Lecture Notes in Mechanical Engineering

Series Editors

Fakher Chaari, National School of Engineers, University of Sfax, Sfax, Tunisia

Francesco Gherardini, Dipartimento di Ingegneria "Enzo Ferrari", Università di Modena e Reggio Emilia, Modena, Italy

Vitalii Ivanov, Department of Manufacturing Engineering, Machines and Tools, Sumy State University, Sumy, Ukraine

Editorial Board

Francisco Cavas-Martínez, Departamento de Estructuras, Construcción y Expresión Gráfica Universidad Politécnica de Cartagena, Cartagena, Murcia, Spain

Francesca di Mare, Institute of Energy Technology, Ruhr-Universität Bochum, Bochum, Nordrhein-Westfalen, Germany

Mohamed Haddar, National School of Engineers of Sfax (ENIS), Sfax, Tunisia

Young W. Kwon, Department of Manufacturing Engineering and Aerospace Engineering, Graduate School of Engineering and Applied Science, Monterey, CA, USA

Justyna Trojanowska, Poznan University of Technology, Poznan, Poland

AF892798

Lecture Notes in Mechanical Engineering (LNME) publishes the latest developments in Mechanical Engineering—quickly, informally and with high quality. Original research reported in proceedings and post-proceedings represents the core of LNME. Volumes published in LNME embrace all aspects, subfields and new challenges of mechanical engineering. Topics in the series include:

- Engineering Design
- Machinery and Machine Elements
- Mechanical Structures and Stress Analysis
- Automotive Engineering
- Engine Technology
- Aerospace Technology and Astronautics
- Nanotechnology and Microengineering
- Control, Robotics, Mechatronics
- MEMS
- Theoretical and Applied Mechanics
- Dynamical Systems, Control
- Fluid Mechanics
- Engineering Thermodynamics, Heat and Mass Transfer
- Manufacturing
- Precision Engineering, Instrumentation, Measurement
- Materials Engineering
- Tribology and Surface Technology

To submit a proposal or request further information, please contact the Springer Editor of your location:

China: Ms. Ella Zhang at ella.zhang@springer.com
India: Priya Vyas at priya.vyas@springer.com
Rest of Asia, Australia, New Zealand: Swati Meherishi at swati.meherishi@springer.com
All other countries: Dr. Leontina Di Cecco at Leontina.dicecco@springer.com

To submit a proposal for a monograph, please check our Springer Tracts in Mechanical Engineering at https://link.springer.com/bookseries/11693 or contact Leontina.dicecco@springer.com

Indexed by SCOPUS. All books published in the series are submitted for consideration in Web of Science.

Shripad Revankar · Kamalakanta Muduli ·
Debjyoti Sahu
Editors

Recent Advances in Thermofluids and Manufacturing Engineering

Select Proceedings of ICTMS 2022

Springer

Editors
Shripad Revankar
Purdue University
West Lafayette, IN, USA

Kamalakanta Muduli
Papua New Guinea University of Technology
Lae, Papua New Guinea

Debjyoti Sahu
KIIT Deemed to be University
Bhubaneswar, India

ISSN 2195-4356 ISSN 2195-4364 (electronic)
Lecture Notes in Mechanical Engineering
ISBN 978-981-19-4387-4 ISBN 978-981-19-4388-1 (eBook)
https://doi.org/10.1007/978-981-19-4388-1

© The Editor(s) (if applicable) and The Author(s), under exclusive license to Springer Nature Singapore Pte Ltd. 2023
This work is subject to copyright. All rights are solely and exclusively licensed by the Publisher, whether the whole or part of the material is concerned, specifically the rights of translation, reprinting, reuse of illustrations, recitation, broadcasting, reproduction on microfilms or in any other physical way, and transmission or information storage and retrieval, electronic adaptation, computer software, or by similar or dissimilar methodology now known or hereafter developed.
The use of general descriptive names, registered names, trademarks, service marks, etc. in this publication does not imply, even in the absence of a specific statement, that such names are exempt from the relevant protective laws and regulations and therefore free for general use.
The publisher, the authors, and the editors are safe to assume that the advice and information in this book are believed to be true and accurate at the date of publication. Neither the publisher nor the authors or the editors give a warranty, expressed or implied, with respect to the material contained herein or for any errors or omissions that may have been made. The publisher remains neutral with regard to jurisdictional claims in published maps and institutional affiliations.

This Springer imprint is published by the registered company Springer Nature Singapore Pte Ltd.
The registered company address is: 152 Beach Road, #21-01/04 Gateway East, Singapore 189721, Singapore

Preface

This International Conference on Thermofluids and Manufacturing Science (KIIT ICTMS 2022) held in the School of Mechanical Engineering of KIIT Deemed to be University, Bhubaneswar, India, during January 24–25, 2022. Philanthropist and educationalist Dr. Achyuta Samanta established KIIT Deemed to be University, formerly Kalinga Institute of Industrial Technology, which is an innovation-driven university located at Bhubaneswar in the state of Odisha, India. KIIT was established in 1992 as an Industrial Training Institution (ITI) which has grown to become KIIT Deemed to be University in 2004. Students from 62 countries study in this coeducational university. All the academic programs are accredited by NAAC of University Grants Commission of India and NBA as per Washington Accord of All India Council for Technical Education, which are benchmarks of quality education. NAAC (Government of India-appointed agency to evaluate universities) has conferred KIIT with the highest grade "A" with a CGPA of 3.36/4. KIIT Deemed to be University acquired the tag of Institution of Eminence (IoE) by Ministry of Education, Government of India. KIIT School of Mechanical Engineering, established in the year 1997, produces graduates, postgraduates and scholars who can meet the requirements of the industry with skill, patience and inclined to do research. Selected consultancy and core research areas of the school include thermofluids and renewable energy, material processing for fuel cell applications, green manufacturing technology, industrial engineering and management. Students of the school bagged international recognition for the development of competition vehicle, robots and aero-models. Research and development initiatives of the school are supported by various private companies and Government of India agencies like ARDB, BRNS, AICTE and DST.

Mechanical engineering deals with many subjects which always thrive to make our lives better. Material and manufacturing, design, thermal science and fluid engineering have been the niche areas in mechanical engineering. However, they are mutually dependent. Thermofluids is the new approach to deal with thermal science, i.e., heat transfer and thermodynamics and fluid engineering together. Application of this area includes combustion and power generation. Researchers in this field broadly aim to achieve a carbon neutral world powered by sustainable energy. Therefore, the research and innovation have been beneficial for the mankind as the efficiency of engine, gas turbine, pumps, refrigerators and thermal power plant improved in last few decades drastically. In the pursuit of reducing emission, the modification in engine exhaust and smoke stack was really challenging but was achieved by the researchers. The use of renewable energy like solar thermal system, solar photovoltaic and wind turbine has increased. Computer simulation has enabled engineers to solve many problems virtually.

Similarly, manufacturing is the backbone of an economy. In view of achieving sustainable development, the energy consumed in the manufacturing sector and transport sector is most relevant. Therefore, any breakthrough in green energy may also show the path of green transport. We need divergent ideas of sustainable growth and sustainable energy to meet. World is facing an unprecedented situation since year 2020. However, advancement of manufacturing technology has enabled us to scale up the production and transportation of medicine, vaccine and medical equipment in pandemic-hit post-COVID-19 world. Rapid prototyping, 3D printing and supply chain management experts have put in all the effort to tackle the situation and contribute to the technical knowledge and ideation. The crisis is far from being over. A combined effort involving and transcending thermal and manufacturing science can solve many more problems including resource planning and management.

KIIT ICTMS 2022 proceedings spans over seven topical tracks, which are well balanced in content, which creates an adequate discussion space for the interesting topics in mechanical engineering. There were 67 oral presentations by participants and the selected among them are written as book chapters and gathered in this book. Efforts taken by peer reviewers contributed a lot to improve the quality of each chapter. They provided constructive critical comments; hence, the quality of chapters was improved. We are grateful to the international/national advisory committee, conveners, session chairs, student volunteers, the institute management and Prof. Bharat Chandra Routara, the conference chair, who selflessly contributed to the success of this conference. Also, we are thankful to all the authors who submitted papers, because of which the conference became a meaningful one. It was the quality of their presentations that really made this conference a great success. Last but not least, we are thankful for the gracious support of Springer for believing us in every step of our journey. Their cooperation was praiseworthy and inspirational.

Bhubaneswar, India

Prof. Shripad Revankar
Prof. Kamalakanta Muduli
Dr. Debjyoti Sahu
Secretary, KIIT ICTMS 2022
debjyoti.sahufme@kiit.ac.in

Contents

Computational Fluid Dynamics

Aerodynamics

Heat Transfer

Combustion Analysis

About the Editors

Dr. Shripad Revankar is Professor of Nuclear Engineering and Director of Multiphase and Fuel Cell Research Laboratory in the School of Nuclear Engineering at Purdue University, West Lafayette, Indiana. He is serving the SNE since 1987. He received his B.Sc., M.Sc. and Ph.D. all in Physics from Karnatak University, India, and M.Eng. in Nuclear Engineering from McMaster University, Canada. He also serves as BK21 Plus Visiting Professor in the Division of Advanced Nuclear Engineering at Pohang University of Science and Technology (POSTECH), South Korea. He has done pioneering work in many niche technical areas that includes advanced reactor systems, reactor safety, reactor thermal hydraulics, composite fuel for advanced nuclear reactors, instrumentation, multi-phase flow and heat transfer, microgravity multiphase flow, direct energy conversion, hybrid power systems, nuclear hydrogen generation, solar energy storage, packed bed reactor, renewable energy, and fuel cell technology. He has published more than 400 peer reviewed technical articles in archival scientific journals and conference proceedings and author/co-author of three books: *Advances in Nuclear Fuels*, InTech publisher 2012, *Fuel Cells-Principles, Design and Analysis*, CRC Press 2014, and *Storage and Hybridization of Nuclear Energy: Techno-economic Integration of Renewable and Nuclear Energy*, Academic Press, 2018.

He has served as the research and educational consultant to academia, national laboratories and industries in Canada, China, Hong Kong, India, South Korea, and USA. He is Chief Editor of *Frontier in Energy-Nuclear Energy* and Chief Editor of *International Journal of Magnetism and Nuclear Science*. He is Life Member of American Nuclear Society (ANS), American Society of Mechanical Engineer (ASME), American Institute of Chemical Engineers (AIChE), Korean Nuclear Society, (KNS), and Indian Society for Heat and Mass Transfer (ISHMT). He is also a member of American Society for Engineering Education (ASEE), Electro Chemical Society (ECS), and American Association for Advancement of Science (AAAS). He was elected as Fellow of ASME in 2008, Fellow of ANS in 2015, and Fellow of AIChE in 2017. He received Technical Achievement Award from American Nuclear Society Thermal Hydraulics Division in 2019 for his significant contribution to reactor thermal hydraulics through experiments and for the modeling of phenomena important in the analysis of nuclear reactor safety and applications.

Dr. Kamalakanta Muduli is presently working as Associate Professor in the Department of Mechanical Engineering, Papua New Guinea University of Technology, Lae, Morobe Province, Papua New Guinea. He has obtained Ph.D. from the School of Mechanical Sciences, IIT Bhubaneswar, Orissa, India. And he has Master's degree in Industrial Engineering. He has over 15 years of academic and research experience in Universities in India and Papua New Guinea. He is a recipient of ERASMUS+ KA107 Award provided by European Union. He has published 51 papers in peer-reviewed international journals most of which are indexed in Clarivate analytics, Scopus and listed in ABDC and presented more than 25 papers in reputed conferences. He appeared as a guest editor for selected special issues of journals and books published by Taylor and Francis, MDPI, CRC Press, Wiley scrivener, and Apple Academic Press. He also has guided three Ph.D. students. His current research interest includes sustainable supply chain management, and Industry 4.0 applications in operations and supply chain management. He is a fellow of Institution of Engineers India. He is also a senior member of Indian Institution of Industrial Engineering and member of ASME.

Dr. Debjyoti Sahu currently serves as Assistant Professor at the School of Mechanical Engineering, KIIT Deemed to be University, Bhubaneswar. He obtained his B.E. in Automobile Engineering from Malnad College of Engineering, Hassan, and M.Tech. in Energy Systems Engineering from BVB College of Engineering and Technology Hubli, Karnataka. He obtained his Ph.D. in Energy Technology from Indian Institute of Technology Guwahati, India. He has taught various subjects of UG and PG courses at the Department of Mechanical Engineering, Amrita Vishwa Vidyapeetham, Bengaluru, during 2008–2018. Now he teaches battery technology and automobile engineering at KIIT Deemed to be University, Bhubaneswar. He has been involved in several experimental research including hydrogen adsorption, metal organic frameworks, refrigeration, and bio-fuel. He along with his research collaborators holds an Australian Innovation Patent on 'Usage of cow urine as an additive with the unusable cooking oil for the emission reduction from diesel engine exhaust'. He has published several journal papers on adsorption, refrigeration, and IC engine performance. He is Life Member of Indian Science Congress and Institution of Engineers, India.

Fluid Dynamics

Numerical Analysis of Gas Entrainment in the Viscous Liquid Pool Developed by a Counter-Rotating Pair of Fully Submerged Rollers

Santosh Kumar Panda and Basanta Kumar Rana

1 Introduction

Entrainment has long been basically known to be one of the most notable characteristics of all two-phase connections. It must be in process of developing the shear-induced turbulent flow that delivers fluid over a fluid–solid interaction. As a result, scientists have always been interested in investigating entrainment dynamics. The entrainment hypothesis generally aims to increase the interfacial area to improve quality and performance. It is used in a variety of engineering scenarios. Hence, a variety of application in process and power sectors, such as internal combustion engine, jet pumps, ejectors and aerators [1, 2], surgical dressings [3, 4], inkjet printers [4–6], duct separators and air conditioner. It enhances the interfacial area, hence improves system performance. Numerous researchers [7–10] analysed that an external rotating field perturbed in a stratified interface because of roller rotations. Lorenceau et al. [11] observed the growth and generation of a sharp-shaped cusp in a highly viscous pool because of the partially submerged rotating roller and developed a relationship between the sharp tip width and radius of curvature and also signified that Capillary number is a function of roller velocity. Decre et al. [12] examined about the entire meniscus profile for asymmetric coating of two rollers (one of which was partially submerged). The formation of free surface cusps for low Reynolds numbers was addressed by Jeong and Moffatt [10], who counter-rotated two parallel mounted cylinders on the same horizontal plane below the free surface and assumed the rollers as vortex dipoles and employing complex variable strategies

S. K. Panda · B. K. Rana (✉)
School of Mechanical Engineering, Kalinga Institute of Industrial Technology, Deemed to be University, Bhubaneswar 751024, India
e-mail: basanta.uce@gmail.com

S. K. Panda
e-mail: pandasantosh131@gmail.com

© The Author(s), under exclusive license to Springer Nature Singapore Pte Ltd. 2023
S. Revankar et al. (eds.), *Recent Advances in Thermofluids and Manufacturing Engineering*, Lecture Notes in Mechanical Engineering,
https://doi.org/10.1007/978-981-19-4388-1_1

to solve the issues. Sharma et al. [13] investigated the revolving movement of a single partially submerged horizontal cylinder between the air and water interface to understand more about gas penetration into liquids. Particle movements were monitored by using simplified particle image velocimetry. For various gas–liquid combinations with half part of roller immersed in liquid, Rana et al. [2] investigated computationally to determine the behaviour of wrapping liquid film in the receding junction and entrained cusp-shaped entrainment in the advancing end portion. Kumar et al. [14] examined entrapment of air filament in a highly viscous liquid pool because of pair of fully submerged counter-rotating solid roller. They looked at the dynamics of bending a filament towards a high-speed spinning roller and discovered an air collection zone that is ejected from the air filament. For various gas–liquid fluid combinations, Panda et al. [15] recently established the effect of horizontal cross-flow in a gaseous medium and rotational inertia on cusp-shaped entrained air in the advancing end and wrapping film thickness in the receding end. The effect of Archimedes number (Ar) on steady-state counter plots, the transient phenomena of liquid tip movement and the breaking of a collapsible gaseous jet were investigated in detail. We have addressed the interfacial structure for air-polybutene pair with variable roller rotations using a computational model (Fig. 1). The influence of different submerged depths (b) due to different roller rotation speeds and the influence of Archimedes number (Ar) on steady-state numerical counters plots were deemed to be the main priorities of this analysis.

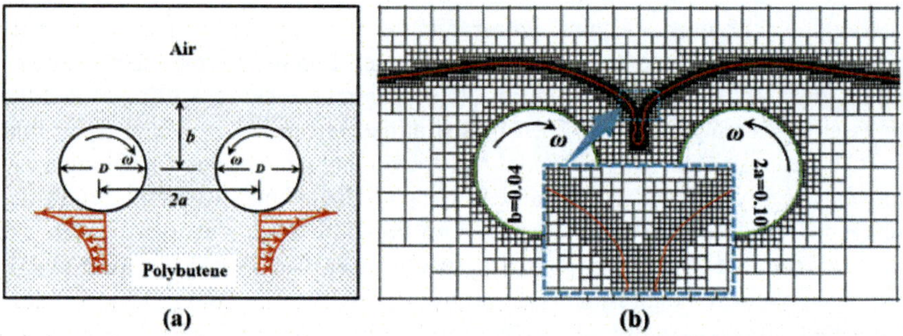

Fig. 1 **a** Schematic illustration of the computational domain having two counter revolving cylinders completely immersed in high viscos polybutene liquid, and **b** illustration of mesh structure of the entire computational domain in an adaptive method

Table 1 Properties of air-polybutene fluid pair for current investigation [14]

Fluid pair	Name of the fluid	ρ (kg/m^3)	μ (Pas)	σ (N/m)
Pair 1	Polybutene	883	12.9	0.0340
	Air	1.225	1.789×10^{-5}	

2 Numerical Methodology

2.1 *Computational Domain*

To investigate the entrainment dynamics around the symmetric rotation (one clockwise and the other counterclockwise) of cylinders, it is decided to choose a $10D \times 10D$ simulation domain with no-slip boundary conditions. Figure 1 shows the simulation model with two rotating solid rollers having 0.06 m diameter each immersed totally in polybutene, and Fig. 1a displays a schematic view of the computational domain with two fully submerged rotating solid rollers in highly viscous polybutene liquid. Here, two equal diameter rollers are positioned horizontally with a spacing of $2a$ and a depth of b from the unperturbed interface. To figure out what entrainment patterns are unique to each case, air-polybutene is selected as the test fluid pair, and the properties of the chosen air-polybutene pair are shown in Table 1. Figure 1b exhibits the adaptive mesh structure which is distributed across the whole computational domain at $b = 0.04$, with finer cells produced in the converging location. The key objective here is to generate a finer mesh in the input domain with high-quality cells but less cells that the calculations become intractable. In converging locations of the air-polybutene interface, the mesh is adaptive and finer which are critical for subsequent calculations.

2.2 *Formulation of Problem*

The entire present investigation is carried out by Gerris solver [16, 17]. It generally utilises VOF-based approach which includes incompressible and density Navier–Stokes solver for identification of entrainment dynamics in a numerical way. Here, momentum and continuity equations along with body and surface forces are exhibited as follows:

$$\rho\left(\frac{\partial u_\alpha}{\partial t} + u_\beta.\frac{\partial u_\alpha}{\partial x_\beta}\right) = -\frac{\partial p}{\partial x_\beta} + \frac{\partial}{\partial x_\beta}\left[\mu\left(\frac{\partial u_\alpha}{\partial x_\beta} + \frac{\partial u_\beta}{\partial x_\alpha}\right)\right] + \sigma k \delta_s n \quad (1)$$

$$\frac{\partial p}{\partial t} + \nabla.(\rho u) = 0 \quad (2)$$

Here, surface tension term generally deals with the depth of penetration inside the liquid pool.

Equation (2) can be changed to the following form outside of the interfacial regime, assuming fluid density remains constant.

$$\frac{\partial u_\alpha}{\partial x_\alpha} = 0 \tag{3}$$

Presently, $\rho = \rho(x, t)$, $u_\alpha = (u, v)$ and $\mu = \mu(x, t)$ are the fluid density, velocity and dynamic viscosity, respectively. Interface deals with Dirac distribution function δ_s (a surface tension term), and σ, k and n are the coefficients of surface tension, curvature and normal to the interface, respectively.

Density and viscosity are delineated by volume fraction term, i.e. $c(x, t)$ for a particular position by considering a generalised fluid as follows:

$$\rho(\widehat{c}) = \widehat{c}\,\rho_{\text{liq}} + (1 - \widehat{c})\rho_{\text{gas}} \tag{4}$$

$$\mu(\widehat{c}) = \widehat{c}\,\mu_{\text{liq}} + (1 - \widehat{c})\mu_{\text{gas}} \tag{5}$$

The liquid density, viscosities and gaseous density, viscosities are denoted by ρ_{liq}, μ_{liq}, ρ_{gas} and μ_{gas}, respectively. In a cell, a filtered value can be calculated by averaging the volume fraction c's four corner values (x, t). The corner values are computed employing bilinear interpolation from cell-centred statistics.

We can derive the appropriate constitutive equation on volume fraction using Eq. (2):

$$\frac{\partial c}{\partial t} + \nabla.(cu) = 0 \tag{6}$$

A staggered spatial grid and a second-order accurate method are being used to discretise the above Eqs. (1–2 and 6) over time. Gerris provides the linear interface for each cell applying polynomial line methodology [17].

3 Validation

Presently, we have compared the current entrained air filament length in angular form ($\theta/2\pi$) to Kumar et al. [14]. (air-polybutene: $\rho_{\text{air}} = 1.225$ kg/m^3, $\mu_{\text{air}} = 1.789 \times 10^{-5}$ Pas, $\rho_{\text{polybutene}} = 883$ kg/m^3, $\mu_{\text{polybutene}} = 12.9$ Pas, $\sigma = 0.034$ N/m, rotational speed ratio (Re_2/Re_1) = 2.8 or rotational speed (ω) = 5 rad/s and diameter of each roller = 0.046 m for numerical modelling assessment). Figure 2 shows that these two numerical simulations relate with each other, indicating that our current investigation

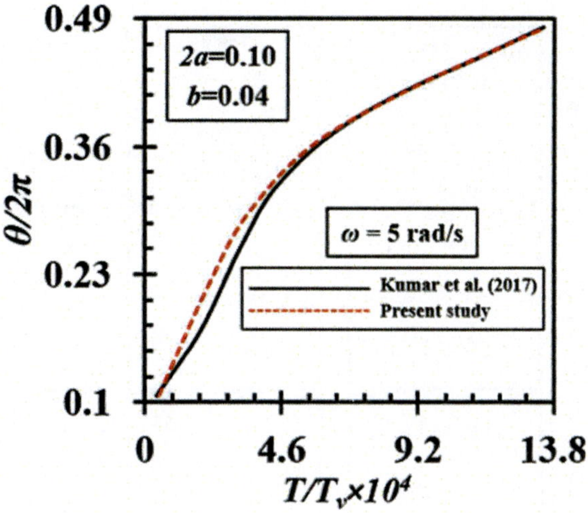

Fig. 2 Comparison of dimensionless entrained air filament length in angular form ($\theta/2\pi$) with Kumar et al. [14] results

or simulations are credible. Here, T stands for physical time (seconds), and $T_V = \mu_l^3/\rho_l\sigma^2$ represents the viscous time scale.

4 Results and Discussion

In this analysis, the converging rotational field generated by fully submerged rollers immersed in polybutene has been analysed for numerous roller rotations, i.e. = 2.5–10 rad/s. On the surface of the solid rollers, a no-slip boundary condition is applied, and the rollers are positioned at a fixed distance from the centre (x-axis), i.e. $2a$ = 0.10, and at various submerged depths (y-axis) (b = 0.035–0.08). The entrainment pattern for each case is elucidated using air-polybutene as test fluid pair (Table 1). For this investigation, the diameter (D) of each roller is chosen to be 0.06 m.

Figure 3 shows the variation of interfacial structure for air-polybutene at various roller rotations (ω = 2.5–10 rad/s) for varied submerged depths (b = 0.035–0.08) at $2a$ = 0.10. Figure 3a–p shows that the depth of penetration gradually reduces as the submerged depth increases (b), due to the lesser influence of rotational inertia as the submerged depth increases. As a result, viscosity-driven air pumping well below cusp opposing surface tension and gravity appears to be greaterat higher submerged depths and declines as submerged depth increases. This type of observations is generally found in industrial applications and gravure printing, etc. Another point to note is that when the roller rotations reduce, the depth of penetration decreases as well. It occurs because a lower rotational field is generated due to the lesser angular speed of rollers, and it gradually increases as the speed of the roller increases. Hence, liquid flux is found to be lower at reduced roller rotations and higher at higher rotations. As

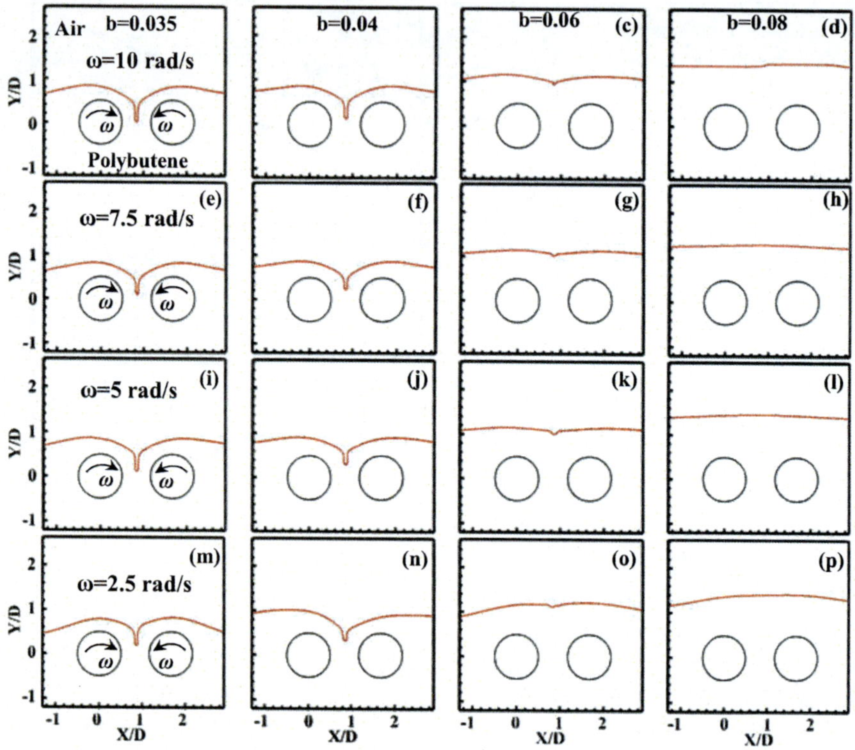

Fig. 3 Steady numerical counters for air-polybutene at varied (a–d) ω=10 rad/s and b=0.035–0.08, (e–h) ω=7.5 rad/s and b=0.035–0.08, (i–l) ω=5 rad/s and b=0.035–0.08, and (m–p) ω=2.5 rad/s and b=0.035–0.08

a result, we can determine that cusp-shaped air entrainment is caused by viscosity-driven capillary pull, which permits air to be pumped against surface tension and gravity.

The Archimedes number $\left(\mathrm{Ar} = \frac{gD^3\rho^2}{\mu^2}\right)$ was kept constant in all of the earlier steady counters (Fig. 3). The current research focuses on the remarkable changes in entrainment structures due to variations in Ar (gravitational pull) with specified $2a$, b and ω rad/s (Fig. 4a). We usually change the Ar = 0.0115–0.0398 with ω rad/s, i.e. 10 rad/s. From Fig. 4a, it can be observed that interface switches away from solid rollers as Archimedes number increases for a fixed $\omega = 10$ rad/s, and likewise, the steady cusp configuration diminishes with gradual increase of Ar for a fixed ω rad/s. Figure 4b depicts the remarkable changes in interface patterns due to various angular speeds of rollers. One may observe from Fig. 4b the depth of penetration increases due to increase in angular speed of rollers. It is due to higher rotational inertia generated at higher speed of rollers. For a fixed roller rotation ($\omega = 10$ rad/s), Fig. 4c depicts that as the Archimedes number increases gradually, the vertical shift of the interface ($y* = y/D$) penetrates less. The density difference between the two fluids causes a

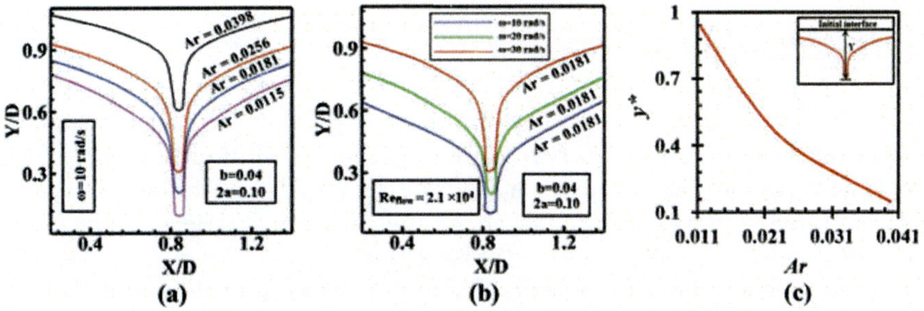

Fig. 4 **a, b** Steady numerical counters for varying Archimedes number (Ar) for fixed roller rotations ($\omega = 10$ rad/s) and varying roller rotations ($\omega = 10$–30 rad/s) and **c** changes in the orientation of the macroscopic boundary as a function of Archimedes number (Ar)

considerable increase in gravitational pull, which diminishes macroscopic interface penetration ($y^* = y/D$). When we change the Ar number range for different fixed ω rad/s, similar kinds of results can be found. Figure 5 illustrates the velocity vectors diagram near the rotational roller for $2a = 0.10$, $b = 0.04$ and $\omega = 10$ rad/s. It is observed that near the roller, the magnitude of velocity vector is higher due to higher rotational inertia and magnitude decreases at bottom below position of rollers and also in away from rollers because of lower rotational inertia (clearly noticed from Fig. 5). The baseline analysis that identifies the flow distribution between two rollers is termed velocity vector analysis. The development of a cusp shape is also revealed by the velocity profile due to various rotational inertia.

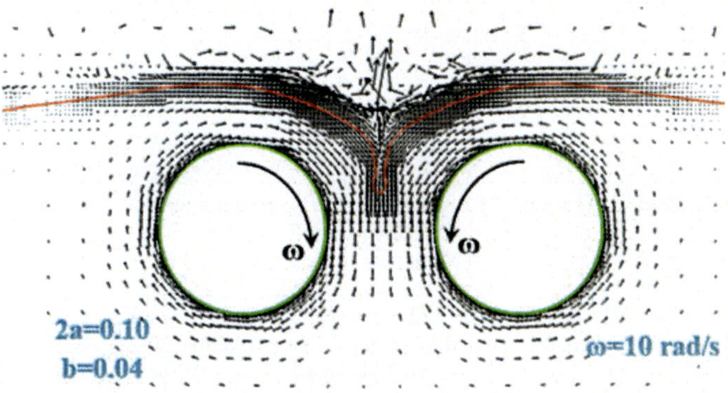

Fig. 5 Velocity vectors around rotating rollers for $2a = 0.10$, $b = 0.04$ and $\omega = 10$ rad/s

5 Conclusions

The influence of different roller rotations and submerged depths on steady numerical contours for air-polybutene has been physically explained. It is quite observed that the length of entrainment increases due to increased speed of rollers, and also depth of penetration increases due to gradual decrease of submerged depth. Variation in Archimedes number (Ar) and how consequence does Archimedes number (Ar) have on the positioning of the macroscopic interface? Both have been thoroughly investigated in parallel efforts. This research is aimed into the effects of fluid properties and various rotational speeds of rollers. Finally, to visualise the flow patterns between solid rollers, a velocity profile has been established.

6 Application

This issue seems to be somewhat related to oceanic mixed layers, which include the transmission of hydrophysical properties between the top and bottom layers. In the ocean, this activity is vital to sustaining the physical climate of coastal areas. The depth of the mixed layer, in this context, is critical in establishing the temperature range of the marine areas.

Acknowledgements The authors would like to express their thankfulness to the Science and Engineering Research Board (SERB), Department of Science and Technology (DST), Government of India, for funding the research project through a research grant (SRG/2019/001438).

References

1. Bagul RK, Pilkhwal DS, Vijayan PK, Joshi JB (2013) Entrainment phenomenon in gas–liquid two-phase flow: a review. Sadhana 38(6):1173–1217
2. Rana BK, Das AK, Das PK (2016) Numerical study of air entrainment and liquid film wrapping around a rotating cylinder. Ind Eng Chem Res 55(46):11950–11960
3. Lutsko JF (2016) Mechanism for the stabilization of protein clusters above the solubility curve: the role of non-ideal chemical reactions. J Phys Condens Matter 28(24):244020
4. Cooper MG (1969) The microlayer and bubble growth in nucleate pool boiling. Int J Heat Mass Transf 12(8):915–933
5. Clanet C, Lasheras JC (1997) Depth of penetration of bubbles entrained by a plunging water jet. Phys Fluids 9(7):1864–1866
6. Eggers J, Villermaux E (2008) Physics of liquid jets. Rep Prog Phys 71(3):036601
7. Joseph DD, Nguyen K, Beavers GS (1984) Non-uniqueness and stability of the configuration of flow of immiscible fluids with different viscosities. J Fluid Mech 141:319–345
8. Joseph DD, Nelson J, Renardy M, Renardy Y (1991) Two-dimensional cusped interfaces. J Fluid Mech 223:383–409
9. Joseph DD (1992) Understanding cusped interfaces. J Nonnewton Fluid Mech 44:127–148
10. Jeong JT, Moffatt HK (1992) Free-surface cusps associated with flow at low Reynolds number. J Fluid Mech 241:1–22

11. Lorenceau E, Restagno F, Quéré D (2003) Fracture of a viscous liquid. Phys Rev Lett 90(18):184501
12. Decré M, Gailly E, Buchlin JM (1995) Meniscus shape experiments in forward roll coating. Phys Fluids 7(3):458–467
13. Sharma Y, Rana BK, Das AK (2015) Rotary entrainment in two phase stratified gas-liquid layers: an experimental study. Int J Chem Mole Nucl Mater Metall Eng 9:1367–1372
14. Kumar P, Das AK, Mitra SK (2017) Bending and growth of entrained air filament under converging and asymmetric rotational fields. Phys Fluids 29(2):022101
15. Panda SK, Rana BK, Kumar P (2021) Competition of roller rotation and horizontal crossflow to control the free surface cusp-induced air entrainment. Phys Fluids 33(11):112114
16. Popinet S (2003) Gerris: a tree-based adaptive solver for the incompressible Euler equations in complex geometries. J Comput Phys 190(2):572–600
17. Popinet S (2009) An accurate adaptive solver for surface-tension-driven interfacial flows. J Comput Phys 228(16):5838–5866

MHD Flow Measurements of Automatic Control Valve of Gas Turbine Engine Subject to Inclined Magnetic Field

S. Ganesh, P. Chandrasekar, and J. Jayaprabakar

1 Introduction

MHD study enhanced in recent years has resulted in its widespread applications in a various industries. Especially in the design of cooling systems with MHD generators and liquid metal. The key thing for the successful research is always the study of MHD channel. In general, magnetohydrodynamics is also called as magneto fluid dynamics [1, 2]. The basic principle which is behind MHD is that currents can be induced by magnetic fields in a moving and electrically conductive fluids, thus the fluid can be polarized and magnetic field will be changed itself. MHD applications vary extensively across multiple disciplines ranging from MHD moving pistons to MHD driven ships, biomedical sensors and actuators. Specifically, the flow of electrically conducting fluids possesses a prominent role in industrial and engineering applications such as hydraulic lifts and electric motors [3].

MR fluid (Magneto-Rheological fluid) is an oil type of fluid having micro particles in it and it is used in carrier fluids it is also called as smart fluids. Under the influence of magnetic field, it acts as a viscoelastic solid where the apparent viscosity is greatly increased. Mainly, the yield stress of the fluid can be properly adjusted by varying the magnetic field intensity [4, 5]. The micro metal particles included in these fluids are in the powder form of iron, carbonyl iron and cobalt alloys and Examples of the carrier liquids are hydrocarbons and silicone-based oils. The basic concept behind MHD and MR fluids is when the metal particles present in the smart fluid are subjected to

S. Ganesh · P. Chandrasekar
Department of Mathematics, Sathyabama Institute of Science and Technology, Chennai, India

J. Jayaprabakar (✉)
Department of Mechanical Engineering, Sathyabama Institute of Science and Technology, Chennai, India
e-mail: jp21tn@gmail.com

© The Author(s), under exclusive license to Springer Nature Singapore Pte Ltd. 2023

S. Revankar et al. (eds.), *Recent Advances in Thermofluids and Manufacturing Engineering*, Lecture Notes in Mechanical Engineering,
https://doi.org/10.1007/978-981-19-4388-1_2

magnetic field; it forms the chain structure and the smart fluid changes into semi-solid state which prevents the movement of the fluid. This transformation is quick, reversible and also the rheological behaviour of the fluid will be changed. Owing to the rheological change of the fluid, the yield stress can be increased significantly by changing the intensity of the magnetic field. The application of this principle in shock absorbers, hydraulic valves. There are number of research on the relationship between the conducting fluids and magnetic field [6, 7].

2 Numerical Modelling

The governing equations are

$$\frac{\partial u}{\partial x} + \frac{\partial v}{\partial y} = 0 \tag{1}$$

$$\rho\frac{\partial u}{\partial t} = -\frac{\partial p}{\partial x} + \mu\left(\frac{\partial^2 u}{\partial x^2} + \frac{\partial^2 u}{\partial y^2}\right) - \sigma_e B_0^2 u \sin^2\alpha \tag{2}$$

$$\rho\frac{\partial v}{\partial t} = -\frac{\partial p}{\partial y} + \mu\left(\frac{\partial^2 v}{\partial x^2} + \frac{\partial^2 v}{\partial y^2}\right) \tag{3}$$

σ is the electrical conductivity and $B_0 = \mu_e H_{0,}$ B_0 is the electromagnetic induction, μ_e being the magnetic permeability and H_0 is the transverse magnetic field.

In Steady State, $\frac{\partial u}{\partial t} = 0$ and $\frac{\partial v}{\partial t} = 0$ then Eqs. (2) and (3) we get,

$$0 = -\frac{\partial p}{\partial x} + \mu\left(\frac{\partial^2 u}{\partial x^2} + \frac{\partial^2 u}{\partial y^2}\right) - \sigma_e B_0^2 u \sin^2\alpha \tag{4}$$

$$0 = -\frac{\partial p}{\partial y} + \mu\left(\frac{\partial^2 v}{\partial x^2} + \frac{\partial^2 v}{\partial y^2}\right) \tag{5}$$

The B.C's [8] are

$u = 0$ on $y = h$ and $y = -h$; $v = v_0\, e^{i\omega t}$ on $y = h$ and $v = -v_0\, e^{i\omega t}$ on $y = -h$ where v_0 the velocity of suction at the channel's walls.

Let $\eta = \frac{y}{h}, u = u(x, y)e^{i\omega t}, v = v(x, y)\, e^{i\omega t}, p = p(x, y)\, e^{i\omega t}$ where ω is the frequency. Equations (1), (4) and (5) becomes

$$\frac{\partial u}{\partial x} + \frac{1}{h}\frac{\partial v}{\partial \eta} = 0 \tag{6}$$

$$0 = -\frac{\partial p}{\partial x} + \mu\left(\frac{\partial^2 u}{\partial x^2} + \frac{1}{h^2}\frac{\partial^2 u}{\partial \eta^2}\right) - \sigma_e B_0^2 u \sin^2\alpha \tag{7}$$

$$0 = -\frac{\partial p}{\partial y} + \mu\left(\frac{\partial^2 v}{\partial x^2} + \frac{1}{h^2}\frac{\partial^2 v}{\partial \eta^2}\right) \tag{8}$$

Let $\nu = \frac{\mu}{\rho}$ = Viscosity in Kinematics, ρ being the fluid's density, μ is the viscosity coefficient and p is the pressure.

The B.C's [9, 10] are rewritten as

$$u(x, 1) = 0, u(x, -1) = 0 \tag{9}$$

$$v(x, 1) = v_0, v(x, -1) = -v_0 \tag{10}$$

Assume that ψ is the stream function.

$$u = \frac{1}{h}\frac{\partial \psi}{\partial \eta} \tag{11}$$

$$v = -\frac{\partial \psi}{\partial x} \tag{12}$$

A stream function of the type can satisfy the equation of continuity.

$$\psi(x, \eta) = (hU(0) - v_0 x) f(\eta) \tag{13}$$

The average entrance velocity at $x = 0$ is $U(0)$.

We get Eq. (13) when adding Eqs. (11) and (12).

$$u = \frac{1}{h}(hU(0) - v_0 x) f'(\eta) \tag{14}$$

$$v = v_0 f(\eta) \tag{15}$$

The differentiation with respect to the dimensionless variable is represented by the prime $\eta = \frac{y}{h}$. Since the fluid is released around the same rate from both walls, v_0 is independent of x.

Equations (14) and (15) are placed in Eqs. (7) and (8) to obtain,

$$-\frac{1}{\rho}\frac{\partial p}{\partial x} = \left(U(0) - \frac{v_0 x}{h}\right) \cdot \left(-\frac{\nu}{h^2} f'''(\eta) + \left(\frac{\sigma_e B_0^2 \sin^2 \alpha}{\rho}\right) f'(\eta)\right) \tag{16}$$

$$-\frac{1}{h\rho}\frac{\partial p}{\partial \eta} = -\frac{\nu \cdot v_0}{h^2} f''(\eta) \tag{17}$$

Differentiating with respect to 'x', we get Eq. (17).

Fig. 1 Design of the plate with valve

$$\frac{\partial^2 p}{\partial x \partial \eta} = 0 \tag{18}$$

Differentiating w.r. to 'η' given Eq. (16), it follows that

$$\frac{-1}{\rho}\frac{\partial^2 p}{\partial x \partial \eta} = \left(U(0) - \frac{v_0 x}{h}\right) \cdot \frac{\mathrm{d}}{\mathrm{d}\eta}\left(-\frac{\nu}{h^2} f'''(\eta) + \left(\frac{\sigma_e B_0^2 \sin^2 \alpha}{\rho}\right) f'(\eta)\right) \tag{19}$$

From Eqs. (18) and (19) becomes,

$$\frac{\mathrm{d}}{\mathrm{d}\eta}\left(-\frac{\nu}{h^2} f'''(\eta) + \left(\frac{\sigma_e B_0^2 \sin^2 \alpha}{\rho}\right) f'(\eta)\right) = 0 \tag{20}$$

Figure 1 shows the simple schematic part model of the model which is to be selected to spray the pure water into the chamber to reduce the pollutants. Before choosing the valve, the optimization and the hydrodynamic behaviour was studied.

3 Graphical Representation

The study is made in order to find the effect of Hartman number $M1$ on axial and radial velocity profiles. The test was conducted for various channel values at $X = 0$, $X = 3$ and $X = 5$ and also for different values of $\omega t = 0$, $\omega t = \pi/4$, $\omega t = \pi/2$, $\omega t = 3\pi/4$, and $\omega t = \pi$. The distance between the plates (h) was kept 1 for all the test conditions where the inlet axial velocity varied from 0.5 to 1. The value of $M1$ was varied from 1 to 5 to determine the influence of $M1$ on the axial velocity. The same

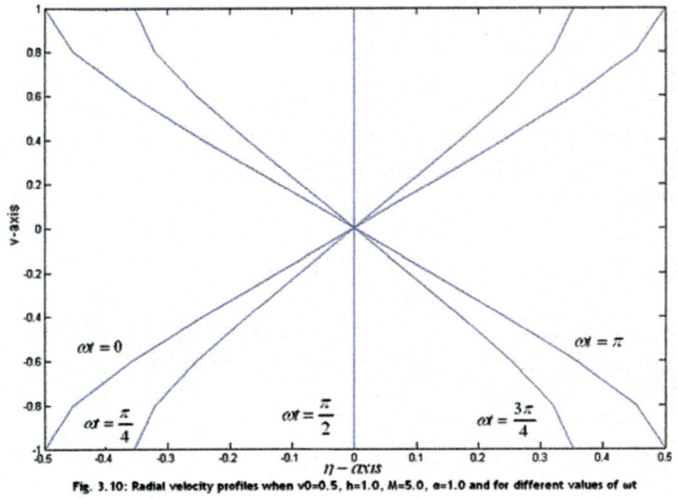

Fig. 2 Profiles of radial velocity when $v_0 = 0.5$, $h = 1.0$, $M_1 = 5.0$, $\alpha = 1.0$ and for a variety of ωt

procedure was followed for radial velocity of the values of 0.5 and 1 with the above said X and values. The value of $M1$ was varied from 1 to 5.

Figure 2 show the value of axial velocity at different inclination of magnetic force (= 0, and) with the same boundary conditions $U0 = 0.5$, $V0 = 0.5$, $M1 = 5.0$ and $h = 1$ but for $X = 1$, $X = 3$ and $X = 5$. It's clear increasing the X value increases the velocity. The velocity increment when the conditions are the same but the $U0 = 1$. The effect of Hartman number on velocity. The boundary conditions are $U0 = 0.5$, $V0 = 0.5$, $M1 = 5.0$, $h = 1$ and $X = 1$ with various magnetic field inclination angles.

4 Results and Discussion

The relationship between the fluid flow and Hartman number is found by doing this analytical method. Hartman number is directly proportional to magnetic force and inversely proportional to viscous force since it's the relationship between them. Thus, the higher the Hartman number, stronger the magnetic force and the smaller the viscous force. The following results were obtained based on this principle. The increase in Hartman number causes a decrease in velocity, which is based on the fact that a stronger magnetic field reduces the flow. It's also been noticed that when the magnetic field's inclination increases, the velocity decreases.

Figure 2 shows the effect of Hartman number and the channel distance X and inclination of magnetic field on velocity profiles. The distance X shows the direct relationship with the velocity profiles. As the value of X increased from 1 to 5, the magnitude velocity profile value was increased. Similarly, when the inlet axial velocity was increased from 0.5 to 1, the same positive effect on the magnitude of axial velocity was noticed. Contrarily, when the Hartman number value $M1$ was increased from 1 to 5, there was a decrement in the magnitude of axial velocity profile [11, 12]. On the other hand, for radial velocity, Hartman number increment showed a little increment in the magnitude of radial velocity profile. When the Hartman number increased from 1 to 5, a marginal increase in the velocity was found (Figs. 3 and 4).

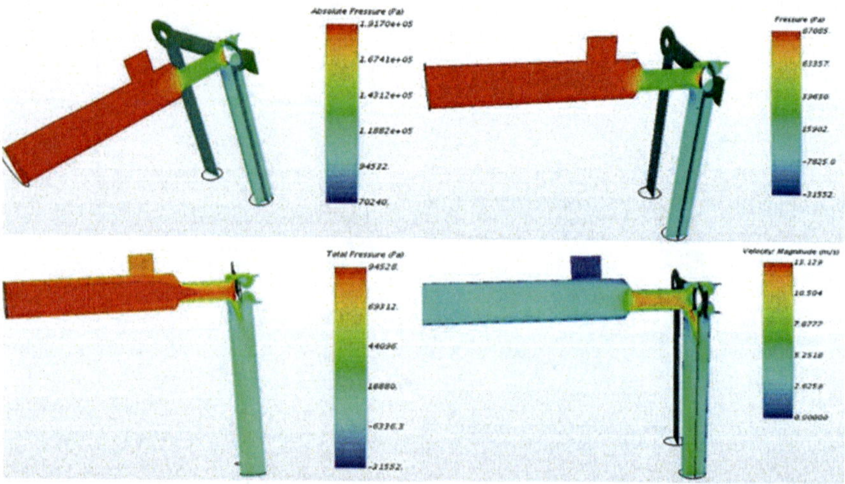

Fig. 3 Plates with 15 mm valve design

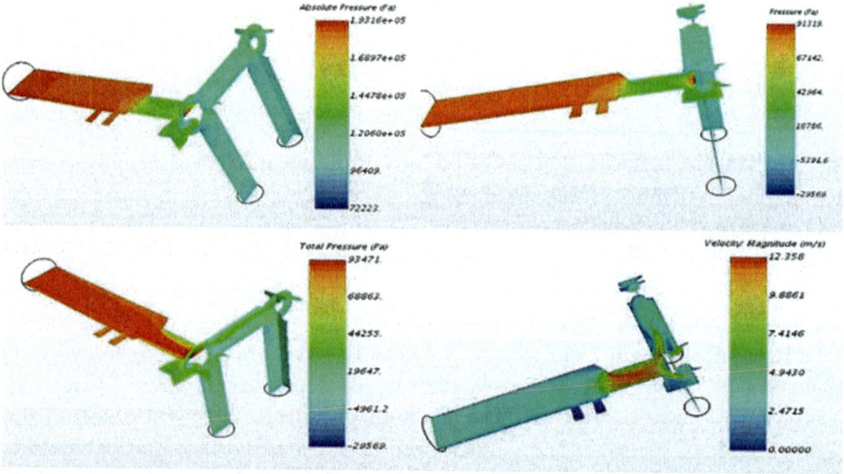

Fig. 4 Plates with 20 mm valve design

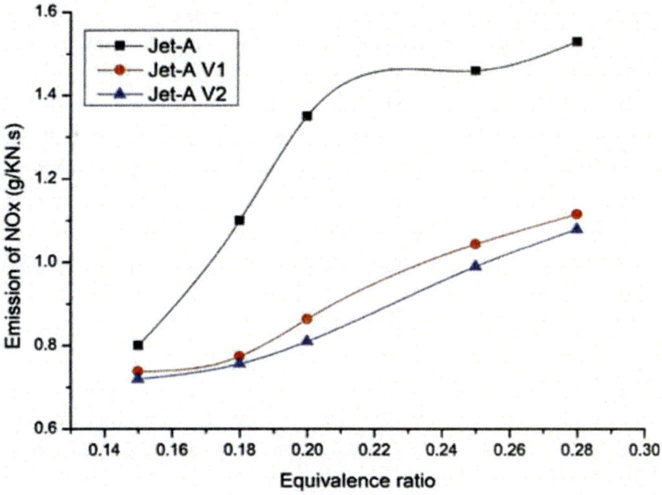

Fig. 5 Emission of NO_x for the modified chamber

Influence of Pure water control valve on NO_x emission: Emission of NO_x was the major drawback for the all transport vehicles. As the demand of the gas turbine engine increases the emission of the pollutants from the engine was a major worry for the environment [13–15]. As the environmental concern grows, several initiates and technology are tested widely to reduce the emission of NO_x. Among them are biodiesel and the modification of the combustion chamber. Here we had introduced the valve which sprays the pure water to reduce the NO_x and other pollutants. Two different configurations of the valves were employed in this current study and compared with the conventional Jet-A fuel. So will the equivalence ratio. The emission of NO_x was increased dramatically. However, in comparison to a standard combustion chamber, the chamber with the valve introduced in the engine reported less NO_x irrespective of the equivalence ratio [16–18]. The average reduction of the NO_x recorded by the Jet-A $V1$ and Jet-A $V2$ were 45% and 52%, respectively as presented in Fig. 5.

5 Conclusions

From the obtained results, it is concluded that reducing the magnetic field allows the fluid to flow freely thus raises the flow's velocity. The Lorentz force acting on the fluid is the explanation for this. When the inclination of the magnetic field is considered, the higher the inclination the lower the velocity is obtained. This is because the magnetic force slows down the motion of the fluid. High velocity is obtained when the angle of inclination is low and the applied magnetic force is low. In other words, increasing the pressure gradient is the cause to the increment in velocity owing to the nature of its force given to the movement of the fluid. In addition, it is evident that the magnitude of velocity profile is increased when the two parallel porous plates move apart from each other. When the plates move closer, it gives negative effect on the velocity profile. Hartman number also relatively reduces the magnitude of axial velocity as it is increased but increases the radial velocity. Further, with the effect of Hartman number increment, axial velocity profile shows linear variation while radial velocity gives nonlinear variation with respect to various ωt. Further, the engine with modified combustion chamber produces less amount of NO_x. Compared to the $V1$, the $V2$ produces lesser amount of NO_x irrespective of the equivalence ratio.

References

1. Bagchi IC (1996) Unsteady flow of viscoelastics Maxwell fluid with transient pressure gradient through a rectangular channel. Ind J Mech Math 4(2)
2. Attia HA (1999) Transient MHD flow and heat transfer between two parallel plates with temperature dependent viscosity. Mech Res Commun 26(1):115–121
3. Delhi Babu R, Ganesh S (2018) The mathematical model for steady magnetohydrodynamic flow between parallel porous plates with an angular velocity. Int J Amb Energy:1–9
4. Ganesh S, Krishnambal S (2007) Unsteady magnetohydrodynamic stokes flow of viscous fluid between two parallel porous plates. J Appl Sci 7(3):374–379
5. Babu RD, Ganesh S (2017) The effects of steady magneto hydrodynamic stokes flow between two parallel porous plates. In: 2017 international conference on innovations in information, embedded and communication systems (ICIIECS). IEEE, pp 1–6
6. Li J, Liu L, Zheng L, Bin-Mohsin B (2016) Unsteady MHD flow and radiation heat transfer of nanofluid in a finite thin film with heat generation and thermophoresis. J Taiwan Inst Chem Eng 67:226–234
7. Ganesh S, Delhi Babu R, Anand VWJ, Chandrasekar P (2018) Magnetohydrodynamic flow of viscous fluid between two parallel porous plates with bottom injection and top suction. Int J Amb Energy:1–4
8. Manigandan S, Vijayaraja K (2018) Flow field and acoustics characteristics of elliptical jet. Aircraft Engineering and Aerospace Technology.
9. Manigandan S, Vijayaraja K (2018) Flow mixing and screech characteristics of elliptical supersonic jet. Int J Turbo Jet-Engines 1
10. Mishra S, Sohret Y, Shukla AK (2019) Conventional and advanced exergy analysis of air-film blade cooled gas turbine cycle. In: Advances in fluid and thermal engineering. Springer, Singapore, pp 283–296
11. Manigandan S, Gunasekar P, Sruthisree N, Aich K, Sathya K, Selvan A, Nithya S (2018) Visualization of vortex flow field around a flat plate with noncircular hole. IOP Conf Ser Mater Sci Eng 310(1):012135
12. Chandra Sekar P, Ganesh S, Anand VWJ, Delhi Babu R (2019) Analytical solution for steady magneto hydrodynamic flow between parallel porous plates with inclined magnetic field. Int J Amb Energy:1–9
13. Coban K, Şöhret Y, Colpan CO, Karakoç TH (2017) Exergetic and exergoeconomic assessment of a small-scale turbojet fuelled with biodiesel. Energy 140:1358–1367
14. Manigandan S, Atabani AE, Ponnusamy VK, Pugazhendhi A, Gunasekar P, Prakash S (2020) Effect of hydrogen and multiwall carbon nanotubes blends on combustion performance and emission of diesel engine using Taguchi approach. Fuel 276:118120
15. Devi PB, Joseph DR, Gokulnath R, Manigandan S, Gunasekar P, Anand TP, Venkatesh S, Vimal MR (2020) The effect of TiO_2 on engine emissions for gas turbine engine fueled with Jatropha, butanol, soya and rapeseed oil. Int J Turbo Jet-Engines 37(1):85–94

16. Manigandan S, Atabani AE, Ponnusamy VK, Gunasekar P (2020) Impact of additives in Jet-A fuel blends on combustion, emission and exergetic analysis using a micro-gas turbine engine. Fuel 15(276):118104
17. Gunasekar P, Manigandan S, Venkatesh S, Gokulnath R, Vimal R, Boomadevi P (2019) Effect of hydrogen addition on exergetic performance of gas turbine engine. Aircraft Engineering and Aerospace Technology
18. Balli O, Sohret Y, Karakoc HT (2018) The effects of hydrogen fuel usage on the exergetic performance of a turbojet engine. Int J Hydrogen Energy 43(23):10848–10858

Study of Laminar Flow Stagnation Problem for a Viscous Fluid with Transverse Magnetic Effect in Axial and Radial Directions

S. Ganesh, V. W. J. Anand, P. Chandrasekhar, M. Prasanna Jeyanthi, J. Jayaprabakar, and Vishwambhar S. Patil

1 Introduction

The Generalized area of research in 2-D problem of flow which is considered to be stagnation in nature for a viscous fluid in both axial and radial directions was discussed by [3]. Ordinary differential equations are obtained from the partial differential equation with boundary conditions by Blasius [3] and were integrated by Hiemenz [3] numerically. Howrath [12] then improved the solution found by Hiemenz [3]. The exact solution studied shows a very close relation to boundary layer solution problems past a body that is arbitrarily shaped and helps to validate the inference that the boundary layer theory in the field of fluid flow is considered to be an important region. Aboul-Hassan and Attia [1] discussed the hydromagnetic unsteady flow of a fluid which is of viscoelastic nature along with temperature-dependent viscosity. Attia [2] discussed the suction and injection effects of flow which is unsteady in nature between the two plates which are parallel and considered with variable properties. Carslaw and Jacger [4] discussed the heat conduction in solids. Delhi Babu and Ganesh [6] analyzed MHD flow in steady state between parallel plates with angular velocity. Ezzat et al. [7] have investigated Magnetohydrodynamic micropolar boundary layer flow problem. Ganesh and Krishnambal [8] analyzed MHD fluid flow in viscous nature which is considered between parallel plates. Ganesh and Krishnambal [9] discussed viscous MHD flow which is unsteady

S. Ganesh (✉) · V. W. J. Anand · P. Chandrasekhar · M. P. Jeyanthi
Department of Mathematics, Sathyabama Institute of Science and Technology, Chennai, India
e-mail: ganesh.maths@sathyabama.ac.in

J. Jayaprabakar
Department of Mechanical Engineering, Sathyabama Institute of Science and Technology, Chennai, India

V. S. Patil
Department of Mathematics, Govt. College of Engineering, Karad, India

© The Author(s), under exclusive license to Springer Nature Singapore Pte Ltd. 2023
S. Revankar et al. (eds.), *Recent Advances in Thermofluids and Manufacturing Engineering*, Lecture Notes in Mechanical Engineering,
https://doi.org/10.1007/978-981-19-4388-1_3

in nature between two parallel plates. Delhi Babu and Ganesh [5] revealed the study of effects in steady MHD flow which is considered between the parallel plates. Ganesh et al. [10] discussed the MHD flow of viscous fluid between two parallel plates with top suction and bottom injection. Gupta [11] discussed the electrically conducting flow of fluid with an accelerated plate with the presence of magnetic effect. Howard and Rachford [12] discussed the geometries of linear and radial flow. Katz [13] discussed about the natural gas flow. Scheidegger [15] studied about the physics of flow through porous media. Mohamed [14] discussed flow of fluid considered through a porous medium. Tzirtzilakis and Xenos [16] studied about the flow in a driven cavity in the Biomagnetic fluid.

The objective of this problem is considered as an extension for electrically conducting flow case with transverse magnetic effect. Due to friction, heat is generated in air molecules during the missile's re-entry. When the missile reenters the earth, heat is generated sufficiently because of the amount of friction of molecules in air. In order to ionize the available air which is found closer to the stagnation point, this is large sometimes due to the heating of viscous. Hence at the wall of the channel, we see that the amount of transfer of heat is reduced due to electrical conductivity of air and a magnetic effect being applied so that the forces which are induced in air retards. The point of stagnation is considered to be the origin of the plate which is infinite and is of 2-D laminar flow where x, is taken as distance in the plane flow of the plate and y is considered as the distance from the given plate, H_0 a magnetic field is applied which is considered to be a constant in the direction of the y axis.

2 Mathematical Formulation

Let the Continuity equation be

$$u_x + v_y = 0 \tag{1}$$

Momentum equations are

$$uu_x + vu_y = -\frac{1}{\rho}p_x + \nu\nabla^2 u - \frac{\sigma_e B_e^2}{\rho}u \tag{2}$$

$$uv_x + vv_y = -\frac{1}{\rho}p_y + \nu\nabla^2 v - \frac{\sigma_e B_e^2}{\rho}v \tag{3}$$

Here σ_e is considered as the electrical conductivity and $B_0 = \mu_e H_o$, where μ_e is magnetic permeability. Here we ignore the effects due to magnetic induction. Assuming the electric field to be zero because of no external forces and effect of the polarization of the field which is ionized is very small for the 2-D condition. Hence Maxwell equations are superfluous. Velocity for the flow in the neighborhood of point is considered to be stagnated at $x = 0$, $y = 0$, and is

$$\left(\frac{\partial u}{\partial x}\right)_{x=\infty} = \beta x \left(\frac{\partial v}{\partial x}\right)_{x=\infty} = -\beta y \tag{4}$$

Pressure gradient p_x which is derived from Eq. (2) with $v = 0$ is written as

$$\frac{-1}{\rho}\frac{\partial p}{\partial x} = \beta'^2 x, \tag{5}$$

Here

$$\beta'^2 = \beta^2 + \frac{\sigma B^2}{\rho}\beta \tag{6}$$

From Eqs. (5) and (6) it is seen that the magnetic field affects the pressure.
Define the Stream function ψ such that

$$\psi = \sqrt{\nu\beta'} \cdot x \cdot F(\eta), \tag{7}$$

where $\eta = \sqrt{\frac{\beta'}{\nu}} \cdot y$, satisfying.

$$u = \psi_y, v = -\psi_x \tag{8}$$

which satisfies the continuity equation.

It follows that

$$u = \beta' x F'(\eta), v = -\sqrt{\nu\beta'} F(\eta). \tag{9}$$

Using Eqs. (5) and (9) in Eq. (2), we have

$$\frac{\partial^3 F}{\partial \eta^3} + F(\eta)\frac{\partial^2 F}{\partial \eta^2} = \left(\frac{\partial F}{\partial \eta}\right)^2 - 1 + C\frac{\partial F}{\partial \eta} \tag{10}$$

where

$$C = \frac{\sigma B^2}{\rho\beta'} \tag{11}$$

$$u = v = 0 \text{ at } y = 0 \text{ and } u = \beta x \text{ at } y = \infty \tag{12}$$

are considered as the boundary conditions.

From the above boundary conditions using Eq. (9).

$$F(\eta)_{\eta=0} = 0,\ F'(\eta)_{\eta=0} = 0;\ F'(\eta)_{\eta=\infty} = \frac{\beta}{\beta'} \tag{13}$$

Using (6) and (11), we have

$$\left(\frac{\beta^2}{\beta'^2}\right) + \left(\frac{C\beta}{\beta'}\right) = 1$$

Therefore, the solution for the above equation is

$$\frac{\beta}{\beta'} = \frac{-C + \sqrt{C^2 + 4}}{2} \tag{14}$$

Thus, we have to solve (10) such that

$$F(\eta)_{\eta=0} = 0,\ F'(\eta)_{\eta=0} = 0;\ F'(\eta)_{\eta=\infty} = \frac{-C + \sqrt{C^2 + 4}}{2} \tag{15}$$

3 Results and Discussion

The non-linear Eq. (10) which actually differs from the equation derived by Blasius due to the term $C \cdot F'$ shows the impact of the magnetic field which is seen in Figs. 4, 5 and 6 is drawn using Matlab. Equation (15) is integrated for three different values of C, i.e., 0, 2, and 3. If $C = 0$, then the solution (i.e., non-magnetic case) agrees well with the solution of Howarth's [12]. Profiles drawn for F, $F^{'}$ and F'' shows the effect of the magnetic field (Figs. 1, 2, 3, 4, 5, 6, 7, 8 and 9) which are drawn using Matlab. It is seen that $F^{'}$ (Figs. 4, 5 and 6) and $\frac{u}{x}$, is constant throughout other than the layer which is of constant thickness. Pressure is decreasing along the direction of flow in the plane and this effects in the constancy of thickness which recompense its normal rise because of viscosity. Hence we see that the increase in magnetic effect results in decrease of $F^{'}$.

Using (9) the skin-friction

$$\tau_\omega = \left(\mu\frac{\partial u}{\partial y}\right)_{y=0}$$

at a station x is given by

$$\left(\frac{\sqrt{\nu}\tau_\omega}{x\mu\beta'^{3/2}}\right) = F''(0). \tag{16}$$

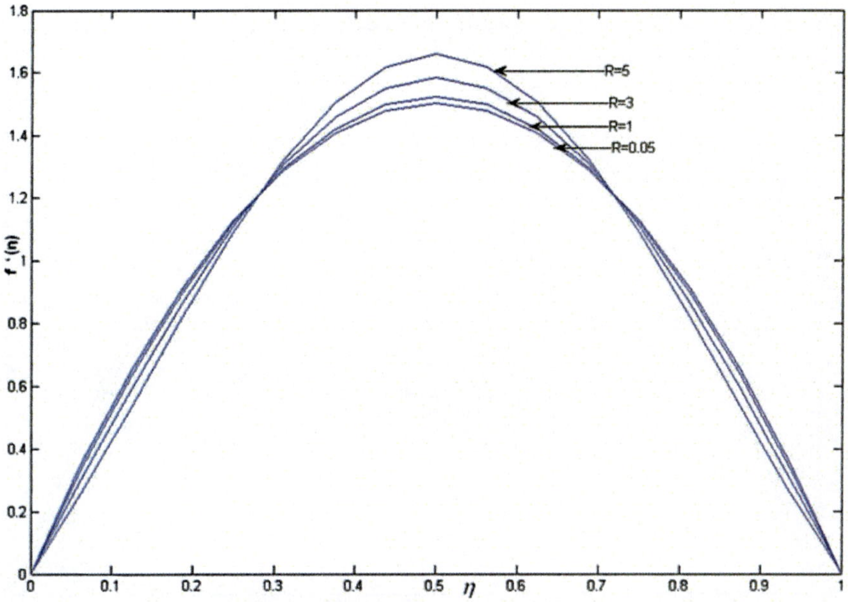

Fig. 1 *F* variation for $C = 0$ and $R > 0$

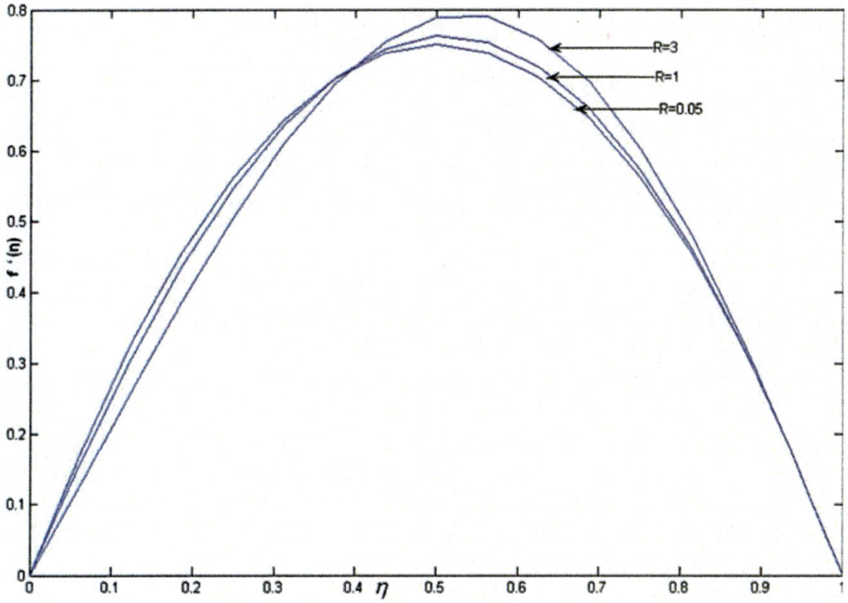

Fig. 2 *F* variation for $C = 2$ and $R > 0$

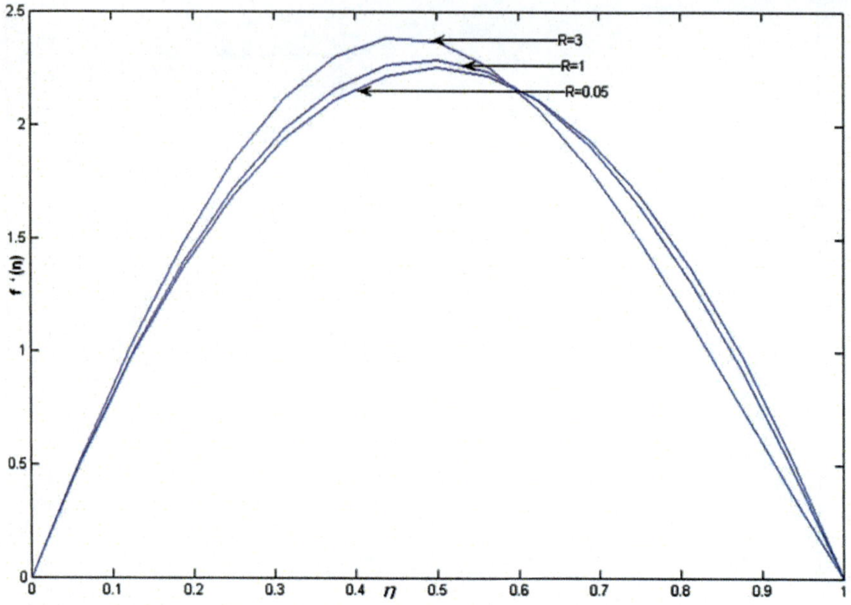

Fig. 3 F variation for $C = 3$ and $R > 0$

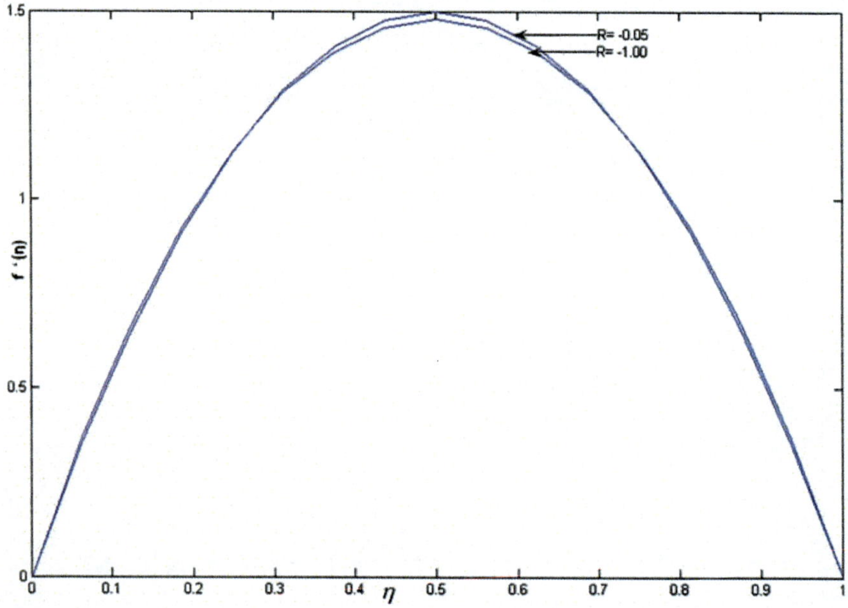

Fig. 4 F' variation for $C = 0$ and $R > 0$

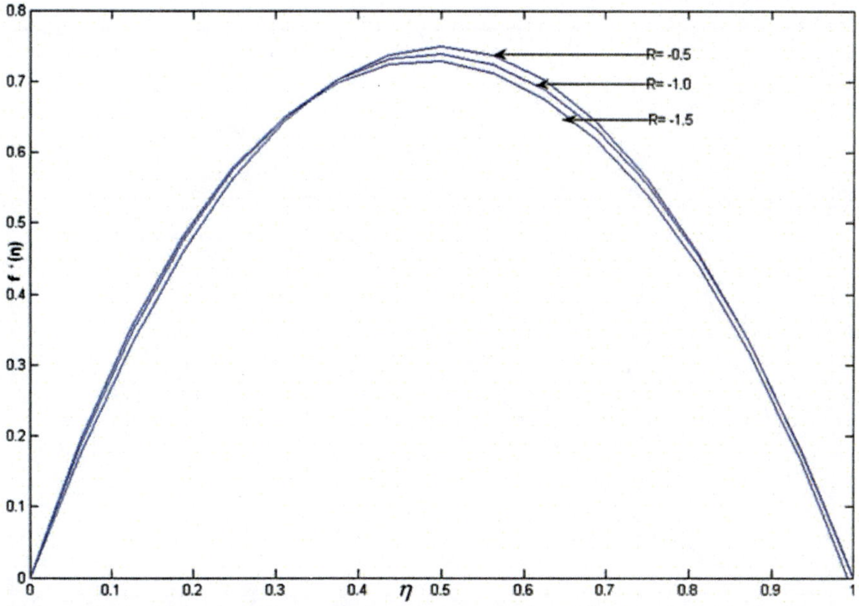

Fig. 5 $F^{'}$ variation for $C = 2$ and $R > 0$

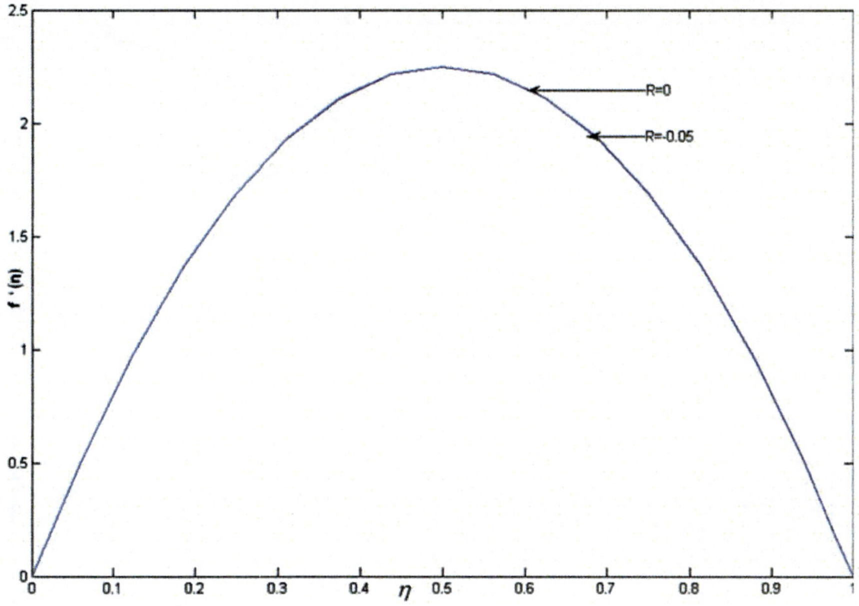

Fig. 6 $F^{'}$ variation for $C = 3$ and $R > 0$

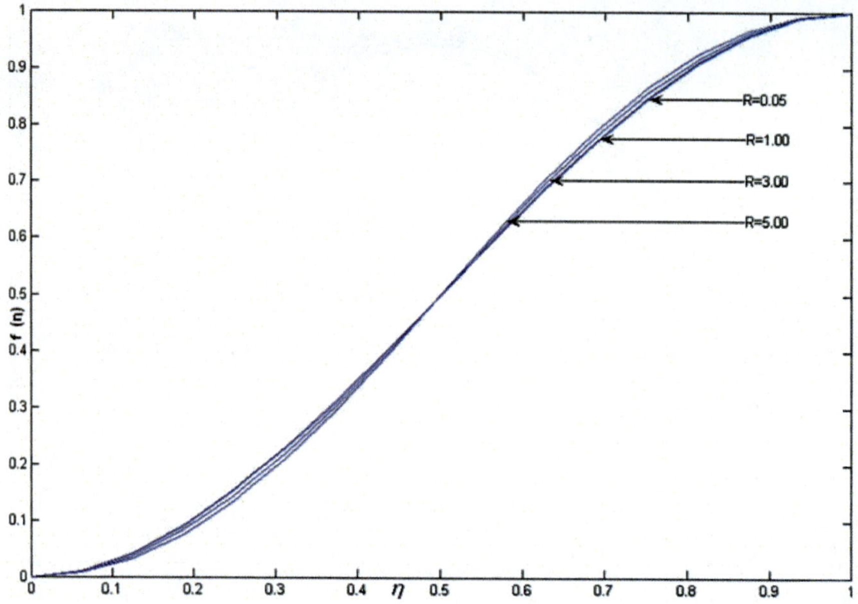

Fig. 7 F'' variation for $C = 0$ and $R > 0$

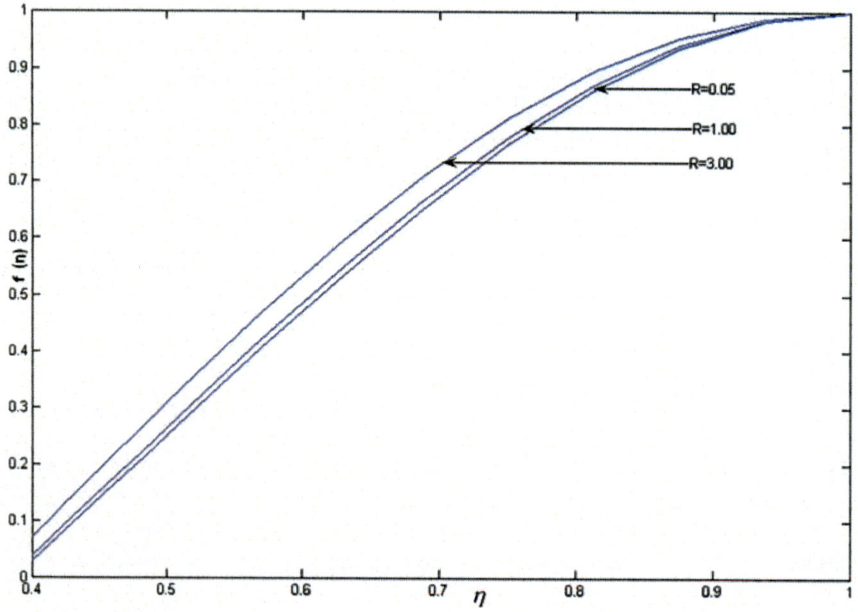

Fig. 8 F''variation for $C = 2$ and $R > 0$

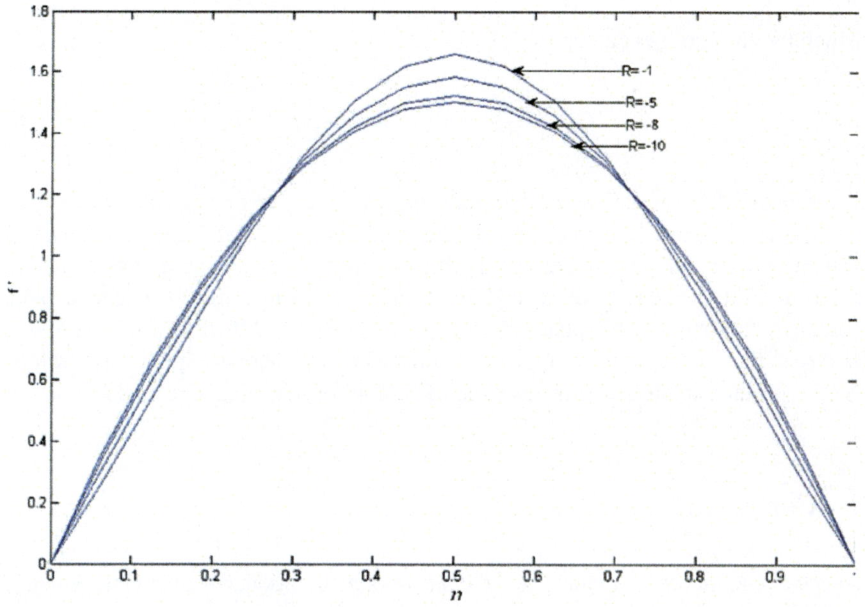

Fig. 9 F'' variation for $C = 3$ and $R > 0$

$F''(0)$ is given as 1.25, 0.71 and 0.60 for $C = 0$, 2 and 3.

We see that the drag coefficient decreases due to the increase in magnetic effect which is seen from Figs. 6, 7, 8 and 9.

Applying Eq. (9) in Eq. (3), we see that

$$v\frac{\partial v}{\partial y} = -\frac{1}{\rho}\cdot\frac{\partial p}{\partial y} + v\cdot\frac{\partial^2 v}{\partial y^2},$$

which when integrated gives

$$\frac{p}{\rho} = \varphi(x) - \frac{v^2}{2} + v\frac{\partial v}{\partial y}, \tag{17}$$

Here $\varphi(x)$ is considered as an arbitrary function. From (5) we get

$$\varphi'(x) = -\beta'^2 x.$$

Hence Eq. (17) changes to

$$\frac{p}{\rho} = C - \frac{\beta'^2 x^2}{2} - \frac{1}{2}v^2 + v\frac{\partial v}{\partial y},$$

which by virtue of (9) becomes

$$\frac{p}{\rho} = C - \frac{\beta'^2 x^2}{2} - \frac{\nu\beta'}{2}F^2(\eta) - \nu\beta' F'(\eta). \tag{18}$$

From the above equation, we determine the pressure given $F(\eta)$ and $F^{'}(\eta)$. Hence we see that the increase in magnetic effect results in decrease of $F^{'}$. Also, it is visible that the drag coefficient decreases due to the increase in magnetic effect and we determine the pressure given F and $F^{'}$. Hence due to increase in magnetic field, velocity component of the plate which is parallel decreases its drag coefficient. It is also seen that the velocity component is constant everywhere except the layer which is of constant thickness and is observed to be true for non-magnetic cases also.

4 Conclusion

Hence due to increase in magnetic field the velocity component of the plate which is parallel decreases its drag coefficient. It is also seen that the velocity component is constant everywhere except the layer which is of constant thickness and is observed to be true for non-magnetic cases also.

References

1. Aboul-Hassan AL, Attia HA (2002) Unsteady hydromagnetic flow of a viscoelastic fluid with temperature-dependent viscosity. Can J Phys 80:1015–1024
2. Attia HA (2005) The effect of suction and injection on the unsteady flow between two parallel plates with variable properties. Tamkang J Sci Eng 8:17–22
3. Blasius H, Hiemenz (1908) Z Angew Math Phys 56:1–37
4. Carslaw HS, Jacger JC. Book on conduction of heat in solids. Oxford University Press
5. Delhi Babu R, Ganesh S (2017) The effects of steady magnetohydrodynamic stokes flow between two parallel porous plates. ICIECS
6. Delhi Babu R, Ganesh S (2018) The mathematical model for steady magnetohydrodynamic flow between parallel porous plates with an angular velocity. Int J Ambient Energy. https://doi.org/10.1080/01430750.2018.1507944
7. Ezzat M, Othman M, Helmy K (1999) A problem of a micropolar MHD boundary layer flow. Can J Phys 77:813–827
8. Ganesh S, Krishnambal S (2006) Magnetohydrodynamic flow of viscous fluid between two parallel porous plates. J Appl Sci (ANSINET) 11:2420–2425
9. Ganesh S, Krishnambal S (2007) Unsteady magnetohydrodynamic stokes flow of viscous fluid between two parallel porous plates. J Appl Sci 7(3):374–379
10. Ganesh S, Delhi Babu R, Anand VWJ, Chandrasekar P (2018) Magnetohydrodynamic flow of viscous fluid between two parallel porous plates with bottom injection and top suction. Int J Ambient Energy:1–4.https://doi.org/10.1080/01430750.2018.1531266
11. Gupta AS (1960) On the flow of an electrically conducting fluid near an accelerated plate in the presence of a magnetic field. J Phys Soc Jpn 15:1894–1897
12. Howard DS, Rachford HH. Trans AIME 207:92

13. Katz DL. Hand book of natural gas engineering. McGraw-Hill
14. Selim MA (1961) Flow through a porous medium. J Appl Sci Res 10:363–368
15. Scheidegger AE. Book on the physics of flow through porous media. The Macmillan Company, New York
16. Tzirtzilakis EE, Xenos MA (2013) Biomagnetic fluid flow in a driven cavity. Meccanica 48(1):187–200

Analysis of Dufour and Soret Effect on Unsteady MHD Prandtl Fluid Flow Past an Expanding/Shrinking Surface

Nalini S. Patil, Vishwambhar S. Patil, J. Jayaprabakar, and S. Ganesh

Nomenclature

a	Expanding/shrinking parameter
A	Fluid consistency index
b	Suction/injection parameter
B	Free-stream velocity
$B(x)$	Magnetic field
B_0	Magnetic field strength
C	Concentration
C_1	Material fluid parameter
C_w	Concentration at the wall
C_∞	Free-stream concentration
c	Constant
c_p	Specific heat at constant pressure
c_s	Concentration susceptibility
Du	Dufour number
D_m	Coefficient of mass diffusivity

N. S. Patil
Department of Mathematics, Pratap College, Amalner 425401, India

V. S. Patil (✉)
Department of Mathematics, Govt. College of Engineering, Karad 415124, India
e-mail: vishme.p@gmail.com

J. Jayaprabakar
Department of Mechanical Engineering, Sathyabama Institute of Science and Technology, Chennai, India

S. Ganesh
Department of Mathematics, Sathyabama Institute of Science and Technology, Chennai, India

© The Author(s), under exclusive license to Springer Nature Singapore Pte Ltd. 2023
S. Revankar et al. (eds.), *Recent Advances in Thermofluids and Manufacturing Engineering*, Lecture Notes in Mechanical Engineering,
https://doi.org/10.1007/978-981-19-4388-1_4

k_1	Mean absorption coefficient
k_t	Thermal diffusion ratio
M	Magnetic parameter
Pr	Prandtl number
q_r	Radiative heat flux
R	Radiation parameter
S_0	Moving surface velocity
Sc	Schmidt number
Sr	Soret number
T	Fluid temperature
T_w	Temperature at the wall
T_∞	Ambient temperature
u_e	Stream velocity
u, v	Velocity components
V_0	Suction/injection velocity of surface

Greek Symbols

τ_{xy}	Shear stress tensor
ε, δ	Prandtl fluid parameter
σ	The electric conductivity of a fluid
σ_s	Stefan–Boltzmann constant
κ	Thermal conductivity of the fluid
η	Similarity variable
ρ	The density of the fluid
λ	Unsteadiness parameter
θ	Dimensionless temperature
ν	Kinematic viscosity
ϕ	Dimensionless concentration

Subscripts

w	Surface condition
∞	Condition at infinity

1 Introduction

Non-Newtonian fluids are not satisfying the linear stress–strain relationship, and hence, mathematical framing of the rheology of such fluids is a little bit complex in nature. But, the wide range of applications in manufacturing, food, biomedical, geophysics, nuclear reactors, chemical engineering, etc., of non-Newtonian fluids attracts academic scholars to explore the physical aspects of such flow phenomenon, and hence, we found different flow models. Among these models, inelastic models are very useful for the investigation of mixed effects of elastic and viscous properties. Hamid et al. [1] reported an investigation of unsteady stagnation point flow of Prandtl fluid past an infinite plate. Akbar et al. [2] obtained a dual solution for MHD stagnation point Prandtl fluid flow past an expanding/shrinking surface. Hayat et al. [3] investigated Cattaneo-Christov double diffusion effect on 3D Prandtl fluid flow. Khan et al. [4] analyzed the stratification and energy generation effect on MHD Prandtl fluid. Recently, Patil et al. [5] reported the study of chemically reacting MHD Prandtl nanofluid past a stretching surface.

Magnetohydrodynamics (MHD) is dealing with the dynamics of electrically conducting fluids and their behavior under the influence of an external magnetic field. Examples of such electrically conducting fluids are liquid metals, salty water, electrolytes, plasma, etc. A wide range of applications of the MHD effect can be seen in power generators, accelerators, compressors, MRI scanning, plasma confinement, crude oil purification, thermonuclear fusion plants, propulsion systems in aircraft, etc. Amanulla et al. [6] reported slip effects on the steady MHD Prandtl–Eyring fluid flow over a permeable sphere. Mandal et al. [7] reported a double diffusion effect on MHD mixed convection on the inclined stretching surface. Rajput et al. [8] investigated unsteady MHD mixed convectional flow along with a vertical permeable plate. Ganesh et al. [9] analyzed unsteady MHD compressible flow over a stainless steel surface. Dessie et al. [10] established the results for viscous dissipation effect on MHD unsteady flow of immersed nanoparticles past stretching device with chemical reaction. Prasad et al. [11] reported the analysis for MHD of suspended nanoparticles in base liquid over a semi-infinite plate with thermo-diffusion, radiation absorption, and chemical reaction effect. Khan and Alqahtani [12] introduced the mathematical framing of MHD nanofluid flow over a surface embedded in porous media. Further, a recent study of the MHD effect with various fluid models can be seen through [13–17].

The heat and mass transfer phenomenon plays a crucial role in many day-to-day activities. The applications can be listed as manufacturing processes, cooling processes, heat insulation, paper production, geothermal activities, reactor engineering, nuclear power generation, solar energy, polymer manufacturing, food industries, glass blowing, ceramic, paint industry, etc. Mostly, in heat and mass transfer problems, researchers ignored the Dufour and Soret effect, due to its small quantity than the effect characterized by Fourier's and Fick's laws. Energy flux developed due to concentration dissimilarities is the Dufour effect, while mass flux generated due to temperature differences is the Soret effect. These effects play a significant role in hydrology, petroleum, reactor engineering, drying process, geosciences, etc. Hayat et al. [18] explored the double diffusion effect on the rotating peristaltic flow of Prandtl flow in flexible walls. Reddy and Chamkha [19] studied the Dufour–Soret and thermophoresis effect on micro-polar nanofluids past a stretching surface. The double diffusion effect on stagnation point flow on a shrinking surface was studied by Bhattacharyya et al. [20]. Bhatti and Rashidi [21] investigated the double diffusion effect on Williamson nanofluid running over porous shrinking/stretching sheets. Soret and Dufour's effect on unsteady heat and mass transmission of Eyring–Powell flow over an expanding porous surface was studied by Layek et al. [22]. Patil et al. [23] observed a double diffusion effect on the MHD Prandtl nanofluid flow which was generated by linear stretching surface. More investigations about thermo-diffusion and diffusion thermo-effect can be seen through the literature [24–28].

The present study investigates the Dufour and Soret effect on heat and mass transmission of unsteady Prandtl fluid flow past an expanding/shrinking surface. The similarity variables are used to transform a highly nonlinear system of PDEs into the system of ODEs. The transformed ODEs are numerically investigated with R–K's fourth-order scheme [5, 16] accompanied by the shooting technique.

2 Problem Formulation

The incompressible viscoelastic non-Newtonian Prandtl fluid flow over an expanding/shrinking sheet is considered for the study. The device is stretched along the x-axis, while the y-axis is kept perpendicular to it. The fluid is moving in $y \geq 0$ region, and the magnetic field of value $B = B_0 x^{-1/2}$ is applied normally to the flow. The flow is in motion with velocity $u = \frac{ax}{1-ct}$. The temperature and concentration at $y = 0$ and $y = \infty$ are T_w, $T_\infty$and C_w, C_∞, respectively, as shown in Fig. 1 [22].

According to the Patil et al. [29], the stress–strain relationship for Prandtl fluid is given by

$$\tau_{xy} = A \sin^{-1}\left(\frac{1}{C_1} u_y\right) \tag{1}$$

where A and C_1 are fluid consistency indices and material fluid parameter.

Expanding $\sin^{-1}\left(\frac{1}{C_1} u_y\right)$ using the Taylor series, we have

$$\sin^{-1}\left(\frac{1}{C_1} u_y\right) \cong \frac{1}{C_1} u_y + \frac{1}{6}\left(\frac{1}{C_1} u_y\right)^3, \left|\frac{1}{C_1} u_y\right| \leq 1 \tag{2}$$

Using common boundary layer approximation, the governing system of equation is given as

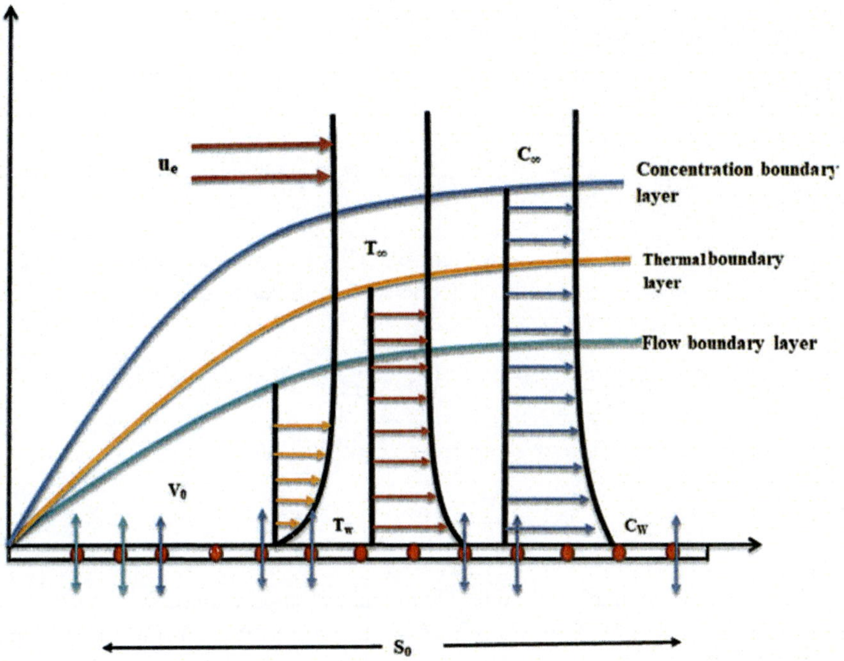

Fig. 1 Physical sketch of the governing flow

$$u_x + v_y = 0 \tag{3}$$

$$u_t + uu_x + vu_y = (u_e)_t + u_e(u_e)_x + \frac{A}{\rho C_1}u_{yy} + \frac{A}{2\rho C_1^3}(u_y)^2 u_{yy} - \frac{\sigma B^2(x)u}{\rho} \tag{4}$$

$$T_t + uT_x + vT_y = \frac{\kappa}{\rho c_p}T_{yy} - \frac{1}{\rho c_p}(q_r)_y + \frac{D_m k_t}{c_s c_p}C_{yy} \tag{5}$$

$$C_t + uC_x + vC_y = D_m C_{yy} + \frac{D_m k_t}{T_m}T_{yy} \tag{6}$$

With boundary conditions

$$u = S_0, v = V_0, T = T_w, C = C_w \text{ at } y = 0 \tag{7}$$

$$u \to u_e, T \to T_\infty, C \to C_\infty \text{ at } y \to \infty \tag{8}$$

Following Patil et al. [5], the radiation heat flux q_r is mathematically expressed as

$$q_r = -\frac{4\sigma_s}{3k_1}(T^4)_y \tag{9}$$

The nonlinear term T^4 in Eq. (9) can be simplified using Taylor's series about T_∞, and after deleting higher-order terms, we get $T^4 \cong 4T_\infty^3 T - 3T_\infty^4$. Hence, Eq. (5) is reduced as

$$T_t + uT_x + vT_y = \frac{\kappa}{\rho c_p} T_{yy} + \frac{16\sigma_s T_\infty^3}{3k_1 \rho c_p} T_{yy} + \frac{D_m k_t}{c_s c_p} C_{yy} \tag{10}$$

Now, the governing nonlinear system of PDEs is remodeled using the group-theoretic technique. For this purpose, we select the similarity variables as

$$\eta = \sqrt{\frac{a}{\nu(1-ct)}}y, v = -\sqrt{\frac{\nu a}{1-ct}} f(\eta), u = \frac{ax}{1-ct} f'(\eta),$$
$$\theta(\eta) = \frac{T - T_\infty}{T_w - T_\infty}, \phi(\eta) = \frac{C - C_\infty}{C_w - C_\infty}, V_0 = -b\sqrt{\frac{\nu a}{1-ct}}$$
$$u_e = \frac{bx}{1-ct}, S_0 = \frac{a^2 x}{1-ct}, B(x) = \frac{B_0}{\sqrt{x}} \tag{11}$$

Employing the similarity variables (11) in the above-mentioned system of equations with boundary conditions in (7) and (8), we get the following transformed ODEs:

$$\left[\varepsilon\left(1+\delta f''^2\right)\right]f''' - \lambda\left[\frac{\eta}{2}f'' + f'\right] + ff'' - f'^2 + B\lambda + B^2 - Mf' = 0 \tag{12}$$

$$\frac{1}{\Pr}\left(1 + \frac{4}{3}R\right)\theta'' + \mathrm{Du}\phi'' - \frac{\lambda}{2}\eta\theta' + f\theta' = 0 \tag{13}$$

$$\frac{1}{\mathrm{Sc}}\phi'' + \mathrm{Sr}\theta'' - \frac{\lambda}{2}\eta\phi' + f\phi' = 0 \tag{14}$$

and modified boundary limits

$$f'(0) = a, f(0) = b, \theta(0) = 1, \phi(0) = 1 \tag{15}$$

$$f'(\infty) = B, \theta(\infty) = 0, \phi(\infty) = 0 \tag{16}$$

where the derivative is taken concerning η.

The physical parameters Pr, Sr, Du, λ, Sc, B, δ, ε and R denote Prandtl number, Soret number, Dufour number, unsteady parameter, Schmidt number, free-stream velocity parameter, fluid parameters, and radiation index which are as follows:

$$\mathrm{Sr} = \frac{D_m k_t (T_w - T_\infty)}{T_m \nu (C_w - C_\infty)}, \mathrm{Du} = \frac{D_m k_t (C_w - C_\infty)}{c_s c_p \nu (T_w - T_\infty)}, \Pr = \frac{\rho c_p \nu}{\kappa}, \lambda = \frac{c}{a},$$

$$\mathrm{Sc} = \frac{\nu}{D_\mathrm{m}}, R = \frac{4\sigma_s T_\infty^3}{kk_1}, \varepsilon = \frac{A}{\rho\nu C_1}, B = \frac{b}{a}, \delta = \frac{a^3x^2}{2C_1^2\nu(1-ct)^3}$$

$$M = \frac{\sigma B_0^2(1-ct)}{a\rho x}.$$

2.1 Numerical Treatment

The transformed Eqs. (12)–(14) including boundary conditions (15)–(16) are extremely complex. Here we integrate the system of initial value problem (IVP) until all boundary limits are fulfilled (see ref. Rajput et al. [30]). Hence, we form the initial value problem (IVP) and solved using shooting scheme.

$$f' = l \tag{17}$$

$$l' = m \tag{18}$$

$$m' = \frac{\lambda\left(\frac{\eta}{2}m + l\right) + l^2 - fm + Ml - (\lambda B + B^2)}{\varepsilon + \varepsilon\delta m^2} \tag{19}$$

$$\theta' = p \tag{20}$$

$$p' = \frac{3\,\mathrm{Pr}\left(\lambda\frac{\eta}{2} - f\right)(p - \mathrm{Du}\,\mathrm{Sc}\,q)}{3 + 4R - 3\,\mathrm{Pr}\,\mathrm{Sc}\,\mathrm{Sr}\,\mathrm{Du}} \tag{21}$$

$$\phi' = q, \tag{22}$$

$$q' = \frac{\left(\lambda\frac{\eta}{2} - f\right)(3\mathrm{ScSr}\,\mathrm{Pr}\,p - 3q - 4Rq)}{3\,\mathrm{Pr}\,\mathrm{Sr}\,\mathrm{Sc}\,\mathrm{Du} - 3 - 4R}. \tag{23}$$

Along with boundary limits,

$$f'(0) = a, f(0) = b, \theta(0) = 1, \phi(0) = 1 \tag{24}$$

$$f'(\infty) = B, \theta(\infty) = 0, \phi(\infty) = 0. \tag{25}$$

3 Result and Discussion

The present section is organized to explore the flow affecting non-dimensional physical quantities such as an expanding/shrinking parameter a, Prandtl number Pr, Dufour number Du, Soret number Sr, magnetic parameter M, radiation parameter R, and Schmidt number Sc on non-dimensional velocity $\left(f'(\eta)\right)$, temperature (θ), and concentration (ϕ). For this purpose, we choose the Prandtl number, $\text{Pr} = 0.21$ for air at 20 °C, $\text{Pr} = 1.06$ for electrolyte solution [31], and $\text{Pr} = 2.0$ for pure water. Throughout the study, we assume $\delta = 0.6$, $\text{Pr} = 2.0$, $R = 2.0$, $\text{Du} = 0.4$, $\text{Sc} = 0.6$, $M = 1.4$, $a = 1$, and $\text{Sr} = 1$. The result obtained is plotted and discussed graphically with restrictions imposed in the subsequent section.

The effect of ε on the velocity profile is illustrated in Fig. 2. The progressive values of the fluid parameter retard the fluid viscosity, so the fluid can smoothly moves over the surface and overall velocity rises slowly. It should be noted that, due to slow motion, particles get settled down and rise boundary layer thickness. Here worth mentioning that the effect δ on the motion, energy, and concentration profiles is very less and hence neglected its graphical interpretation.

The performance of the energy profile against the expanding/shrinking parameter is portrayed in Fig. 3. The rising values of a provide more time for the fluid to cool down, and the fluid releases the heat which ultimately leads to a decline in the thermal boundary layer thickens. Hence, progressive values of expanding/shrinking parameters decline the fluid temperature. The effect of suction/injection restriction on velocity and concentration profiles is illustrated in Figs. 4 and 5, respectively. As

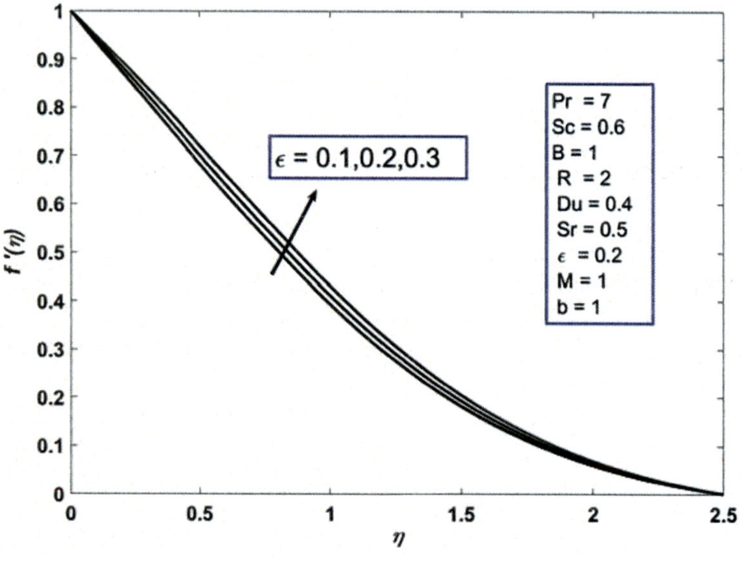

Fig. 2 Velocity profile for distinct values ε

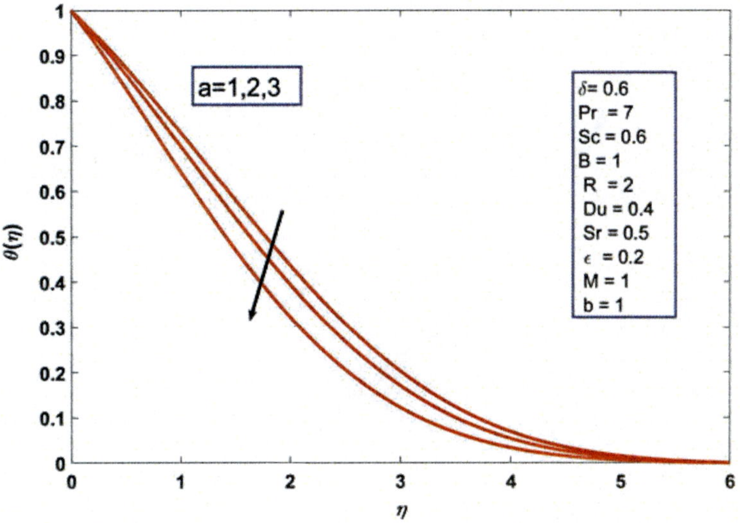

Fig. 3 Temperature profile for fluid parameter a

we know, $b > 0$ represents suction and $b < 0$ is injection. Here we consider the suction, and we observed that for accelerating values of b, the fluid moves fast and hence respective growth is captured in Fig. 4. Due to the rapid movement of particles, they cannot be settling down on the stretching surface and are unable to augment to form a thick boundary layer. Hence, the corresponding concentration profile declines for higher values of b as shown in Fig. 5. The effect of the Prandtl number on the temperature profile is sketched in Fig. 6. Prandtl number characterizes the thermal boundary layer thickness over the velocity boundary layer. The higher value leads to lower thermal diffusivity, and fluid temperature declines.

The effect of radiation parameters on the temperature profile can be observed in Fig. 7. For higher values of radiation parameters, more energy is entered into the flow system. This boosts up the kinetic energy of fluid particles which accelerates the thermal boundary layer.

The effect of the Dufour number Du on temperature and concentration profiles is sketched through Figs. 8 and 9, respectively. We know that energy flux induced due to concentration gradient is the Dufour effect. Hence, for increasing values of the Dufour number, more energy in the system is noted which ultimately increases the thermal boundary layer. The reverse effect is observed on the concentration profile. Hence, concentration acts as a decreasing function of the Dufour number as shown in Fig. 9. Further, the generation of mass flux due to energy differences is the Soret effect, and its influence on non-dimensional concentration profile can be seen in Fig. 10. From Fig. 10, it is clear that for the rise in Soret number, the mass flux induces inflow which boosts the boundary layer thickness. Hence, a typical rise in concentration is captured. Schmidt number (Sc) characterizes the ratio of thermal diffusivity to mass diffusivity. The larger the values, Schmidt number signifies the

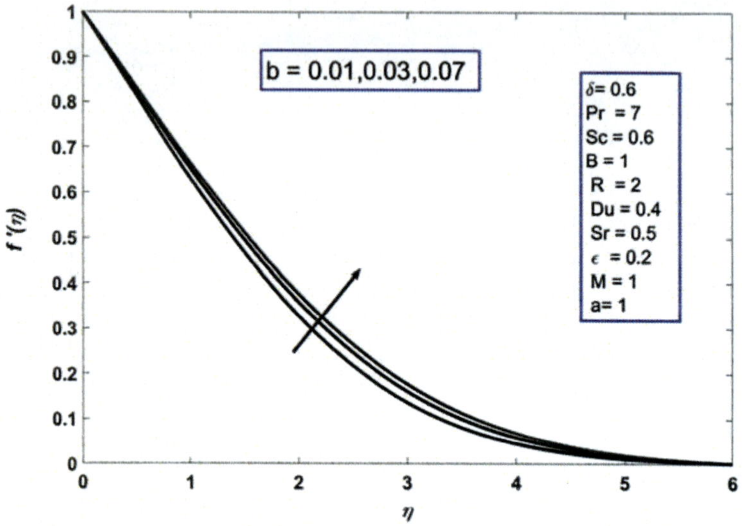

Fig. 4 Velocity profiles for distinct values fluid parameter b

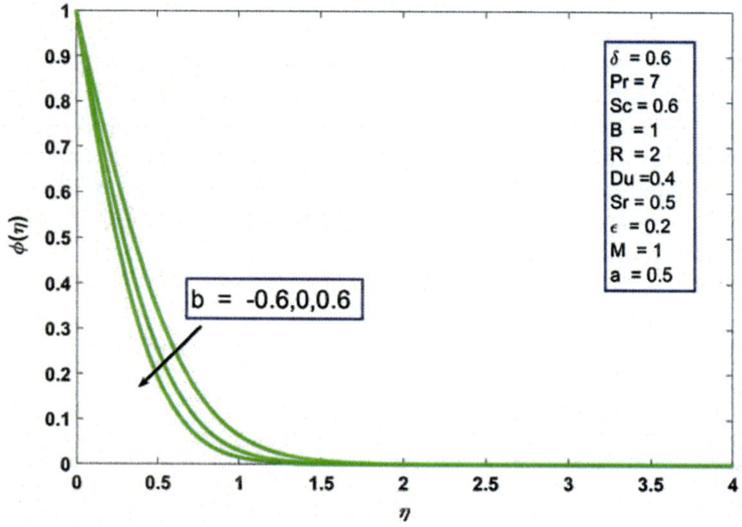

Fig. 5 Concentration profiles for distinct stretching parameter b

high thermal diffusion rate and low mass diffusion. Hence, the concentration declines gradually as observed in Fig. 11.

The effect of magnetic parameters on velocity profile is illustrated in Fig. 12. For magnetic parameter $M \neq 0$, fluid particles experience drag due to induced Lorentz force. The drag motivates the particle to slow down and enhances the fluid viscosity

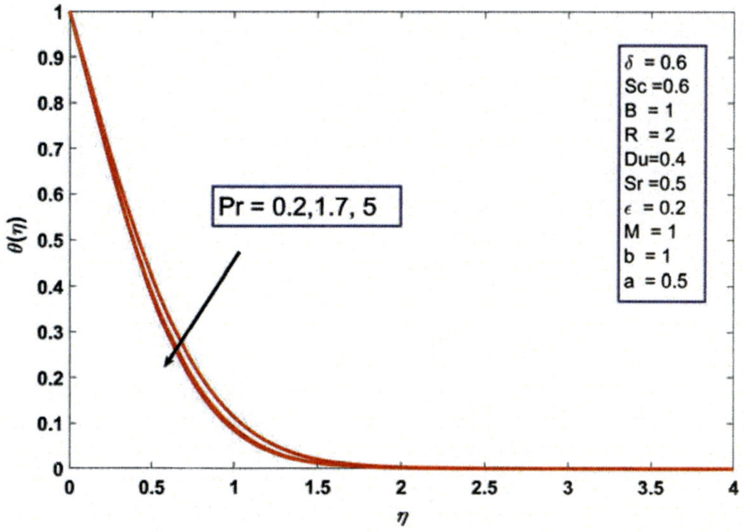

Fig. 6 Temperature profiles for distinct Prandtl number Pr

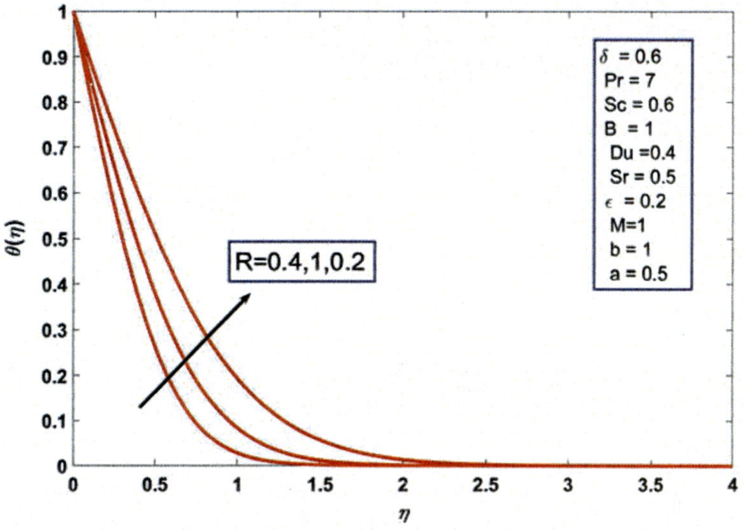

Fig. 7 Temperature profiles for distinct radiation parameter R

which ultimately declines the flow motion. Thus, the applied magnetic field controls the fluid motion over a surface. Figure 13 explores the effect of free-stream velocity on concentration. For higher values of B, the fluid particles are accelerated and move faster over the surface, so that unable to augment in the boundary region. This leads to a decline in concentration profile.

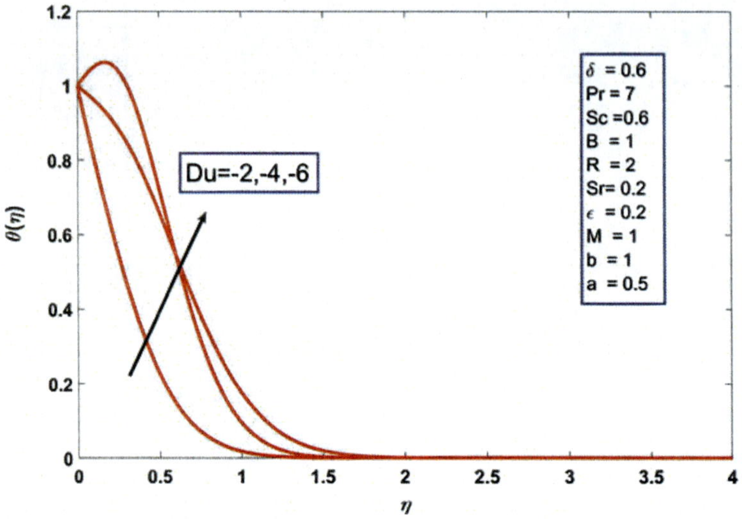

Fig. 8 Temperature profiles for distinct Dufour number Du

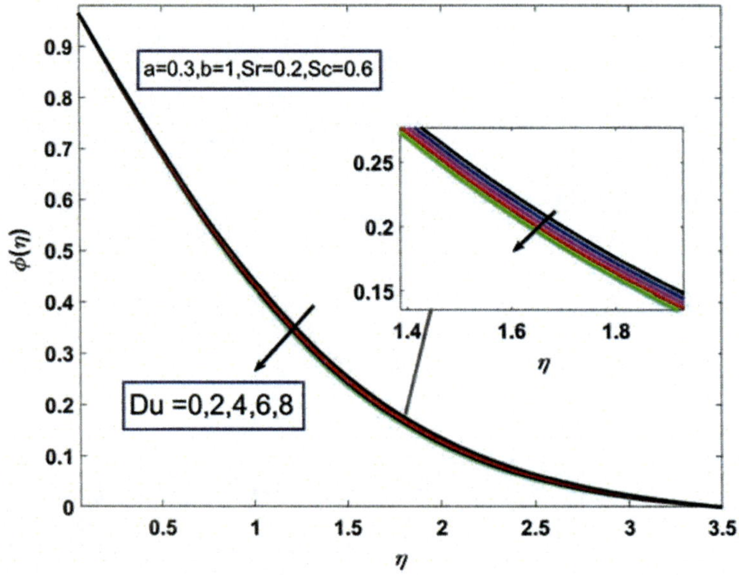

Fig. 9 Concentration profiles for distinct Dufour number Du

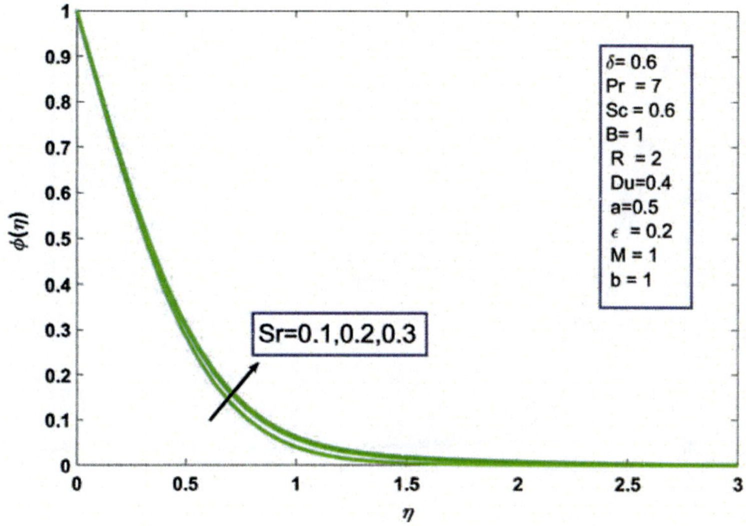

Fig. 10 Concentration profiles for distinct Soret number Sr

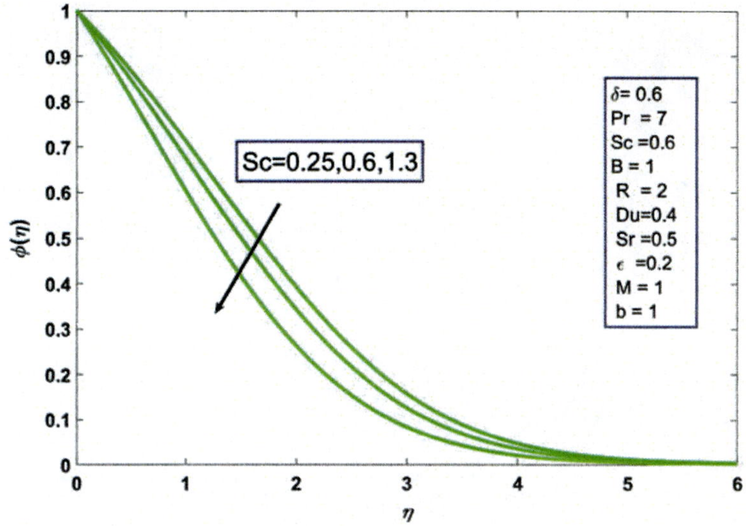

Fig. 11 Concentration profiles for distinct Schmidt number Sc

It is worth mentioning here that concentration exceeds for larger values of Soret numbers. The boundary layer thickness decay was noticed for the distinct values of Dufour number. It is hoped that the current analysis will serve as an adequate endorsement to the field of knowledge.

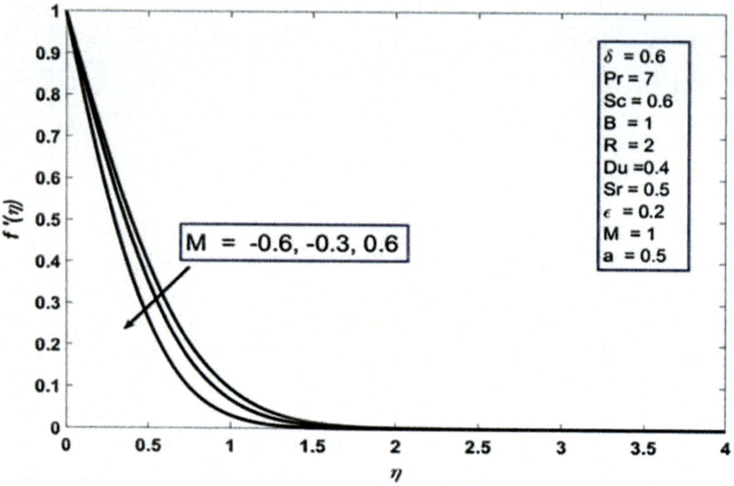

Fig. 12 Velocity profile for *M*

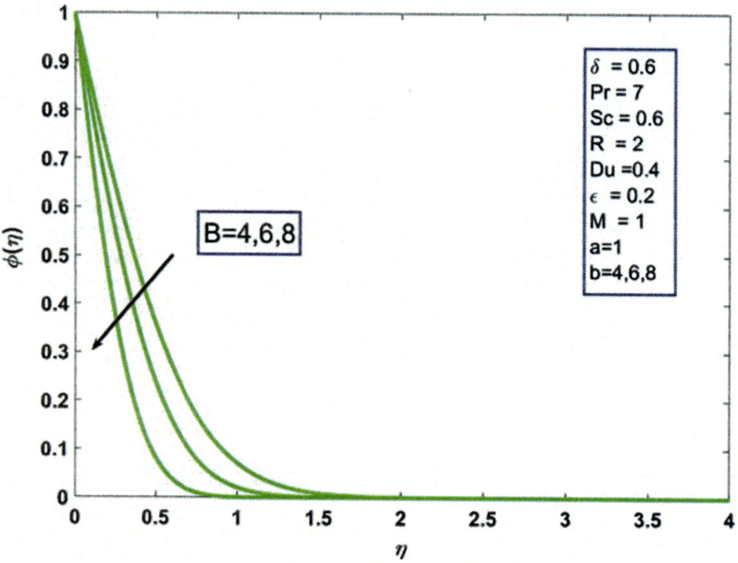

Fig. 13 Concentration profiles for *B*

4 Conclusion

The flow development has been identified by affiliated parameters, their consequences on the different physical aspects have been examined, and the results acquired are acknowledged graphically. The outcomes of the present study are as follows:

- The motion is decreasing function of an external magnetic field. However, it is an increasing function suction/injection parameter.
- The energy profile accelerates for the higher amount of radiation parameter, but it shows reverse nature for the higher amount of expanding/shrinking parameter, Prandtl number.
- The concentration declines for a higher amount of suction/injection parameter, Dufour number, Schmidt number, and free-stream velocity. It increases for higher values of Soret number.

References

1. Hamid M, Zubair T, Usman M, Khan ZH, Wang W (2019) Natural convection effects on heat and mass transfer of slip flow of time-dependent Prandtl fluid. J Comput Des Eng 6(4):584–592
2. Akbar NS, Khan ZH, Haq RU, Nadeem S (2014) Dual solutions in MHD stagnation-point flow of Prandtl fluid impinging on shrinking sheet. Appl Math Mech 35(7):813–820
3. Hayat T, Aziz A, Muhammad T, Alsaedi A (2018) Three-dimensional flow of Prandtl fluid with Cattaneo–Christov double diffusion. Res Phys 9:290–296
4. Khan I, Hussain A, Malik MY, Mukhtar S (2020) On magnetohydrodynamics Prandtl fluid flow in the presence of stratification and heat generation. Phys A 540:123008
5. Patil AB, Humane PP, Patil VS, Rajput GR (2021) MHD Prandtl nanofluid flow due to convectively heated stretching sheet below the control of chemical reaction with thermal radiation. Int J Amb Energy 24:1–13. https://doi.org/10.1080/01430750.2021.1888803
6. Amanulla CH, Saleem S, Wakif A, AlQarni MM (2019) MHD Prandtl fluid flow past an isothermal permeable sphere with slip effects. Case Stud Thermal Eng 14:100447. https://doi.org/10.1016/j.csite.2019.100447
7. Mandal B, Bhattacharyya K, Banerjee A, Verma AK, Gautam AK (2020) MHD mixed convection on an inclined stretching plate in Darcy porous medium with Soret effect and variable surface conditions. Nonlin Eng 9(1):457–469
8. Rajput GR, Patil VS, Jadhav BP (2017) MHD mixed flow of unsteady convection with radiation over a vertical porous plate: lie group symmetry analysis. Appl Comput Mech 11(2):1–12. http://hdl.handle.net/11025/26592
9. Ganesh S, Patil VS, Babu RD, Chandrasekar P, Anish M (2021) Unsteady magnetohydrodynamic flow of a compressible liquid through a stainless steel plate of porous medium. Mater Today Proc 44:3924–3928
10. Dessie H, Kishan N (2015) Unsteady MHD flow of heat and mass transfer of nanofluids over stretching sheet with a non-uniform heat/source/sink considering viscous dissipation and chemical reaction. Int J Eng Res Africa 14:1–12. https://doi.org/10.4028/www.scientific.net/JERA.14.1
11. Prasad PD, Kumar RVMSSK, Varma SVK (2018) Heat and mass transfer analysis for the MHD flow of nanofluid with radiation absorption. Ain Shams Eng J 9(4):801–813. https://doi.org/10.1016/j.asej.2016.04.016
12. Khan I, Alqahtani AM (2019) MHD nanofluids in a permeable channel with porosity. Symmetry 11(3): 378. https://doi.org/10.3390/sym11030378
13. Rajput GR, Patil VS, Prasad JK (2018) MHD flow of Powell-Eyring nanofluid containing nanoparticles and gyrotactic microorganisms over a stretched surface. Rev Compos Mater Av 28(3):405

14. Patil VS, Patil AB, Ganesh S, Humane PP, Patil NS. Unsteady MHD flow of a nano Powell-Eyring fluid near stagnation point past a convectively heated stretching sheet in the existence of chemical reaction with thermal radiation. Mater Today Proc 44:3767–3776
15. Rajput GR, Jadhav BP, Patil VS, Salunkhe SN (2021) Effects of nonlinear thermal radiation over magnetized stagnation point flow of Williamson fluid in porous media driven by stretching sheet. Heat Transfer 50(3):2543–2557
16. Humane PP, Patil VS, Patil AB (2021) Chemical reaction and thermal radiation effects on magnetohydrodynamics flow of Casson–Williamson nanofluid over a porous stretching surface. Proc Inst Mech Eng Part E J Process Mech Eng 2021:09544089211025376
17. Patil AB, Patil VS, Humane PP, Patil NS, Rajput GR (2021) Thermally and chemically reacted MHD Maxwell nanofluid flow past an inclined permeable stretching surface. Proc Inst Mech Eng Part E J Process Mech Eng 2021
18. Hayat T, Zahir H, Tanveer A (2018) Soret and Dufour effects on MHD peristaltic flow of Prandtl fluid in a rotating channel. Res Phys 8:1291–1300. https://doi.org/10.1016/j.rinp.2018.01.058
19. Sudarsana Reddy P, Chamkha AJ (2016) Soret and Dufour effects on MHD heat and mass transfer flow of a micro-polar fluid with thermophoresis particle deposition. J Naval Arch Marine Eng 13(1):39–50. https://doi.org/10.3329/jname.v13i1.23974
20. Bhattacharyya K, Layek GC, Seth GS (2014) Soret and Dufour effects on convective heat and mass transfer in stagnation-point flow towards a shrinking surface. Phys Scr 89(9):095203. https://doi.org/10.1088/0031-8949/89/9/095203
21. MM Bhatti, MM Rashidi (2016) Effects of thermo-diffusion and thermal radiation on Williamson nanofluid over a porous shrinking/stretching sheet. J Mole Liquids
22. Layek GC, Mandal B, Bhattacharyya K (2020) Dufour and Soret effects on unsteady heat and mass transfer for Powell-Eyring fluid flow over an expanding permeable sheet. J Appl Comput Mech 6(4):985–998
23. Patil AB, Patil VS, Humane PP, Shamshuddin MD, Jadhav MA (2022) Double diffusive time-dependent MHD Prandtl nanofluid flow due to linear stretching sheet with convective boundary conditions. Int J Model Simul 1–15. https://doi.org/10.1080/02286203.2022.2033499
24. Sheri SR, Modugula P (2017) Heat and mass transfer effects on unsteady MHD flow over an inclined porous plate embedded in porous medium with Soret-Dufour and chemical reaction. Int J Appl Comput Math 3:1289–1306. https://doi.org/10.1007/s40819-016-0177-4
25. Alao FI, Fagbade AI & Falodun BO (2016) Effects of thermal radiation, Soret and Dufour on an unsteady heat and mass transfer flow of a chemically reacting fluid past a semi-infinite vertical plate with viscous dissipation. J Nigerian Math Soc 35(1):142–158. https://doi.org/10.1016/j.jnnms.2016.01.002
26. Idowu AS, Falodun BO (2020) Effects of thermophoresis, Soret-Dufour on heat and mass transfer flow of magnetohydrodynamics non-Newtonian nanofluid over an inclined plate. Arab J Basic Appl Sci 27(1):149–165. https://doi.org/10.1080/25765299.2020.1746017
27. Reddy PS, Chamkha AJ (2016) Soret and Dufour effects on unsteady MHD heat and mass transfer from a permeable stretching sheet with thermophoresis and non-uniform heat generation/absorption. J Appl Fluid Mech 9(5):2443–2455
28. Ullah I, Khan I, Shafie S (2017) Soret and Dufour effects on unsteady mixed convection slip flow of Casson fluid over a nonlinearly stretching sheet with convective boundary condition. Sci Rep 7:1113. https://doi.org/10.1038/s41598-017-01205-5
29. Patil VS, Patil NS, Timol MG (2015) A remark on similarity analysis of boundary layer equations of a class of non-Newtonian fluids. Int J Non-lin Mech 71:127–131. https://doi.org/10.1016/j.ijnonlinmec.2014.10.022
30. Rajput G, Patil V, Prasad J (2017) Hydromagnetic bioconvection flow in the region of stagnation-point flow and heat transfer in non-Newtonian nanofluid past a moving surface with suction: similarity analysis. Int J Heat Technol 35:25–31. https://doi.org/10.18280/ijht.350104
31. Hossain MS, Samand MA, Mohebujjaman M (2013) Heat and mass transfer of an MHD forced convection flow along a stretching sheet with chemical reaction, radiation and heat generation in presence of magnetic field. Res J Math Stat 5(1–2)

A Numerical Method to Analyze the Steady Conjugate Heat Transfer Around a Slab Suspended in Air

Basanta Kumar Rana, Jitendra Kumar Patel, Manoj Ukamanal, Prakash Ghose, Swarup Kumar Nayak, and Achinta Sarkar

1 Introduction

Natural convection heat transfer has been the easiest and convenient way of cooling various engineering heated components. This sort of heat transfer is common in a wide range of industrial applications. Because of the minimal price, easy operation and lack of additional technology that made the design easier, cost-effective and noise-free, this form of heat exchange is widely used in a variety of manufacturing applications. As a result, numerous investigators sought to examine natural convection analyses using different geometries [1–7]. Additionally, heat losses from slab is highly necessary as well as crucial among a variety of scientific uses, such as furnace sidewalls, structures to conserve thermal energy, food storage chambers, etc. Kimura et al. [8] carried out tests and analytic analyses upon the vertically heated slabs employing freshwater as base fluid and discovered three different regimes. Salt [9] performed an unsteady heat transfer study in a multi-layer composite laminates with a sudden change in base fluid. Barletta [10] defined the resilience of free convection inside a porous medium slab and reformed the investigation by evaluating previous studies. Saeid [11] studied analytically the constant conjugated heat transfer from porous enclosures having limited wall thickness for a broad variety of thermal conductivities and wall thickness to height ratios to create the porous structure. Tari and Mehrtash [12] reported natural convection through heat sinks employing plate fins and evaluated with prior scientific investigations. Dash and Dash [13] exhibited natural convection in the laminar zone around the thick vertical alignment tube. They further continued to demonstrate the cooling studies and Nu relationship as just a function of geometric and thermodynamic variables. Vynnycky and Mitchell [14]

B. K. Rana (✉) · J. K. Patel · M. Ukamanal · P. Ghose · S. K. Nayak · A. Sarkar
School of Mechanical Engineering, KIIT Deemed to be University, Bhubaneswar, Odisha 751024, India
e-mail: basanta.uce@gmail.com

© The Author(s), under exclusive license to Springer Nature Singapore Pte Ltd. 2023

S. Revankar et al. (eds.), *Recent Advances in Thermofluids and Manufacturing Engineering*, Lecture Notes in Mechanical Engineering,
https://doi.org/10.1007/978-981-19-4388-1_5

demonstrated steady free convection inside a vertical porous slab having permeable borders using 2-D steady-state equations. Behera et al. [15] studied natural convection overheated three distinct bend geometries (straight, round and parabolic) inside this laminar flow regime and established the optimal X/L for greatest heat loss for bend plates; additionally bends in the plate somewhat lowers thermal performance.

A thorough analysis of the current research indicates that a significant amount of research had been conducted to investigate spontaneous convection in thick slabs with porous media. Furthermore, there is no research detailing the conjugate heat transfer measurement all around heated thick 2-D slabs. As a result, attempts are being taken to understand the conjugate heat transfer study all around slab for various heat conductivity, aspect ratios and Rayleigh number.

2 Problem Description

The computing domain has been depicted schematically in Fig. 1a, with the thick heated slab confronted here to ambient atmosphere. L and b represent the slab's length and breadth, accordingly. The current problem is a 2-D model. The isothermal boundary condition has been placed on the slab's right-hand side that is kept at a constant temperature of 326 K. Meanwhile, the remaining of the slab's three surfaces are subjected to a coupled boundary condition, the temperature attained is influenced by thermal conductivity and slab's width. Figure 1b depicts the design of mesh within the domain. The temperature of quiescent air is assumed to be 300 K. The smallest-sized grids are evidently formed near to the slab's surfaces in order to acquire the thin boundary layer, while somewhat coarse cells are generated distant from slab's surface.

3 Numerical Methodology

Flow domain is assumed to be two-dimensional, steady and laminar. Governing differential equations of continuity, momentum and energy are to be solved and are written below:

Continuity equation:

$$\nabla \cdot \vec{V} = 0 \tag{1}$$

Momentum equation:

$$\rho\left(\vec{V} \cdot \nabla\right)\vec{V} = -\nabla p + \mu \nabla^2 \vec{V} + \rho \vec{g} \tag{2}$$

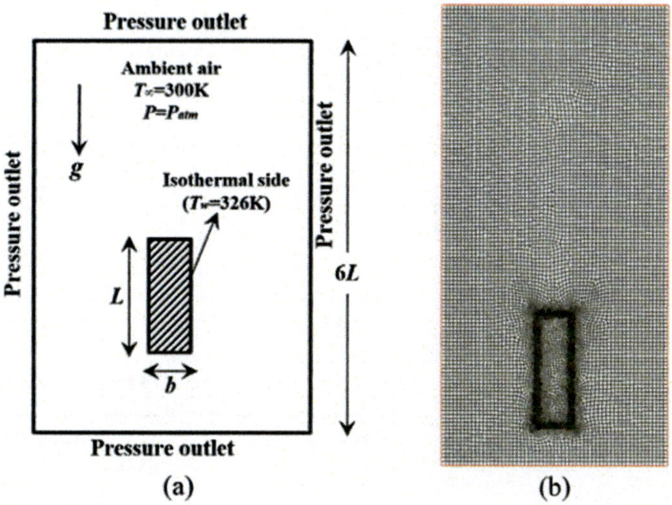

Fig. 1 **a** Layout sketch of the working domain including the slab suspended in atmospheric air. **b** Design of the mesh

Energy equation in fluid domain:

$$\left(\vec{V}\cdot\nabla\right)T=\alpha\nabla^2\vec{V} \tag{3}$$

Energy equation in solid slab:

$$\nabla^2T=0 \tag{4}$$

where, $\vec{V}=\hat{\imath}u+\hat{\jmath}v$ is the fluid velocity.

The solution includes the Boussinesq approximations to represent the free convection phenomena. To analyze variables from Eqs. (1) to (4), boundary conditions must be fulfilled. The slabs and sub-domain sides are subjected to the following boundary conditions.

Right wall of slab (isothermal surface): $T=T_w, u=v=0$.

At solid–fluid interface: $k_{\text{solid}}(\partial T_{\text{solid}}/\partial n)=k_{\text{fluid}}(\partial T_{\text{fluid}}/\partial n)$.

At solid–fluid interface temperature continuity: $T_{\text{solid}}=T_{\text{fluid}}$.

Pressure outlet boundary condition on the edges of the computational domain: $p=p_{\text{atm}}, T=T_\infty$.

The implementation of relevant non-dimensional numbers are highly necessary to describe the underlined physics thoroughly, which are mentioned below:

Rayleigh number, $\text{Ra}=\left\{g\beta L^3(T_w-T_\infty)\right\}/\alpha\nu$.

Nusselt number: $\text{Nu}=hL/k$, where $h=Q/(\text{slab surface area}\times\Delta T)$.

The current study is supported by the test findings of Kimura et al. [8], who evaluated both the top and bottom faces to be insulated. The analysis shows that the

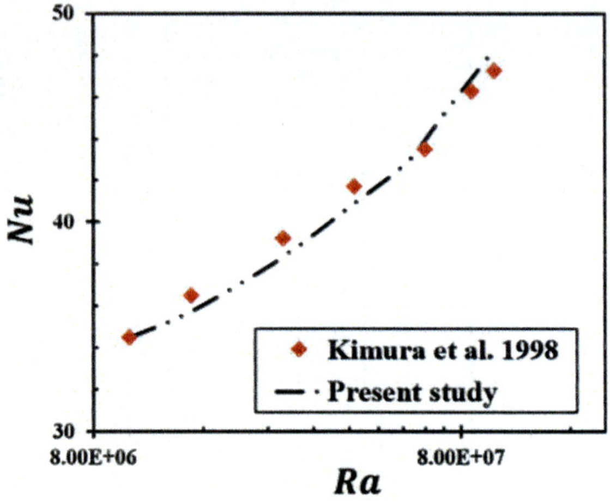

Fig. 2 Comparison of present work with the experimental study of Kimura et al. [8]

current research and the previous analysis are really in good conformity, as can be seen in Fig. 2.

4 Results and Discussion

A numerical analysis is conducted to report the detailed characterization of conjugate heat transfer around a 2-D thick rectangular-shaped slab exposed to quiescent air. Isothermal condition is maintained at one edge of the slab and subsequently, heat is being transferred to rest three edges due to the presence of a temperature gradient. A wide range of Rayleigh number ($10^5 \leq \text{Ra} \leq 10^8$) is employed to investigate this study.

4.1 Pattern of Thermal Plume

One of the key criteria for extracting the underlying physics over such an examination is the explication of thermal and flow field surrounding the heated slab. The very similar tendencies were exhibited by Rana et al. [5], Dash and Dash [13] and Rana [7]. Figures 3 and 4 clearly depict the distribution of thermal plumes inside a steady-state arrangement round the block suspended across an air medium over varied aspect ratios. Simulations were conducted with two distinct slab elements, concrete and wood, that are often used in real-time applications. Since right side of the slab being perceived isothermal, the temperature upon the face remains constant. Furthermore, the other sides were considered to be non-isothermal, therefore the temperature distribution is entirely determined by thermal conductivity of the slab, thickness of the slab as well as the Rayleigh number.

The temperature gradient within the concrete slab is seen to be a function of both Rayleigh number and thermal conductivity. In the condition of reduced thermal conductivity, the thermal gradients in between right and left sidewalls are noticeably significant. Once the temperature gradient near the sidewall is higher, then heat removal rate is significantly larger for the non-isothermal surfaces. Additionally, the temperature of non-isothermal sidewalls are expected to be highest with increasing Ra values for specific values of L/b and slab materials since the strength of the

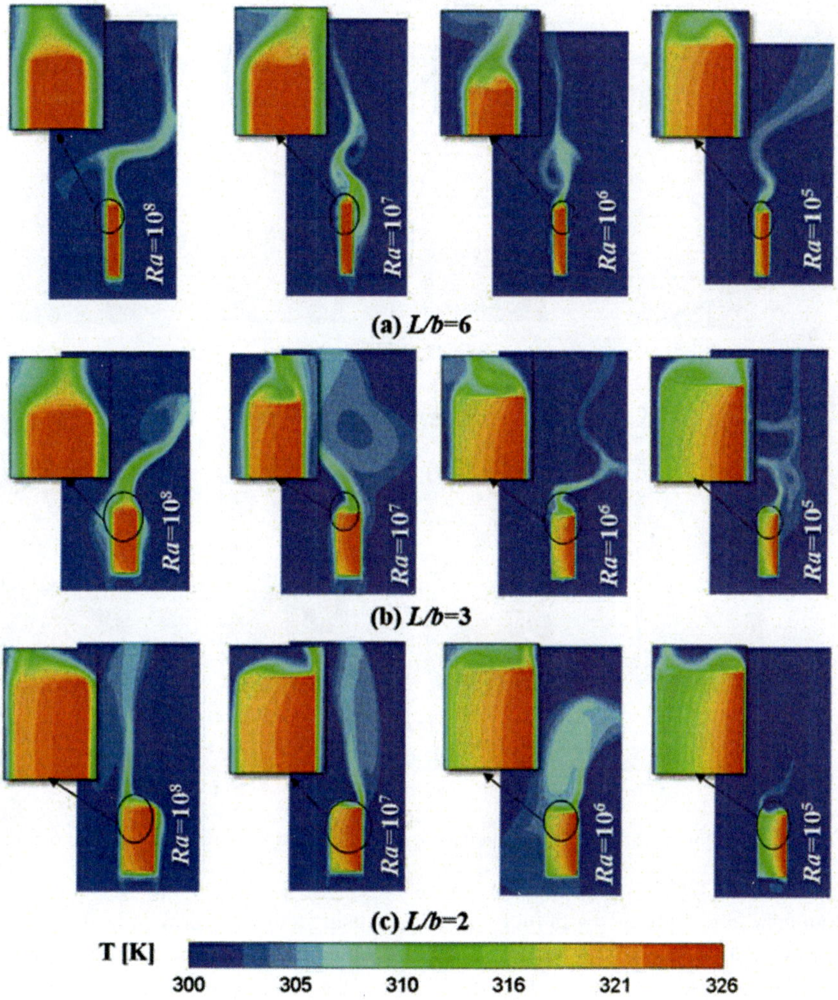

Fig. 3 Distribution of thermal plumes out from concrete slab ($k = 2.25$ W/mK) for various L/b ratios

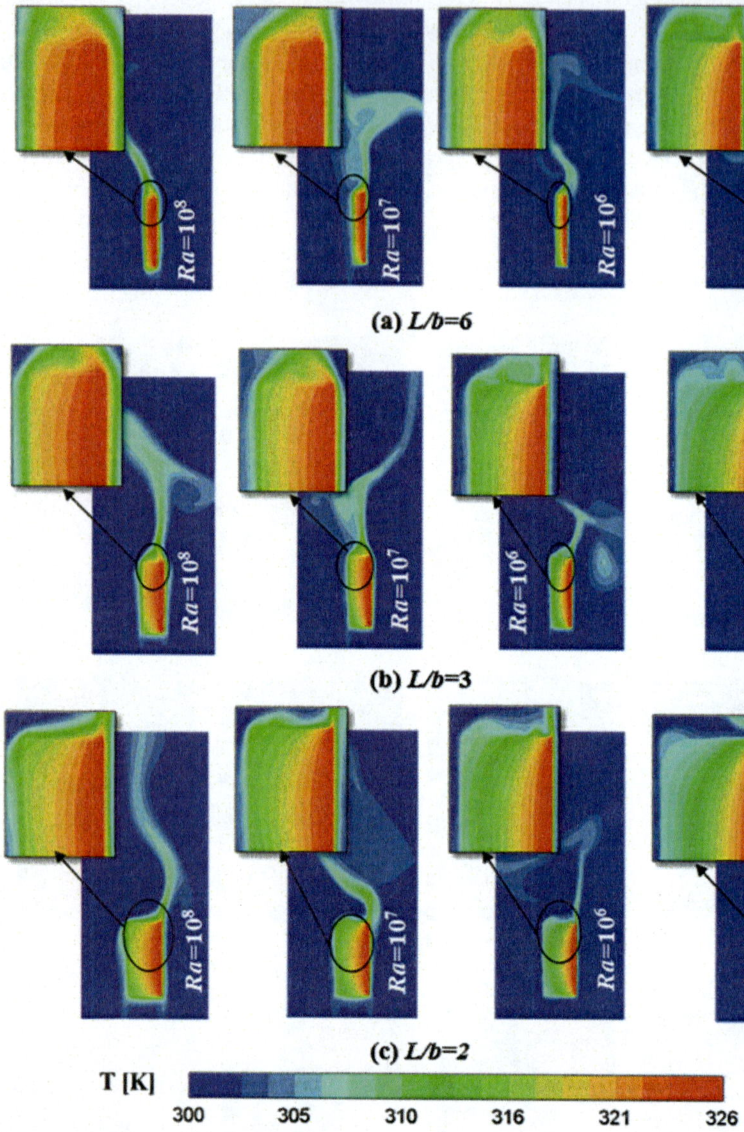

Fig. 4 Distribution of thermal plumes out from wood slab ($k = 0.12$ W/mK) for various L/b ratios

boundary field increases with increasing Ra, causing a upward flow of heated liquid towards the surface would become quicker. As a result, it may be predicted that Q for non-isothermal surfaces is boosted at higher Ra values compared to reduced Ra values. The slabs aspect ratio is really essential and has a large impact on 'Q'. The length of the slab was fixed throughout this experiment, and the change was done for thickness (b), as shown in the diagram. The heat transfer rate from non-isothermal surfaces was indeed observed to be decreased as the aspect ratio diminishes, because the temperature gradient near to non-isothermal sidewalls has been noticed to be relatively low, leading to reduced advection heat exchange from sidewall at lower *L*/*b* comparable to significantly larger *L*/*b* for a fixed Ra and *k*. There is a noticeable variance in the design of thermal plume for varying aspect ratios of slabs. Furthermore, the temperature distribution significantly sliced downward at non-isothermal sidewalls as in case of wood over concrete caused by (*k*) of the slab materials, as seen in Figs. 3c and 4c.

4.2 Effect of Aspect Ratios and Thermal Conductivity on Nusselt Number

Heat removal rate from slabs surface is the most significant parameters in this type of study that is very useful in developing mechanical devices. The heat transfer rate can be expressed through a well non-dimensional number defined as the Nusselt number. Figure 5 depicts the relationship between the surface averaged Nu and the Ra number over various aspect ratios and slab materials. In case of concrete and wood slabs, it was discovered that Nu rises gradually as Ra enhances. When contrasted with wood, the growth of Nu for concrete seems to be considerably higher. Additionally, above Ra $= 10^7$, the heat transfer rate keeps increasing. Furthermore, the Nu pattern is definitely observed to rise with increasing aspect ratio since convective heat transfer from non-isothermal surfaces rises exponentially.

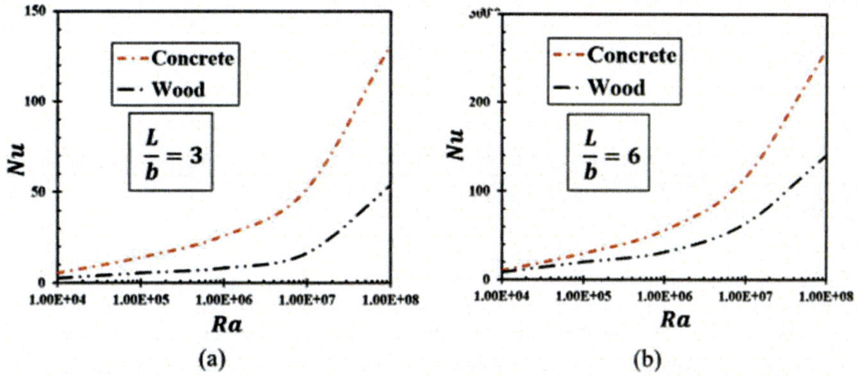

Fig. 5 Influence of aspect ratio and thermal conductivity of the material on the Nusselt number

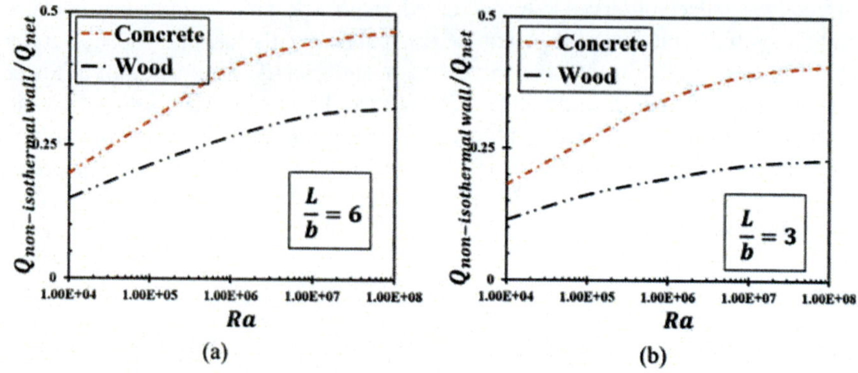

Fig. 6 Influence of L/b and k on the contribution of rate of heat transfer from the non-isothermal surfaces

4.3 *Effect of Aspect Ratios and Thermal Conductivity on* Q_{con}

We have additionally tried to determine the significance of heat loss from non-isothermal surfaces to the net heat loss rate. The similar findings were observed by other researchers [7, 13] in past in their respective investigations. The contribution of heat loss rate (Q_{con}) can be determined by using following mathematical expression: $Q_{\text{con}} = Q_{\text{non-isothermal walls}}/Q_{\text{net}}$. It is noticed that Q_{con} rises notably with the continuous growth of Rayleigh's number. Moreover, Q_{con} is found to be greater when the concrete material is used compared to wood because of higher value of k (Fig. 6).

5 Conclusion

To comprehend the laminar conjugate heat transfer around vertically heated thick slabs submerged inside the stationary air, a computational research has been conducted. Variations of several input significant factors, including Rayleigh number, length to width ratio, and thermal conductivity of slabs, are used to illustrate the comprehensive study. The primary findings emphasized are as follows:

I. Heat dissipation rate from non-isothermal materials surface gets increased for higher Ra values compared to lower Ra values for specific aspect ratios and thermal conductivity. Likewise, a larger value of thermal conductivity corresponds to a higher value of heat removal rate from non-isothermal materials surfaces.

II. Heat dissipation rate from non-isothermal surfaces are observed to be decreased when the aspect ratio declines because as temperature distribution adjacent to non-isothermal sidewalls are observed to be smaller, resulting in lesser advection heat transfer from the walls.

III. The Nusselt number rises when the aspect ratio of non-isothermal materials enhances as advection heat transfer rises substantially.
IV. Owing to the high (k) of the concrete material, the heat exchange contribution ratio for concrete was significantly larger than that for wood.

References

1. Hassan KE, Mohamed SA (1970) Natural convection from isothermal flat surfaces. Int J Heat Mass Transf 13(12):1873–1886
2. Jia H, Gogos G (1996) Laminar natural convection heat transfer from isothermal spheres. Int J Heat Mass Transf 39(8):1603–1615
3. Liu J, Zhao CJ, Liu H, Lu WQ (2018) Numerical study of laminar natural convection heat transfer from a hemisphere with adiabatic plane and isothermal hemispherical surface. Int J Therm Sci 131:132–143
4. Acharya S, Dash SK (2021) Natural convection heat transfer from a vertical hollow cylinder with surface holes. J Heat Transf 143(4):042602
5. Rana BK, Singh B, Senapati JR (2021) Thermofluid characteristics on natural and mixed convection heat transfer from a vertical rotating hollow cylinder immersed in air: a numerical exercise. J Heat Transf 143(2):022601
6. Rana BK, Senapati JR (2021) Entropy generation analysis and cooling time estimation for a rotating vertical hollow tube in the air medium. J Heat Transf 143(4):042101
7. Rana BK (2022) Conjugate steady natural convection analysis around thick tapered vertical pipe suspended in the air. Sādhanā 47(1):1–16
8. Kimura S, Okajima A, Kiwata T (1998) Conjugate natural convection from a vertical heated slab. Int J Heat Mass Transf 41(21):3203–3211
9. Salt H (1983) Transient conduction in a two-dimensional composite slab-I. Theoretical development of temperature modes. International journal of heat and mass transfer 26(11):1611–1616
10. Barletta A (2015) A proof that convection in a porous vertical slab may be unstable. J Fluid Mech 770:273–288
11. Saeid NH (2007) Conjugate natural convection in a porous enclosure: effect of conduction in one of the vertical walls. Int J Therm Sci 46(6):531–539
12. Tari I, Mehrtash M (2013) Natural convection heat transfer from inclined plate-fin heat sinks. Int J Heat Mass Transf 56(1–2):574–593
13. Dash MK, Dash SK (2020) Natural convection heat transfer and fluid flow around a thick hollow vertical cylinder suspended in air: a numerical approach. Int J Therm Sci 152:106312
14. Vynnycky M, Mitchell SL (2022) On steady natural convection in a vertical porous slab with permeable boundaries. J Fluid Mech 935:A21
15. Behera S, Acharya S, Dash SK (2020) Natural convection heat transfer from linearly, circularly and parabolically bent plates: a study of shape effect. Int J Therm Sci 150:106219

Low Reynolds Number Flow Past Array of Staggered Cylinders

Ankit Kumar, Biranchi Narayana Das, Srikant Panigrahi, Pooja Chaubdar, and Atal Bihari Harichandan

1 Introduction

The studies of flow past single or multiple cylinders having circular shapes are of utmost practical importance in engineering and science. The geometries seem simple but when the fluid flow is investigated and analyzed, many complex fluid flow equations come into play depending upon various flow fields downstream and upstream of the cylinder. The features of the separated wake are incredibly difficult to analyze analytically and so, the numerical simulations and experiments come into the analysis. When studying the flow past a cylinder of rectangular shape, the flow separation occurs at the corners but in case of cylinders of circular shape prediction, the point of flow separation complicates the investigation. When investigating the flow behaviors past the circular cylinders, the flow field becomes very much complex to understand and predict the future flow behavior. Having multi cylinders in a setup further complicates the problems and hence the analysis becomes further complex and the wake, in this case, becomes different than the single-cylinder setup.

The geometrical configuration of multiple cylinders placed side by side has garnered a lot of attention from the researchers. Experimental investigations have been carried out for flow past two cylindrical bodies in staggered and tandem configurations [1–3]. The floe interference between the cylinders in different configurations has been studied systematically keeping the distances between the centers lower than five times diameter of cylinder. Experiments have also been carried out keeping the distance between the two cylinders as 3.47 times the diameters for flow

A. Kumar
Department of Aeronautical Engineering, National Formosa University, Huwei, Taiwan

B. N. Das · S. Panigrahi · P. Chaubdar · A. B. Harichandan (✉)
Aerospace Engineering, School of Mechanical Engineering, KIIT Deemed to be University, Bhubaneswar 751024, India
e-mail: atal.harichandanfme@kiit.ac.in

© The Author(s), under exclusive license to Springer Nature Singapore Pte Ltd. 2023

S. Revankar et al. (eds.), *Recent Advances in Thermofluids and Manufacturing Engineering*, Lecture Notes in Mechanical Engineering,
https://doi.org/10.1007/978-981-19-4388-1_6

past two tandem cylinders [4]. The authors observed two different flow field patterns switching randomly in time. The avg. time interval for two field patterns is dependent strongly on turbulent intensity. Numerical simulations for flow past a periodic array of staggered cylindrical bodies have been performed and reported [5]. The authors have computed the uniperiodic flows with every cell of the periodicity consisting of either two or ten cylinders. Flow past a pair of cylindrical bodies have been investigated at Reynolds number 100 and 1000 in staggered and tandem arrangement configuration [6]. A finite volume scheme has been reported for the simulation. In the literature, most of the numerical simulations have been reported for Reynolds numbers between the range of 100 and 1000 but, the experimental results have been reported for Reynolds numbers more than 10^4. This helps in understanding the flow behaviors and patterns and the results are of immense importance in understanding the underlying phenomena.

Finite element method is used for solving 2*D* Navier Stokes equation to study the vortex shedding phenomenon from two cylindrical bodies having varying diameters for different gap ratios and the angular location of smaller cylinders [7]. Results show that at smaller gap ratio, only one wake is observed after the two cylinders when the gap ratios are medium, interactions between the vortex shedding from larger cylinder and shedding from smaller cylinder is observed and when the hap ratios are large, the interaction between shedding becomes very weak. When the gap ratio is smaller, single-wake mode of shedding is achieved, at medium gap ratio, interaction mode is achieved and at larger gap ratios two wake mode is observed. The frequency of the Vortex shedding for larger cylinder is most of the time lower than that of single cylinder at same Re. The authors also reported that diameter ratio of smaller and the larger cylinders would also impact the hydrodynamic forces on cylinders and vortex shedding from two cylinders. A 2*D* high-order unstructured solver form simulating compressible flows with spectral difference scheme for viscous and inviscid flows has been developed and validated [8]. The solver is used for simulating the laminar flow past two cylinders kept side by side at varying spacing *S*, (center to center/diameter). When $s = 1.1$, the flow pattern of wake resembles vortex street of the single body and when $s = 1.4, 1.5, 1.7, 2$, the authors predicted that the asymmetric flow pattern for the flow around the pair of cylinders and separation points of lower and upper cylinders aren't synchronized. At $s = 2.5$ and 3, the anti-symmetric in-phase flow patterns have been predicted for $\mathrm{Re} = 100$. The unsteady flow of lower and upper cylinders separates at the locations of same degrees on surfaces of cylinder. When the distance is further increased between the cylinders when $s = 3.4$ and 4, the symmetric anti-phase flow pattern is observed. The points of separation of lower and the upper cylinder are symmetrical across centerline of gap flow. A solver based upon cartesian staggered grid FVM has been used for simulating viscous, unsteady, incompressible, and 2*D* flow past two transverse circular cylindrical bodies in laminar and turbulent flow [9]. The Reynolds number and the gap ratios have been varied for investigating various parameters and their effect on the flow behavior characteristics and hydrodynamic forces which act on the bodies. A biased flow pattern is found behind the two cylinders when T/D is kept 1.5 and Reynolds number is 100 which is bistable. Wide wake and the narrow regions develop behind every cylinder which varies over time. The coefficient of lift and drag forces also varies irregularly with passing time because of proximity and the gap flow deflects toward cylinder having high frequency resulting the cylinder to experience increased drag. As gap is increased to $T/D = 3$ and 4, flip-flopping pattern is found to be diminishing at for simulation at Reynolds number 100. At $T/D = 3$, streamlines and the shedding shows in-phase asymmetric pattern while, for $T/D = 4$, synchronized anti-phase vortices shedding after the lower and upper cylinders. The flow is deflected from the lower cylinder toward the upper one as the lower cylinder experiences a higher drag. When the Reynolds number is 200 and the gaps are high, shedding of completely periodic vortex is found at $T/D = 4$ and the flow patterns are found to be symmetric. At $T/D = 1.5$, a repulsive force develops between the cylinders and the wake patterns after the cylinders is found to be similar to that as behind one single buff body. Two identical cylinders of circular shape placed side by side at oblique and right attack angles have been investigated by computing $3D$ Navier Stokes equation through Petrov–Galerkin finite element method [10]. The effect of gap ratios and flow attack angle between two cylinders of circular shape on vortex shedding flow and the hydrodynamic forces of cylinders have been investigated. When the gap ratios are small $G/D = 0.5$, 1, the biased flow is observed for both angles at $\alpha = 0°$, 45° which leads to narrow and wide wake after cylinders. The coefficient of forces at $\alpha = 0°$ has been compared with $\alpha = 45°$ at same Reynolds number. Flow around two staggered cylinders has been investigated by varying the distance between the cylinders at angles of incidence (α) at Reynolds number 100–800 using spectral element method [11]. The authors have identified six flow pattern behaviors namely Induced separation (IS), Shear layer reattachment (SLR), Vortex impingement (VI), Vortex pairing and enveloping (VPE), Synchronized vortex shedding (SVS), Vortex pairing splitting and enveloping (VPSE) which transform from one flow pattern to another by varying the angle of incidence, Reynolds number or the distance between cylinders. At smaller angles of incidence up to 30°, IS, SLR, and VI flow patterns are observed holding only one von Karman vortex shedding in the wake and from 30° to 50° angles of incidence, VPE and VPSE flow patterns are observed presenting two synchronized von Karman vortices. Using Immersed Boundary Method has been investigated numerically for flow over 3 circular identical cylinders arranged in an equilateral triangle arrangement [12]. The spacing ratio and the Reynolds numbers have been varied and their effect has been studied on strouhl numbers, hydrodynamic forces, and the flow structures. The authors have observed nine different flow structures, flip-flopping, single bluff-body, steady symmetric, deflected, steady asymmetric, in-phase, anti-phase, hybrid, and fully developed in-phase co-shedding flow. Flow over cylindrical body under influence of $1D$ gusty flow has been investigated which impinges on the cylinders through CFRNUS scheme [13].

2 Governing Equations and Computational Domain

The governing equations of viscous incompressible flow in two dimensions are two components of momentum equations and continuity equations which are mixed parabolic-elliptic in nature. These equations in non-dimensional primitive variables are expressed as:

Continuity equation:

$$\frac{\partial u}{\partial x} + \frac{\partial v}{\partial y} = 0$$

Momentum equations:

X-momentum:

$$\frac{\partial v}{\partial t} + \frac{\partial (x^2)}{\partial x} + \frac{\partial (uv)}{\partial y} = -\frac{\partial p}{\partial x} + \frac{1}{\mathrm{Re}}\left(\frac{\partial^2 u}{\partial x^2} + \frac{\partial^2 u}{\partial y^2}\right)$$

Y-momentum

$$\frac{\partial v}{\partial t} + \frac{\partial (uv)}{\partial x} + \frac{\partial (v^2)}{\partial y} = -\frac{\partial p}{\partial y} + \frac{1}{\mathrm{Re}}\left(\frac{\partial^2 v}{\partial x^2} + \frac{\partial^2 v}{\partial y^2}\right)$$

where, u and v are the velocity components in x- and the y-directions, respectively, Re is Reynolds number, p is ratio of pressure and density and t is non-dimensional time. At the beginning of simulation process, a uniform free-stream velocity and the pressure field are prescribed in every cell of the flow domain has been given below:

$$\left.\begin{array}{l} u[i] = u_\infty \\ v[i] = v_\infty \\ p[i] = p_\infty \end{array}\right\} \text{for all the triangular cells in flow domain}$$

The cylindrical body is introduced suddenly to uniform free-stream flow. For rectangular computational domain the outlet, inlet, bottom, and top boundaries have been kept away from the surfaces. Dirichlet boundary condition has been applied at outer boundaries and no-slip condition has been applied on the body.

The geometrical description of the three-cylinder configurations having equal or unequal diameters has been defined with two parameters namely, longitudinal gap, L, and the transverse gap, T depending upon center to center distances as shown in Fig. 1. The inflow boundary has been kept at ten times of cylinder diameter from center of the cylinder at front and outflow boundary is kept at 25 times of cylinder diameter downstream from the rear cylinders center. The bottom and top boundaries are kept at 15 times of the cylinder diameter from the center of top and bottom cylinders.

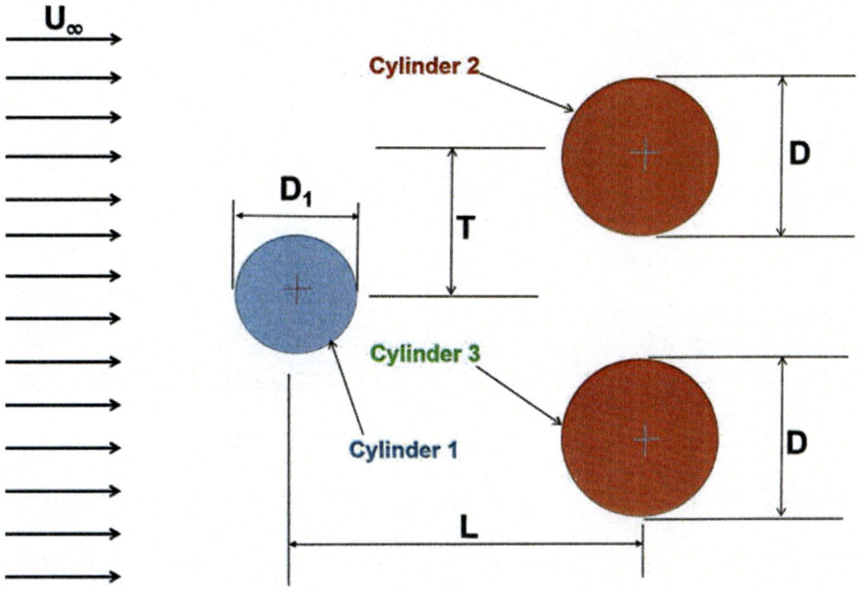

Fig. 1 Problem description

3 Grid Independence Study and Code Validation

Unsteady flow past a single cylinder at Reynolds number 100, 200 has been computed on triangular cell grid having 14,878 nodes and 29,464 cells in which 160 nodes are on the surface of the body surface. The mesh has been selected depending upon the results derived through the grid independence test out of four different grids. The convergence of the Pressure-Poisson equation has been ensured by keeping the global maximum residual of the equation to 10^6 for all the numerical simulations. The simulations have been carried out on CFR scheme and up to 400 s (Fig. 2; Table 1).

Figure 3 shows vorticity contours and the streamlines at an instantaneous time $t =$ 400 s and Fig. 4 shows the cl and cd versus time in seconds acting on the cylindrical

Fig. 2 Close-up view of the mesh for single circular cylinder

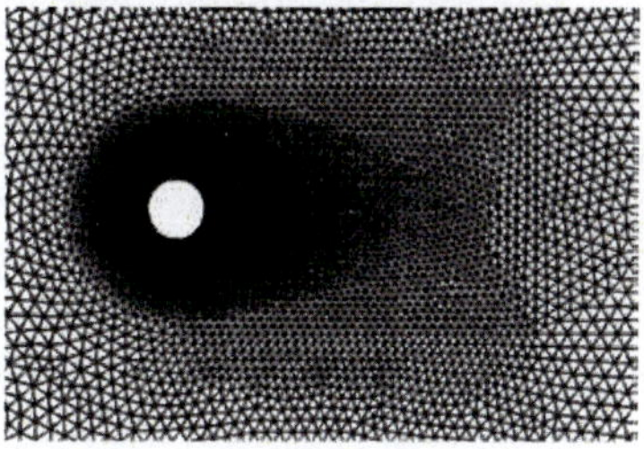

Table 1 Grid independence study for at Re = 100

Number of nodes body	Coefficient of drag (CD)	Coefficient of lift (CL)	Strouhal number (St)
(Grid 1) 80	1.113 ± 0.022	± 0.182	0.165
(Grid 2) 120	1.186 ± 0.016	± 0.22	0.163
(Grid 3) 160	1.361 ± 0.011	± 0.279	0.162
(Grid 4) 200	1.382 ± 0.011	± 0.282	0.161

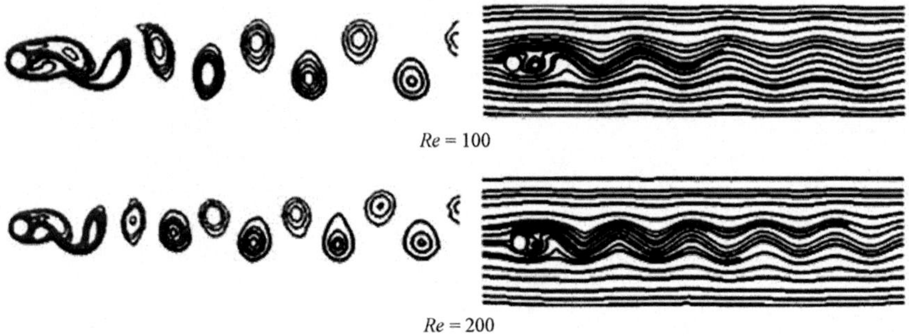

Fig. 3 Vorticity contours and the streamlines around a single circular cylinder

body. The plots show a clear periodicity in coefficient of drag and lift which shows that the vortex shedding from the cylinder's rear surface is periodically taking place. Table 2 shows mean values and the amplitudes of coefficient of drag and lift and Strouhal nos. of the present results and are compared with the already published results. The results from the current research are found in agreement with other reports published by other researchers. The authors now then applied the current unstructured CFR scheme [14–16] to study flow around multiple cylinders in staggered arrangements.

4 Results and Discussion

1. Cylinders of equal diameters in staggered arrangement

Flow around three circular cylinders in staggered arrangement has been simulated by varying the longitudinal and transverse gap at Reynolds number 100. Three staggered cylinders having the arrangement of transverse gap $T = 0.7D$, $1.5D$, and $3D$ and the longitudinal gap $L = 3D$ at Reynolds number 100. Figure 5 shows the streamlines and the vorticity contours for the different gap ratios at Re = 100. The vorticity contours show that the flow behaviors are impacted by the transverse distance between downstream cylindrical bodies and the longitudinal distance between downstream and the upstream cylinders. The total combined effect of both the transverse and longitudinal

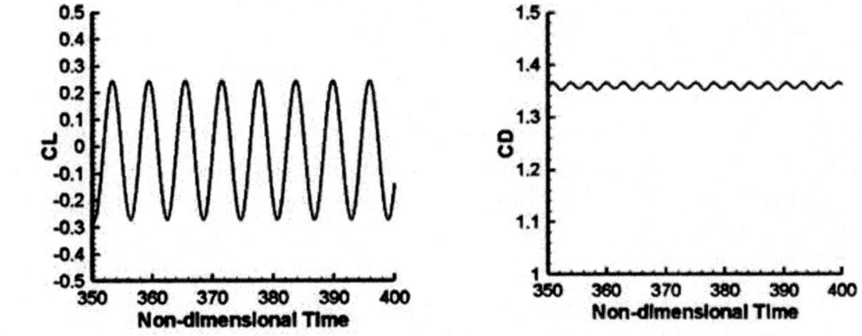

$Re = 100$

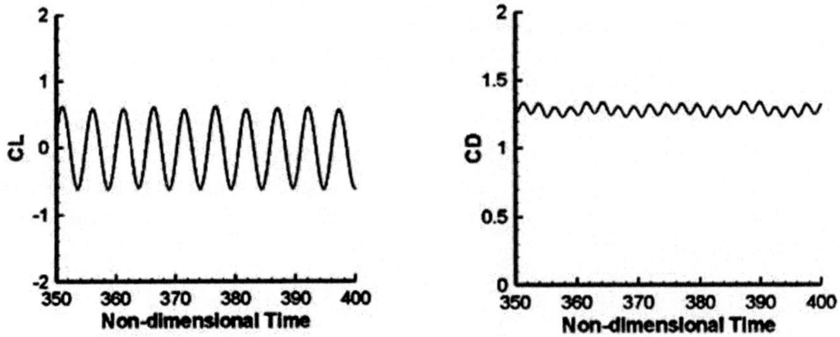

$Re = 100$

Fig. 4 Lift and drag coefficients of flow around a single circular cylinder

Table 2 Parameters for flow over single circular cylinder at Re = 100 and 200

Parameters	CD		CL		St	
Re	100	200	100	200	100	200
Braza [17]	1.365 ± 0.0149	1.39 ± 0.049	± 0.249	± 0.76	0.160	0.200
Ding [18]	1.369 ± 0.010	1.349 ± 0.049	± 0.288	± 0.658	0.159	0.196
Present	1.351 ± 0.011	1.31 ± 0.049	± 0.279	± 0.603	0.160	0.192

distance helps in determining the nature of interaction of the flow. A single-wake pattern of the bluff body is observed when the downstream cylinders are placed in closer proximity ($T = 0.7D$) as shown in Fig. 5. The vortex shedding is suppressed from gap sides of both downstream cylindrical bodies (inner vortices) completely as the wake pattern is not affected as the gap flow is very weak. The resultant flow behaves as there is only one body with increased length which is accompanied with a sheet of the vortices shed from free-stream side (outer vortices) is existing. When $T = 1.5D$, a synchronized in-phase wake pattern is seen having two parallel in-phase streets which are symmetric w.r.t. horizontal centerline of upstream cylinder. This behavior of wake pattern is observed near the wake of downstream cylinders when the longitudinal distance is increased slightly but in downstream, vortex interaction from the upstream cylinder (steady and an elongated wake pattern) and downstream cylinder lead to ceasing of inner vortices and a single wider vortex sheet is generated. When the transverse gap is further increased, the merging of the vortices decreases as the mutual interaction of the wakes of the cylindrical bodies in the staggered arrangements further weakens. The two cylinders in downstream have exhibited synchronized in-phase wake patterns but the single cylinder in upstream is exhibiting a wake pattern of steady nature which then elongated between the wake patterns of the downstream cylinders in rear position. Figure 6 shows the plots for coefficient for lift and drag versus time at Reynolds number = 100 with cylinders arranged in a staggered configuration having $L = 3D$ and $T = 1.5D$. At $L = 3D$, the coefficient of lift force becomes negative for the downstream cylinders with more negative value for lower cylinder in the arrangement when compared to the upper cylinder. This occurs because of the total combined effect of the critical transverse and longitudinal gap with the former gap playing a significantly dominating role. For smaller longitudinal gap ($\leq 3D$) between the downstream and upstream cylindrical bodies, the upstream cylinder is going to have a great impact on wake behavior patterns of downstream cylindrical bodies because of a stronger and more complex vortices interaction in the wake of the cylinders.

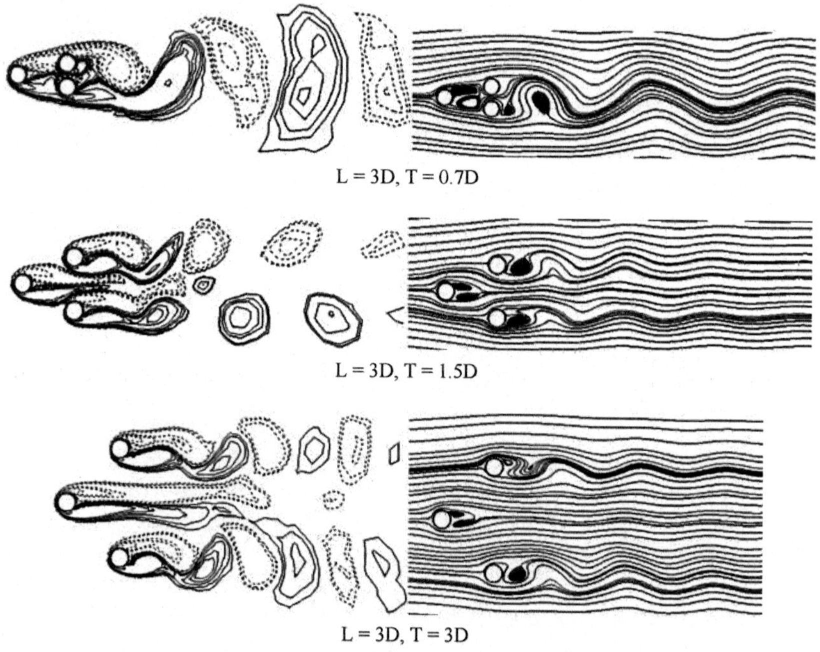

Fig. 5 Vorticity contours, the streamlines for flow over 3 cylinders in staggered arrangement at Re = 100

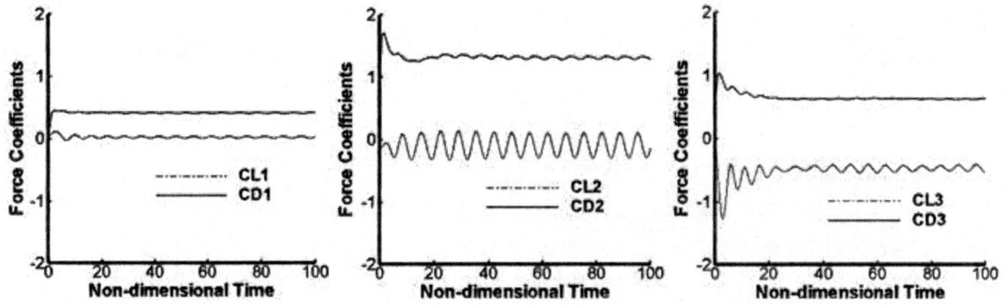

L = 3D and T = 1.5D

Fig. 6 CL and CD of flow over 3 cylinders in staggered arrangement at Re = 100

2. Cylinders of unequal diameters in staggered arrangement

Three cylinders are arranged in staggered arrangement having $L = 3D$ and $T = 1.5D$ to perform numerical simulation at Reynolds number 100. The downstream cylinders are having equal diameters and the upstream cylinders are of diameter $0.5D$ and $1.5D$ for the current investigation as shown in Fig. 7. Figure 8 shows vorticity contours and the streamlines for flow over three cylinders of unequal diameters in staggered arrangement at an instantaneous time $t = 100$ s. The vorticity pattern shows that the downstream of the rear cylinders are found to be mostly depending on transverse gap between them. A synchronized in-phase wake pattern is found in both cases which have been considered for the transverse gap. This wake pattern is found to be existing up to few centimeters in longitudinal distance in the rear cylinder but in downstream, the vortices interaction from downstream and upstream cylinder leads to disappearance of the inner vortices and a single wider vortex sheet is generated. The interaction of shedding of the vortices is found to be stronger when $D_1 = 1.5D$ than $D_1 = 0.5D$ as the case in which $D_1 = 1.5D$, the upstream cylinder is having a large wake which then mixes with the vortices shed from downstream cylinder intensely. This leads to formation of a wake pattern of synchronized nature relatively quickly when compared to the case in which $D_1 = 0.5D$ which has weaker vortices in wake of the upstream cylinder. Time history of the coefficients of forces has been shown in Fig. 9 which shows a variation in temporal behavior of coefficient of forces depending on the dimensions. The upstream cylindrical body which is smaller becomes steady after showing initial periodicity but for the larger upstream cylindrical body, the periodic variations have sustained. This phenomenon occurs as the effective Reynolds number would be able to scale with dimension of cylinder and for $0.5D$ cylinder, the effective Reynolds no. would be 50. At this Re, the shedding barely starts in the cylinder. In case where $T = 1.5D$, the downstream cylinders have always exhibited variation of coefficients of forces.

The RMS values of coefficients of forces for different cylinder arrangements at Re = 100 are highlighted in Table 3. The results show that the upstream cylinder experiences the least RMS drag and lift forces and the maximum RMS forces are experienced by downstream cylinders. The lower downstream cylinders show lesser drag and lift force than the upper ones. But when compared to the isolated cylinder,

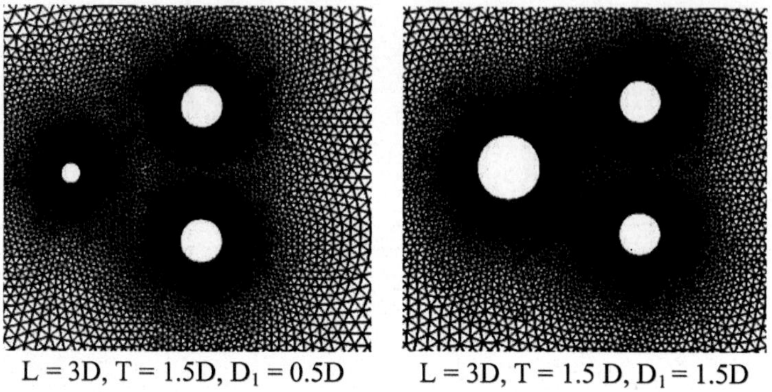

Fig. 7 Close-up view of mesh for three staggered cylinders of unequal diameters

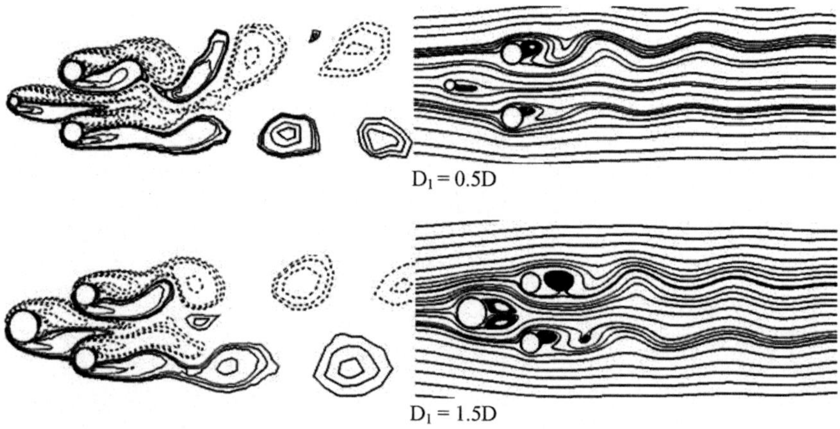

Fig. 8 Vorticity contours, streamlines of flow over 3 cylinders of unequal diameters in staggered arrangement at $Re = 100$

much lesser forces have been experienced. As the transverse gap would increase, these values might approach the values of the isolated cylinder.

5 Conclusion

The numerical analysis of three cylinders in different arrangements has been studied through a finite volume solver using CFRUNS scheme. The grid independence test has been conducted to choose the most optimum grid and has been validated with existing reported results in literature. In the first arrangement, the transverse gap is being varied and the mutual interaction of the wakes of different cylinders has

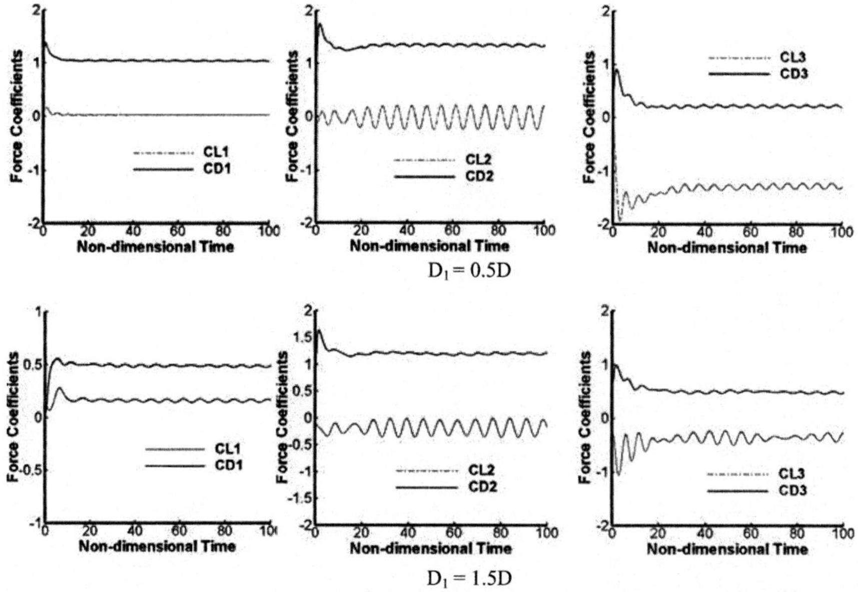

Fig. 9 Coefficients of lift and drag of flow over 3 cylinders of unequal diameters at Re = 100

Table 3 RMS values of coefficients of forces for different cylinder arrangements at Re = 100

Cylinder arrangement		CL_{rms}	CD_{rms}
Isolated cylinder		0.2079	0.00479
Isolated cylinder [8, 19]		0.2200	0.0051
$D_1 = D$	1st cylinder	0.0117	0.0056
	2nd cylinder	0.1428	0.0159
	3rd cylinder	0.0377	0.0061
$D_1 = 0.5D$	1st cylinder	0.0005	0.0029
	2nd cylinder	0.1629	0.0161
	3rd cylinder	0.0389	0.0137
$D_1 = 1.5D$	1st cylinder	0.0163	0.0125
	2nd cylinder	0.1261	0.0135
	3rd cylinder	0.0494	0.0122

been studied keeping the diameters of the cylinders equal. In the next arrangement, the diameters of the cylinders have been varied and the wake interaction has been analyzed. The authors have observed that the cylinders lying in the wake of upstream cylinder experiences forces that are highly unsteady in nature. The RMS values of the coefficients of forces are lower than the respective values for the isolated cylinder. As the transverse gap is increased the values might approach those of an isolated cylinder. Both the longitudinal and the transverse distance between the cylinders have an impact on the characteristics of flow behavior.

References

1. Zdravkovich MM (1977) Review of flow interference between two circular cylinders in various arrangements. J Fluids Eng (ASME) 99:618–633
2. Chen SS (1987) Flow-induced vibrations of circular cylindrical structures. Hemisphere Publication, Washington, DC
3. Kiya M, Arie H, Tamura HM (1980) Vortex shedding from two circular cylinders in staggered arrangement. J Fluids Eng Trans (ASME) 102:166–173
4. Kiya M, Mochizuki O, Ido Y, Suzuki T, Arai T (1992) Flip-flopping around two bluff bodies in tandem arrangement. Springer, Berlin, pp 15–18
5. Johnson AA, Tezduyar TE, Liou J (1993) Numerical simulations of flows past periodic arrays of circular cylinders. Comput Mech 11:371–383
6. Mittal S, Kumar V, Raghuvanshi A (1997) Unsteady incompressible flows past two cylinders in tandem and staggered arrangement. Int J Numer Meth Fluids 25:1315–1344
7. Zhao M, Cheng L, Teng B, Liang D (2005) Numerical simulation of viscous flow past two circular cylinders of different diameters. Appl Ocean Res 27:39–55
8. Liang C, Premasuthan S, Jameson A (2009) High-order accurate simulation of low-Mach laminar flow past two side-by-side cylinders using spectral difference method. Comput Struct 87:812–827
9. Hesam SM, Navid N, Behzad GD (2011) Numerical simulation of flow over two side by side circular cylinders. J Hydrodyn 23(6):792–805
10. Thapa J, Zhao M, Cheng L, Zhou T (2015) Three-dimensional simulations of flow past two circular cylinders in side-by-side arrangements at right and oblique attacks. J Fluids Struct 55:64–83
11. Hsu LC, Chen CL, Ye JZ (2017) A study of flow patterns for staggered cylinders at low Reynolds number by spectral element method. J Mech Sci Technol 31(6):2765–2780
12. Chen W, Ji C, Alam MM, Williams J, Xu D (2020) Numerical simulations of flow past three circular cylinders in equilateral-triangular arrangements. J Fluid Mech. https://doi.org/10.1017/jfm.2020.124
13. Parekh CJ, Roy A, Harichandan AB (2018) Numerical simulation of incompressible gusty flow past a circular cylinder. Alex Eng J 57(4):3321–3332
14. Harichandan AB, Roy A (2010) Numerical Investigation of low reynolds number flow past two and three circular cylinders using unstrucutred grid CFR scheme. Int J Heat Fluid Flow 31(2):154–171
15. Harichandan AB, Roy A (2012) Numerical investigation of flow past single and tandem cylindrical bodies in the vicinity of a plane wall. J Fluids Struct 33:19–43
16. Harichandan AB, Roy A (2012) CFR: a finite volume approach for computing incompressible viscous flow. J Appl Fluid Mech 5(3):39–52
17. Braza M, Chassaing P, Ha MH (1986) Numerical study and physical analysis of the pressure and velocity fields in the near wake of a circular cylinder. J Fluid Mech 163:79–130
18. Ding H, Shu C, Yeo KS, Xu D (2007) Numerical simulation of flows around two circular cylinders by mesh-free least square-based finite difference methods. Int J Num Methods Fluids 53:305–332
19. Liang C, Papadakis G, Luo X (2009) Effect of tube spacing on the vortex shedding characteristics of laminar flow past an inline tube array: a numerical study. Comput Fluids 38:950–964

Optimization in Fluid Mixing in Microchannels: A Review

Swagatika Acharya, Vijay Kumar Mishra, and Jitendra Kumar Patel

Nomenclature

Symbols	**Description**
Re	Reynolds number
θ_c	Circulation time
V	Volume flow rate
Q	Pumping capacity
P	Power consumption
N	Speed of impeller
D	Diameter of impeller
$Nmin$	Minimum agitation speed
t	Time
ρ	Density
$\Delta\rho$	Density difference between continuous and dispersed phase, kg/m^3
μ_c	Continuous phase viscosity, N s/m^2
μ_d	Dispersed phase viscosity, N s/m^2
σ	Interfacial tension, Nm^{-1}
C_0, a_0	Constants depend on impeller types and their locations
U	Fluid flow velocity
L	Length of the plate
ν	Viscosity

S. Acharya (✉) · V. K. Mishra · J. K. Patel
School of Mechanical Engineering, KIIT Deemed to be University, Bhubaneswar 751024, India
e-mail: mishra.vdm@gmail.com

© The Author(s), under exclusive license to Springer Nature Singapore Pte Ltd. 2023

S. Revankar et al. (eds.), *Recent Advances in Thermofluids and Manufacturing Engineering*, Lecture Notes in Mechanical Engineering,
https://doi.org/10.1007/978-981-19-4388-1_7

1 Introduction

1.1 Fluid Mixing

Fluid mixing is a very crucial activity and has various applications such as food handling, drug creation, substance designing, biotechnology, agri-synthetic arrangements, paint fabricating and water sanitization, etc. [1]. Muskett [1] identified various objectives to be achieved by mixing, like liquid estimation, stage blending, mass and hotness move between the liquids, controlling the liquid speeds, and so forth. It is advised in this work that in the event of blending two fluids, one of the fluids must be made broken into tiny beads. Further, at that point, by utilizing the Laser Doppler Anemometry strategy, the speeds of the fluids must be noticed and choppiness must be made to upgrade the blending effectiveness.

Richard et al. [2] used homogeneity property to determine the effectiveness of mixing. To accomplish homogeneity in blending, two miscible liquids for mixing were considered. Thickness of the liquids was used to study mixing interaction in various systems. Different systems for different fluid were used and studied; Fierce system was picked for low thickness liquids, temporary system for respectably gooey liquids, and laminar systems for exceptionally thick liquids. Prior information about blender and characteristics of liquids is essential before carrying out mixing experiments. Classification of mixing process is shown in Fig. 1 [1, 2].

Low viscosity in turbulent, moderate in transitional, and high in laminar regimes, respectively, were the characteristics of the fluids used in the blending experiment. Improvement in the homogeneity of the final product was observed. Dispersion and emulsion are the two different processes that are applicable to the immiscible fluids to be mixed. In dispersion, when the two fluids get closed to each other, one fluid breaks down to small droplets and merged with the second one and mixing is done. An emulsion is a kind of colloid shaped by consolidating two fluids that typically don't blend. In an emulsion, one fluid contains a scattering of the other fluid.

Fig. 1 Types of mixing process

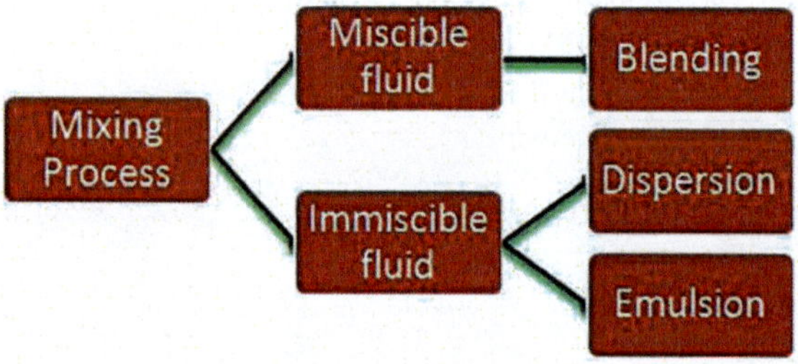

1.2 Mixing Factors

The fluid mixing performance depends upon the factors like mixing time, circulation time, mixing speed and power consumption, etc. [3]. The diameter and speed of the impeller are the two parameters based on which the mixing time depends. Circulation time is the time taken to transport a huge mass of fluid from one place to another within the container. The more the mixing speed the better will be the performance. The energy required for the blending process per unit time is called as the power consumption for the same.

$$\theta_c = \frac{V}{Q} \tag{1}$$

$$N_{\min} = C_o D^{a_o} (\mu_c)^{1/9} (\mu_d)^{-1/9} \sigma^{0.3} (\Delta P)^{0.25} \tag{2}$$

$$P \propto N^3 D^5 \tag{3}$$

Here θ_c, V, and Q denote the circulation time, discharge, and pumping capacity of the fluid, respectively. $N_{\min}$ is the minimum mixing speed whereas P, N, and D are the power consumption, speed, and diameter of the impeller, respectively.

2 Microfluidic Devices

In order to have an efficiently mixed fluid [4, 5], the mixing need to be done at the molecular level which is called as micromixing. For this purpose, microfluidic devices are required. Over last few decades, there is a huge demand for the Lab-On-a-Chip technique which has a great impact in the field of biotechnology, medicine, engineering, chemical industries, etc. It is a device of very small size in which several laboratory functions can be performed. A microfluidic chip is a bunch of miniature channels carved or shaped into a material (glass, silicon, or polymer). The miniature channels shaping the microfluidic chip are associated together to accomplish the ideal elements. This organization of miniature channels caught into the microfluidic chip is associated with the outside by data sources and yields penetrated through the chip, as an interface between the large scale and miniature world.

The aim of microfluidic mixing [6] is to achieve a thorough and rapid mixing of multiple samples in microscale devices. As the Reynolds number of the fluids flowing in the microchannel is very small, turbulent mixing does not occur in this case.

$$R_e = \frac{UL}{\nu} \tag{4}$$

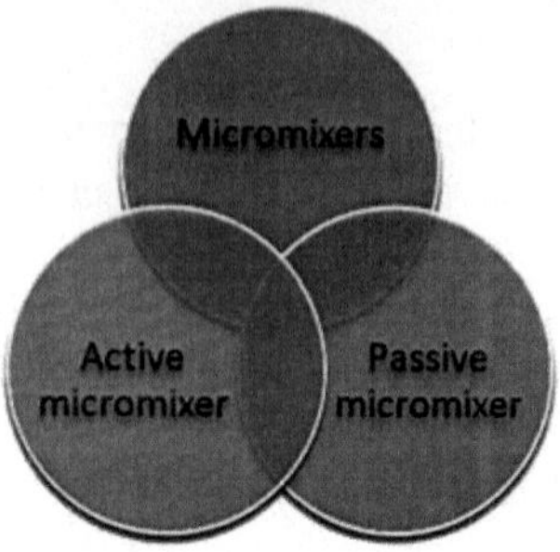

Fig. 2 Types of micro-mixers

Here diffusive species mixing assumes a significant part however it is an intrinsically sluggish interaction. Thus, the point of microfluidic mixing plans is to improve the blending productivity to such an extent that a careful mixing execution can be accomplished inside more limited blending channels, which can decrease the trademark size of microfluidic gadgets. Besides, the improvement of proficient mixing plans is fundamental for expanding the throughput of microfluidic frameworks and to understand the idea of miniature complete investigation frameworks and lab-on-a-chip frameworks. Expanding the contact region between the species to be mixed is one of the most productive methods for upgrading the diffusive-mixing impact. Likewise, past investigations introduced mixing plans by taking care of the examples of premium through discrete by means of openings, cantilever plate-valves or multi-diverts in the microfluidic gadget. An effective methodology is to expand the contact region between the mixing species by planning the microchannel designs with the goal that the species are collapsed on numerous occasions as they stream along the blending channel.

Since microfluidic gadgets are compelled to depend on lethargic diffusive-mixing process, there is a need in the field to find new methods by which mixing effectiveness can be upgraded. Shockingly, numerous laminar streams can possibly accomplish turbulent mixing, and microfluidic gadgets are presently being manufactured to exploit this marvel. Classification of micro-mixers is shown in Fig 2. Microfluidic mixing technique can be classified as either "active", where an external energy force is applied to integrate the sample species, or "passive", where the contact area and contact time of the species samples are increased through specially-designed microchannel configurations.

2.1 Active Micro-Mixer

In case of active micro-mixer, the mixing can be enhanced by stirring the fluid flow with the use of some external energy source. These types of micro-mixers can be categorized [7–9] as pressure, acoustic, thermal, magnetic, and electrical field-driven mixers. Various types of active micro-mixers are shown in Fig. 3.

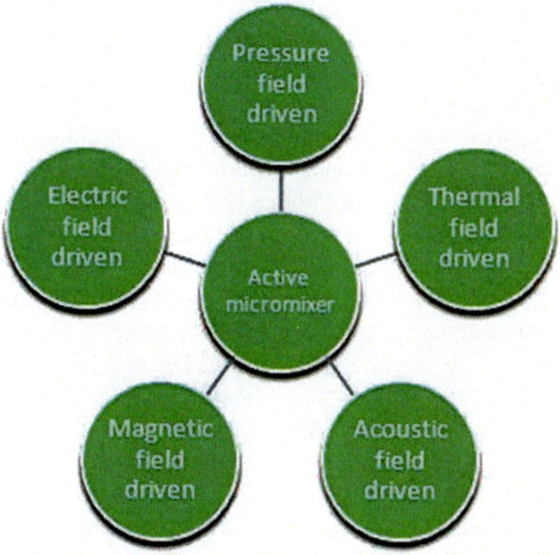

Fig. 3 Types of active micro-mixers

A functioning micro-mixer [10] utilizes unsettling influence created by an outside field for the mixing and the mixing system is fast. Using the quick mixing of the attractive mix bar, predefined convergences of arrangements can be created and gathered at the power source of the micro-mixer by ascertaining and applying fitting stream rates at the delta. The microfluidic arrangement comprises of a strain regulator, stream sensors, and a micro-mixer. The arrangements in the micro-mixer chamber are mixed constantly utilizing an attractive mix bar causing immediate mixing of the liquids. The tension-driven stream combined with stream sensors empowers exact command over the stream pace of the liquids entering the micro-mixer.

In case of pressure-driven active micro-mixer [11–13], disturbance is created by using an external micro pump that regulates the flow of the fluids inside the channel. By repeating the process of allowing and restricting the flow movement, turbulence is created which helps in better mixing. Mixing can be achieved effectively in acoustic type of micro-mixer with the help of an actuator which is used to stir the fluids in a flexible surface to generate air bubbles which in turn get energized and help in mixing. In thermal driven micro-mixer [14, 15], thermal power is used to raise the diffusion coefficient so that turbulence gets generated and mixing is done. In case of magnetic as well as electric types of micro-mixers, magnetic and electric fields are created, respectively which in turn helps in better fluid mixing.

2.2 *Passive Micro-Mixer*

In active micro-mixer [16] the intricacy of design and size is expanded, which requires extra human resources. Thus, the likelihood of coordinating active microfluidics with

LOCs applications has dropped off. To counter these disadvantages, the development of the test is accomplished either utilizing liquid properties or latent systems with practically no outer supporting force sources. Thus, latent microfluidics has been taken on and used generally in current examination undertakings to keep away from the utilization of outside supporting force gadgets. This strategy is basic, simple to fabricate, and needn't bother with any actuators or outer force supplies, as it utilizes fundamental research center instruments, as micropipettes, and clinical gadgets, for example, needle siphons. Passive micro-mixer [17] is one of the microfluidic gadgets. It uses no energy input with the exception of the component (pressure head) used to drive the liquid stream at a consistent rate.

2.2.1 Trends in Passive Micro-Mixer

In last few years (2000–2018), so many research works [16] have been done in passive micro-mixers and numbers of article are published on different grounds. Fig 4 represents the trends in passive micro-mixers. The different trends are pressure-driven, capillary, surface tension, vacuum-based, hydrostatic, gravity-driven, and osmosis passive micro-mixer. It is clear from the figure that pressure-driven, capillary, surface tension, vacuum-based, hydrostatic micro-mixers are the latest trends. Currently, numbers of research works are going on the same and articles are published.

Passive micro-mixer is classified into four main categories: Fig. 5 shows different types of passive micro-mixers.

The mechanism of sequential passive micro-mixer is to split the fluid streams and then recombine them in order to have better mixing efficiency. In this process, the mixing contact area of the fluids gets increased and hence the mixing time gets reduced. In droplet micro-mixer [18], the fluids need to be broken into tiny droplets. After the droplets formed, mixing occurs. Chaotic advection [19] means the movement of the substance in the transverse direction of the fluid flow. The advantage

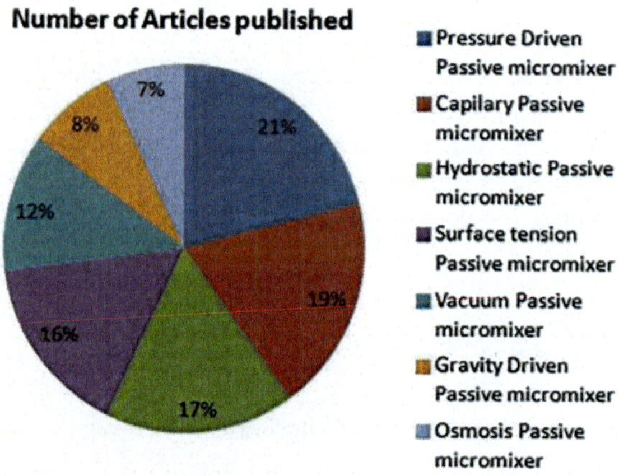

Fig. 4 Approaches in passive micro-mixers

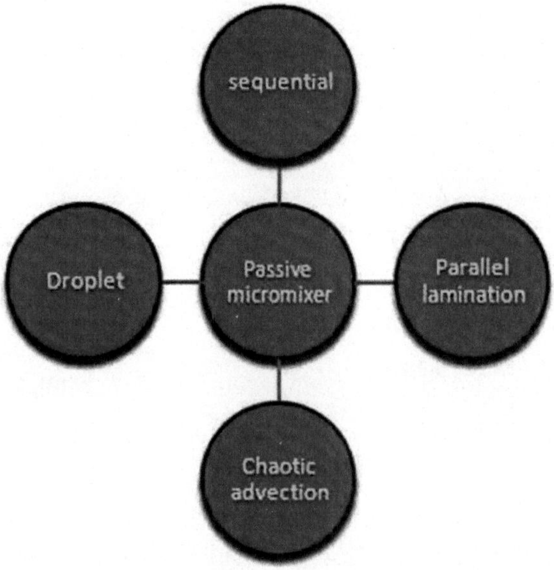

Fig. 5 Types of passive micro-mixers

of this method is that transverse components get generated which creates a better mixing surface area for mixing. In case of parallel lamination [20, 21] type of passive micro-mixer the inlet of the channel can be divided into numbers of parts but there will be a single outlet. The fluid streams coming from the inlet passage get merged and after mixing come out of the channel. The working principle of this type of micro-mixer is to improve the contact surface area by minimizing the inner diffusion duration of the fluids. This type of micro-mixer can be of two types.

2.2.2 T Type Passive Micro-Mixer

It is [22] T-shaped, hence named so. Here the inlet channel is divided into two segments. The angle between the two inlet channels is 1800. The outlet channel is perpendicular to the inlet one. Navier-Stokes equation, continuity equation, and convection-diffusion equation are required to describe the fluid flow in T channel.

$$\rho.\nabla u = 0 \tag{5}$$

$$\rho\frac{du}{dt} + \rho u(u.\nabla) = -\nabla p + \mu\nabla^2 u \tag{6}$$

$$\frac{dc}{dt} + u\nabla.C = D\nabla^2 C \tag{7}$$

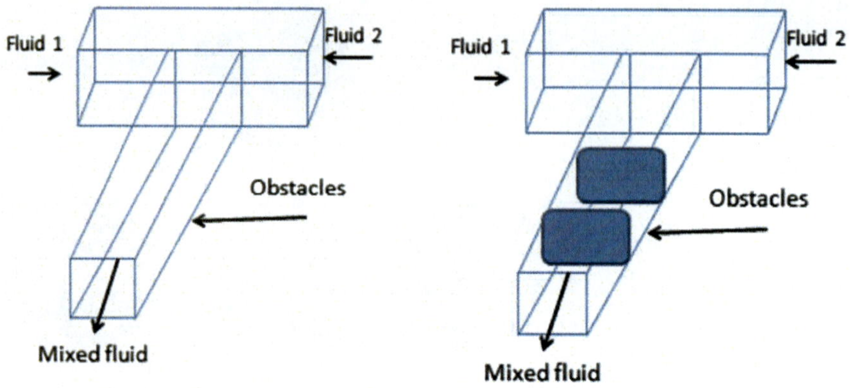

Fig. 6 T-shaped microchannel without and with obstacles

As we know, the fluid flow in microchannels mainly depends on diffusion process. Due to low Reynolds number turbulence is not created which hampers the speed of fluids and as a result, the mixing is not up to the level. In order to enhance the fluid mixing efficiency, turbulence needs to be created and that can be done with the presence of obstacles in the flow channel. Hence different types of obstacles are placed in the flow channel at different positions. Figure 6 represents the T type micro-mixer without and with obstacles.

It is found that by using either obstacle in the flow channel or baffle plates, turbulence is getting created which helps in improving the fluid velocity and in turn the mixing efficiency. As a result, the mixing quality is also getting enhanced.

2.2.3 Y-type Passive Micro-Mixer

It is one of the easiest models [23] for fluid mixing. Here the inlet channel is divided into two segments that are inclined at some angle to each other. It consists of a single outlet passage for the mixed fluid to come out. Two different fluids enter to the channel, they get mixed and the mixed one comes out of that microchannel. As we know that the Reynolds number is very low in this type of channel, hence the velocity profile is set up not a long way from the inlet channel. Here also the mixing efficiency can be improved by using obstacles in the flow passage. Figure 7 represents the Y-type microchannel without- and with- obstacles.

3 Optimization in Fluid Mixing

For fluid mixing purposes, number of devices have been designed and developed. But there is huge demand for the highly efficient device for optimizing the fluid

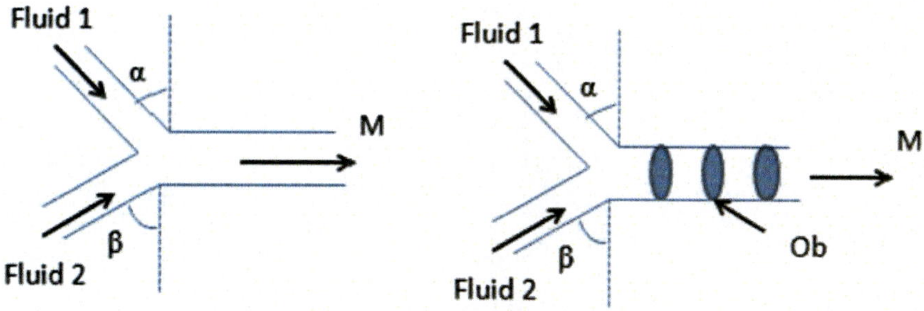

Fig. 7 Y shaped microchannel without and with obstacles

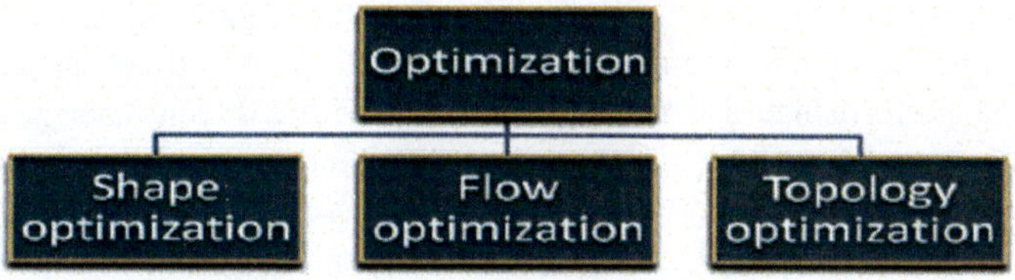

Fig. 8 Optimization approach in micro-mixers

mixing process. In this regard, Computational Fluid Dynamics (CFD) is the most accurate way to predict the flow behavior within the system which consists of efficient as well as effective designing process. In fluid mechanics, optimization relates to satisfy the objective functions like measuring the performance and the designing parameters. The optimization can be carried out in three different manners like shape optimization, size optimization, and topology optimization, etc. and are shown in Fig. 8.

3.1 *Shape Optimization*

Shape optimization [24] mainly relates to the shape of the microchannel so that the mixing efficiency can be enhanced at low designing cost. This concept is mostly applied to the static mixers where no moving part is there for the mixing purpose. Here the flow domain includes two inlet channels, a single outlet channel, and a mixing channel having numbers of baffles evenly spaced. Baffle plates are used in order to improve the mixing efficiency by creating turbulence in fluid flow. In this work, seven baffles are used to achieve maximum efficiency. Figure 9 shows the schematic of the experimental setup.

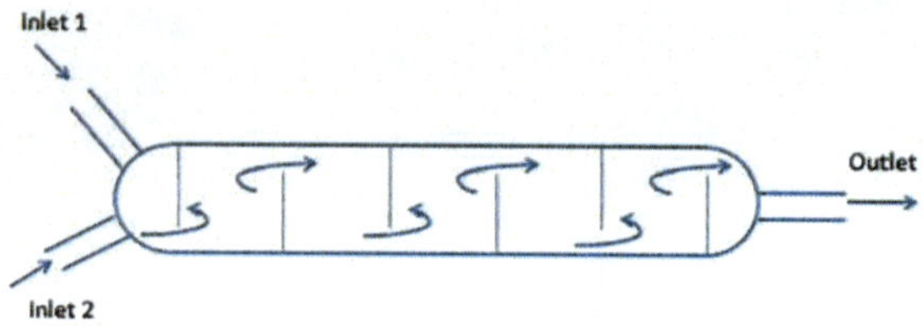

Fig. 9 Static mixer with baffle plates

3.2 Flow-Based Optimization

The best [25] presentation in the field of fluid mixing can be firmly coupled to the flow conditions. Designing and optimizing flow development requires definite information about the genuine unique circumstance. The flow analysis and optimization result in improved performance and mixing efficiency with reducing energy consumption, maximum volume flow rate, minimum maintenance and running cost, etc. Flow-based optimization can be accomplished by for example coupling of computational liquid elements (CFD) and optimization programming. The contribution of the advancement calculation is usually the recreation results relating to the different design points. Such recreations are typically directed in batch mode to kill the inactive time. In the batch mode, the solver arrangement is predefined and a similar setup is applied to all the design points. When there are huge changes in design, which prompt flow shift in power, solver would not have the option to react to the changes. To upgrade the appropriateness of CFD-based optimization, the smart CFD solver is adjusted to examine the physical science in each design point and afterward allocate fitting actual models to arrange the solver.

3.3 Topology Optimization

Topology optimization [26] is a numerical technique that enhances material format inside a given design space, for a given arrangement of loads, limiting conditions, and requirements fully intent on expanding the presentation of the framework. Topology optimization is not the same as shape optimization as this can attain any shape inside the design space, rather than managing predefined designs. The conventional topology optimization definition utilizes a finite element method (FEM) to assess the design execution.

Topology optimization has a wide scope of uses in aviation, mechanical, biosynthetic, and structural designing. At present, designs for the most part use topology optimization at the idea level of a design cycle. Because of the free structures that normally happen, the outcome is regularly hard to make. Hence the outcome arising out of topology optimization is frequently adjusted for manufacturability. Adding requirements to the detailing to build the manufacturability is a functioning field of exploration. Sometimes results from topology optimization can be straightforwardly produced utilizing added substance fabricating; topology optimization is consequently a vital piece of plan for added substance producing.

4 Conclusion

Fluid mixing is understood with the help of works covering basic concepts to advanced experiments in microchannels. Various kinds of systems for improving mixing process in miniature mixers have been covered in this survey. It is suggested that each plan has its own pros and cons. The most appropriate method for mixing needs to be selected based upon the particular application as a primary concern. There is no single mixing idea satisfying all the desired goals. Various aspects of mixing devices such as creation cost, execution, etc. are covered. Subsequent to going through the above papers we can conclude that if a mixed sort hindrance is put in either T molded or Y formed stream channel, then the mixing effectiveness can be upgraded without expanding the tension drop.

References

1. Muskett MJ (1986) The measurement of fluid mixing processes. Mater Des 7(4):188–191. https://doi.org/10.1016/0261-3069(86)90120-2
2. Grenville RK, Nienow AW (2004) Blending of miscible liquids. Handb Ind Mixing, pp 507–542. https://doi.org/10.1002/0471451452
3. Ghotli RA, Raman AAA, Ibrahim S, Baroutian S (2013) Liquid-liquid mixing in stirred vessels: a review. Chem Eng Commun 200(5):595–627. https://doi.org/10.1080/00986445.2012.717313
4. Convery N, Gadegaard N (2019) 30 years of microfluidics. Micro Nano Eng 2:76–91. https://doi.org/10.1016/j.mne.2019.01.003
5. Asghar C, Hossein F, Aghayie H, Reza V (2015) Numerical simulation of time- dependent electro-osmotic micro-mixer for laboratory-on-a-chip applications. Res J Recent Sci 4(2):83–90
6. Seok G, Seok J, Beom CC, Sang K, Lee H (2014) Applications of micromixing technology. Analyst 135(3):460. https://doi.org/10.1039/b921430e
7. Bayareh M, Ashani MN, Usefian A (2020) Active and passive micro-mixers: a comprehensive review. Chem Eng Process 147. https://doi.org/10.1016/j.cep.2019.107771
8. Hejazian M, Nguyen NT (2017) A rapid magnetofluidic micro-mixer using diluted ferrofluid. Micromachines 8(2):37. https://doi.org/10.3390/mi8020037
9. Ward K, Fan ZH (2015) Mixing in microfluidic devices and enhancement methods 25. https://doi.org/10.1088/09601317/25/9/094001
10. Sprogies T, Köhler JM, Groß GA (2008) Evaluation of static micro-mixers for flow-through extraction by emulsification. Chem Eng J 135:199–202. https://doi.org/10.1016/j.cej.2007.07.032
11. Lee HY, Voldman J (2007) Optimizing micro-mixer design for enhancing dielectrophoretic microconcentrator performance. Anal Chem 79(5):1833–1839. https://doi.org/10.1021/ac061647q

12. Fang W, Yang J (2009) A novel microreactor with 3D rotating flow to boost fluid reaction and mixing of viscous fluids. Sens Actuators B 140(2):629–642. https://doi.org/10.1016/J.SNB.2009.05.007
13. Rao LT, Goel S, Dubey SK, Javed A (2019) Performance investigation of T-shaped micro-mixer with different obstacles. J Phy 1276. https://doi.org/10.1088/1742-6596/1276/1/012003
14. Santana HS, Júnior JLS, Taranto OP (2015) Numerical simulations of biodiesel synthesis in microchannels with circular obstructions. Chem Eng Process 98:137–146. https://doi.org/10.1016/j.cep.2015.10.011
15. Oualha K, Amar MB, Michau A, Kanaev A (2017) Observation of cavitation in exocentric T-mixer. Chem Eng J 321:146–150. https://doi.org/10.1016/j.cej.2017.03.111
16. Narayanamurthy V, Jeroish ZE, Bhuvaneswari KS, Bayat P, Premkumar R, Samsuri F, Yusoff MM (2020) Advances in passively driven microfluidics and lab-on-chip devices, a comprehensive literature review and patent analysis. RCS Adv 10(20):11652–11680. https://doi.org/10.1039/d0ra00263a
17. Wang C-T, Chen Y-M (2011) Flow mixing of double two-inlet Y-Type micro channel with optimal layout of obstacles. J Mech 27(02):N1–N4. https://doi.org/10.1017/jmech.2011.22
18. Huanming X, Jiawei W, Zhiping W (2018) A comparative discussion of different designs of passive micro-mixers: specific sensitivities of mixing efficiency on Reynolds numbers and fluid properties. Microsyst Technol 24(2):1253–1263. https://doi.org/10.1007/s00542-017-3496-4
19. Sudarsan AP, Ugaz VM (2005) Fluid mixing in planar spiral microchannels. Lab Chip 6(1):74–82. https://doi.org/10.1039/b511524h
20. Chiu PH, Chang CC, Yang RJ (2013) Electro kinetic micromixing of charged and non- charged samples near nano–microchannel junction. Microfluid Nanofluid 14(5):839–844. https://doi.org/10.1007/s10404-012-1116-2
21. Gambhire S, Patel N, Gambhire G, Kale S (2016) A review on different micro-mixers and its micromixing channel microchannel. Mater Sci, pp 409–413. https://doi.org/10.14741/Ijcet/22774106/spl.4.2016.83
22. Raza W, Hossain S, Kim KY (2020) A review of passive micro-mixers with a comparative analysis. Micromachines 11(5). https://doi.org/10.3390/mi11050455
23. Rudyak V, Minakov A (2014) Modeling and optimization of Y-type micro-mixers. Micromachines 5(4):886–912. https://doi.org/10.3390/mi5040886
24. Alexias P, Giannakoglou KC (2020) Shape optimization of a two-fluid mixing device using continuous adjoint. Fluids 5(1). https://doi.org/10.3390/fluids5010011
25. Li L, Cheng Z, Lange CF (2018) CFD-based optimization of fluid flow product aided by artificial intelligence and design space validation. Math Prob Eng https://doi.org/10.1155/2018/8465020
26. Andreasen CS, Gersborg AR, Sigmund O (2009) Topology optimization of microfluidic mixers. Int J Numer Methods Fluids 61:498–513. https://doi.org/10.1002/fld.1964

A Simple Solution of Third Grade Fluid Flow Problem: Cascade-Forward Type ANN Approach

Swagatika Acharya, Vijay Kumar Mishra, Jitendra Kumar Patel, Gaurav Gupta, Mrityunjay K. Sah, Pinky Shah, and Debjyoti Sahu

Nomenclature

A	Fluid parameter (Third grade)
Ac	Area of cross section
A_1, A_2, A_3, A_n	Kinematic tensor
$a_0, a_2, a_4, a_6, a_8, b_0, b_2, b_4, b_6, b_8, b_{10}, b_{12}, c_1, c_2$	Constants
Br	Brinkman number
C_p	Specific heat, kJ/kgK
c_i	iTh constant
D/Dt	Material derivative
f	Body force per unit volume
g, $\boldsymbol{v}, \tilde{v}$	Functions
h	Half depth of channel, m
k_{th}	Fluid thermal conductivity
L	Channel length
l_1, l_2	Length constants
N	Pressure gradient (Non-dimensional)
$\alpha_1, \alpha_2, \beta$ $\beta 1, \beta 2, \beta 3$	Material constants
ρ	Fluid density, kg/m^3
μ	Dynamic viscosity, PaS

S. Acharya · V. K. Mishra (✉) · J. K. Patel · M. K. Sah · P. Shah · D. Sahu
School of Mechanical Engineering, Kalinga Institute of Industrial Technology, Bhubaneswar 751024, India
e-mail: mishra.vdm@gmail.com

G. Gupta
Vellore Institute of Technology, Vellore 632014, India

© The Author(s), under exclusive license to Springer Nature Singapore Pte Ltd. 2023

S. Revankar et al. (eds.), *Recent Advances in Thermofluids and Manufacturing Engineering*, Lecture Notes in Mechanical Engineering,
https://doi.org/10.1007/978-981-19-4388-1_8

Θ	Temperature (non-dimensional), K
Θ_N	Temperature (non-dimensional) for Newtonian fluid
p^*	Pressure
Q	Flow rate
q	Ratio of heat fluxes
q_1, q_2	Heat flux at the plates
R	Residual
S	Sum of the residual square
T^*	Temperature
$T_m{}^*$	Bulk mean temp.
$T_{wl}{}^*$	Temperature of the lower wall
u	Velocity in axial direction (non-dimensional)
u_N	Velocity of Newtonian fluid (non-dimensional)
$\boldsymbol{u}^*$	Velocity in axial direction
u_0	Average velocity
V^*	Velocity vector
Φi	Base function
w_i	*i*Th weight function
$\boldsymbol{\tau}$	Stress
CFANN	Cascade forward artificial neural network
LSM	Least square method
SCG	Scaled conjugate gradient
Nu	Nusselt number

1 Introduction

Various engineering problems of practical interest are difficult to solve [1, 2], either due to large time consumption and/or absence of mathematical relation between the input and the output. Such problems can be solved easily by employing artificial neural network (ANN) [3–7]. As compared to other approach, ANN method offers following advantages in solving a problem: (a) complex mathematical relations are not required (b) after training of network, results for any new case can be obtained within seconds (c) complex problems with multiple input and out parameters can be handled easily (d) easy to apply optimization and inverse analysis (e) experimental and/or numerical and/or analytical data can be used. ANN is nature inspired mathematical tool, where simple mathematical operations are performed by neurons (processing elements). Large numbers of neurons when interconnected with each other, can perform difficult tasks like approximating any random function. ANN can be used to solve wide range of engineering problems, including non-Newtonian fluid flow problems. Non-Newtonian fluid has some unique properties like shear thinning and shear thickening [8–10].

In contrast of Newtonian fluids, fluids such as a non-Newtonian fluid don't follow Newton's law of fluid flow, and their shear stress depends non-linearly on the strain rate. These characteristics of non-Newtonian fluids can be most appropriately modeled by third grade fluid model. By using semi analytical method such as least square method (LSM) [11–14], governing equations can be solved under specified boundary conditions. Hence, the generated data can be fed as pair of input and output, to ANN for training. After that the trained ANN can solve any small alterations of the problem within short time.

Under ANN, a cascade-forward type network has connection of layers to all the downstream layer of the network i.e. input layer is not only connected to the first hidden layer but also to the subsequent hidden and out layers. Such high level of connectivity between the layers leads to high level of accuracy of the network for low number of neurons in the hidden layer. Obviously, care should be taken to avoid over fitting in such cases. In the present work, a cascade-forward type ANN (CFANN) is employed and demonstrated to solve a fluid flow problem with non-Newtonian fluid. The problem is solved for velocity and temperature, by using LSM. Velocity along with temperature distribution (along the width of the channel) is computed by LSM for 14 values of a parameter relevant in non-Newtonian fluids called third grade fluid parameter, and then applied in the CFANN during the training stage. The trained CFANN is then used to retrieve temperature distribution corresponding to a new velocity distribution. Different number of neurons in the hidden layer is tested to retrieve the temperature distribution, and performance is compared and analyzed. For all such cases, scaled conjugate gradient (SCG) algorithm is employed for training of the network. Present work intends to drastically reduce the complexity and time required to solve non-Newtonian based fluid flow problems. CFANN type network along with LSM is not reported in open literature to compute complete temperature distribution for a fluid flow problem with non-Newtonian fluid.

Work in this paper is presented as:

- development of input–output pair of data by solving the governing equations together applying appropriate boundary conditions
- development of CFANN with different hidden neurons number
- Training of the CFANN with the generated data, and monitoring of the CFANN by tools such as histogram, performance curves etc.
- using the CFANN for retrieval of temperature
- analysis or the result and identification of most suitable CFANN.

Present approach intends to solve small alterations of the problem in this work with least effort, without depending upon any other approaches like LSM etc. Such simplicity and time effective approach may find applications in industries associated with non-Newtonian fluids like paints, grease, polymer etc.

2 Formulation

A non-Newtonian fluid is moving through two plates in parallel arrangement, each of length L are distance H distance apart (see Fig. 1). The flow direction dimension is taken large as compared to dimensions in other directions ($L > > H = W$; say W is width). Both the bottom and the top plates are supplied with uniform and constant heat flux, q1 and q2 respectively. Present analysis assumes fluid properties to remain constant. The fluid flow is laminar and in-compressible, and in steady state. Further, the flow is also assumed to be fully developed, both in terms of hydro-dynamically as well as thermally.

The necessary equations governing the flow are presented and generalized with suitable non-dimensional terms. The solution methodology subjected to proper boundary conditions is also presented (see Sect. 2.1). Various governing equations used in the analysis are given through Eq. (1)–Eq. (24).

Continuity equation can be represented as:

$$\nabla \cdot V^* = 0 \tag{1}$$

In Eq. (1), V^* is the velocity vector. Similarly, the conservation of Momentum Equation can be expressed as:

$$\rho \frac{DV^*}{Dt} = \nabla \cdot \tau + f \tag{2}$$

Body forces are not considered in conservation of momentum equation (Eq. 2). Under third grade fluid model, the non-Newtonian fluid is modeled by following stress tensor and strain rate tensor relation:

$$\tau = -pI + \mu A_1 + \alpha_1 A_2 + \alpha_2 A_1^2 + \beta_1 A_3 + \beta_2(A_1 A_2 + A_2 A_1) + \beta_3(trA_1^2)A_1 \tag{3}$$

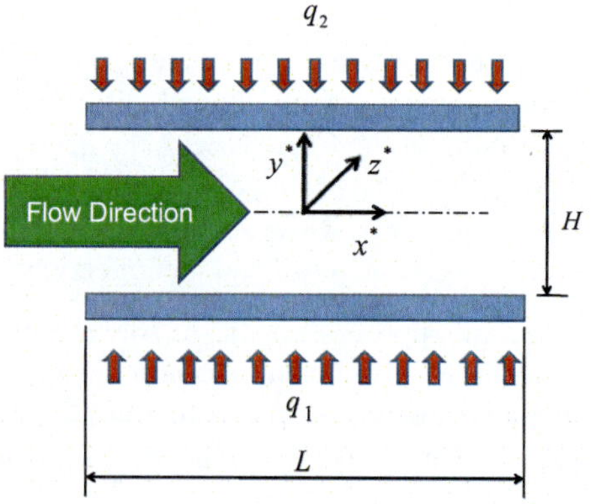

Fig. 1 Fluid flows through two parallel plates (both heated uniformly)

Kinetic tensor can be expressed by Eqs. (4) and (5) respectively.

$$A_1 = (\mathrm{grad}V^*) + (\mathrm{grad}V^*)^{\mathrm{Transpose}} \tag{4}$$

$$A_n = \frac{\mathrm{d}A_{n-1}}{\mathrm{d}t} + A_{n-1}(\mathrm{grad}V^*) + (\mathrm{grad}V^*)^{\mathrm{Transpose}} A_{n-1}, \quad n = 1, 2, 3 \tag{5}$$

As per convention the energy equation is as expressed in Eq. (6).

$$\rho c_p \frac{\mathrm{d}T^*}{\mathrm{d}t} = \tau : \mathrm{grad}(V^*) - \nabla.(-k_{\mathrm{th}} \nabla T^*) \tag{6}$$

On the basis of hydro-dynamically developed flow assumption, the simplified form for the velocity vector can be expressed as in Eq. (7).

$$v^* = \left[u^*\left(y^*\right), 0, 0\right] \tag{7}$$

Momentum conservation equation gets modified after using the above velocity vector, and can be expressed by Eqs. (8) and (9) and Eq. (10) along x, y and z axes respectively (x^*, y^*,$z^*$are dimensional coordinates).

$$\frac{\partial p^*}{\partial x^*} = \mu \frac{\mathrm{d}^2 u^*}{\mathrm{d}y^{*2}} + 2(\beta_2 + \beta_3) \frac{\mathrm{d}}{\mathrm{d}y^*}\left(\frac{\mathrm{d}u^*}{\mathrm{d}y^*}\right)^3 \tag{8}$$

$$\frac{\partial p^*}{\partial y^*} = (2\alpha_1 + \alpha_2) \frac{\mathrm{d}}{\mathrm{d}y^*}\left(\frac{\mathrm{d}u^*}{\mathrm{d}y^*}\right)^2 \tag{9}$$

$$\frac{\partial p^*}{\partial z^*} = 0 \tag{10}$$

Moving forward with same procedure, energy equation can also be modified as given in Eq. (11).

$$\rho c_p u^* \frac{\partial T^*}{\partial x^*} = \mathrm{kth}\left(\frac{\partial^2 T^*}{\partial x^{*2}} + \frac{\partial^2 T^*}{\partial y^{*2}}\right) + \mu\left(\frac{\mathrm{d}u^*}{\mathrm{d}y^*}\right)^2 + 2(\beta_2 + \beta_3)\left(\frac{\mathrm{d}u^*}{\mathrm{d}y^*}\right)^4 \tag{11}$$

Here, kth denotes fluid thermal conductivity (eventually a third grade fluid). However, the next term on the right side of the Eq. (11) is representative of viscous dissipation. The third term exists solely because of parameter of the third grade fluid (Eq. 11). Moreover, it can be simplified for Newtonian fluid, if values for β_2 and β_3 are taken to be zero. With the assumption of thermally developed condition, the temperature distribution can be expressed in non-dimensional form, $\theta = \frac{T^* - T^*_{w_l}}{T^*_m - T^*_{w_l}}$.

Here, T_m^* and $T_{w_l}^*$ are the bulk mean temperature of the fluid and the upper plate, respectively. Under the assumption of constant heat flux with uniform strength throughout the length of the plate, following expressions can be written:

$$\frac{\partial T^*}{\partial x^*} = \frac{\mathrm{d}T_m^*}{\mathrm{d}x^*} = \text{const.}, \quad \frac{\partial^2 T^*}{\partial x^{*2}} = 0 \tag{12}$$

The modified energy equation of Eq. (11) can be further simplified by using expressions given in the Eq. (12) as follows:

$$\rho c_p u^* \frac{\partial T^*}{\partial x^*} = k\text{th}\left(\frac{\partial^2 T^*}{\partial y^{*2}}\right) + \mu\left(\frac{\mathrm{d}u^*}{\mathrm{d}y^*}\right)^2 + 2(\beta_2 + \beta_3)\left(\frac{\mathrm{d}u^*}{\mathrm{d}y^*}\right)^4 \tag{13}$$

2.1 *Boundary Conditions*

Due to no-slip conditions prevailing on the surface of both the plates, a mathematical expression can be written for dimensional axial velocity as follows:

$$u * (-\mathrm{h}) = 0, \; u * (\mathrm{h}) = 0 \tag{14}$$

Application of uniform and constant heat flux on both the bottom and the top plate leads to following expressions:

$$k_{th}\left(\frac{\partial T^*}{\partial y^*}\right)_{-h} = \mp q_1, \; k\text{th}\left(\frac{\partial T^*}{\partial y^*}\right)_h = \pm q_2 \tag{15}$$

For the purpose of generalization, the temperature is made non-dimensional as:

$$\theta = \frac{T^* - T_{w_l}^*}{q_1 h / k\text{th}} \tag{16}$$

Further, for the purpose of generalizing the solution, velocity and distance are made non-dimensional as:

$$u = \frac{u^*}{u_0}, \; y = \frac{y^*}{h} \tag{17}$$

After making it non-dimensional, the conservation equations of momentum and energy becomes:

$$\frac{\mathrm{d}^2 u}{\mathrm{d}y^2} + 6A\left(\frac{\mathrm{d}u}{\mathrm{d}y}\right)^2 \frac{\mathrm{d}^2 u}{\mathrm{d}y^2} = N \tag{18}$$

$$\beta u = \frac{\mathrm{d}^2\theta}{\mathrm{d}y^2} + \mathrm{Br}\left(\frac{\mathrm{d}u}{\mathrm{d}y}\right)^2 + 2\mathrm{ABr}\left(\frac{\mathrm{d}u}{\mathrm{d}y}\right)^4 \tag{19}$$

$$A = \frac{\beta_2 + \beta_3}{\mu}\left(\frac{u_0}{h}\right)^2, \ \mathrm{Br} = \frac{\mu u_0^2}{q_1 h}, \ u_0 = \frac{1}{N}\frac{\mathrm{d}p^*}{\mathrm{d}x^*}\frac{h^2}{\mu}, \ \beta = \frac{\rho c_p u_0}{q_1}\frac{\mathrm{d}T^*}{\mathrm{d}x^*} \tag{20}$$

Boundary conditions for the velocity field are:

$$u(-1) = 0, u(1) = 0 \tag{21}$$

Similarly, boundary conditions in non-dimensional form for the energy equation can be expressed as:

$$\theta(-1) = 0 \tag{22}$$

$$\frac{\mathrm{d}\theta}{\mathrm{d}x}(-1) = -1 \tag{23}$$

$$\frac{\mathrm{d}\theta}{\mathrm{d}x}(1) = \pm\frac{q_2}{q_1} = q \tag{24}$$

2.2 *Solution Involving LSM*

In LSM, if differentiation of any function $v(y)$ gives $g(y)$, and use of any linearly independent base function ϕ_i with some combination can approximate the function $v(y)$. Mathematically $v(y)$ can be expressed in terms of linear combination of base function ϕ_i, as:

$$v \cong \tilde{v} = \sum_{i=1}^{n} c_i \phi_i \tag{25}$$

If differentiation of $v(y)$ gives $g(y)$, then residue can be obtained as:

$$R(y) = D\{\tilde{v}(y)\} - g(y) \neq 0 \tag{26}$$

Now, for the entire domain, the residue can be reduced to zero.

$$\int_x R(y) w_i(y) \mathrm{d}y = 0; \ i = 0, 2, 3 \ldots, n \tag{27}$$

Setting the weights as equal as the residue, Eq. 27 simplifies as,

$$S = \int_y R(y)R(y)\mathrm{d}y = \int_y R^2 \mathrm{d}y \tag{28}$$

To minimize S, the expression need to be used is:

$$\frac{\partial S}{\partial c_i} = 2\int_y R(y)\frac{\partial R}{\partial c_i}\mathrm{d}y = 0 \tag{29}$$

The above-mentioned governing equations are solved by employing LSM. The non-linear nature of Eqs. (18) and (19), makes them difficult to solve. The solution becomes challenging, if the complexity of the problem is increased any further. Benefits with ANN approach here is, only some cases need to be solved by LSM, whereas other alterations can be solved by ANN.

3 Results and Discussions

The LSM method explained in the previous section, is used to get the solution; and then the result obtained is compared with the solution reported in the literature [14], through exact method. The velocity distribution for two different values of third grade fluid parameter, is used for the comparison; and the validation is reliable as per Fig. 2a. The solution from the LSM method is also extended for Newtonian fluid, and the result is compared with the exact solution as shown in Fig. 2b. Thus the method of LSM is able to give reliable result for both the cases of Newtonian fluid and the non-Newtonian fluid.

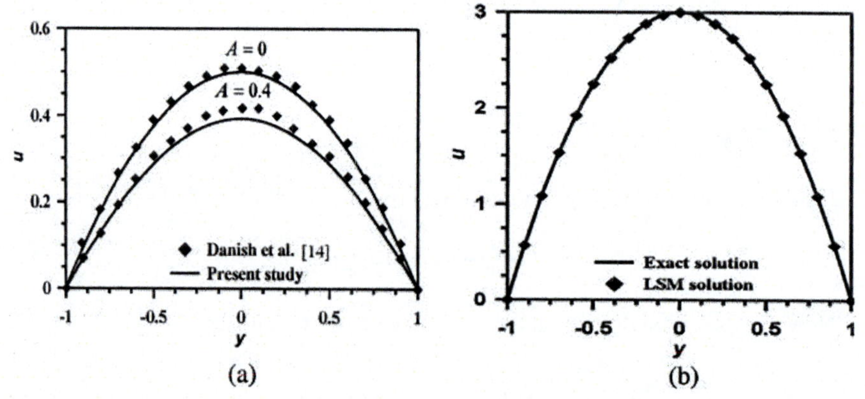

Fig. 2 Comparison of present LSM solution with the exact solution reported in the literature [14] **a** velocity distribution for the case of third grade fluid parameter A **b** velocity distribution for the case of Newtonian fluid

3.1 CFANN Deployment

After validating the results of LSM, CFANN is employed for solving the problem. Input data required for the training of the CFANN, is generated by computing velocity and temperature distribution for fourteen different values of A (third grade fluid parameter). Data generated for training purpose is in the form of input (velocity distribution) and output (temperature distribution) pairs. The input–output pair of data is used to feed the CFANN during training stage and the performance of training phase is monitored with different means. Also, effect of number of neurons in the hidden layer (2, 4 and 8) of results of CFANN are analyzed and compared. One such CFANN model is shown in Fig. 3, where velocity distribution (201 points) is used as input and temperature distribution (201 points) is fed as target. Total pairs of input and output pairs are 14. In each pair, data for 201 points are taken. One hidden layer with two neurons is used. Specialty of CFANN is high connectivity between neurons of input, hidden and the output layer. This level of connection means the network will be able to give reliable result with moderate number of neurons. Tan-sigmoid activation function based hidden layer and pure linear function based output layer are used here.

Flowchart of the details of the steps taken in use of CFANN approach is shown in Fig. 4. Starting with the employment of LSM (least square method) for solving the governing equations under suitable boundary conditions. Then, proceeding to compute velocity and temperature distribution for various values (14 numbers) of A. Later on, the data is used to train the CFANN. Finally, analyzing the developed CFANN model with regression coefficient and other means, the analyzed CFANN model is deployed for development of temperature distribution for any new velocity distribution.

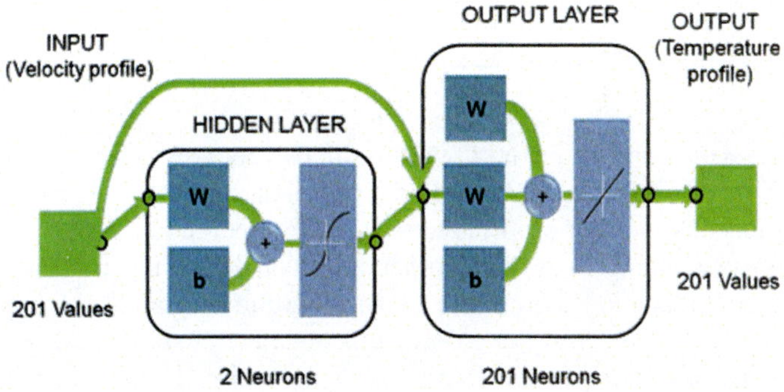

Fig. 3 CFANN model for development of temperature distribution with two number of neurons in the hidden layer

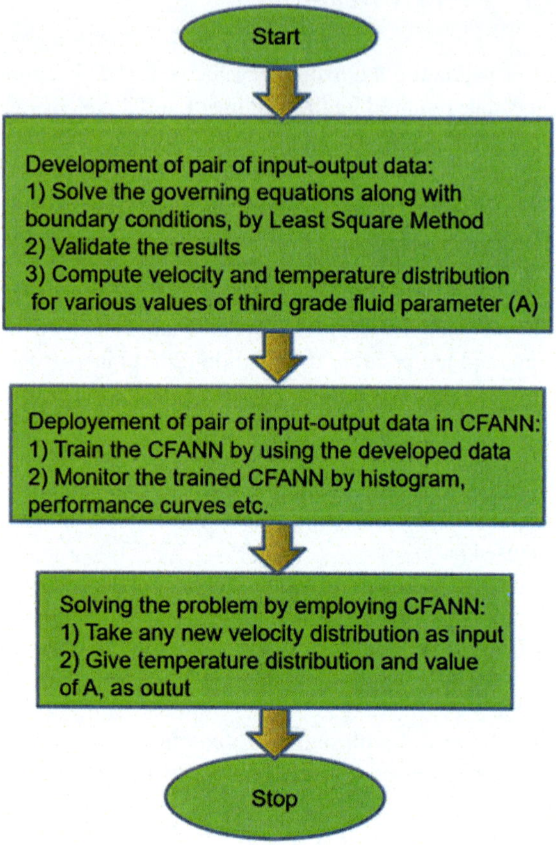

Fig. 4 Flowchart of CFANN approach for solving a fluid flow problem with third grade fluid

3.2 Performance Curves

Training, validation and testing for CFANN with two neurons in the hidden layer are shown in Fig. 5. It is observed that the slopes of validation and testing curves acquire similar trend, thus indicate proper training of CFANN. Gradient plot and val fail plots are as shown in Fig. 5; these are important for observing the growth of training. Gradient should attain low value close to the best validation point, which is correct for the present case. Also, val-fail plot indicates number of successive iterations for which the value of mean square error (mse) is not declining. It is observed from Fig. 6, that at iteration/epoch 136, the mse does not decline continuously for 6 number of iteration; and hence the training is stopped.

In Fig. 7, the histogram of the input data in CFANN, is represented. Interestingly, maximum number of data fall near ZEL (zero error line), and gradually decreases farther away. Sufficient amount of data on the ZEL indicates proper training for the cases corresponding to the input data. Whereas, data away from the ZEL indicates perfect training of the CFANN for new cases (for which data is not used in training).

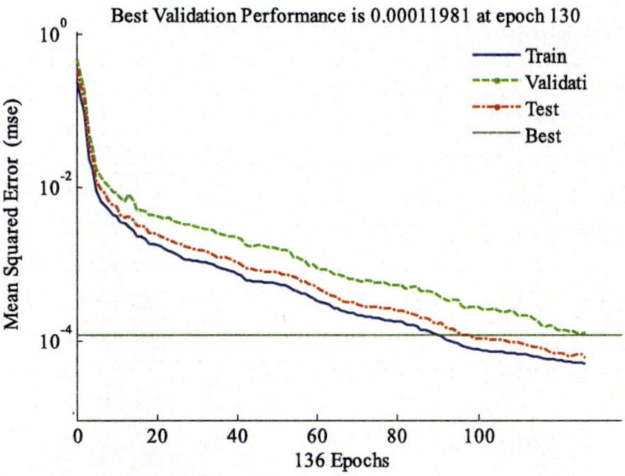

Fig. 5 Monitoring of training stage of CFANN by performance curves

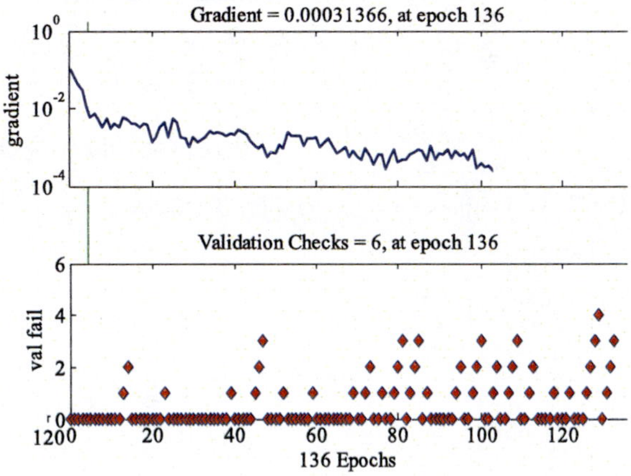

Fig. 6 Monitoring of training stage of CFANN by gradient and val fail plots

Moreover, the CFANN model suggests a good distribution of data through all the stages (training, validation, testing) across the left and right side of ZEL. Another important way to predict the performance of CFANN for any practical situation, is the regression coefficient; and is shown in Fig. 8. The ideal case (output = target) is shown by the dashed curve, whereas actual case predicted by the CFANN model is shown by solid line in Fig. 8. Approximately one value of regression coefficient means that the model has learned to get a linear trend among the output and the target.

After successfully developing the CFANN model, a new velocity distribution is fed as input; and corresponding temperature distribution is given by the CFANN as output. Various number of neurons (2, 4 and 8) are explored in the hidden layer, and accuracy of the results from CFANN are compared in Fig. 9. The difference between the actual temperature distribution and the one obtained from CFANN is plotted against the width of the channel. All cases are capable to give the temperature within accuracy of 2.5%. Also, the accuracy with 4 and 8 neurons is higher than that

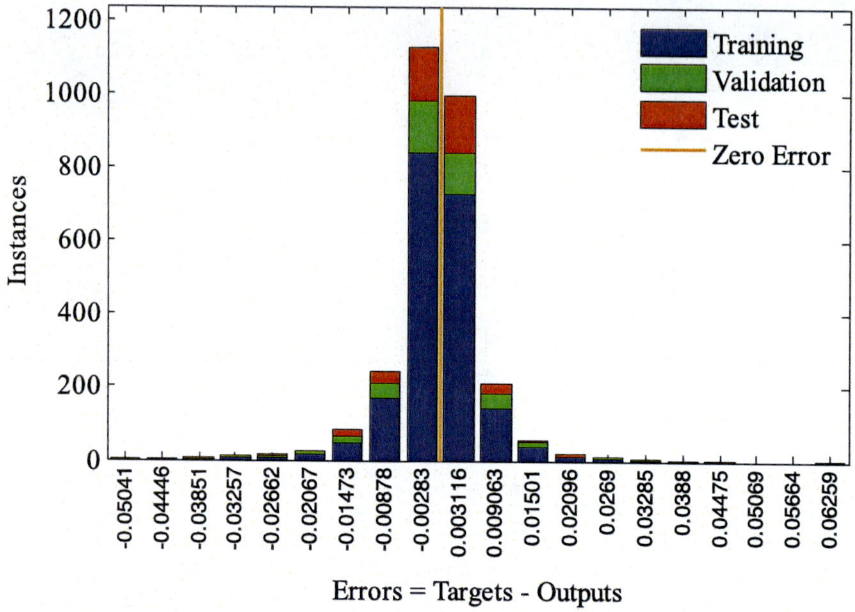

Fig. 7 Monitoring of training stage of CFANN by histogram

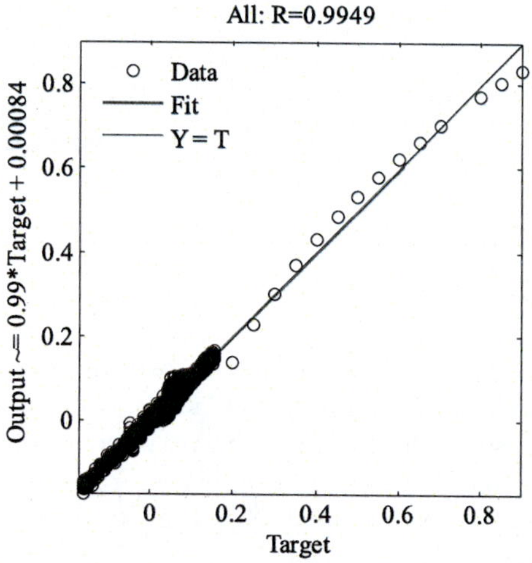

Fig. 8 Monitoring of training stage of CFANN by regression coefficient

with 2 number of neurons. Clear comparison of 4 and 8 neurons can't be made (see Fig. 9) quantitatively, due to almost similar plots of noisy data. Time consumption for different number of neurons is shown in Fig. 10. Time consumption for 4 and 8 neurons is of the same order, while time consumption with 2 neurons is less. Very less time consumption with two neurons is obvious, as less neurons lead to less computation. The present CFANN is also explored to retrieve the value of A (third grade fluid parameter). Table 1 shows the retrieved values of the A by CFANN. CPU time consumption along with accuracy of retrieval of A, is shown for three different cases (2, 4 and 8 neurons in hidden layer) [11, 12].

At Table 1, it is summarized that the accuracy of retrieval of A, is high with 4 neurons as compared to 8 neurons. Also it is important to note that, while accuracy with 2 neurons is higher than 8 neurons (as per Table 1) but Fig. 9 suggest that 2

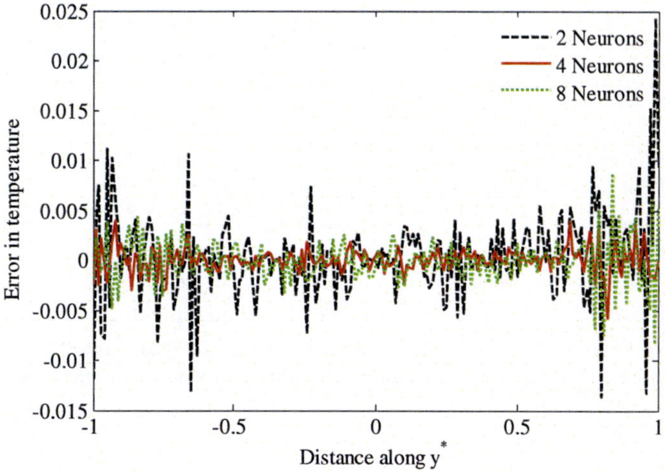

Fig. 9 Plot of error in temperature distribution developed by CFANN for different number of neurons in the hidden layer

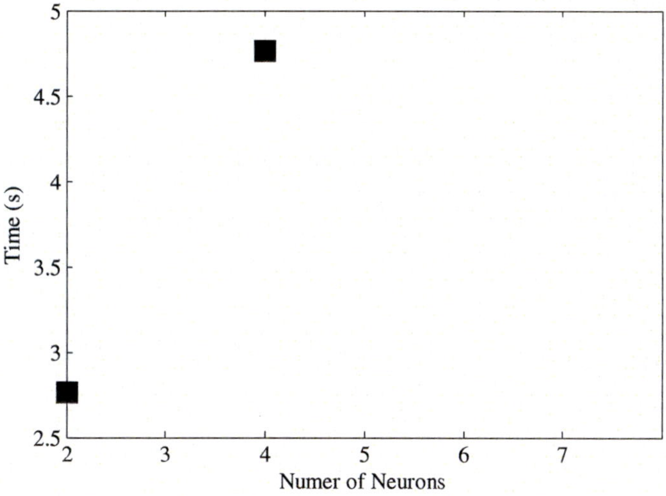

Fig. 10 Time consumption in training the three CFANN model

Table 1 Comparison of retrieval of A by CFANN for different number of neurons in the hidden layer

S. No	Number of neurons	Absolute error in retrieved value of A (%)	Time (s)
1	2	1.52	2.76
2	4	1.74	4.76
3	8	6.32	4.74

neurons gives comparatively high error at some points. Thus over all, 4 neurons is best for the CFANN.

For all the cases, scaled conjugate algorithm is used for training; and found to be very fast with high accuracy for the present problem.

In summary, the profiles of the temperature and the velocity are calculated by LSM for fourteen values of third grade fluid parameter, A. This data of velocity and temperature profile can be easily fed into CFANN for training. The training has to be monitored utilizing four important plots: performance plot, gradient and val-fail plot, histogram and regression plot [15]. After training, a velocity profile has to be fed as input, and output can be the temperature. Thus the performance of CFANN in generating the temperature profile can be compared for various neurons in the hidden layer also.

4 Conclusions

With appropriate boundary conditions set, the equations governing the flow problem involving non-Newtonian fluid are solved by employing semi analytical technique i.e., least square method. CFANN is demonstrated here in easiest possible way to solve fluid flow problem with non-Newtonian fluid. Further, attempt is made to retrieve the value of third grade fluid parameter (A) from a given velocity profile. The accuracy and time consumption in retrieving the value of A, is comparable for different number of neurons in the hidden layer. It can be proposed that, four hidden layer neurons is most suitable from time and accuracy point of view, as it retrieves the value of A within high accuracy of 1.75%.

References

1. Mishra VK, Mishra SC, Basu DN (2017) Simultaneous estimation of four parameters in a combined mode heat transfer in a 2-D rectangular porous matrix with heat generation. Numer Heat Transf-A 71(6):677–692
2. Mishra VK, Mishra SC, Basu DN (2016) Simultaneous estimation of properties in a combined mode conduction–radiation heat transfer in a porous medium. Heat Transf—Asian Res 45(8):699–713

3. Dumek V, Druckmuller M, Raudensk M, Woodbury KA (1993) Novel approaches to the ICHP: neural networks and expert systems, inverse problems in engineering: theory and practice. In: 1993 proceedings of the first international conference on inverse problems in engineering, ASME no I00357 275–282
4. Jambunathan K, Hartle S, Ashforth-Frost S, Fontama VN (1996) Evaluating convective heat transfer coefficients using neural networks. Int J Heat Mass Transf 39:2329–2332
5. Sablani SS (2001) A neural network approach for non-iterative calculation of heat transfer coefficient in fluid-particle systems. Chem Eng Process 40:363–369
6. Amani M, Amani P, Bahiraei M, Wongwises S (2019) Prediction of hydrothermal behavior of a non-Newtonian nanofluid in a square channel by modeling of thermophysical properties using neural network. J Therm Anal Calorim 135:901–910. https://doi.org/10.1007/s10973-018-7303-y
7. HemmatEsfe M, Bahiraei M, Mahian O (2018) Experimental study for developing an accurate model to predict viscosity of CuO–ethylene glycol nanofluid using genetic algorithm based neural network. Powder Technol 338:383–390. https://doi.org/10.1016/j.powtec.2018.07.013
8. Al Mukahal FH, Wilson SK, Duffy BR (2015) A rivulet of a power-law fluid with constant width draining down a slowly varying substrate. J Non-Newton Fluid Mech 224:30–39
9. Jalil M, Asghar S (2013) Flow of power-law fluid over a stretching surface: a Lie group analysis. Int J Non-Linear Mech 48:65–71
10. Tso CP, Francisca JS, Hung YM (2010) Viscous dissipation effects of power-law fluid flow within parallel plates with constant heat fluxes. J Non-Newton Fluid Mech 165:625–630
11. Hatami M, Ganji DD (2013) Heat transfer and flow analysis for SA-TiO2 non-Newtonian nano fluid passing through the porous media between two co-axial cylinders. J Mol Liq 188:155–161
12. Pourmehran O, Rahimi-Gorji M, Gorji-Bandpy M, Ganji DD (2015) Analytical investigation of squeezing unsteady nanofluid flow between parallel plates by LSM and CM. Alex Eng J 54:17–26
13. Fakour M, Vahabzadhe A, Ganji DD, Hatami M (2015) Analytical study of micro polar fluid flow and heat transfer in a channel with permeable walls. J Mol Liq 204:198–204
14. Danish M, Kumar S, Kumar S (2012) Exact analytical solutions for the Poiseuille and Couette-Poiseuille flow of third grade fluid between parallel plates. Commun Non Linear Sci 17:1089–1097
15. Acharya S, Mishra VK, Patel JK (2021) Generation of temperature profile by cascade-forward type artificial neural network in flow of non-Newtonian third grade fluid through two parallel plates with viscous dissipation. AIP Conf Proc 2341:030026. https://doi.org/10.1063/5.0051820

Computational Fluid Dynamics

Formation and Development of Cusp-Induced Air Entrainment Caused by Fully Submerged Rotary Solid Roller: A Numerical Investigation

Santosh Kumar Panda and Basanta Kumar Rana

1 Introduction

The most commonly noticeable phenomenon in the multiphase flow community is entrainment, which deals with the identification of interfacial interactions in terms of discrete entity transport across the boundary between the two fluid frames. It has a wide variety of applications, such as the incursion of microbial species, the evolution of enormous macroscopic entities within a substance and various things that happen in day-to-day life. It mostly emanates from the physical periphery between phases, which avows for the particular transmission of embedded entities based on the strength of the distinct influence. As a result, the entrained bodies exhibit particular characteristics and portray their dynamic behaviour, which is regulated by the medium. As a result, numerous researchers [1–4] have concentrated their efforts on the transformation and growth of the interface because of two immiscible fluids and also the progressive entrainment of one phase into another. Wilson [5] claimed that the suggested analysis can be performed on the horizontal roller coating mechanism. The immiscible gas–liquid interaction is the most widely encountered phenomenon of interfacial interactions. Engineering devices namely adhesive tapes [1, 2], industrial appliance which deals with roll coaters using gravure printing technique for coating of paper, inkjet printers [2–4], jet pump, chemical reactors, nuclear reactors and geothermal energy plants are mostly based on the entrainment phenomenon. The imposition of roller/rollers (in between the gas–liquid interface or in the liquid phase) at the interfacial level is the most common approach for establishing steady entrainment as termed as rotary entrainment [1–13]. For the determination of an equivalent

S. K. Panda · B. K. Rana (✉)
School of Mechanical Engineering, KIIT Deemed to be University, Bhubaneswar 751024, India
e-mail: basanta.uce@gmail.com

S. K. Panda
e-mail: pandasantosh131@gmail.com

© The Author(s), under exclusive license to Springer Nature Singapore Pte Ltd. 2023
S. Revankar et al. (eds.), *Recent Advances in Thermofluids and Manufacturing Engineering*, Lecture Notes in Mechanical Engineering,
https://doi.org/10.1007/978-981-19-4388-1_9

coating viscosity, Tharmalingam and Wilkinson [1] can be informative. According to the researchers, the equivalent coating viscosity is also proportional to roller speed. Cerro and Scriven [6] established a film profile equation for a non-Newtonian fluid with a power-law fit. By balancing viscous, gravitational and inertial forces, they developed a theoretical understanding of liquid creep. Tekic and Jovanovic [7] established a nonlinear method which was based on two-dimensional approach for analysis of dragged liquid flux generated due to partially submerged revolving cylinder, matching the results of Tharmalingam and Wilkinson [1]. Many researchers [3, 4 and 8] observed the interactions of immiscible fluids around the rotating rollers to predict the interfacial dynamics of immiscible fluid pairs. The vortex dipole theory was used by Jeong and Moffatt [8] to investigate cusped interfaces using analytical formations. Based on their analytical calculation, they observed cusp-shaped interface between counter-rotating fully submerged rollers in their experiments. Furthermore, Jeong [9, 10] extended the same arrangement by putting the vortex dipoles in random locations to see how field inclination and spacing affect the result. In order to interpret the film generation process, Yu et al. [2] used the volume of fluid (VOF) technique to estimate the film thickness across a horizontal revolving roller. Gundlach et al. [11] described the free surface flow between two concentric cylinders with a high viscous liquid through numerical analysis. They compared their findings to previous experiments and discovered that as the viscosity of the liquid increases, the rebuilding of the interfaces basically changes. The cusp-shaped air entrainment and the liquid wrapping dynamics in a stratified layer near the solid clockwise rotating cylinder were studied by Rana et al. [12]. They investigated the transient phenomena of wrapping at the receding junctions and entrainment at the advancing end to show how liquid characteristics affect interfacial structures. The influence of an asymmetric rotating field generated by counter-rotating cylinder and diametric asymmetry conditions had been identified by Kumar et al. [13]. They had demonstrated how liquid characteristics affect interfacial arrangements. Panda et al. [14] recently reported the impact of horizontal cross-flow in a gaseous medium and rotational inertia on liquid wrapping film thickness in the receding end and cusp-shaped entrained air in the advancing end for a wide assortment of gas–liquid fluid pairs. The effect of Archimedes number (Ar) on steady-state counter plots, the transient phenomena of liquid tip movement and the breaking of a collapsible gaseous jet, were displayed. A survey of the literature in this field reveals that several experimental, computational and analytical research have been conducted to better understand the complicated gas–liquid interfacial properties produced by the rotational field. Despite this, the formation and development of cusp-induced entrainment caused by a fully submerged solid roller is still need to address and researchers must act quickly to illustrate the gaps in the existing literature. By exhibiting the unique entrainment phenomenon, we have designed a numerical model to study the interfacial configuration for a fully submerged horizontal roller. This issue appears to be quite identical to oceanic mixed layers, in which hydrophysical properties are traded between top and bottom layers. In the ocean, this phenomenon is important in regulating the physical climate of coastal cities. Two fully submerged rollers are widely used in many industrial applications such as jet pumps, ejectors, chemical reactors and other devices that use the entrainment principle or favour increasing the interfacial area to optimise the performance. This problem, on the other hand, is unfavourable in common industrial processes like pouring molten glass or metal, which usually results in a loss of strength due to gas entrapment. The influence of various submerged depths (*h/D*) and rotating inertia (ω rad/s) on steady-state counters plots, as well as steady air entrainment thickness (H), are the main goals of this investigation.

2 Numerical Methodology

2.1 Model Description

We conducted two-dimensional numerical simulations on a revolving horizontal cylinder positioned fully in the liquid medium. To track the reorientation of the interface, we implemented a grid-based Volume of Fluid (VOF) framework. Here, we used the open-source freeware Gerris solver to simulate this study using an adaptive, multilayer and hierarchical model (Popinet [15, 16]). The surface of the solid cylinder is employed with the no-slip boundary condition as shown in Fig. 1a, which represents the schematic perspective of the simulation domain. A computational domain of 0.2 m × 0.2 m was chosen for all physical periphery, with no slip and no penetration criteria. For this investigation, we have chosen the diameter of the cylinder equal to 0.06 m and all simulations are performed by using the clockwise rotation of the roller. Figure 1b depicts adaptive mesh structure distribution across the domain and as a result, the finer cells formed at the interface. These cells generally ensured the execution of extensive calculations near the interface. With stratified air, highly viscous polybutene is selected as the test liquid and their relevant properties are shown in Table 1.

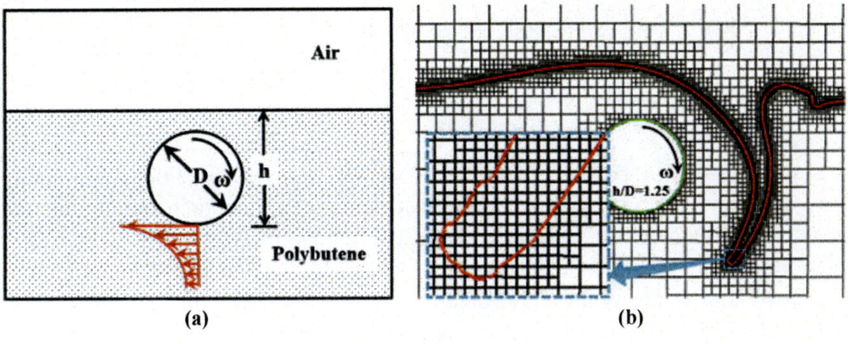

Fig. 1 **a** Distribution of adaptive cells throughout the computational region at h/D = 1.25, **b** schematic illustration of the simulation domain with a single clockwise rotating fully immersed horizontal solid roller at h/D = 1.25

Table 1 Properties of air-polybutene pair for the present simulation

Fluid pair	Name of the fluid	ρ (kg/m^3)	μ (Pa.s)	σ (N/m)
Pair 1	Polybutene	883	12.9	0.0340
	Air	1.225	1.789×10^{-5}	

We have chosen a non-dimensional number namely the Capillary number ($Ca = \frac{\mu\omega D}{2\sigma}$) for our examination. Where μ, ω and σ are the dynamic viscosity of liquid (Pa. s), rotational inertia (rad/s) and σ is the surface tension of gas–liquid pair. The relative effect of viscous drag forces and surface tension forces acting across gas–liquid interfaces are described by the capillary number.

2.2 *Formulation of Problem*

The entire present investigation are carried out by Gerris solver (Popinet [15, 16]). It generally utilises VOF based approach which includes incompressible, density Navier–stokes solver for identification of entrainment dynamics in a numerical way. Here, momentum, continuity equations along with body and surface forces are exhibited as follows:

$$\rho\left(\frac{\partial u_\alpha}{\partial t} + u_\beta.\frac{\partial u_\alpha}{\partial x_\beta}\right) = -\frac{\partial p}{\partial x_\beta} + \frac{\partial}{\partial x_\beta}\left[\mu\left(\frac{\partial u_\alpha}{\partial x_\beta} + \frac{\partial u_\beta}{\partial x_\alpha}\right)\right] + \sigma k \delta_s n \tag{1}$$

$$\frac{\partial p}{\partial t} + \nabla.(\rho u) = 0 \tag{2}$$

Here, Surface tension term generally deals with the depth of penetration inside the liquid pool.

Equation (2) can be changed to the following form outside of the interfacial regime, assuming fluid density remains constant.

$$\frac{\partial u_\alpha}{\partial x_\alpha} = 0 \tag{3}$$

Presently, $\rho = \rho(x, t)$, $u_\alpha = (u, v)$ and $\mu = \mu(x, t)$ are the fluid density, velocity and dynamic viscosity, respectively. Interface deals with Dirac distribution function δ_s (a surface tension term), σ, k and n are the coefficient of surface tension, curvature and normal to the interface, respectively.

Density and viscosity are delineated by volume fraction term, i.e. c(x, t) for a particular position with considering a generalised fluid as:

$$\rho(\widehat{c}) = \widehat{c}\rho_{liq} + (1 - \widehat{c})\rho_{gas} \tag{4}$$

$$\mu(\widehat{c}) = \widehat{c}\mu_{liq} + (1 - \widehat{c})\mu_{gas} \tag{5}$$

The liquid density, viscosities and gaseous density, viscosities are denoted by ρ_{liq}, μ_{liq}, ρ_{gas} and μ_{gas}, respectively. In a cell, a filtered value can be calculated by averaging the volume fraction c's four corner values (x, t). The corner values are computed employing bilinear interpolation from cell-centred statistics.

We can derive the appropriate constitutive equation on volume fraction using Eq. (2):

$$\frac{\partial c}{\partial t} + \nabla.(cu) = 0 \tag{6}$$

A staggered spatial grid and a second-order accurate method are being used to discretise the above Eqs. (1–2 and 6) over time. Gerris provides the linear interface for each cell applying polynomial line methodology [16].

3 Validation

We displayed the variation of entrained cusp-shaped air film thickness in the advancing end of a solid roller of diameter 0.06 m rotating at an angular speed of 20 rad/s between the air-lubricating oil interface ($\rho_{\text{air}} = 1.225$ kg/m^3, $\mu_{\text{air}} = 1.789 \times 10^{-5}$ Pa.s, $\rho_{\text{lubricating}} = 876.4$ kg/m^3, $\mu_{\text{lubricating}} = 0.136$ Pa.s, $\sigma = 0.035$ N/m), With a submergence ratio of 50%, (roller is equally immersed in air and lubricating oil). Here, we have observed the current model demonstrates an ideal agreement with the previous result as shown in Fig. 3 and this provides us with the necessary conviction for future investigation (Fig. 2).

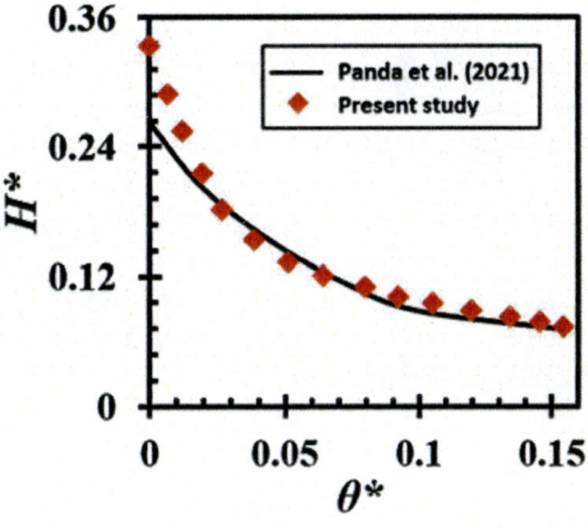

Fig. 2 Comparison of non-dimensional entrained air thickness (H^*) with outcomes of Panda et al. [14]

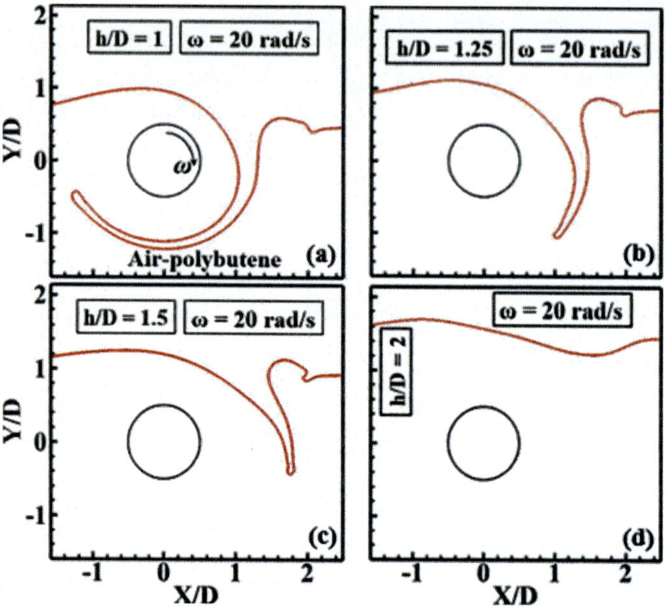

Fig. 3 Deviation of steady numerical contours of air-polybutene pair with $\omega = 20$ rad/s for various submerged depth ($h/D = 1$–2)

4 Results and Discussion

The development of cusp-shaped air entrainment due to clockwise rotating fully submerged solid roller on highly viscous polybutene for various submerged depths (h/D) and angular speed of solid roller (ω rad/s) is explained in this section. We varied the submerged depth (h/D) from 1 to 2 and the angular speed (ω) from 5 to 30 rad/s for this investigation. This research is looking into the effect of different submerged depths on non-dimensional air entrainment thickness ($H^* = H/D$) and the deviation of entrained air filament length (θ^*) at different rotational speeds (Ca).

Figure 3a–d demonstrates the variations steady contour results for air-polybutene at various submerged depth ($h/D = 1$–2) for a fixed roller rotation, i.e. $\omega = 20$ rad/s. Figure 3 shows that for a given roller rotation of 20 rad/s, the length of air-entrained filament is greater at $h/D = 1$ (Fig. 3a) and much less at $h/D = 2$ (Fig. 3d). The reason is quite obvious that the rotational effect of the roller is more prominent when the interface is nearer to it and this effect reduces with a gradual increase of submerged depth. As a result, viscosity-driven air pumping below the cusp against surface tension and gravity is found to be more for higher submerged depth and decreases with increases of submerged depth. So the length of air entrainment is found to be more at $h/D = 1$ (roller just touches the interface), it becomes decreases with the increase of submerged depth and is found to be very less at $h/D = 2$. One more observation is that the cusp-shaped air filament shift gradually away from roller with increases of *h/D* reveals lesser rotational effect due to increase of submerged depth.

The effect of roller rotations on the steady numerical counter for $h/D = 1.25$ is shown in Fig. 4a–d. It can be seen that for higher rotation, i.e. $\omega = 30$ rad/s, the length of cusp-shaped air entrainment is longer and very less for low rotation, i.e. 5 rad/s. It is due to the higher rotational field is generated because of the higher angular speed of the roller and it becomes lesser with the gradual decrease of speed of the roller. As a result, more liquid flux is picked at higher roller rotation and it is found to be lesser in lesser rotations. From this analysis, we may conclude that cusp-shaped air entrainment is a consequence of viscosity-driven capillary pull for pumping of air counter to surface tension and gravitational force.

The dimensionless entrained air filament thickness ($H* = H/D$) and entrained air filament length ($\theta * = \theta/360°$) is an important parameters of this present investigation. Figure 5a highlights the effect of submerged depth (*h/D*) on non-dimensional entrained thickness ($H^* = H/D$) at $\omega = 20$ rad/s for air-polybutene. The width of the air filament is observed to be more at $h/D = 1$ and width gradually dismissed with increases of *h/D* and at the same time filament length is found to be significantly higher in cases of $h/D = 1$ and vary less at $h/D = 2$ (also observed from Fig. 3) due to the higher amount of resisting viscous force is developed which leads to widening of the air filament. However, the percentage reduction in $H*_{min}$ from $h/D = 1$ to 2 is approximately 36%. With the same time, at $h/D = 1$, the percentage increase

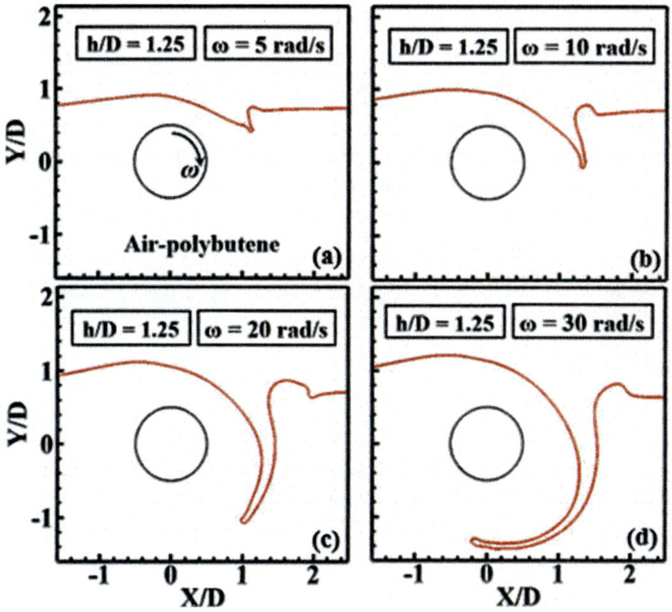

Fig. 4 Deviation of steady numerical contours of air-polybutene pair with $h/D = 1.25$ for various roller rotations ($\omega = 5$–30 rad/s)

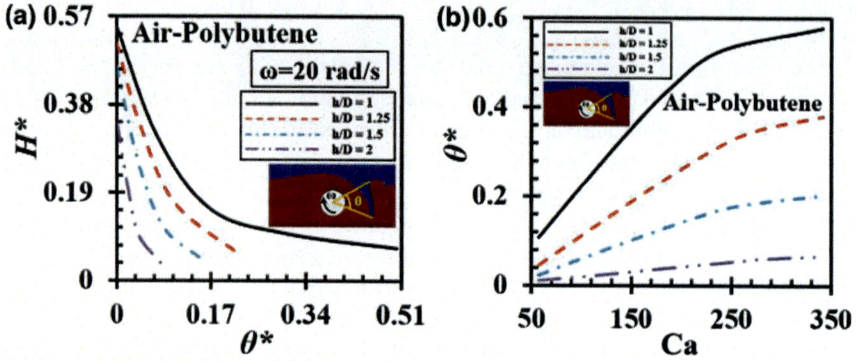

Fig. 5 **a** Effect of *h/D* on non-dimensional air entrainment thickness (H^*), **b** Variation of entrained air filament length at different rotational speeds (*Ca*) for air-polybutene

in air filament length is observed to be greater and lesser with increases of *h/D* because more momentum gets transferred to the highly viscous liquid from the roller surface leading to deeper penetration of filament inside the heavier fluid. Figure 5b depicts the variation of entrained air filament length at different rotational speeds (*Ca*) for air-polybutene. One can be observed that the length of air filament gradually increases with the increase of *Ca*, i.e. roller rotation. As a result, the strength of the jet (centrifugal force) increases due to the increase of *Ca* causing greater air filament length. Another finding is that the length of the air filament is greater (as Ca increases) at $h/D = 1$ and much shorter at $h/D = 2$ (also observed from Fig. 4).

Figure 6 depicts velocity vectors around a fully submerged rotational roller with an angular speed of 20 rad/s and $h/D = 1.25$. This is considered a crucial deliberation in the current investigation. The general consideration that regulates the flow distribution near the horizontal roller is entitled velocity vectors analysis. It also recognises the formation of a cup-shaped structure in the advancing end caused by roller rotation.

5 Conclusions

The effect of submerged depth (*h/D*) and roller rotations on constant numerical contours of air-polybutene has been described physically. The influence of various submerged depths on steady air entrainment thickness (H^*), length of air entrainment (θ^*) and velocity vectors near solid roller have been comprehensively examined in a parallel effort (which was also explained by Panda et al. [14] for partially submerged roller). This study also reveals the influence of viscosity, surface tension, gravitational force and roller angular speed. Finally, near the rotating roller, a velocity vector is provided to visualise the flow patterning at the advancing junctions.

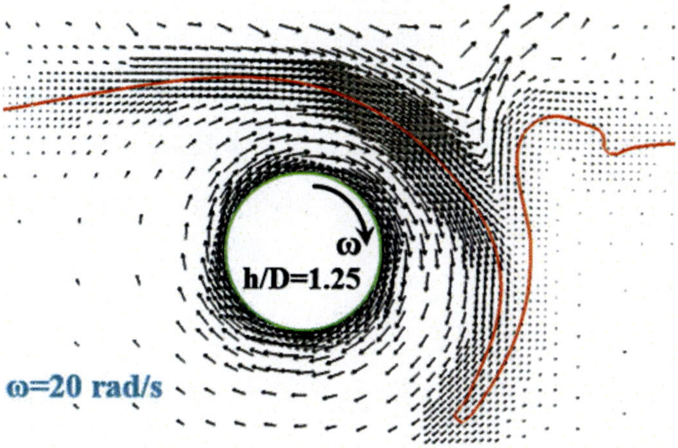

Fig. 6 Velocity vectors near a rotational roller for $h/D = 1.25$ and $\omega = 20$ rad/s

Acknowledgements The authors would like to express their gratitude to the Science and Engineering Research Board (SERB), Department of Science and Technology (DST), Government of India, for funding the research project through a research grant (SRG/2019/001438).

References

1. Tharmalingam S, Wilkinson WL (1978) The coating of Newtonian liquids onto a rotating roll. Chem Eng Sci 33(11):1481–1487
2. Yu SH, Lee KS, Yook SJ (2009) Film flow around a fast rotating roller. Int J Heat Fluid Flow 30(4):796–803
3. Joseph DD, Nguyen K, Beavers GS (1984) Non-uniqueness and stability of the configuration of flow of immiscible fluids with different viscosities. J Fluid Mech 141:319–345
4. Joseph DD, Nelson J, Renardy M, Renardy Y (1991) Two-dimensional cusped interfaces. J Fluid Mech 223:383–409
5. Wilson SD (1982) The drag-out problem in film coating theory. J Eng Math 16(3):209–221
6. Cerro RL, Scriven LE (1980) Rapid free surface film flows. An integral approach. Ind Eng Chem Fundam 19(1):40–50
7. Tekić MN, Jovanović S (1982) Liquid coating onto a rotating roll. Chem Eng Sci 37(12):1815–1817
8. Jeong JT, Moffatt HK (1992) Free-surface cusps associated with flow at low Reynolds number. J Fluid Mech 241:1–22
9. Jeong JT (1999) Formation of cusp on the free surface at low Reynolds number flow. Phys Fluids 11(3):521–526
10. Jeong JT (2010) Two-dimensional Stokes flow due to a pair of vortices below the free surface. Phys Fluids 22(8):082102
11. Gundlach J, Lübke M, Wünsch O (2014) Numerical investigation of the free-surface flow within an annulus. PAMM 14(1):681–682
12. Rana BK, Das AK, Das PK (2016) Numerical study of air entrainment and liquid film wrapping around a rotating cylinder. Ind Eng Chem Res 55(46):11950–11960

13. Kumar P, Das AK, Mitra SK (2017) Bending and growth of entrained air filament under converging and asymmetric rotational fields. Phys Fluids 29(2):022101
14. Panda SK, Rana BK, Kumar P (2021) Competition of roller rotation and horizontal crossflow to control the free surface cusp-induced air entrainment. Phys Fluids 33(11):112114
15. Popinet S (2003) Gerris: a tree-based adaptive solver for the incompressible Euler equations in complex geometries. J Comput Phys 190(2):572–600
16. Popinet S (2009) An accurate adaptive solver for surface-tension-driven interfacial flows. J Comput Phys 228(16):5838–5866

A Study on Stability Condition for an Inconsistent Formulation in High-Density Ratio Flow

Swagatika Acharya, Jitendra Kumar Patel, Manoj Ukamanal, Prakash Ghose, Swarup Kumar Nayak, Achinta Sarkar, Basanta Kumar Rana, and Vijay Kumar Mishra

1 Introduction

The researchers place a high priority on stability and precision while designing numerical algorithms. Previously few important algorithms, namely volume of fluid (VoF) [1] and level set [2], etc. have been introduced in the realm of multifluid flows/multiphase. The numerical problems linked to high-density ratio flows have long plagued the community of multiphase. The non-conservative technique to treating the governing equations, as detailed in [3], could be the root of instability for such flows. The failure to adequately describe the physics (science) could also be due to the uneven transfer of mass and momentum fluxes. Rudman [4] sought to tackle issue by deriving density flux from mass flux, which guarantees mass and moment transport stability. Desjardins and Moureau [5] apply the same consistent technique to the analysis of high-density ratio flows. This idea was later incorporated in the coordinated framework by reported by Bussmann et al. [6]. Raessi and Pitsch in their paper [7] provided a consistent mass and momentum formulation in the VoF framework by adopting an identical method in the discrete advection and momentum equations.

Interestingly, some experts in the literature have effectively implement the concept of consistency in the level set structure [8, 9]. The algorithms in the problem formulation and accompanying simulation of high-density ratio flows must be consistent, as evidenced by the literature. To understand the influence of density ratio on solution stability, Patel and Natarajan [10] conducted a systematic investigation of consistent and inconsistent formulation in range of test circumstances. In spite of high-density ratio, they discovered that the erratic formulation can produce steady outcomes in a

S. Acharya · J. K. Patel (✉) · M. Ukamanal · P. Ghose · S. K. Nayak · A. Sarkar · B. K. Rana · V. K. Mishra
School of Mechanical Engineering, KIIT Deemed to be University, Bhubaneswar, Odisha 751024, India
e-mail: jitendra.patelfme@kiit.ac.in

© The Author(s), under exclusive license to Springer Nature Singapore Pte Ltd. 2023

S. Revankar et al. (eds.), *Recent Advances in Thermofluids and Manufacturing Engineering*, Lecture Notes in Mechanical Engineering,
https://doi.org/10.1007/978-981-19-4388-1_10

select circumstances. Hence, it is recommended that in addition to the density ratio, the effects of additional simulation factors such as Reynolds number (Re), viscosity ratio and Froude number be rigorously investigated. In this vein, authors try to establish a standard by combining all factors of the simulation as well as high-density ratio for which a satisfactory outcome can be produced through an inconsistent formulation.

2 Numerical Methodology

In the current investigation authors used two dimension to resolve incompressible, laminar Navier–Stokes equation, which state,

$$\nabla.\vartheta = 0 \tag{1}$$

$$\frac{\Delta(\rho\vartheta)}{\Delta T} + \nabla(\rho\vartheta) = -\nabla p + \frac{1}{Re}\nabla \times \mu\nabla\vartheta + \frac{\rho}{Fn} + F_s \tag{2}$$

Pressure $= p$; Density $= \rho$ and Viscosity $= \mu$. In accordance with the physics of lighter fluid, Froude and Reynolds numbers are determined as: $F_n = U^2/gL$ and $Re = \rho_l UL/\mu_l$. To implement the surface tension force (F_s) as: $F_s = \frac{k\nabla\emptyset}{W_n}$ the continuum surface force technique is employed. Where 'k' signifies the predicted interface curvature using the Ubbink model [11] along with modified Green Gauss gradient calculation method [12]. The Weber number (W_n), which is non-dimensional, is defined as follows: $W_n = \rho_l LU/\sigma$. It should be noted that in balanced framework all forces, i.e. gravity, pressure and surface tension are employed. To compute the mixture density as well as viscosity in this study, formulation of single fluid is adopted as:

$$\rho = \varphi + (1 - \varphi)^{\rho_2}/_{\rho_1} \tag{3}$$

$$\mu = \varphi + (1 - \varphi)^{\mu_2}/_{\mu_1} \tag{4}$$

To find out the volume fraction (φ), authors employ the V_f method, which involves solving an advection equation using velocity of local fluid as illustrated below:

$$\frac{D\varphi}{DT} = \frac{\Delta\varphi}{\Delta T} + \vartheta.\nabla\varphi = 0 \tag{5}$$

Along with the CUIBS scheme [13] of high resolution, an implicit technique is used to solve the advection problem in order to obtain the volume fraction of the faces. A general momentum equation was solved rather than X and Y momentum equation [14]. For the current issue, the normal momentum equation was solved through inexact Newton Krylov solver implicitly similar to advection equation (PetSc libraries [15]). In both space and time, the current approach has been determined to be second-order accurate. Authors implement the same CUIBS scheme [13] in the convective section of Eqs. (2) and (5) at discrete levels to establish consistency, as described in [10]. Inconsistency may be accounted for in the analysis by employing alternate methods in Eqs. (2) and (5), namely the upwind and CUIBS schemes.

3 Results and Discussion

To examine the consistency and inconsistency of formulations, numerical analysis is performed. We also look into the impact of test settings on consistency in inconsistent formulation. The test under this part is performed repeatedly for an update set of settings to investigate the impact further the results are reported at the end.

3.1 Water Rising in a Tank

A rectangular tank of (2 × 10) cross-sectional area is taken as a computation realm, further which is sub-structured into 100 × 500 discrete elements. In order to perform the computational analysis, various conditions applied to the tank such as Neumann boundary condition, uniform velocity field and slip condition were applied at the tank's top, bottom and side walls, respectively. Firstly, a heavy fluid supplied to the tank from its bottom portion till the level rises to 0.5, after which same fluid begins to enter into the tank with a steady velocity of $U = 1$ therefore the level of fluid rises. A higher *Re* of 1000, density of 1000 and viscosity ratio of 100 taken for the computation, to ensure the consistency thoroughly. The simulation is run in a time step of $t = 10^{-4}$ for a Froude number (Fr) = 1.

Figure 1 depicts the interface position with respect to velocity vectors for inconsistent formulation whereas Fig. 2 illustrates the consistent formulation for the same. In contrast to the preceding study, deformation of the interface is not observable in inconsistent formulation; yet, high-intensity spurious currents are formed throughout the interface. As demonstrated in Fig. 1, the intensity increases with time and by time $t = 2$, currents are 10–15 times greater than mean flow, which is undesirable. In Fig. 2, on other hand, velocity vectors are uniform across the field and stay constrained throughout the simulation. Because uniform velocity fields in the consistent formulation indicate that no additional spurious currents are formed owing to gravity, the balanced force method may be executed properly.

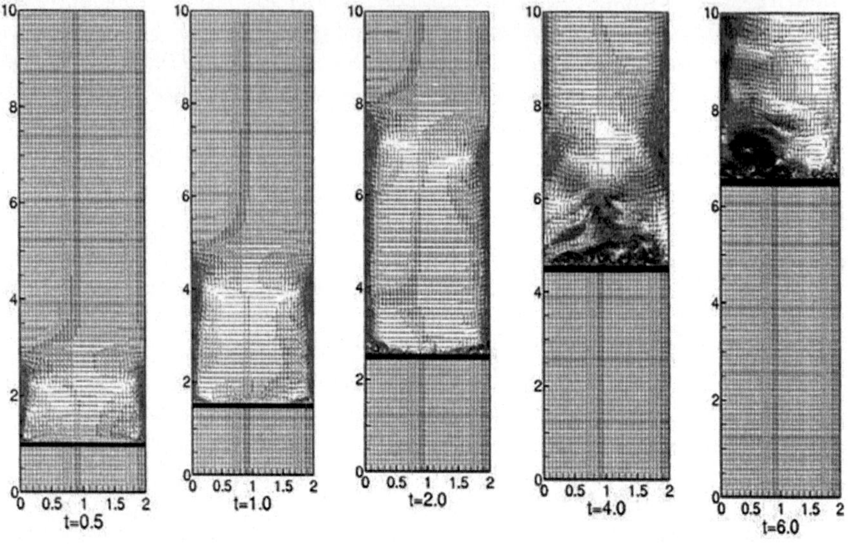

Fig. 1 Time plots of water rising in tank for density ratio 1000, viscosity ratio 100 with $R_n = 1000$ and $F_n = 1$ using the inconsistent formulation

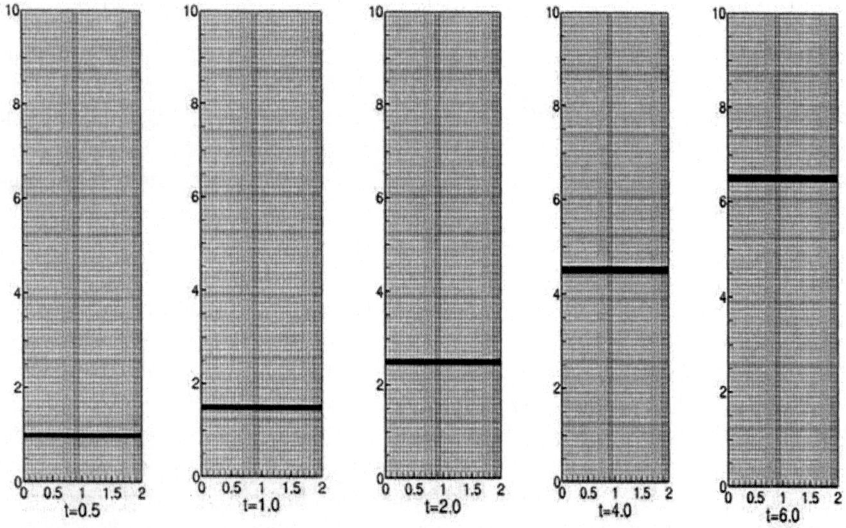

Fig. 2 Time plots of water rising in tank for density ratio 1000, viscosity ratio 100 with $R_n = 1000$ and $F_n = 1$ using the consistent formulation

3.2 Translating Droplet

This test scenario gives a good environment for analysing the creation of spurious currents which may cause interface distortion, allowing the algorithm's usefulness to be extensively analysed. Uniform grid resolution of 0:02 based rectangular channel used in the present investigation, in which a droplet with a diameter D = 1 is used. Uniform velocity and Neumann boundary are specified at the entrance and outflow, respectively. Whereas at the top and bottom wall slip condition is applied. The fluid within the region, along with droplet, is set to the identical inlet velocity of U = 1 at the start. As among the droplet and surrounding fluid relative velocity is zero, which is conceptually the identical scenario as the static droplet.

The highest velocity inaccuracy compared to the background velocity U = 1 is shown in Fig. 3. Three sets of density and viscosity ratio are simulated with Reynold number = 10 and Weber number = 1, with gravitational influence ignored. In all circumstances, regardless of density and viscosity ratios, the consistent formulation produces very minimal velocity inaccuracy, as predicted. Inconsistent formulation, on the other hand, results in velocity inaccuracies of the same magnitude as the background velocity at high-density ratios, distorting the interface as shown in Fig. 4. For low-density ratio, i.e. 10, the velocity inaccuracy is on the order of 10^{-2}, which is comparable to the consistency formulation. The test case is replicated with a higher density ratio of 1000 and a higher viscosity ratio of 100. Despite the huge density ratio, the velocity inaccuracy in simulation is lesser than 10^{-2}, contradicting the original finding. This implies that the viscosity ratio seems to have a favourable influence on the consistency of the inconsistent formulation.

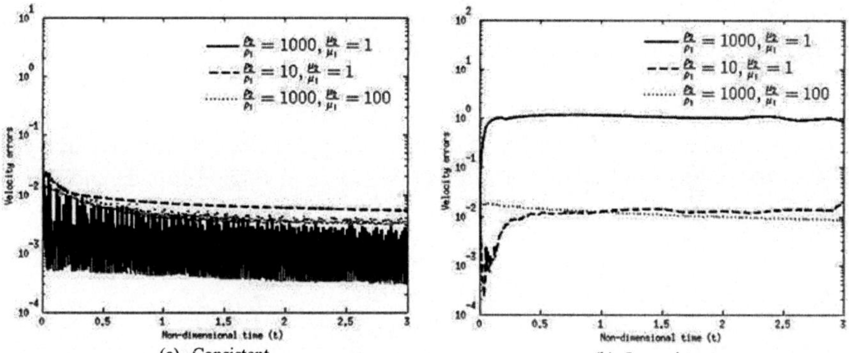

(a) Consistent (b) Inconsistent

Fig. 3 Maximum velocity error relative to the background velocity for **a** consistent and **b** inconsistent formulation for $R_n = 10$, $W_e = 1$

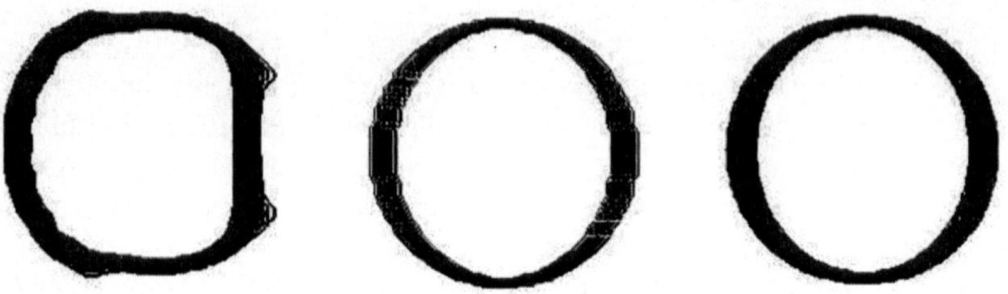

(a) $\rho_2/\rho_1 = 1000$, $\mu_2/\mu_1 = 1$ (b) $\rho_2/\rho_1 = 10$, $\mu_2/\mu_1 = 1$ (c) $\rho_2/\rho_1 = 1000$, $\mu_2/\mu_1 = 100$

Fig. 4 Shape of the droplets after non-dimensional time (t) = 2.5 employing inconsistent formulation for $R_n = 10$ and $W_e = 1$

3.3 Stability Analysis for an Inconsistent Formulation

The aforementioned test examples demonstrate that a consistent formulation may yield reliable results regardless of flow circumstances. The success or failure of an inconsistent formulation, on the other hand, is not primarily determined by the density ratio. Other flow characteristics such as viscosity ratio, Reynolds number and convection velocity all play a crucial part in maintaining stability.

Table 1 summarises the test cases performed in this study. The inconsistency of the system's ability to deliver tangible outcomes is noted over the last column, in which satisfactory and undesirable are indicated, respectively. This shows that not only does a large density ratio impair stability, but so does Reynolds no. The numerical studies in the translating droplet test show that raising the viscosity ratio has a positive effect on precisely capturing the form of the droplet in translation. A thorough examination of these factors would be beneficial in developing a criterion for determining if the inconsistent formulation is capable of producing a stable result. Future research will look at the influence of additional simulation factors including Froude and Weber numbers.

Table 1 Test parameters using inconsistent formulation

Test case	S.I	ρ_2/ρ_1	μ_2/μ_1	Re	F_n	W_e	Remark
Water rising	1	1000	100	1000	1	–	Unstable
	2	1000	100	10	1	–	Stable
Translating droplet	1	1000	1	10	–	1	Unstable
	2	10	1	10	–	1	Stable
	3	1000	100	10	–	1	Stable

4 Conclusions

In high-density ratio flows, consistency is crucial for maintaining the algorithms' resilience. We explored an inconsistent formulation in interfacial flows in this paper, which included simulation characteristics such as density ratio, viscosity ratio, Reynolds number and Froude number, among others. Simulation factors such as the Reynolds number, Froude number and viscosity ratio are just as significant as the density ratio.

References

1. Hirt CW, Nichols BD (1981) Volume of fluid (VOF) method for the dynamics of free boundaries. J Comput Phys 39:201–225
2. Sussman M, Smereka P, Osher S (1994) A level set approach for computing solutions to incompressible two-phase ow. J Comput Phys 114:146–159
3. Bell JB, Marcus DL (1992) A second-order projection method for variable-density flows. J Comput Phys 101:334–348
4. Rudman M (1997) Volume tracking method for interfacial ow calculation. Int J Numer Meth Fluids 24:671–691
5. Desjardins O, Moureau V (2010) Methods for multiphase flows with high density ratio. Center Turbulent Res Summer Programm, pp 313–322
6. Bussmann M, Kothe DB, Sicilian JM (2022) Modeling high density ratio incompressible interfacial flows. In: ASME conference proceedings, pp 707–713
7. Raessi M, Pitsch H (2012) Consistent mass and momentum transport for simulating incompressible interfacial flows with large density ratios using the level set method Comput. Fluids 63:70–81
8. Ghods S, Herrmann M (2013) A consistent rescaled momentum transport method for simulating large density ratio incompressible multiphase flows using level set methods. Phys Scr
9. Nangia N, Griffith BE, Patankar NA, Bhalla APS (2019) A robust incompressible Navier-Stokes solver for high density ratio multiphase flows. J Comput Phys 390:548–594
10. Patel JK, Natarajan G (2017) A novel consistent and well-balanced algorithm for simulations of multiphase flows on unstructured grids. J Comput Phys 350:207–236
11. Ubbink O (1997) Numerical prediction of two fluid systems with sharp interfaces. Ph. D. thesis, Imperial College London, London, UK
12. Patel JK, Natarajan G (2018) A cost-effective curvature calculation approach for interfacial flows on unstructured meshes. Int J Numer Methods Fluids 88(7):347–362
13. Patel JK, Natarajan G (2015) A generic framework for design of interface capturing schemes for multi-fluid flows. Comput Fluids 106:108–118
14. Natarajan G, Sotiropoulos F (2011) IDeC(k): a new velocity reconstruction algorithm on arbitrarily polygonal staggered meshes. J Comput Phys 230(17):6583–6604
15. Balay S, Abhyankar S, Adams M, Brown J, Brune P, Buschelman K, Eijkhout V, Gropp W, Kaushik D, Knepley M (2014) PETScusers manual revision 3.5, Argonne National Laboratory (ANL). http://www.mcs.anl.gov/petsc

Computational Modelling for Cyclone Separators Performance Evaluation

Prakash Ghose, Tarak Kumar Sahoo, Swarup Kumar Nayak, Basanta Kumar Rana, Jitendra Kumar Patel, Achinta Sarkar, and Manoj Ukamanal

1 Introduction

Due to ease of fabrication, cyclone separators were indeed commonly used in process industry for segregate dust and debris from the gas. This is also less expensive than other separating processes. There in cement production, multi-stage cyclone separators have substituted the lengthy kilns significantly to reduce kiln length. This employs a counter flow heating method for separate CO_2 from calcium carbonate dust particles [1].

The gas and particulate combination are delivered tangentially further into cyclone separator. As a result, a powerful vortex develops, and indeed the pressure along the cyclone walls increases, resulting in the formation of a negative pressure zone inside and outside the centre core. Fewer relatively low-mass particles are suspended mostly by the vortex core and emerge well with gas, which itself is undesirable. As a result, it has an impact on collecting efficiency. CFD is a tool that can be used to make accurate predictions about the efficiency of cyclone separators. Wang et al. [2] studied the effect of input duct angle upon that effectiveness of a typical cyclone separator. They discovered that when dust particle diameter grows, collection efficiency enhances, and at 15 ° positive input duct angle, collection efficiency is maximised (around 80%). Azadi et al. [3] analytically examined the influence of cyclone size and velocity input to cyclone separator upon system efficiency. It is been observed that as cyclone size rises, so does the cut-off diameter and perhaps the pressure decrease. Misiulia et al. [4] investigated the influence of output pipe design on vortex disintegration. Investigators noted that a substantial vortex disintegration was detected with a standard output pipe layout, but not with a well-designed outlet pipe. Therefore, this has no effect on the cyclone separator's efficiency. Venkatesh et al. [5] examined the staging effects of

P. Ghose (✉) · T. K. Sahoo · S. K. Nayak · B. K. Rana · J. K. Patel · A. Sarkar · M. Ukamanal
School of Mechanical Engineering, KIIT Deemed to Be University, Bhubaneswar 751024, India
e-mail: pghosefme@kiit.ac.in

© The Author(s), under exclusive license to Springer Nature Singapore Pte Ltd. 2023
S. Revankar et al. (eds.), *Recent Advances in Thermofluids and Manufacturing Engineering*, Lecture Notes in Mechanical Engineering,
https://doi.org/10.1007/978-981-19-4388-1_11

cyclone separators on separation efficiency. For particle diameter of 27 μm as well as larger, a single stage configuration achieves 100% efficiency. Particulate less than 27 μm were not segregated successfully, however, for particle diameter 7 μm and larger, four staging arrangements result in 100% efficiency. Venkatesh et al. [6] studied the effectiveness of squared cross-sectional cyclone separators theoretically and practically. The Taguchi technique was used to optimise the size factors of a square type cyclone separator. Bogodage et al. [7] evaluated the effect of collecting hoper geometric arrangement on total cyclone separators efficiency. It is been stated that a short hoper significantly improves in contrast to a long hoper in terms of grade performance. Rafiee and Sadeghiazad [8] examined the efficiency of vortex tubes and observed that the heavy particulates of an aircraft vehicle's exhaust stream are effectively separated using the cyclone method.

The collection efficiency of a cyclone separator including particulates of varied sizes has been discussed in this study for a variety of gas mass flow rates. For the model, an established cyclone separator from NUVAKO VISTAS CORP. Ltd. was being implemented. Throughout this computational work, calcium carbonate particles are utilised as the particulate phase and hot exhaust gas have been used as a continuous phase. This investigation also examines the pressure variation across the cyclone and the flow velocity as well.

2 Physical Geometry and Computational Modelling

In this investigation, a NUVAKO VISTAS CORP. Ltd. twin cyclone separator has been employed. The simulation has been carried out up to the Y-joint of downdraft tube. Figure 1a depicts the intake, gas output, and particulate outflow. As seen in Fig. 1b, the whole computational volume is discretised with 1.8 million hexahedral unstructured cells. For continuous phase, the conservation of continuity and momentum equations in Reynolds averaged form has been resolved. To calculate the eddy viscosity, a realisable k-ε turbulent model has been used. In the Lagrangian frame of reference, the particle phase is solved. The particle–particle interaction, on the other hand, is not taken into account for this work.

Table 1 illustrates the thermophysical characteristics of the gas phase. Table 2 displays the cases that have been executed. The cyclone separator performance is assessed using three different gas mass flow rates, while the particle mass flow rate remains constant across all situations. As in second case, the input mass flow boundary condition seems to be the actual operational plant's mass flow rate condition. The particle sizes examined for modelling vary from 1 to 35 μm, and the dispersion of particulate matter in the blend is assumed to be equal across all situations.

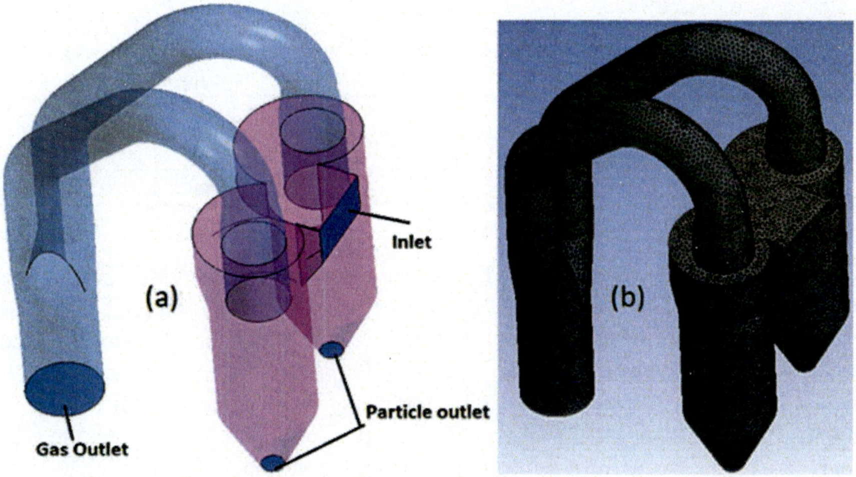

Fig. 1 **a** Physical geometry, **b** meshing

Table 1 Thermophysical characteristics of the gas and particles

Density	1400 kg/m^3
Gas density @ 300 °C	0.58 kg/m^3
Viscosity @ 300 °C	2.81×10^{-5} kg/m s

Table 2 Operating conditions

Cases	Air flow rate (kg/sec)	Particle mass flow rate (kg/sec)
Case-1	90	54.3
Case-2	57	54.3
Case-3	20	54.3

3 Results and Discussion

The computational work is validated against the real-time experimental pressure data along the axis of the cyclone at different height. The readings are taken at 2, 4, and 6 m height from the bottom along the axis of the cyclone. From table, it can be seen that there is insignificant difference observed between simulated and experimental readings. Hence, various computational models used in this simulation are well validated. The similar trends are also observed in the work of Allahmer and Dabbas [11] (Table 3).

Figure 2 shows intense near the cyclone separator's wall. At and across the axial location of cyclone separator, a negative pressure vortex core has been detected. The centrifugal force is significant in this phenomenon. Since centrifugal force consistently keeps driving the gas towards the cyclone's wall, pressure builds inside this

Table 3 Comparison between experimental and CFD pressure (Pa) all along cyclone's axis at various heights

Axial distance from bottom (y) in m	Pressure (Pa) experiment	Pressure (Pa) CFD
2	0.8	0.6
4	− 6.0	− 8.0
6	− 28.0	− 34.0

region. This has been noticed that when the gas mass flow rate increases, core vortex turns out to be stronger. As a result, there seems to be a significant decrement in pressure there in centre of the cyclone, whereas pressure builds up more towards the cyclone separator's wall. It can be seen from the axial velocity contour in Fig. 3 is that axial velocity near the wall is significantly greater than central core. It is indeed clear that it adheres to the forced vortex principle. The velocity near the wall increases when the gas flows tangentially. Regarding the mass flow rate impact, it is noticed that as the mass flow rate of gas increases, so does the axial velocity around the wall and close to the deep tube increases. The velocity, meanwhile, seems to be nearly negligible in the centre section of the central bottom part and the dip tube of the cyclone. As a result, it facilitates the movement of the accumulated particles downward. Figure 4 depicts the vector for velocity at different cyclone cross-sections.

The difference between the intake mass flow rate of particles and the mass flow rate of the collected particle via the bottom aperture of the cyclone has been used to determine collection efficiency. Figure 5 shows that the particle residence period is longer for larger-sized particles. It demonstrates the impact of centrifugal force on separation. Whenever particle size increases, this becomes heavier, and so the centrifugal force increases. Consequently, it travels a long period of time in the cyclone. Lightweight particles, however, move for a shorter period because of reduced centrifugal force. Contrastingly, the gravitational force has a significant impact on particles residence time. The heavier particles travel relatively at lower velocity, the residence time of the particle will gradually decrease.

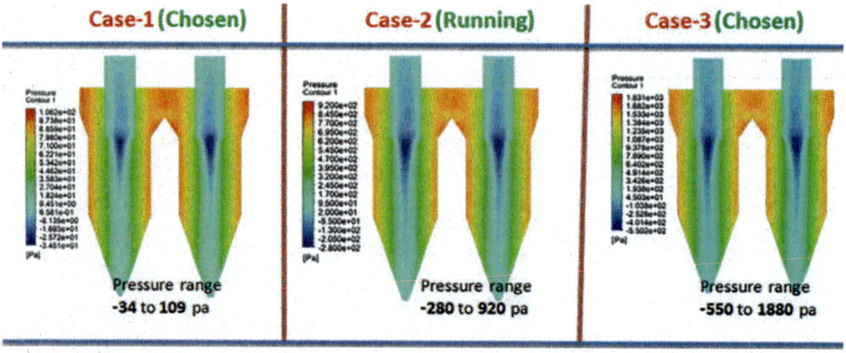

Fig. 2 Pressure variations in the centre plane for various gas mass flow rates

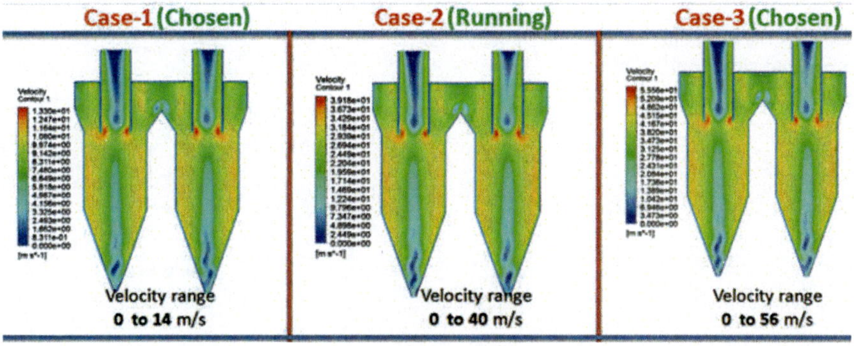

Fig. 3 Axial velocity variations in the centre plane for various gas mass flow rates rate

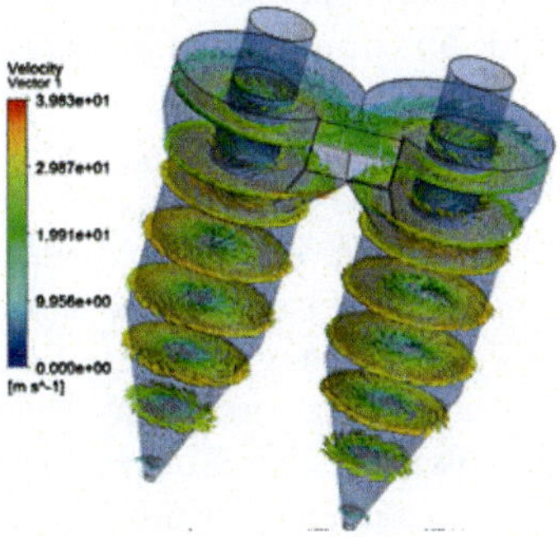

Fig. 4 Velocity vector on various planes (running case)

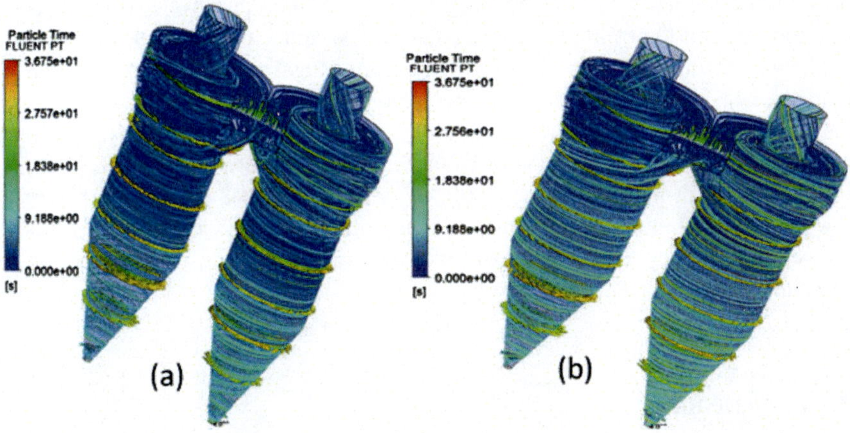

Fig. 5 Particles residence time for **a** one micron, **b** ten micron

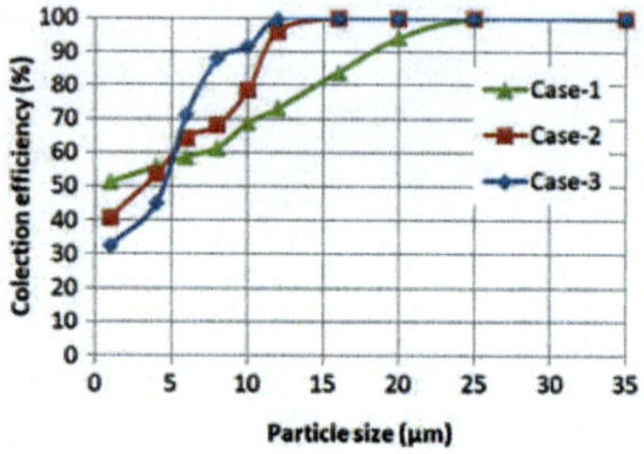

Fig. 6 Collection efficiency of the particles for different mass flow rates

The collection efficiency of the particles is depicted in Fig. 6. Because of the reduced centrifugal force, heavy particle cannot be segregated adequately when the gas mass flow rate (gas velocity) declines. Owing to the low-centrifugal force, they were unable to retain their position near the cyclone walls, instead being stuck there in central vortex core. In case-1 (lowest gas flow velocity), particle larger than 25 micron is completely captured just at bottom of cyclone separator. It can be deduced that particle size >25 micron will have 100% collection efficiency. However, lighter particles ≤5 micron, on the other hand, are collected efficiently over larger particles in greater mass flow condition cases. This is owing to the weak vortex core's inability to retain lighter particles. Whenever the mass flow rate increases, an extremely significant negative pressure so at centre of the vortex retains, the lighter particles owing to reduced centrifugal force. Because of increased centrifugal force for heavier particles, this is not really possible. The increased centrifugal force holds them forever close to the cyclone walls.

According to the findings, whenever the mass flow rate surpasses the desired value, an immense pressure drop develops all along axis, and then, an elevated pressure zone develops around the cyclone separator's wall. Owing to high-mass flow rate, flow velocity tends to rise; therefore, an intense centrifugal force consistently tries to hold the heavier particles nearer to the cyclone walls. As a consequence, heavier particles drop down very slowly and are not confined in the core vortex. However, lighter particles ≤5 μm get confined within core. Considering particles ≥12 μm in diameter, with mass flow rate of 90 kg/s results in 100% collection efficiency. For particles ≥16 microns in diameter, a mass flow rate of 57 kg/s results in 100% collection efficiency. The collection efficiency for particles with a diameter of 5 μm does not really vary with velocity.

4 Conclusion

It is significant to design a cyclone separator since the geometric configuration and mass flow rate of the dust and gas combination have a major influence on the cyclone separator's collection efficiency. Computational modelling was used to investigate the influence of mass flow rate of the gas and the particle size on the separation efficiency of a particular cyclone separator. It is concluded that if the proportion of heavier particles inside a mixture is considerable high, one must use a higher mass flow rate of gas, and if the proportion of small size particles are high, then one should use a lower mass flow rate of gas (gas velocity).

References

1. Kashani E, Mohebbi A, Heidari MG (2018) CFD simulation of the preheater cyclone of a cement plant and the optimization of its performance using a combination of the design of experiment and multi-gene genetic programming J. Powder Technol 327:430–441
2. Wang S, Li H, Wang R, Wang X, Tian R, Sun Q (2019) Effect of the inlet angle on the performance of a cyclone separator using CFD-DEM. J Adv Powder Technol 30:227–39
3. Azadi M, Azadi M, Mohebbi A (2010) A CFD study of the effect of cyclone size on its performance parameters. J Hazard Mater 182:835–841
4. Misiulia D, Antonyuk S, Andersson AG, Lundström TS (2020) High-efficiency industrial cyclone separator: a CFD study J. Powder Technol 364:943–953
5. Venkatesh S, Sakthivel M, Saranav H, Saravanan N, Rathnakumar M, Santhosh K (2020) Performance investigation of the combined series and parallel arrangement cyclone separator using experimental and CFD approach J. Powder Technol 361:1070–1080
6. Venkatesh S, Kumar RS, Sivapirakasam S, Sakthivel M, Venkatesh D, Arafath SY (2020) Multi-objective optimization, experimental and CFD approach for performance analysis in square cyclone separator. J Powder Technol 371:115–29
7. Bogodage SG, Leung AY (2015) CFD simulation of cyclone separators to reduce air pollution J. Powder Technol 286:488–506
8. Rafiee SE, Sadeghiazad M (2017) Experimental and 3D CFD analysis on optimization of geometrical parameters of parallel vortex tube cyclone separator. J Aerosp Sci Technol 63:110–22
9. Launder BE, Spalding DB (1983) Numerical prediction of flow, heat transfer, turbulence and combustion: Elsevier, pp 96–116
10. Ghose P, Datta A, Mukhopadhyay A (2016) Effect of prediffuser angle on the static pressure recovery in flow through casing-liner annulus of a gas turbine combustor at various swirl levels J Journal Therm Sci Eng Appl 8
11. Alahmer A, Al-Dabbas M (2014) Modeling and simulation study to predict the cement portland cyclone separator performance J. Emirates J Eng Res 19:19–25

Computational Modeling of a Gas-Fired Walking Beam Reheat Furnace for Billet Heating and Scale Formation Prediction

Prakash Ghose, Tarak Kumar Sahoo, Rishitosh Ranjan, Manas Kumar Padhy, and Kunja Bihari Sahu

1 Introduction

Reheating furnace is widely used in steel plants to heat billets from room temperature to the desired temperature [1]. After heating, the billets are sent to the rolling mill. In walking beam type furnace the billets are either continuously fed into the furnace at an ultra-low speed or hold slabs in different heating zones for a particular time. Thus the slabs are heated up to very high temperature due to a higher residence time in the furnace. However, the residence time depends upon the size of the furnace, thickness of the slab, amount and type of fuel used and many other factors. Basically, two types of heating methods are used such as; electrical heating and heating by fuel combustion [2]. In the combustion heating process, various types of fuels are used. Garcia et al. [3] used natural gas as fuel. They investigated the effect of burners' positions on heating performance computationally. Liu et al. [4] used a mixture gas of coke oven gas and blast furnace gas as fuel. Mainly it has been focused on the effect of fuel flow rate and its inlet conditions on furnace performance. Chakraborty et al. [5] used solid fuel (coal) for furnace heating. In their work, billet temperature investigation across the furnace has been performed. Euler–Lagrange multiphase approach has been used for the simulation. Various approaches such as; High viscosity laminar flow model, Dynamic mesh model and Source term model are adopted by different researchers for computational modeling of walking beam heating inside the furnace. Chang et al. [6] investigated the performance of all three computational approaches and mentioned their respective merits and demerits. However, it has been mentioned that both High viscosity laminar flow model and Source term model predict better

P. Ghose (✉) · T. K. Sahoo · R. Ranjan · K. B. Sahu
School of Mechanical Engineering, KIIT Deemed to Be University, Bhubaneswar 751024, India
e-mail: pghosefme@kiit.ac.in

M. K. Padhy
Centurion University, Bhubaneswar 751009, India

© The Author(s), under exclusive license to Springer Nature Singapore Pte Ltd. 2023
S. Revankar et al. (eds.), *Recent Advances in Thermofluids and Manufacturing Engineering*, Lecture Notes in Mechanical Engineering,
https://doi.org/10.1007/978-981-19-4388-1_12

than the Dynamic mesh model. Tang et al. [7] used mixed fuel and used Dynamic mesh model for simulation. They mainly focused on the effect of the walking pattern of billets on heating. Moreover, in order to reduce the computational time, a 2D self-developed code is used, where; extracted data of the 3D model is incorporated in 2D code. In another work, Tang et al. [1] varied the fuel flow rate in different heating zones to study the temperature uniformity on plate surface with these variances. Wang et al. [8] mainly focused on burner positions and the number of burners for the same fuel flow rate on furnace overall efficiency. They reported that 6 number of burners settings gives higher efficiency than 13 numbers of burner settings. Han and Chang [9] investigated the optimum residence time to achieve desired preheat temperature through performance efficiency calculations.

Scaling formation during reheating is a major challenge in rolling plants. Due to contact of CO, CO_2 and H_2O in combustion products and high billet surface temperature, the surface of the billet is oxidized and a thin layer of oxide scale is formed on the billet surface. Due to scaling formation, a significant amount of metal loss is observed. Jung et al. [10] and Landfahrer et al., [11] investigated the scale formation on slab heating. However, they only considered parabolic law to predict the scale thickness and neither linear law nor the effect of product species concentration on scale formation has been considered. Liu et al. [12] considered both parabolic law and linear law for scale thickness prediction. They incorporated the effect of combustion product species in the rate constant formulation itself. On the other hand Schluckner et al., [13] investigated the scale formation considering both the laws and species concentration.

In this work, simulation has been performed within a small-scaled combustion-based furnace to predict the slab heating, when the slab is continuously moving within the furnace for different slab velocities. The effect of slab velocity on scaling thickness has also been evaluated.

2 Physical Model and Computational Modeling

The schematic of the furnace is shown in Fig. 1. Four numbers of burners are placed as shown in the figure. Three different slab velocities are chosen in such a way that the slab crosses the entire furnace in 60, 30 and 15 min. For these cases, fuel–air velocity is kept as 1 m/s. Another study is also done by varying the air–fuel mixture mass flow rate. For these cases, the slab residence time is kept as 15 min. All operating conditions are given in Table 1.

For this work, High viscosity laminar flow model is adopted for simulation. The entire solid domain is considered as a fluid domain and all thermophysical properties are set for low carbon steel except viscosity. However viscosity term for solid is insignificance, but the motion to the solid slab is given by considering it as a high viscous fluid. In order to nullify the boundary layer formation near wall of the slab, the slab interface boundary is given the same as slab velocity. Moreover, the buoyancy effect is not considered at all. This is a conjugate problem, where all three modes of

Fig. 1 Schematic of furnace

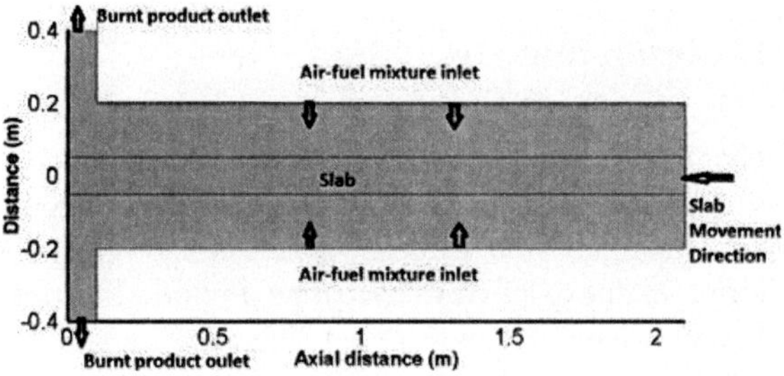

Table 1 Operating conditions

Cases	Methane air equivalence ratio	Air–fuel mixture velocity (m/s)	Slab velocity (m/s)	Slab inlet temperature (K)	Air–fuel mixture inlet temperature (K)
Case-1	1	1	0.0002332	400	300
Case-2		1.5	0.0002332		
Case-3		2	0.0002332		
Case-4		1	0.0001166		
Case-5		1	0.0000583		

heat transfer such as; convection, conduction and radiation are considered. Discrete ordinate radiation model [14] is considered to incorporate the radiative heat transfer from flue gas to slab surface. WSGGM is used to evaluate the absorption coefficient. Here methane is used as fuel and the premixed fuel equivalence ratio is maintained as 1. For the gas phase, all the thermophysical properties are constant except specific heat. In order to evaluate scale thickness an Arrhenius rate constant is used as follows [15];

$$x^2 = 6.1e^{(-169452/R_u T)}$$

where x is the scale thickness, R_u is the universal gas constant and T is the slab surface temperature. However, the above correlation predicts only Wusite (FeO) formation as scale. Magnetite (Fe_3O_4) and Hematite (Fe_2O_3) are not considered as scale, because both of them have only 5% contribution to the total scaling [13].

The simulation has been carried out with the help of ANSYS Fluent 19.2. Since computational capabilities of today's systems are very good, therefore particularly 2D models with very fine mesh can be simulated within very less time. However, the entire domain has meshed with a large number of elements (85,312). For mass momentum coupling, SIMPLE algorithm is used. The second-order upwind scheme is used for spatial discretization for all the governing equations. A convergence criterion for energy is set to 10^{-6} and for the rest it is 10^{-3}.

3 Result and Discussion

It has been observed that when the slab velocity is very less, a higher slab temperature is observed within the portion of the slab that is exposed to the hot gas phase zone. It is due to higher slab residence time. As slab velocity increases, the slab temperature decreases due to less residence time. Figure 2 illustrates the temperature field of both fluid and slab zone at different slab velocities. With slab velocity 0.0002332 m/s, 0.0001166 m/s and 0.0000583 m/s the entire slab crosses the furnace in 15, 30 and 60 min, respectively.

The desired temperature of the slab at furnace discharge should be higher than around 1473 K [13]. The temperature across the slab at the outlet (furnace discharge) is shown in Fig. 4. It has been observed that throughout the slab cross-section, the temperature is above the desired temperature when the slab residence time is 60 min. The discharge slab temperature across the slab decreases with increase in slab velocity due to less residence time. However, according to the size and fuel flow rate of this furnace 15 min of residence time is sufficient to achieve desired discharge temperature as shown in Fig. 4. The surface temperature along the length of the slab is shown in

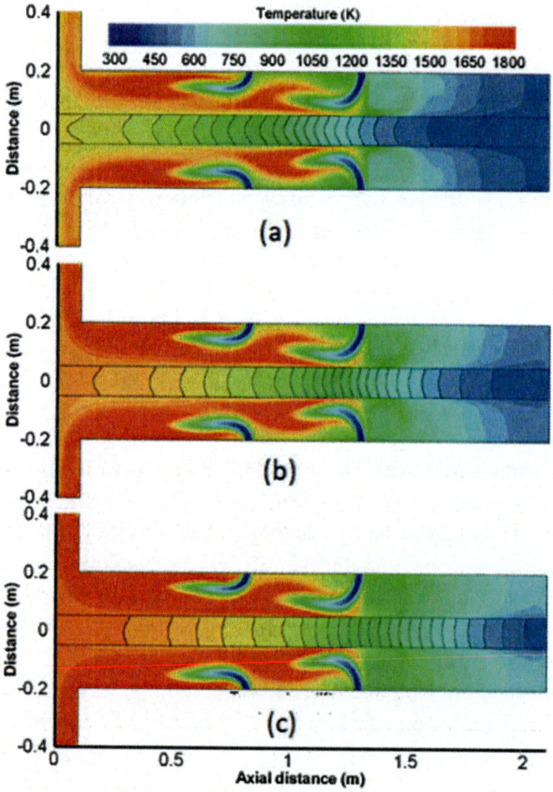

Fig. 2 Temperature field at slab velocity **a** 0.0002332 m/s **b** 0.0001166 m/s **c** 0.0000583 m/s. For all cases fuel–air velocity is kept as 1 m/s

Fig. 6. With lower slab velocity, slab surface temperature became high, which has been observed from this figure. As discussed lower slab velocity causes higher slab residence time. Hence heat transfer from the hot flue gas becomes higher. However radiative heat transfer plays a major role in furnace heating.

Figure 3 illustrates the effect of air–fuel mass flow rate on slab heating. A higher amount of fuel in combustion provides higher heat energy. Therefore it has been observed that with increase in air–fuel velocity or mass flow rate, slab temperature becomes high. In order to demonstrate exact values of slab temperature distribution across the slab at the outlet (furnace discharge) for various air–fuel mixture velocities, Fig. 5 has been prepared. From both Figs. 4 and 5, it has been noticed that at the center of the slab temperature is minimum and maximum value exist at slab surface. This temperature drop is caused by the thermal resistance of the solid slab.

The scale thickness formed on slab surface in millimeters is shown in Fig. 7. From the figure, it has been observed that for lower slab velocity, scale thickness is highest throughout the slab surface as compared to higher slab velocity cases. Higher plate surface temperature caused by lower slab velocity enhances the scale formation rate. Moreover due to higher slab residence time, scale (FeO) formation rate increases due

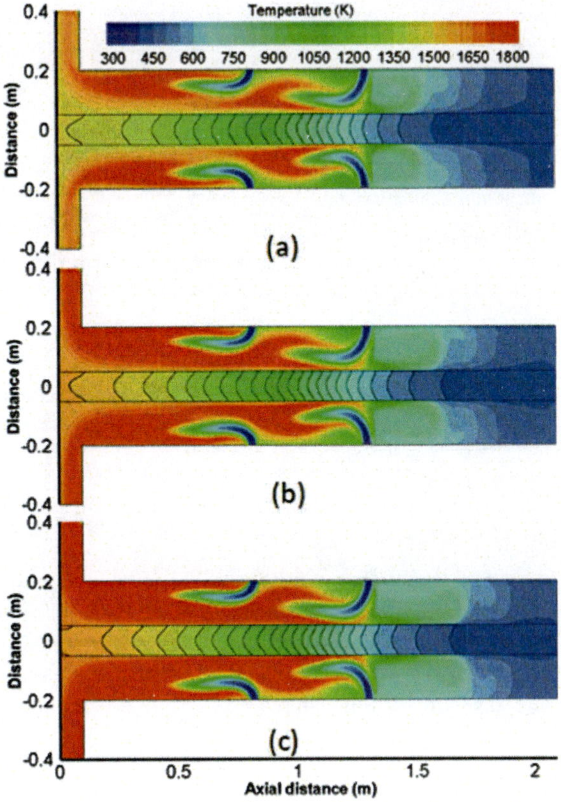

Fig. 3 Temperature field at air–fuel mixture velocity **a** 1 m/s **b** 1.5 m/s **c** 2 m/s. For all cases slab velocity is kept as 0.0002332 m/s

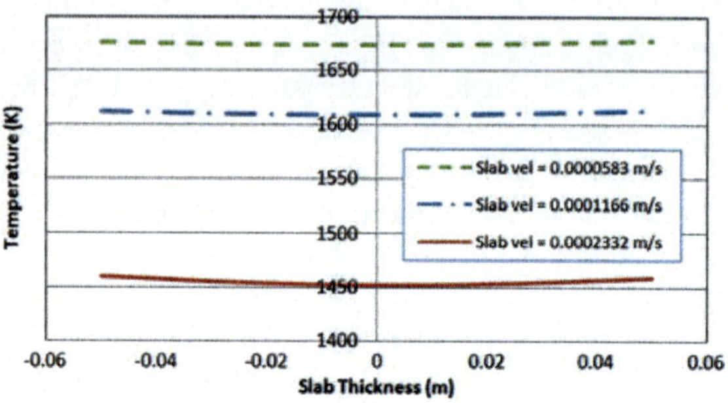

Fig. 4 Cross sectional slab temperature at slab outlet position for different slab velocity (air–fuel inlet velocity = 1 m/s)

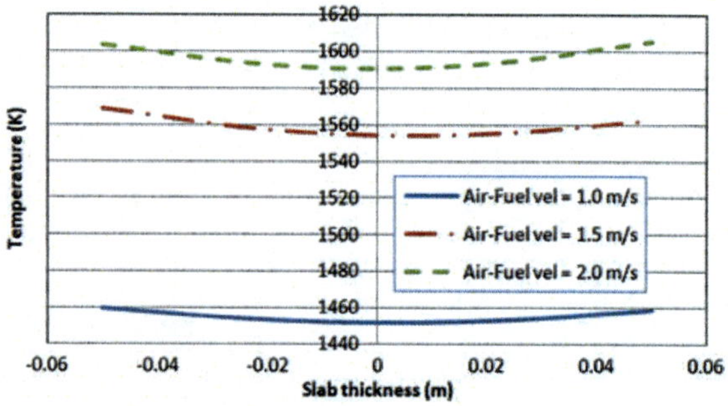

Fig. 5 Cross sectional slab temperature at slab outlet position for different air–fuel inlet velocity (slab velocity = 0.0002332 m/s)

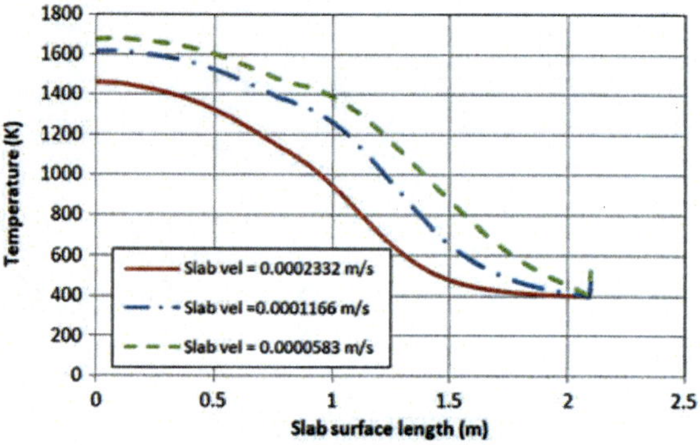

Fig. 6 Slab surface temperatures along the slab length

to higher slab surface exposure time to oxygen content species (CO_2, O_2, H_2O and CO). As slab velocity increases due to lesser slab residence time, scale formation rate decreases, hence scale thickness becomes minimum. It can be noticed that with 60 min slab residence time, scale thickness is around 5.6 mm at the discharge end, whereas with 15 min slab residence time it is around 2.25 mm (Figs. 7).

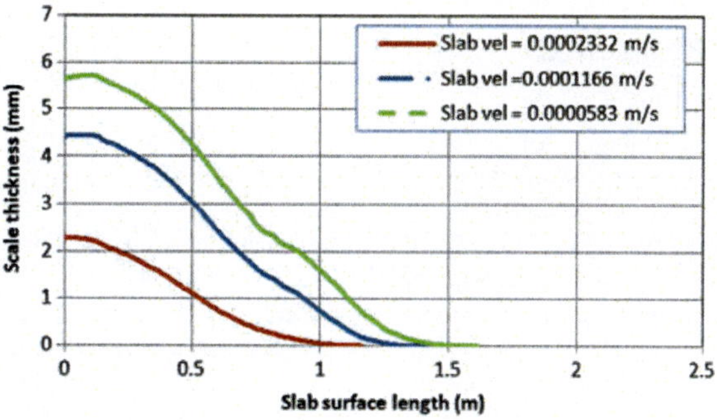

Fig. 7 Scale thicknesses along the slab for different slab velocities

4 Conclusion

The influence of slab velocity and premixed methane-air flow rate on slab heating in the walking beam reheat furnace was investigated through computational simulation. The slab is assumed to be fluid having all thermophysical properties are of steel, whereas viscosity is chosen as an arbitrary higher value. The interfacial walls between the combustion zone and slab zone are also moved at the same velocity as the slab. Hence boundary layer formation is eliminated and the fluid slab behaves like a solid. Continuity, momentum, species transport and energy equations are solved in both combustion zone and slab zone simultaneously. The discrete ordinate (DO) radiation model is used for participating media consideration in the combustion zone and the eddy dissipation model is used to determine the combustion reaction rate.

From the investigation, it has been observed that when the slab crosses the furnace in 15 min (slab velocity of 0.0002332 m/s), the minimum temperature across the slab at the discharged end is higher than 1450 K for fuel–air mixture velocity of 1 m/s. Therefore, the slab velocity of 0.0002332 m/s is adequate to obtain the desired temperature for rolling. When the slab velocity decreases further, the slab temperature became much higher than the desired temperature at the discharged end due to higher slab residence time and that also causes the thicker scale formation. With a slab velocity of 0.0002332 m/s when air–fuel inlet velocity increases from 1 to 1.5 m/s and 2 m/s the slab temperature at discharge end become much higher than the desired temperature required for rolling. Therefore an unnecessary use of fuel will be there if the slab velocity will be higher than 1 m/s. Moreover, higher thickness scales form due to the higher slab surface temperature.

References

1. Tang G, Wu B, Bai D, Wang Y, Bodnar R, Zhou CQ (2018) CFD modeling and validation of a dynamic slab heating process in an industrial walking beam reheating furnace. J Appl Therm Eng 132:779–789
2. Singh VK, Talukdar P (2013) Transfer m. Comparisons of different heat transfer models of a walking beam type reheat furnace. 47:20-6
3. García AM, Colorado AF, Obando JE, Arrieta CE, Amell AA (2019) Effect of the burner position on an austenitizing process in a walking-beam type reheating furnace. J Appl Therm Eng 153:633–645
4. Liu Y, Wang J, Min C, Xie G, Sundén B (2020) Performance of fuel-air combustion in a reheating furnace at different flowrate and inlet conditions. J Energ 206:118206
5. Chakraborty S, Rajora A, Singh SP, Talukdar P (2017) Heat transfer and discrete phase modelling of coal combustion in a pusher type reheating furnace. J Appl Therm Eng 116:66–78
6. Chang JH, Oh J, Lee H (2020) Development of a roller hearth furnace simulation model and performance investigation. J Int J Heat Mass Transf 160:120222
7. Tang G, Wu B, Bai D, Wang Y, Bodnar R, Zhou CQ (2017) Modeling of the slab heating process in a walking beam reheating furnace for process optimization. J Int J Heat Mass Transf 113:1142–1151
8. Wang J, Liu Y, Sundén B, Yang R, Baleta J, Vujanović M (2017) Analysis of slab heating characteristics in a reheating furnace. J Energ Convers Manage 149:928–936
9. Han SH, Chang D (2012) Optimum residence time analysis for a walking beam type reheating furnace. J Int J Heat Mass Transf 55:4079–4087
10. Jang JH, Lee DE, Kim MY, Kim HG (2010) Investigation of the slab heating characteristics in a reheating furnace with the formation and growth of scale on the slab surface. J Int J Heat Mass Transf 53:4326–4332
11. Landfahrer M, Schluckner C, Prieler R et al (2019) Numerical and experimental investigation of scale formation on steel tubes in a real-size reheating furnace. J Int J Heat Mass Transf 129:460–467
12. Liu X, Worl B, Tang G et al (2019) Numerical simulation of heat transfer and scale formation in a reheat furnace. J Steel Res Int 90:1800385
13. Schluckner C, Gaber C, Demuth M, Forstinger S, Prieler R, Hochenauer C (2018) CFD-model to predict the local and time-dependent scale formation of steels in air-and oxygen enriched combustion atmospheres. J Appl Therm Eng 143:822–835
14. Sahu AK, Ghose P (2020) Computational modelling on pulverized coal combustion in a 5 kw burner to study no reduction through co-flow methane used as secondary fuel. J Comput Therm Sci: An Int J 12
15. Chen R, Yeun W (2003) Review of the high-temperature oxidation of iron and carbon steels in air or oxygen. J Oxid Metals 59:433–468

Aerodynamics

Computational Study of Combustion Process in a Gas Turbine Engine

S. K. Muduli, R. K. Mishra, and P. C. Mishra

1 Introduction

Combat and advanced trainer aircraft operate in all weather conditions and from sea level to very high altitude to fulfill various missions. Thrust requirements during combat operation of these aircrafts necessitate the gas turbines engines to work at elevated levels of pressure and temperatures [1, 2]. Aero engine fulfills the thrust demand at any combination of altitude and forward speed when all its modules, systems, and accessories behave efficiently and as predicted when working independently and jointly [3]. This necessitates the performance characterization of each engine module such as fan, compressor, combustion chamber, turbine, jet nozzles, and various systems from sea level static condition to high altitude high Mach number operating conditions [3]. In this regard, the annular combustor of a low-bypass turbofan engine is extensively analyzed using computational tool to study the combustion process which will be useful to establish its performance parameters. The major performance parameters are overall pressure loss, combustion efficiency, and exit temperature distributions which are controlled by the air flow distribution in the various zones of combustion chamber, fuel atomization and mixing, ignition, and combustion processes [4, 5]. This paper deals with computational study of combustion process and its dependency on operating conditions. Detailed study of flow field and temperature field inside the flame tube has been carried out for understanding the effect of combustor inlet conditions covering various operating altitudes.

S. K. Muduli (✉) · P. C. Mishra
School of Mechanical Engineering, KIIT Deemed to be University, Bhubaneswar, Odisha 751024, India
e-mail: sk.muduli7@gmail.com

R. K. Mishra
RCMA (GTRE), Bangalore, India

© The Author(s), under exclusive license to Springer Nature Singapore Pte Ltd. 2023

S. Revankar et al. (eds.), *Recent Advances in Thermofluids and Manufacturing Engineering*, Lecture Notes in Mechanical Engineering,
https://doi.org/10.1007/978-981-19-4388-1_13

CFD analyzes are widely used for gas turbine combustor design and optimization in spite of complex combustor geometries and grid generation technology required for a reasonable accuracy. In this study, ANSYS CFX solver has been used to solve the governing equations by using Reynolds-averaged Navier–Stokes equations on unstructured meshes with 4.25 millions elements. In the present case, rate of fuel flow, fuel spray cone angle, and particle sizes are used as inputs. Design of each combustor is unique due to complex combustion aerodynamics and heat transfer. Hence, computational approaches are used for preliminary design and actual testing for final validation and freezing of the design [6, 7]. Though a substantial amount of work has been carried out in this field in the past [4, 5, 8, 9], the present study will help to further augment the knowledge level as it covers the off-design operating points.

1.1 Configuration of Combustion Chamber

An annular combustor with configuration shown in Fig. 1 has been used in the present study. The compressed air from compressor with low-diffused speed enters the combustor.

After the pre-diffuser from the dump diffusion zone, air flow is splitted to three flows, i.e., outer annulus flow, inner annulus flow, and core or dome flow. Swirlers are positioned around duplex atomizers to supply diffused and swirled air mass flow for stabilized combustion. The chamber has three zones, i.e., primary, intermediate, and dilution. These zones are having air injection holes on both outer and inner liners. Tiny liner cooling holes are provided on both outer and inner liners to protect the liner material from thermal erosion. The combustion chamber used in this study has been

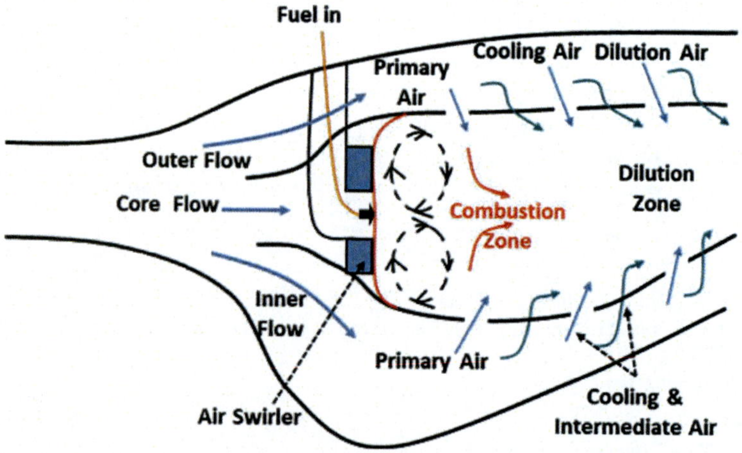

Fig. 1 Configuration of an annular type combustion chamber without center line

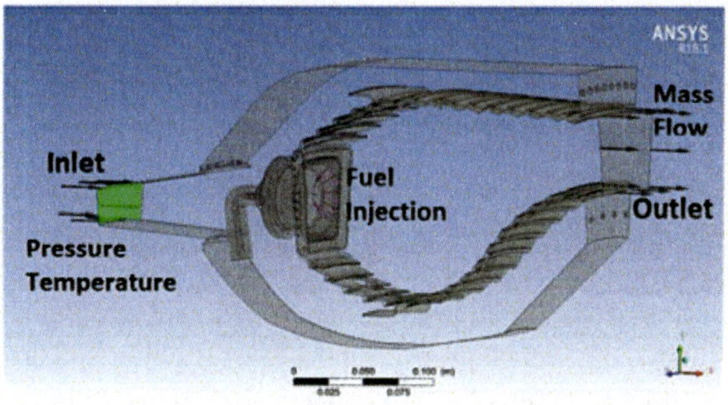

Fig. 2 Combustor model showing the boundary conditions used for computational analysis

designed to operate at a pressure of 2430 kPa and outer temperature of 845 K with overall fuel–air ratio of about 0.025. The primary zone equivalence ratio approaches a value nearly 1.1.

1.2 Computational Model

Gas turbine combustors are generally modeled to have in sector which can represent the complete combustion system. Annular combustors having large inner and outer liners having multiple atomizers are modeled in sector containing one or two atomizers depending on the size of the combustor and computational time required. [5]. It is generally assumed that the flow field and results in other sectors of the combustor are identical to that obtained in the sector analyzed. In the present work, an annular sector combustion chamber of 13^0 sector is modeled and placed as Fig. 2. The primary air injection ports, dilution ports, cooling holes, and all major components in the flow path of the combustion chamber are modeled from the inlet to the exit. The combustor assembly under study having added design features such as mounting brackets and flanges, small filets with different manufacturing tolerances were simplified during modeling and mesh generation.

1.3 Numerical Solution Procedure

Reynolds-averaged Navier–Stokes equations on unstructured meshes were solved using ANSYS CFX with velocity and pressure equations as a single implicit system. Jet A1 ($C_{12}H_{23}$) fuel was injected as a discrete phase using cone injection model from the atomizer. Spray cone angle, fuel particle size, and its temperature were specified during this analysis. Eddy-dissipation combustion model was assumed where mixing time is defined by the turbulent kinetic energy k and dissipation rate ε [8–10].

1.4 Boundary Conditions

Inlet and outlet boundary conditions such as total pressure, temperature, and total mass flow, respectively, are specified. Lagrangian particle tracking method is used for fuel flow rate with spray cone angle as input which are shown in Fig. 2. Fuel droplet sizes are assumed to follow the Rosin–Rammler distribution which describes the volume fraction of droplets as a function of droplet size [11, 12]. The distribution is divided into ten equal volumetric sizes. For the fuel droplet size distribution, a Rosin–Rammler exponent of 1.8 is used. Due to the wide variation of combustor operating pressure and temperature over operating envelope, the air and fuel properties also vary and affect the atomizer characteristics. To account this, Lefebvre's correlation for pressure swirl atomizer given in Eq. (1) is used to estimate the fuel particle mean drop size [13, 14].

$$\mathrm{SMD} = 4.52\left(\frac{\sigma \mu_L^2}{\rho_A \Delta P_L^2}\right)^{0.25} (t \cos\theta)^{0.25} + 0.39\left(\frac{\sigma \rho_L}{\rho_A \Delta P_L}\right)(t \cos\theta)^{0.75} \quad (1)$$

where σ is the surface tension in kg/s^2, μL is dynamic viscosity of fuel kg/ms, ρ_A and P_L are air and fuel densities in kg/m^2, ΔP is the pressure drop across the atomizer in N/m^2, θ is half-spray cone angle, and t is the film thickness. The axial velocity of droplets at the point of injection is computed from the fuel flow rate and the estimated fuel film thickness. Periodic boundary conditions are imposed on side walls with adiabatic combustor walls [15]. Iterations are carried out till the residual error converges to 1e^{-5} for mass, momentum, and species concentration and an order of 1e^{-6} for energy.

1.5 Grid Sensitivity Analysis

Sensitivity study of grid model was carried out by numerical analysis at constant inlet pressure of 555 kPa and temperature of 666 K keeping combustor fuel–air ratio of 0.025. Analysis was carried out with three different grid models of 2.5, 4.25, and 6.1 million cells. Variation in pressure loss ($\%\Delta P$) and EGT was estimated for different grid sizes and found that the variations in values are negligible beyond 4.25 million grid sizes. The order of deviations found, i.e., about 7.5% in pressure loss and less than 2% in exit temperature are generally acceptable for validating the CFD code and are similar to reference [16]. The 4.25 million grid size has been used in this computational study to provide a good representation of global flame structure and to produce reasonably accurate local flame temperatures which is also computationally affordable.

Table 1 Combustor operating conditions for computational analysis

S. no	Altitude in km	Flight Mach mo	P_{inlet} in kPa	T_{inlet} in K	Fuel–air ratio	SMD, μm
1	0	0	2400	820	0.025	24
2	4	0	1450	750	0.025	30
3	6	0.4	1100	715	0.025	34
4	8	0.4	827	680	0.025	39
5	10	0.4	676	664	0.025	43
6	12	0.6	555	666	0.025	48

2 Results and Discussion

During the study of flow field analysis, velocity distribution in various zones, location and size of recirculation zone, Mach numbers along the liner and pressure distributions are studied, whereas temperature distribution and jet fuel fraction are included as a part of temperature field study. Computational analysis has been carried out varying between sea level static conditions up to 12 km with 0.6 Mach number covering the entire operating range of the combustor for which it is designed, the details of which are presented in Table 1. Considering the properties of fuel, i.e., Jet $A1$, the fuel particle mean drop size is estimated and is included in Table 1 [17]. The drop size varied from 24 μm at sea level (24 Bar) to 48 μm at 12 km altitude (5.55 Bar) for the atomizers used in the present combustor.

3 Flow Field Analysis

Primary zone of combustor anchors the flame and provides required time, speed, and turbulence to the air–fuel mixture for complete burning. It also provides toroidal flow reversal which is required for continuous ignition of fuel–air mixture. This flow reversal and recirculation zone are caused by both swirling air produced by the swirlers at the front-end dome of the combustor liner and air jets injected to the primary zone through ports on both outer and inner liner walls. These air flow patterns in the primary zone are not only important for achieving complete combustion, but also to maintain flame stability in a continuously flowing air/gas stream over the entire operational envelope [18, 19]. Cooling air left out after supply for combustion is entered through dilution holes for liner wall cooling and to maintain an acceptable temperature distribution tolerable to downstream hot end components. The mixing is important in reducing the gas temperature to meet the life requirements of high-pressure turbine blades and nozzle guide vanes. Air through various cooling passages keeps the wall temperatures within the tolerable limits. Because of inaccessibility of placement of sensors inside the liner, flow field studies are generally carried out by flow visualization techniques and computational methods. Cooling and dilution

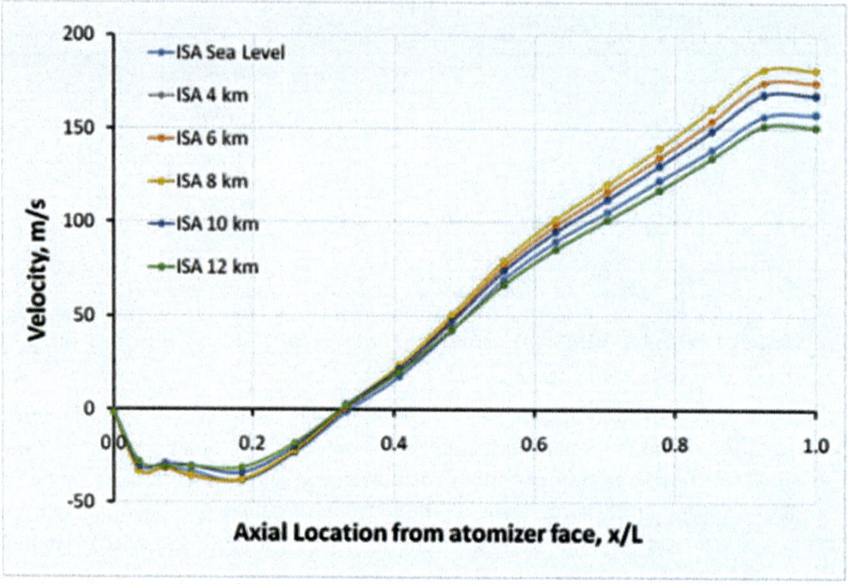

Fig. 3 Mean centerline velocities in x-direction (u) along the combustor length for different altitude conditions

flows along the combustor liner are estimated using the results of 1D flow network analyzes which gives a split of 38% and 30% air flow in outer and inner annulus, respectively, and balance 32% through the core of the combustor. The mean stream velocity (u) along the length of the combustor at the center line is shown in Fig. 3. Negative value indicates the presence of central recirculation zone which spreads up to a length of 33% from the face of the atomizer $x/L = 0$. It further becomes positive after addition of primary air. A recirculation zone is created with the help of swirler which helps to stabilize and hold the flame inside the combustor.

Though the velocity distribution inside the liner in primary zone is identical up to $x/L = 0.5$ for all conditions and not affected by the altitude, the exit velocities at higher altitudes are relatively less than that at lower altitudes. The proportion of air distribution in primary zone is maintained at all altitudes causing the recirculation zone size non-affected. But as the absolute pressure drop across the liner is decreased at higher altitude, the velocity is reduced.

3.1 *Temperature Field Analysis*

The temperature field inside the combustor liner is analyzed to study the initiation of combustion and achievement of peak temperature in the primary zone and distribution

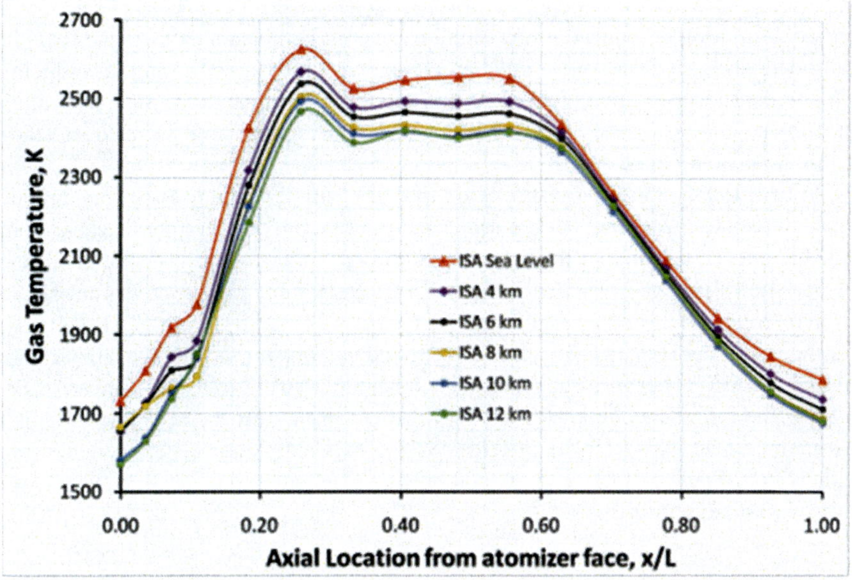

Fig. 4 Maximum gas temperatures along the centerline of the combustor liner for different altitude conditions, fuel–air ratio = 0.025

along the liner length till the exit plane. A noticeable flame zone is established in the entire primary zone covering the dome height as shown in Fig. 4.

It is evident that the peak temperature as well as the exit gas temperature is maximum for sea level condition where both pressure and temperatures are maximum causing a faster chemical reaction. Combustion products thus produced have a thorough mixing with fresh incoming mixture that takes place by high-kinetic energy in the incoming jets [20]. Heat release rate reaches maximum at equivalence ratio of the mixture at nearly 1–1.05 and is a function of mixture as well as fuel initial temperature [21]. Beyond this point, there is a fall in the concentration of oxygen and un-burnt fuel. This offset the reaction rate resulting in reduction in heat release rate and gas temperature beyond primary zone till the exit at each altitude and also with increase in altitude even though overall fuel–air ratio of 0.025 is maintained for all operating conditions.

3.2 *Flame Stability*

Flame stability is studied at the highest altitude of operation, i.e., at 12 km. To examine the size of flame zone and gas temperatures in combustor primary zone, steady-state analysis carried out at different overall fuel–air ratios, reducing the fuel flow rate [3]. The peak temperature occurs at about 26% of length from front end, and temperature level is reduced in the dilution zone till combustor exit. As the fuel–air ratio decreases, there are no significant change up to FAR 0.015. Further reducing FAR to 0.010, combustion takes place and reaches near adiabatic flame temperature in primary zone but it is quenched with the primary jet air and temperature starts falling in primary zone itself.

At FAR 0.005, there is a small region of flame with temperature of about 2400 K close to the atomizer face, but the center liner maximum temperature reaches about 2100 K only at the face of the atomizer and keeps on decreasing thereafter till the exit as shown in Fig. 5. The temperature levels and temperature contours culminate to an inference of the existence of a stable flame at 12 km altitude with FAR = 0.005. Further reducing FAR to 0.003, flame is formed at atomizer face with a temperature of about 1900 K, but it is quenched with fresh air entering in to the primary zone. During the full-scale combustor testing in aero-thermal test facility, flame was not extinguished at fuel–air ratio of 0.003 as the exit temperature started increasing indicating existence of flame in the combustion chamber and it sustained when fuel flow was increased further.

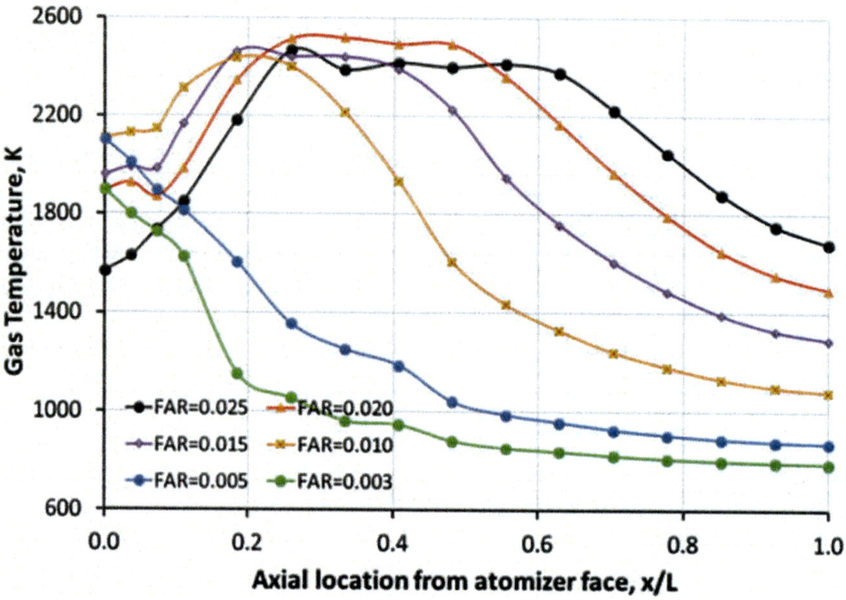

Fig. 5 Gas temperatures along the centerline of the atomizer for different fuel–air ratio for 12 km altitude

3.3 Effect of Particle Size of Fuel on Combustion

Particle size of fuel plays an important role in atomization and influences the combustion as well as the temperature field in the combustor [18, 22]. Though the mean particle size has been considered based on Lefebvre's correlation, there is possibility that the actual fuel particle size may vary w.r.t predicted one due to factors such as variation in spray cone angle, film thickness, pressure, and temperature. The time for evaporation varies exponentially with increase in particle size resulting in delay of heating up of air–fuel mixture and further ignition which affects the combustion flame due to quenching of fresh incoming air. To study the effect of particle size on combustion temperature along the liner length; it is increased by 50% from the predicted size at sea level inlet conditions of 2400 kPa and 820 K. The effect of particle size on maximum gas temperatures along the centerline of the atomizer is found very marginal at sea level condition as shown in Fig. 6. Initial gas temperature from the atomizer face up to 10% length is observed be less due to rise in particle size due to delay in evaporation and heat release. Since the operating pressure and temperature are already in very high level, the effect of fuel particle size is not noticed beyond $x/L = 0.10$.

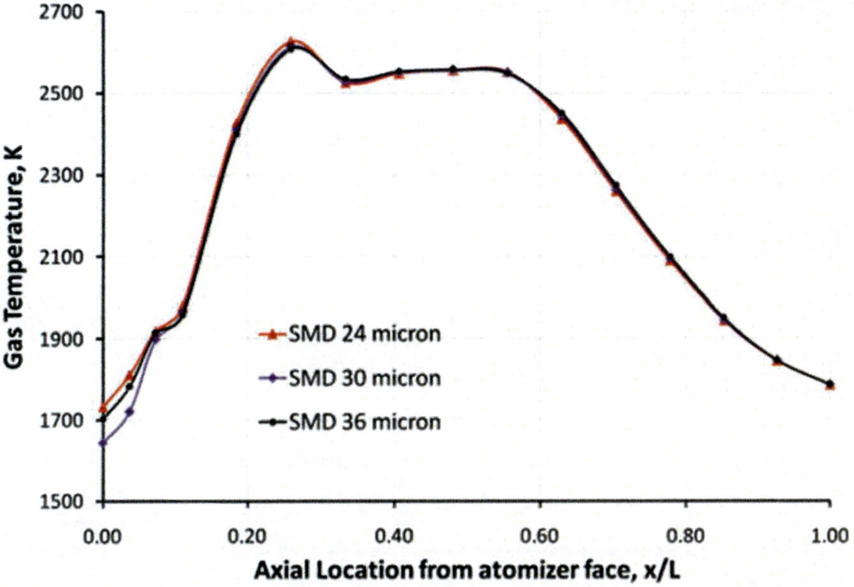

Fig. 6 Fuel particle size and its effect on maximum gas temperatures along the centerline of the atomizer at sea level conditions

3.4 Effect of Spray Cone Angle on Combustion

Fuel is usually sprayed into the combustor at suitable cone angles based on type of atomizer. Fuel atomization improves with increase in spray cone angle which increases the exposure of fuel droplets and further mixing with incoming air [23]. Pressure-swirl atomizers with a nominal half cone angle of 45°are used in the present study for all simulations. Temperature distribution is also studied by varying half cone angle between 30° and 60° and is presented in Fig. 7. Lower as well as higher spray cone angles compared to the mean value has reduced the initial flame temperature close to the atomizer face up to $x/L = 0.10$, beyond which there is nearly no effect till the combustor exit. Lower cone angle increases penetration of spray from atomizer face having a blanket of fresh low-temperature mixtures around causing delay in heat release. Higher cone angle exposes fuel particles to more fresh air near atomizer at lower temperature which reduces the gas temperature near the atomizer face as shown in Fig. 7 [24].

The following observations are noteworthy in this study:

- The size and location of recirculation zone in combustor are nearly insensitive with altitude. The length of recirculation zone is found to be about 33% from atomizer face at all altitudes for the combustor under study.

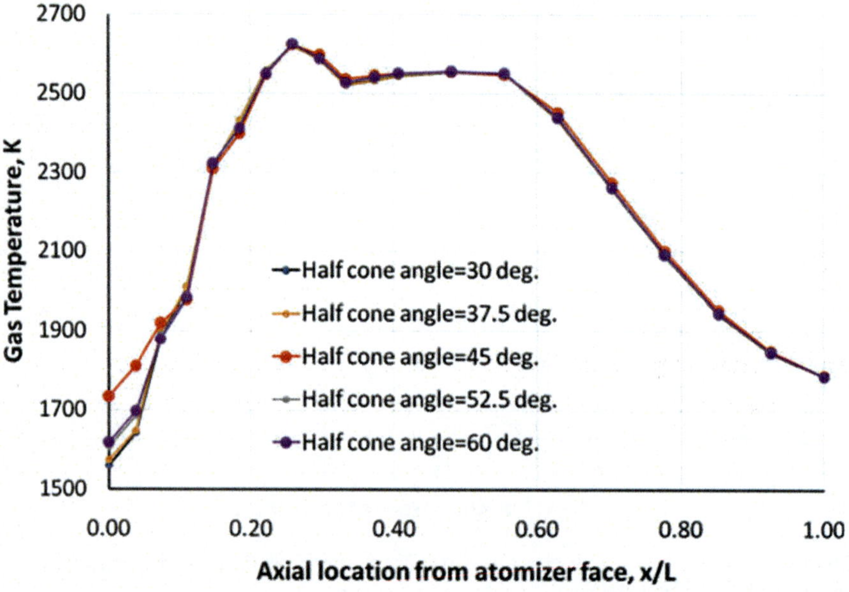

Fig. 7 Spray cone angle and its effect on maximum gas temperatures along the center line of the atomizer at sea level conditions of $P = 2400$ kPa, $T = 820$ K, fuel–air ratio = 0.025, and fuel particle size = 24 μm

- The centerline mean velocities in x-direction (u) in dilution zone as well as at combustor exit decrease with increase in altitude.
- The maximum gas temperature along the combustor length initially increases and attains peak value at a length of about 26% from atomizer face for all conditions. The peak temperature and exit temperature levels decrease with altitude. Temperature levels remain nearly uniform between primary and dilution injection ports and then drop till the exit plane.
- At low-fuel–air ratio of 0.01, combustion takes place and reaches near adiabatic flame temperature in primary zone but it is quenched with the primary jet air and temperature starts falling in primary zone itself. A small flame region is formed close to the atomizer face but it is quenched in primary zone with fuel–air ratio below the above value.
- Effect of particle size on maximum gas temperatures is felt within 10% axial length from atomizer face. Initial gas temperature near atomizer face is observed to decrease with rise in particle size and thereafter almost nil effect on maximum gas temperatures along the atomizer center line.
- Spray cone angle beyond the mean value (on both sides) affects the gas temperature within 10% length from atomizer face. The peak temperature in other zones has been found to be nearly insensitive with variation in spray cone angle.

4 Conclusion

The study of various performance parameters has been carried out in an operational annular combustor. Selected parameters are assessed on standalone test bed, and no further parameters details are available to the operator. However, as the combustor under study operates with aircraft in the designated operating envelope, hence, the study has been carried out to assess parameters in the operating envelope so that it can be utilized to understand the combustion process throughout the operating envelope and issues related to the combustion during operation can be resolved with these inputs.

References

1. Cohen H, Rogers GFC, Saravanamuttoo HIH, Gas turbine theory, 5th edn. Wiley, New Delhi: Dorling Kinderslay (India) Pvt. Ltd. Licensees of Pearson Education in South Asia
2. Giampaolo T (2003) The gas turbine hand book: principles and practices, 2nd edn. Fairmont Press, Inc., Lilbum, GA 30047, USA
3. Walsh PP, Fletcher P (2004) Gas turbine performance. Blackwell Publishing Inc., Malden, MA
4. Lefebvre AH (1998) Gas turbine combustion. Taylor & Francis, Philadelphia
5. Mellor AM (1990) Design of modern turbine combustors. Academic Press, London
6. Mishra RK, Muthuveerappan N, Bhat MN, Jan SA. Radial temperature profile improvement in an annular gas turbine combustor. In: 2nd National Conference on Air Breathing Engines, Trivandrum, India, December

7. Mishra RK, Navindgi RD, Muthuveerappan N, Bhat MN, Jan SA (1998) Performance optimization of a short annular aero gas turbine combustor. In: 4th National Conference on Air Breathing Engines, Bangalore, India, December
8. Li LP, Liu T (2006) Combustion and cooling performance in aero-engine annular combustor. J Appl Thermal Eng 26:1771–1779
9. Armand S, Chen M (2002) A combustion study of gas turbines using multi-species/reacting computational fluid dynamic. In: Proceedings of ASME Turbo Expo 2002, GT-2002-30105
10. Stopford PJ (2002) CFD modelling of industrial burners and furnaces. In: 6th European Conference on Industrial Furnaces and Boilers, Sintra, Portugal, April
11. Khavkin YI (2003) Theory and practice of swirl atomizers. Taylor & Francis
12. Crowe CT, Schwarzkopf JD, Sommerfeld M, Tsuji Y (2011) Multiphase flows with droplets and particles. CRC Press
13. Lefebvre AH (1987) The prediction of Sauter mean diameter for simplex pressure-Swirl atomizers. Atomization Spray Tech 3:37–51
14. Wang XR, Lefebvre AH (1987) Mean drop sizes from pressure-swirl nozzles. J Propulsion Power 3(l):11–18
15. Turrell MD, Stopford PJ, Syed KJ, Buchanan E (2004) CFD simulation of the flow within and downstream of a high-swirl lean premixed gas turbine combustor. Turbo Expo: Power for Land, Sea Air 41669:31–38
16. Srinivasa Rao M, Sivaramakrishna G (2009)Performance improvement of an aero gas turbine combustor. ASME GT2009–59928
17. Handbook of aviation fuel properties, CRC Report no. 530, prepared by Coordinating Research Council Inc. Atlanta, Georgia, published by Society of Automotives Engineers, Warrendale, PA, USA, 1983
18. Mishra RK, Kishore Kumar S, Chandel S (2014) Effect of fuel particle size on the stability of swirl stabilized flame in a gas turbine combustor. Int J Turbo Jet Engines. 32:129–141. https://doi.org/10.1515/tjj-2014-0027
19. Mishra RK, Kishore Kumar S, Chandel S (2015) Lean blow-out studies in a Swirl stabilized annular gas turbine combustor. Int J Turbo Jet-Engines 32(2):117–128
20. Longwell JP, Weiss MA (1955) High temperature reaction rates in hydrocarbon combustion. J Ind Eng Chem 47(8):1634–1643
21. Glassman I (1996) Combustion. Academic Press, New York
22. Reddy KU, Mishra DP (2008) Studies on spray behavior of a pressure swirl atomizer in transition regime. J Propul Power 24(1):74–80
23. Reitz RD, Bracco FV (1979) On the dependence of spray angle and other spray parameters on nozzle design and operating conditions. In: SAE Paper 790494
24. Mishra RK, Kishore Kumar S, Chandel S (2016) Effect of spray cone angle on flame stability in an annular gas turbine combustor. Int J Turbo & Jet-Engines 33(1):35–44

Effect of Diffuser, Rake, and Slant Angles on the Aerodynamics of a Simple Ahmed Body

Karthik Nayakar, Ramayana, K. Supradeepan, and P. S. Gurugubelli

1 Introduction

Aerodynamics of automotive bodies is very important to study for design and other such aspects for the general automotive industry, the aerodynamics characteristics have great influence on fuel economy and emission level [1–3]. New technologies are been developed to reduce aerodynamic drag and hence reduce fuel consumption because of high prices of fuel and also due to the fact that it is a non-renewable fuel. The research that is being pursued nowadays is more leaning toward the improvement of the quality and comfort of the aerodynamics of the vehicle. The most widely researched body in the automotive part is mainly the Ahmed body; it was first devised by S. Ahmed in 1984 [4–9]. It has significant impact on research in automotive industry. This is a bluff body, the skin friction of bluff bodies is negligible, but due to the presence of form drag, there is a lot drag and lift acting on the body [1–3, 8–11]. At the rear of car large turbulent wakes and eddies are formed.

The presence of such eddies and vortices at rear of car increases drag and leads to flow separation and also reduces the downforce acting on the body [7, 10–14]. By changing the shape of body and by adding a few aero devices improve and optimize the aerodynamics of vehicle. Vehicles with lower drag coefficient are more efficient. But reducing drag drastically can reduce the downforce in turn leading to loss of traction which will increase the chances of road accidents. 85% of the road vehicle drag comes from the fact that it is a bluff body [7]. The big developments from the automotive industry come from the study done with the flow on the underside of the Ahmed body, which is the diffuser, this aerodynamic device is mainly seen and used in high-performance sports vehicles which have very low clearance, hence such a

K. Nayakar · Ramayana · K. Supradeepan · P. S. Gurugubelli (✉)
Computing Lab, Department of Mechanical Engineering, BITS Pilani Hyderabad Campus, Hyderabad, Telangana 500078, India
e-mail: pardhasg@hyderabad.bits-pilani.ac.in

© The Author(s), under exclusive license to Springer Nature Singapore Pte Ltd. 2023

S. Revankar et al. (eds.), *Recent Advances in Thermofluids and Manufacturing Engineering*, Lecture Notes in Mechanical Engineering,
https://doi.org/10.1007/978-981-19-4388-1_14

device is very effective in such vehicles. Our main focus in this study is to show how effective the diffuser is in an Ahmed body which directly links to all the road going vehicle [10].

The objective of the diffuser is to provide a transition which ensures that the eddies formed can be reduced, which improves the drag on the body. As mentioned previously, the diffuser in addition to reducing drag force, improved the down- force acting on the body improving the handling of the vehicle [10, 12, 15]. Hence, researchers are studying these aerodynamic devices that can be used on any road going vehicle to improve the performance of normal vehicles. This is because of ground effect. As air flows around the car, it flows with a high velocity under the car; while air flow above the car is at higher pressure [11, 16]. As low-pressure air reaches rear portion of vehicle, it expands due to the addition of a diffuser and a venturi-like effect takes place where the pressure is increases and speed of air is reduced.

Due to this pressure difference of air above and underneath the car, a vacuum is created. This sucks the air out of car from rear enhancing the effect of the diffuser [6, 7, 11]. It is also seen that by changing the angle of the entire body in addition to the diffuser also improves the performance slightly [2, 7, 11, 17]. But most of the researches have no inclusion of angle of rake. In this study conducted on Ahmed body using ANSYS Fluent, the diffuser angle is varied from 0° to 12.5°, after selecting an optimum angle of the diffuser, the changes that the rake angle and slant angles have on the Aerodynamic performance of the Ahmed body is also observed. Using CFD by changing parameters such as the diffuser angle, rake angle, and slant angle are observed and the improvements are compared with respect to a simple Ahmed body.

The main reason behind studying the various drag and lift characteristics on the Ahmed body is to provide useful data for further research and development of other techniques to improve the energy efficiency on a simple Ahmed body. The results are verified analytically, and observations are made as to why drag and lift forces varied as seen in these cases [2, 12, 18]. The optimum values for diffuser, rake, and slant angles were achieved for minimum drag and lift forces.

2 Methodology

The diffuser acts like a expander [7, 19] which in turn maintains low-pressure zone under the car by increasing the velocity of the air beneath the car. This increases the downforce acting on the car [8]. Figure 1 presents a typical diffuser on an Ahmed body, here the diffuser looks as mentioned before, like an expander.

The rake generates a venturi-like effect and helps the air passing below the car to accelerate, thus creating a suction effect and increasing the downforce on the car [10, 15, 19]. Figure 2 shows that the slant angle at the back breaks the streamline flow over an Ahmed body (boundary layer separation) and thus destroying the vacuum created at the back and thus reducing the form drag [11, 12].

These effects when all combined together will naturally increase the down- force on the car and decrease the drag acting on the car. ANSYS packages are used for

Fig. 1 Ahmed body with diffuser angle of 25°

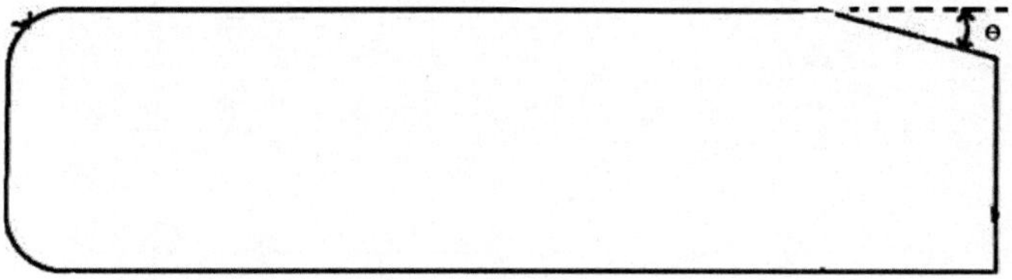

Fig. 2 θ is the slant angle here

design creation, meshing, and simulation. Simulations are performed on a 2D Ahmed body to obtain accurate results. To look into the diffuser of the Ahmed body, 10 different diffuser angles are simulated, as seen in Figs. 3a and b for angles ranging from 0° to 12.5° and a gap interval of 2.5° each.

For studying rake angle on the Ahmed body, the variation on 5 rake angles, i.e., 0, 1.5, 3.75, 5, 7 is observed. And finally, the behavior of 5 different slant angles ranging from 25° to 35° with a gap of 2.5° is reported.

2.1 *Design and Meshing*

2.1.1 Geometry Description

A standard Ahmed body is considered as shown in the Fig. 4 the dimensions of the computational domain being 11,544 mm × 3194.5 mm (l b). For the different diffuser, slant and rake angles separate designs with the desired changes are created and simulated.

2.1.2 Mesh Generation

A mesh with a y+ of 4 was assumed which was suitable for the K-omega SST turbulence model. Figure 5 focuses around the diffuser and the wake of the Ahmed body. The detailed view of the mesh is as shown;

(a) 10 degree diffuser angle

(b) 12.5 degree diffuser angle

Fig. 3 Ahmed body with varied diffuser angles

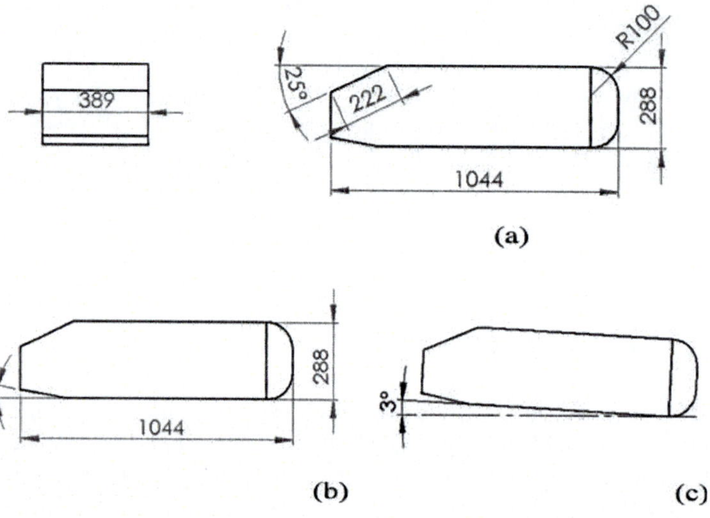

Fig. 4 Typical Ahmed body with **a** slant angle 25°, **b** diffuser angle 10°, and **c** rake angle 3° (all dimensions in mm)

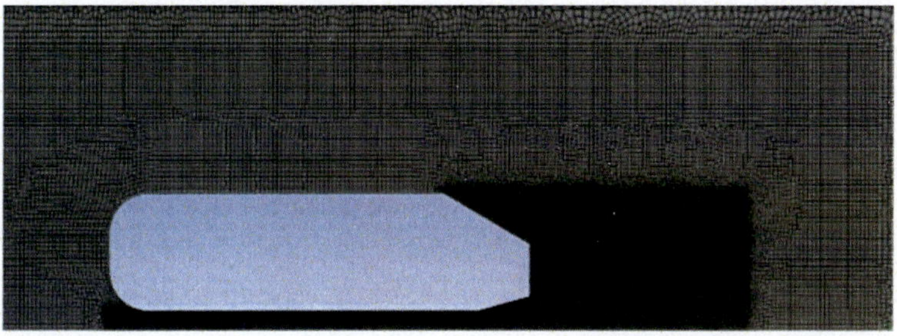

Fig. 5 Close-up view of the computational mesh around the Ahmed body

2.1.3 Setup and Boundary Conditions

The simulation was carried out at a Reynolds number of 1.8 10^8, and a K-omega SST turbulence model was chosen to study the characteristics close to the Ahmed body. The velocity of air in the domain is 25 m/s (90 km/h), and the surface of the Ahmed body is given a no slip boundary condition [10, 12]. An outflow boundary condition is given to the outlet, and a free stream condition is given to prevent formation of any boundary layer in front of the Ahmed body.

3 Results and Analysis

The simulations results obtained were carried out with ANSYS Fluent 2021 Academic version. A K-omega SST model is employed to observe the near wall fluctuations and also study the changes due to the body far away from it. A SIMPLE scheme was used for the entire simulation process, and the slant angle was maintained at 30° for the study of variations of diffuser and rake angles and the height of the Ahmed body from the ground remained constant throughout our simulations.

3.1 Effect of Diffuser Angle Variation

The diffuser angle has been varied from 0 to 12.5°, with a difference of 2.5°, by varying the diffuser length accordingly for a Reynolds number of 1.8 10^8, slant angle of 25°, and keeping rake angle at 0°. The main objective of the diffuser is to create a negative lift force or downforce as many call it [2, 3, 6, 13]. The pressure difference between the top and bottom surfaces of the Ahmed body creates this downforce effect due to the differences in flow speed around them, respectively. This phenomenon can better be described by the Bernoulli's principle in fluid dynamics [11, 13]. From Fig. 6a, it can be observed that as the diffuser angle is increased the downforce acting on the Ahmed body increases till it reaches a maximum downforce

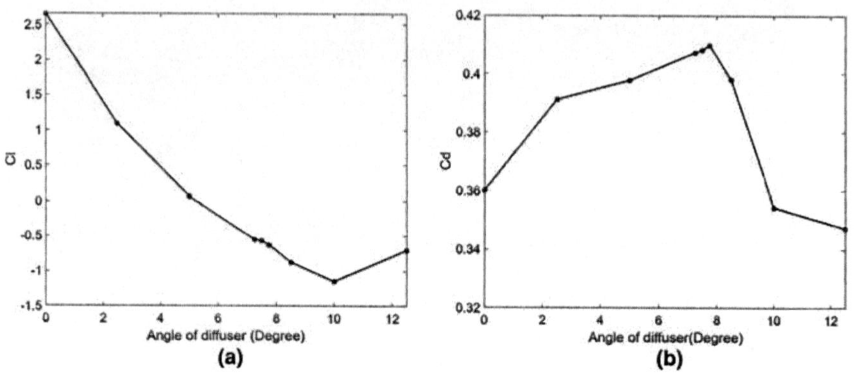

Fig. 6 Effect of varying the diffuser angle on Cl and C_d for a Reynolds number of 1.8 10^8, slant angle of 25°, and rake angle of 0°

for the diffuser angle of 10°. For diffuser angle greater 10°, the downforce acting on the Ahmed body starts to decrease again. The increase in the downforce for diffuser angles 0 10° can be attributed to the flow separation at the diffuser that creates a low-pressure region beneath the body, thereby increasing the downforce acting on it. The reduction in the downforce for the diffuser angles greater than 10° is due to the merging of circulation regions below and above the Ahmed body. As a result, it is seen that the optimal diffuser angle will also strongly depend on the slant angle to increase downforce.

Figure 6b presents the effect of diffuser angle on the drag experienced by the Ahmed body. It can be observed that as the diffuser angle is increased, there is a sharp increase in the drag experienced by the body. Similar to the downforce, this sharp increase can again be attributed to the flow separation below the body due to the diffuser angle, thereby resulting in a low-pressure region behind the Ahmed body and increasing the drag. This increase in drag continues up to diffuser angle of 8°. However, for diffuser angles greater than 8°, the drag starts to fall sharply. This drop in drag can further be explained by the reduction in the size of the wake behind the body.

As mentioned earlier, there is an influence of slant angle on the drag and lift forces acting on the body. Further, the effect of rake angle is investigated by fixing our diffuser angle at 10°.

3.2 *Effect of Rake Angle Variation*

The rake angles have been varied from 0° to 7° gradually, for a Reynolds number of 1.8 10^8 by fixing the diffuser and slant angles at 10° and 25°, respectively, throughout this process. It is observed that an increase in rake angle increases the downforce on the Ahmed body significantly. This could again be an effect of higher flow speeds

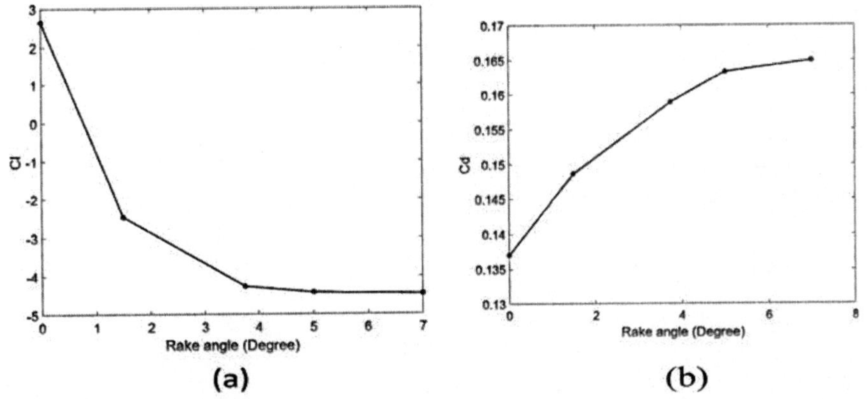

Fig. 7 Effect of varying the rake angle on Cl and Cd for a Reynolds number of 1.8 10^8, slant angle of 25°, and a diffuser angle of 10°

of air in the underside of the Ahmed body, where the flow is faster due to the rake angle and the diffuser at the rear-end of the Ahmed body. Figure 7 summarizes the effect of rake angle variations on the Cl and C_d. It is observed that as the rake angle is increased, the downforce and the drag increase exponentially. Therefore, the improvement in the downforce due to the change in rake angle for the Ahmed body comes at a cost of the drag force.

3.3 Effect of Slant Angle Variation

For the final changes, the slant angle of the Ahmed body is varied from 25° to 35°, with a difference of 2.5°, for a Reynolds number of 1.8 10^8, diffuser angle of 10°, and keeping rake angle at 0°. By varying the slant angle, it is reported that the lift force has a very different trend compared to the usual variation as seen in other cases. There is decrease in lift force from 25° to 27.5°, but then the lift force increases up to 30° with another drop in lift force at 32.5° and finally a slight increase at 35°.

The plots in Fig. 8 show very abnormal changes in the Cl and Cd values, this could be an anomaly in the observations, however, this trend can be explained with the pressure differences between the top and bottom surfaces of the Ahmed body, as the angle increased, the flow separation also increases, at 30° the flow separation at the top surface and the pressure difference between the top and bottom surfaces of the Ahmed body do not induce any drag. Similar observations have been reported in [4, 14–16].

When the slant angle was made 32.5°, it was again observed that the flow separation also increased, but this time, since the wake region at the rear of the Ahmed body slightly reduced, this caused a lift induced drag to act on our Ahmed body, hence observing those trends of lift and drag forces.

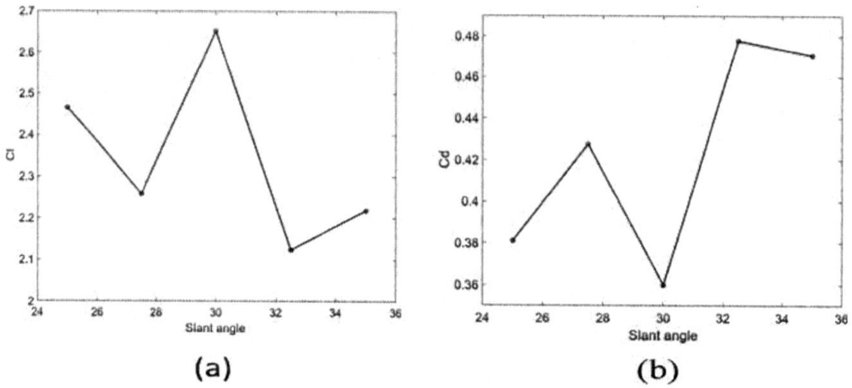

Fig. 8 Effect of varying the slant angle on Cl and Cd for a Reynolds number of 1.8 10^8, diffuser angle of 10°, and rake angle of 0°

4 Conclusions

The conclusive remarks on the aerodynamic effects report that the inclusion of diffuser and rake angles to an Ahmed body improves the aerodynamic drag coefficient and hence downforce increases. By varying the diffuser angle, it is seen that drag increases up to the 8° mark, and then, a large drop is observed. An increase in the downforce is reported up until the 10° mark, where minimum drag and maximum downforce are obtained, which is considered the most optimal or the best diffuser angle. The coefficient of drag observed at 10° is even lesser than that without a diffuser. For varying rake angles, it is observed that the drag coefficient increases up to a point after which it remains constant. And similarly, it is seen that the lift coefficient also decreases up to the same mark. While observing the changes on the slant angle, due to the flow separation, the drag coefficient drops at the 30^0 and resumes a normal trend thereafter. The future scope in the studies of diffusers will include optimizing the shape of the diffuser, the length, and width of the diffuser to obtain maximum downforce. Some examples of using energy efficiency in cars are rear wings designed along with a diffuser to generate downforce. There have also been studies which improve the downforce on the vehicle with a diffuser and the exhaust system combination working in tandem to improve the energy efficiency of the vehicle overall.

References

1. Marklund J (2013) Performance of an automotive under- body diffuser applied to a sedan and a wagon vehicle. SAE Int. J. Passeng. Cars Mech. Syst 1(6)
2. Xingjun H, Rui Z, Zhiming Z (2011) Influence of different diffuser angle on sedan's aerodynamic characteristics. Physics Procedia 22:239–245

3. Guilmineau E (2007) A computational study of flow around a simplified car body. J Wind Eng Ind Aerodyn 96(1207)
4. Strachan RK, Lawson KK (2004) CFD and experimental study of an ahmed reference model. In: SAE 2004 World Congress, Detroit, Michigan, USA, 8–11 March
5. Ahmed SR, Faltin G (1984) Some salient features of the time-averaged ground vehicle wake. In: SAE International Congress and Exposition, Detroit, Michigan, USA, February 27–March 3
6. Baxendale AJ, Graysmith JL, Haynes T (1994) Comparisons between CFD and experimental results for the Ahmed reference mode. In: Proceedings of the Royal Aeronautical Society Conference on Vehicle aerodynamics, London, UK, 18–19 July
7. Bayraktar I, Baysal O (2001) Experimental and computational investigation of an Ahmed body for ground vehicle aerodynamics. In: SAE Paper 2001-01-2742
8. Spohn A, Gillieron P (2002) The flow separations generated by a simpli- fied geometry of an automotive vehicle In In: IUTAM Symposium: Unsteady Separated Flows, April 8–12, Toulouse, France
9. Sims-Williams DB, Duncan B (2003) The Ahmed model unsteady wake: Experimental and computational analyses. In: SAE World Congress, Detroit, Michigan, March 3–6
10. Kohri I, Yamanashi T (2014) Study on the transient behaviour of the vortex structure behind ahmed body. SAE Int J Passeng Cars Mech Syst 7(2)
11. Kalyan DK, Paul AR (2013) Computational study of flow around a simplified 2d ahmed body. IJESIT 2(3):148–153
12. MB A, Patil P, Singh TP (2018) Parametric investigation of effect of diffuser angle on the flow characteristics of an ahmed body. Int J Engineering Res Tech 7
13. Strachan RK, Knowles K (2007) The vortex structure behind an ahmed reference model in the presence of a moving ground plane. Experiments Fluids 42(9):659–669
14. Wang Z, Hu X, Li S (2021) Study on vehicle drag reduction simulation based on suzen–huang model. Physics of Plasmas 28(063512)
15. Bayraktar S, Bilgili YO (2018) Effects of under body diffuser on the aerody- namics of a generic car. Int J Automotive Eng Techn 7(2):99–109
16. Tural Tunay BS, Ozbo V (2014) Effect of slant angles on the flow charac- teristics of ahmed body. Experimental Thermal Fluid Sci 57:165–176
17. Minguez M, Serre E (2008) High order large-eddy simulation of flow over the "ahmed body" car model. Phys Fluids 20(095101)
18. Musa MN, Osman K (2012) Computational analysis of flow field around ahmed car model passing underneath a flyover. AIP Conference Proceedings, 1440(190)
19. Fan Y, Xia C, Cadot O (2020) Experimental and numerical analysis of the bi-stable turbulent wake of a rectangular flat-backed bluff body. Phys Fluids 32(105111)

Aerodynamic Cooling of Electronic Pin Fins

Ankit Kumar, Biranchi Narayana Das, Srikant Panigrahi, Pooja Chaubdar, and Atal Bihari Harichandan

1 Introduction

Cooling of electronics has arisen as challenging problem from beginning of electronic industry. The demand of electronic components has increased faster like a clock speed and simultaneously smaller physical enclosures are preferred among all. Similarly, heat fluxes inside electronic packages increase with increasing clock speeds and decrease with chip sizes. This situation leads to increase demands of removing heat flux rapidly. In twenty-first century, thermal management of microelectronic components plays an important role in the development of electronics industries. Sparrow et al. [1] experimentally investigated and compared the performance of inline and the staggered configurations for heat transfer and the pressure drop. Lower heat transfer coefficient and pressure drop in inline pin fins as compared to staggered pin fins was observed. Jonson and Moshfegh [2] experimentally studied the performance of heat sinks with pin type, strip type and plate type fins for inline as well as staggered arrangements.

The performances of pin fin heat sinks were analyzed by considering circular and square configurations for dissimilar fin heights (H) of 1.0, 1.5 and 2.0 cm and unvarying width (B) of 5.28 cm, respectively. Ryu et al. [3] have experimentally studied fluid flow and heat transfer for square pin fins of 101.43 mm × 101.43 mm with varying velocities from 1 m/s to 5 m/s. They also studied the effect of different pin fin parameters like spacing, height and pitch. Jang and Tzeng [4] investigated heat transfer and the pressure drop phenomenon with square pin fins attached to the

A. Kumar
Department of Aeronautical Engineering, National Formosa University, Huwei, Taiwan

B. N. Das · S. Panigrahi · P. Chaubdar · A. B. Harichandan (✉)
Aerospace Engineering, School of Mechanical Engineering, KIIT Deemed to be University, Bhubaneswar 751024, India
e-mail: atal.harichandanfme@kiit.ac.in

© The Author(s), under exclusive license to Springer Nature Singapore Pte Ltd. 2023

S. Revankar et al. (eds.), *Recent Advances in Thermofluids and Manufacturing Engineering*, Lecture Notes in Mechanical Engineering,
https://doi.org/10.1007/978-981-19-4388-1_15

rectangular channel experimentally. They observed better performances of the square pin fins than the circular pin fins at the same pumping power. They recommended optimum pitch for the inline configuration to be X_T (transverse pitch) = 2 and X_L (longitudinal pitch) = 1.5 whereas the staggered pin fin array showed good results at $X_T = 1.5$ and $X_L = 1.5$.

Soodphakdee et al. [5] studied various heat sinks geometries. A periodical two-dimension flow and isothermal heat transfer surfaces were assumed to simplify the problem. It was found that the staggered configuration provided high heat transfer compared to inline configuration and rounded geometries are performing better than the similar shapes of sharp-edged fin. Elliptical pin fin outperforms at low values of pressure drop and rounded pin fin at high values. Yang and Peng [6] performed a computational simulation of the plate round pin fin heat sinks made up of circular pin fins sandwiched between the plate fins. The result of the configuration on the pressure drop and heat transfer of heat sink was studied. Thermal resistance of the plate circular pin fin heat sink was found to be lesser than plate-fin heat sink with higher value of pressure drop.

Wang et al. [7] have experimentally and numerically investigated thermal and hydraulic phenomena by considering different shaped pin fins like elliptical, circular and the drop shaped with similar cross-sectional areas in the staggered arrangements. The enhancement of heat transfer of drop-shaped fins was lower than circular pin fins for Reynolds numbers varying from 4200 to 8400. But friction losses inside drop-shaped fins were less than circular and elliptical pin fins enabling drop-shaped fins to be used as alternatives of circular pin fins. Abdoli et al. [8] studied experimentally various types of micro pin fins and analyzed the effect on the performances of heat sinks, using modified hydrofoil, hydrofoil, symmetry and circular convex shapes for the micro pin fins of staggered arrangements. It was noticed that the pressure drop is decreased by 30.4% in case of hydrofoil pin fin, 47.3% in convex pin fin when compared with the circular cross-section. The modified hydrofoil pin fin lessens the avg. temperature by 6.4 ^{0}C when compared with hydrofoil pin fin. Sajedi et al. [9] investigated, the method to maximize heat transfer by increasing area with help of splitter and simultaneously reducing pressure drop to avoid flow separation at the wake of pin fin by effects of splitter on the thermal and the hydraulic performances of pin fin heat sinks. An increase is observed in hydraulic as well as thermal performances of the pin fin for circular pin fins and square pin fins using splitter. The pressure drop and the thermal resistance decrease by 36.8% and 13.4% for the circular pin fin with splitter. Still, the pressure drop decreases by 8.5% and the thermal resistance reduce by 23.8% for square pin fins.

Cooling of electronics components has been studied widely by many researchers considering heat sink concept using analytical model, numerical simulation and novel experimental techniques. The enhancements of heat transfer in heat sinks has been the main objectives of all these analysis. The forced and free convection studies have been done by considering different shapes and geometries of fins. Also influence of many geometric parameters like fin length, fin height, spacing between two fins, etc. and heat transfer coefficient for thermal performance has been considered. A pin fin heat sink have been a very common method to enhance the heat transfer due to increase in surface area and coefficient of heat transfer. Also forced convection air cooling is key method to tackle power dissipation and heating issue in electronic industry. However, investigation of staggered pin fins with aerodynamic shapes has enough scope to predict optimum design criteria of microelectronic components with high-pressure drop and heat transfer. In present research, the authors intend to investigate effect on pin fin by means of different aerodynamic shape geometry considering rectangular pin fin, conic pin fin and rounded leading edge pin fin for staggered arrangement to predict the best possible pin fin shape for optimum performance in terms of thermal and hydraulic parameters.

2 Computational Method

2.1 Governing Equations

Mass, momentum and the energy conservation equations in an incompressible flow system are considered for the current investigation.

- Equation of conservation of mass (continuity equation):

$$\frac{\partial u}{\partial x} + \frac{\partial v}{\partial y} = 0 \tag{1}$$

where u is the velocity component along x-direction whereas v is velocity component along y-direction.

- Equations of momentum conservation in steady form:

$$u\frac{\partial u}{\partial x} + v\frac{\partial v}{\partial y} = -\frac{1}{\rho}\frac{\partial p}{\partial x} + \frac{\mu}{\rho}\left(\frac{\partial^2 u}{\partial x^2} + \frac{\partial^2 u}{\partial y^2}\right) + F_b + F_s \tag{2}$$

$$u\frac{\partial u}{\partial x} + v\frac{\partial v}{\partial y} = -\frac{1}{\rho}\frac{\partial p}{\partial y} + \frac{\mu}{\rho}\left(\frac{\partial^2 v}{\partial x^2} + \frac{\partial^2 v}{\partial y^2}\right) + F_b + F_s \tag{3}$$

F_b is body force and F_s is surface force.

- Energy equation

$$C_p\left(u\frac{\partial T}{\partial x} + v\frac{\partial T}{\partial y}\right) = \frac{K}{\rho}\left(\frac{\partial^2 T}{\partial x^2} + \frac{\partial^2 T}{\partial y^2}\right) \tag{4}$$

2.2 Problem Definition

Three different arrangements of pin fin with staggered configuration were simulated. Figure 1 shows the staggered rectangular pin fin configuration. 2-D heat transfer and the fluid flow characteristic in rectangular microchannel heat sink have been studied considering air as the cooling fluid. The length and width of the rectangular channel are 154.3 mm and 4 mm, respectively. An even temperature of 423 K is maintained at heat sink wall. The occurrence of thermal dissipation in the micro channel would have been due to natural convection of air. Figure 2 shows the staggered pin fin arrangement with rounded leading edge and Fig. 3 shows the staggered pin fin arrangement with conic shape.

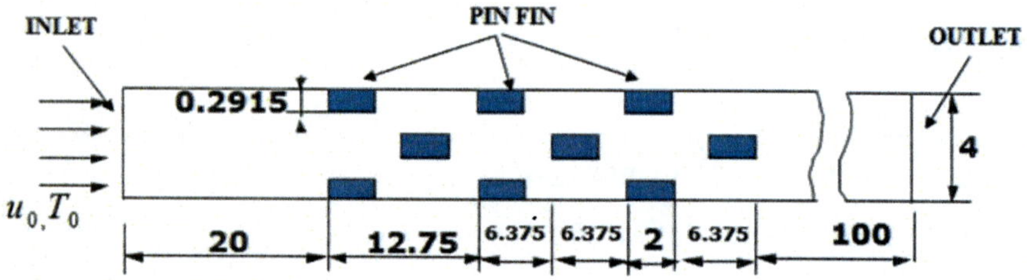

Fig. 1 Schematic diagram of staggered rectangular pin fin heat sink using micro channel

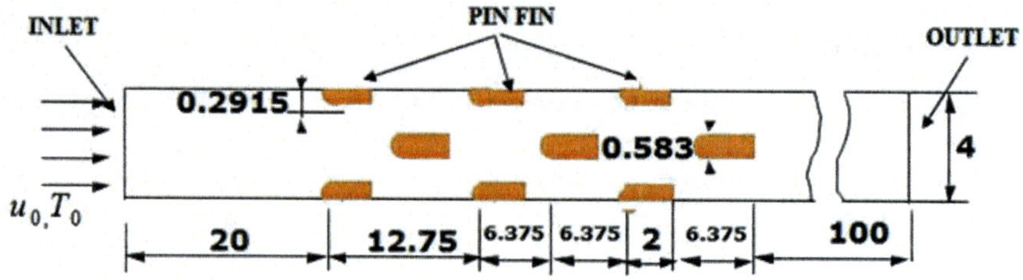

Fig. 2 Schematic diagram of rounded leading edge pin fin heat sink using micro channel

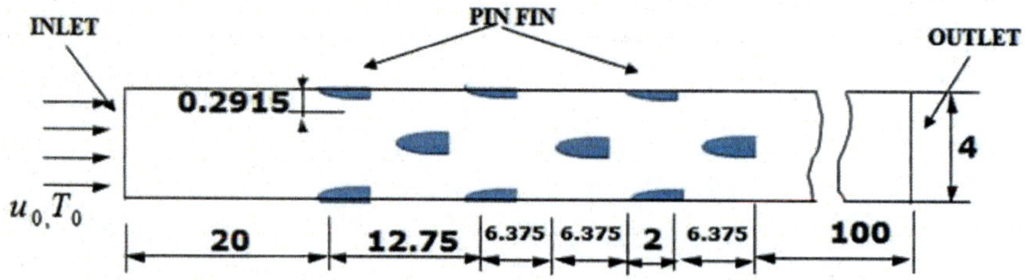

Fig. 3 Schematic of conic pin fin heat sink using micro channel

The thickness and fin's length have been chosen to be 0.583 mm and 34.3 mm, respectively from the commercially existing standard dimensions of heat sinks. The boundary conditions don't influence the flow field near the fin due to large computational domain. The initial boundary conditions of uniform velocity and temperature of 5 m/s, $T_0 = 300$ K are applied. At the fin walls no-slip condition and at heat sink walls a constant, uniform temperature, $T_w = 423$ K is considered. At outlet boundary, outflow is applied. 10e-5 is applied as the convergence criteria for present numerical simulation. Special care has been adopted in selecting a very fine triangular mesh to ensure insensitiveness of estimated flow parameters like pressure drop and heat transfer rate to more mesh-refinement.

3 Results and Discussion

ANSYS Fluent has been used to study the effect on aerodynamically shaped pin fins in a heat sink device. The enhancement of pressure drop and heat transfer plays an important role in escalating aerodynamic efficiency. In current research work, numerical investigation of the heat sink has been carried out to optimize the fin geometries. A 2-D computational domain is used for simulating the effect of hydraulic and thermal parameters. 2nd order upwinding scheme is considered for discretization of derivative terms in momentum and the energy equations along with SIMPLE algorithm being used for the pressure–velocity coupling.

It is observed that pressure drop and the heat transfer values don't vary significantly with grid refinement. So, for the remaining geometries, calculation of pressure drop and the heat transfer, the fine non-uniform mesh was used. The results of QF and P are greater for staggered configurations than for standard configurations. Alike results have also been expressed by researchers like Soodphakdee et al. [5], Leon et al. [10], Sathyamurthy et al. [11].

Figures 4, 5, and 6 represent that the pressure and temperature contours for different pin fin surfaces arranged in staggered manner having an inlet velocity of 5.0 m/s. In staggered rectangular pin fin, maximum pressure is achieved at the front edge of fin. However, in case of rounded front edge pin fin and conic pin fin, maximum pressure it not at front edge due to aerodynamic shape provided at the front edge. Due to this reason, the pressure drop is reduced in case of rounded leading edge pin fin and conic pin fin. Thus, it has been considered as a positive aspect for pumping power.

From temperature contour, it is seen that the temperature around pin fin is also reduced in case of rounded pin fin and conic pin fin. The QF, P, pressure drop and surface heat transfer obtained in present numerical model are presented in Table 1.

$$\mathrm{QF} = \frac{Q}{\left(\Delta P + D_p\right) \times D \times u_0} \tag{5}$$

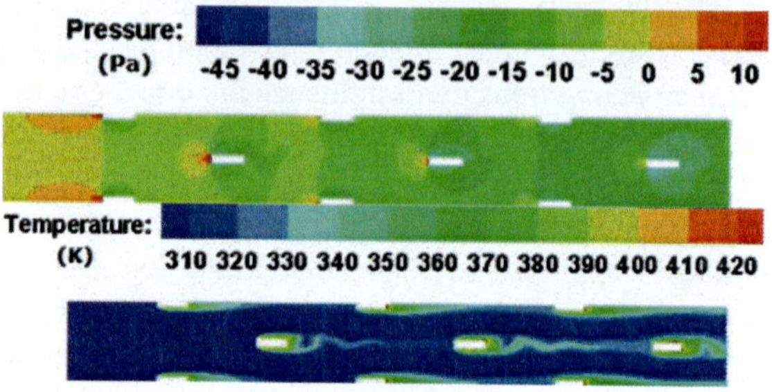

Fig. 4 Temperature and pressure contours for staggered rectangular pin fin

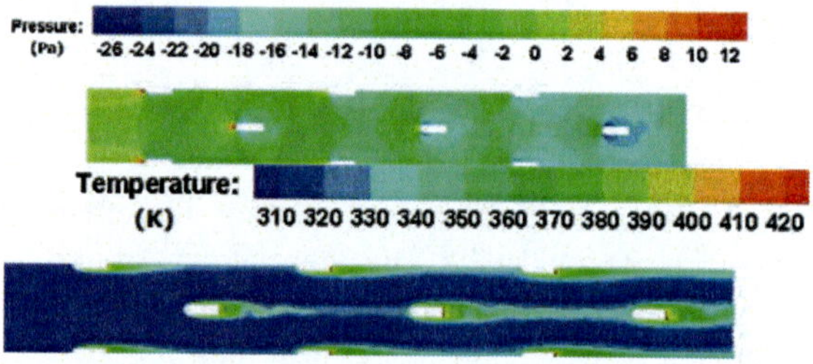

Fig. 5 Temperature and pressure contours for staggered rounded leading edge pin fin

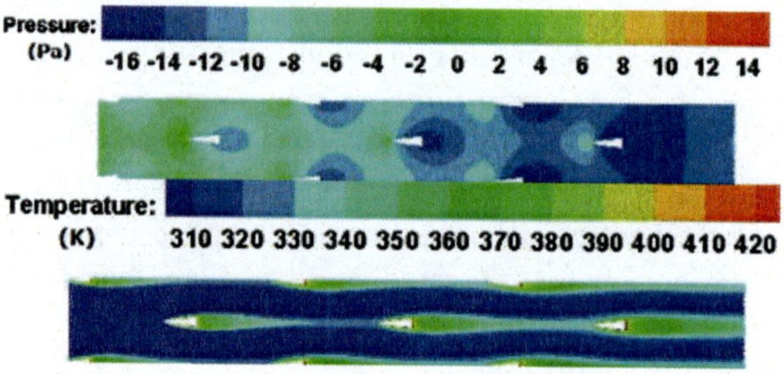

Fig. 6 Temperature and pressure contours for staggered conic pin fin

Table 1 Analytical expressions obtained for pin fin heat sink models

Pin fin model	u_0 (m/s)	Q (W)	Δp (Pa)	P (W)	QF
Staggered rectangular pin fin	5.0	550.41	10.43	0.6429	767.30
Staggered rounded leading edge pin fin	5.0	569.16	9.27	0.6139	926.97
Staggered conic pin fin	5.0	567.27	8.59	0.5970	950.18

$$P = \left(\frac{1}{\rho}\right) \times \dot{m} \times \left(\Delta P + D_p\right) \tag{6}$$

where Q, ΔP, D_p, ρ, D and u_o are heat transfer rate, pressure drop, P dynamic pressure at inlet, air density, width at the inlet of models and inlet velocity.

Table 1 shows different hydraulic and thermal parameters obtained from numerical simulation and analytical equations (5 and 6). The simulation is done with same incoming velocity 5 m/s and Table 2 summarizes the percentage variation of these parameters. The numerical simulation shows that compared to staggered rectangular the value of quality factor is increased by 20.8% for staggered rounded leading edge pin fin and by 23.83% for staggered conic pin fin. Also value of heat transfer in increased by 3.4% and 3.06% for the staggered rounded leading edge pin fin and the staggered conic pin fin, respectively compared to staggered rectangular pin fin. The simulations are also done for velocity range from 2.0 m/s to 8.0 m/s.

Figure 7 shows the varying pumping power and the quality factor (QF) for the rounded leading edge pin fin and conic pin fin at different flow inlet velocities. The staggered rounded leading edges pin fin has better QF with low P and the pressure drop for the inlet flow velocity higher than 4 m/s. Also for velocity ranges from 2 to 4 m/s, the staggered conic fin performs superior in terms of thermal as well as hydraulic parameters.

Table 2 Variation of functional parameters for heat sinks

Model	u_0 (m/s)	Difference in Q (%)	Difference in Δp (%)	Difference in P (%)	Difference in QF (%)
Staggered rectangular pin fin	5.0	–	–	–	–
Staggered rounded leading edge pin fin	5.0	+3.4	−11.12	−4.59	+20.80
Staggered conic pin fin	5.0	+3.06	−17.64	−7.13	+23.83

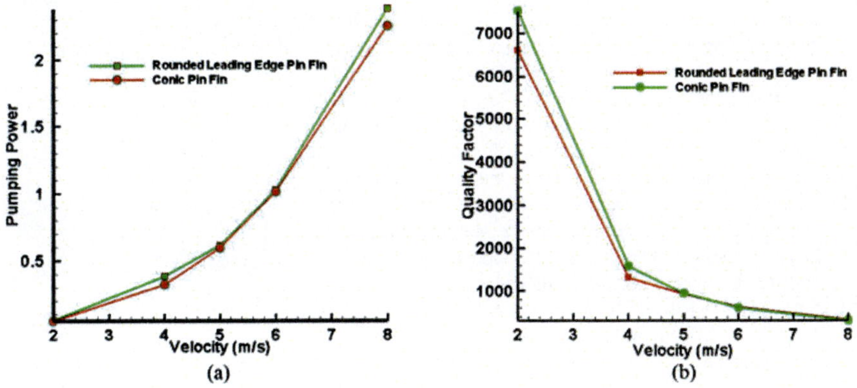

Fig. 7 P and QF for rounded leading edge pin fin and conic pin fin at various flow inlet velocities

4 Conclusions

The investigation of heat transfer and pumping power for different pin fin shapes with staggered configuration has been simulated using finite volume method. These simulations was performed for different values of flow velocities. Three different shapes of pin fin, namely, rectangular, rounded leading edge and conic for staggered configurations are considered. The main objective of authors was to find optimum shapes of pin fin in staggered configuration in terms of P and QF. From this research, it is concluded that staggered rounded leading edge pin fin has better QF with low P and the pressure drop for the inlet flow velocity higher than 4 m/s. However, for velocity ranges from 2 to 4 m/s, the staggered conic fin performs better in terms of thermal as well as hydraulic parameters. Hence, author suggests two best shapes of pin fins in staggered arrangement for heat sinks to reduce the problem of heat dissipation inside microelectronic components in form of rounded leading edge pin fins and conic pin fins.

References

1. Sparrow E, Ramsey J, Altemani C (1980) Experiments on in-line pin fin arrays and performance comparisons with staggered arrays. J Heat Transfer 102:44–50
2. Jonsson H, Moshfegh B (2001) Modeling of the thermal and hydraulic performance of plate fin, strip fin, and pin fin heat sinks—Influence of flow bypass. IEEE Trans: Components Packaging Tech 24:142–149
3. Ryu J, Choi D, Kim S (2002) Numerical Optimization of the thermal performance of a microchannel heat sink. Int J Heat Mass Transf 45:2823–2827
4. Jeng T, Tzeng S (2007) Pressure drop and heat transfer of square pin-fin arrays in in-line and staggered arrangements. Int J Heat Mass Transf 50:2364–2375
5. Soodphakdee D, Behnia M, Copeland D (2001) A comparison of fin geometries for heatsinks in Laminar forced convection: Part I—Round, elliptical, and plate fins in staggered and in-line configurations. Int Microelectronics Packaging Soc 24:1063–1072
6. Yang Y, Peng H (2009) Investigation of planted pin fins for heat transfer enhancement in plate fin heat sink. Microelectron Reliab 49:163–169
7. Wang F, Zhang J, Wang S (2012) Investigation on flow and heat transfer characteristics in rectangular channel with drop-shaped pin fins. J Propulsion Power Res 1:64–70
8. Abdoli A, Jimenez G, Dulikravich G (2015) Thermo-fluid analysis of micro pin-fin array cooling configurations for high heat fluxes with a hot spot. Int J Therm Sci 90:290–297
9. Sajedi R, Osanloo B, Talati F, Taghilou M (2016) Splitter plate application on the circular and square pin fin heat sinks. Microelectronic Reliability 62:91–101
10. Leon O, De Mey G, Dick E, Vierendeels J (2004) Staggered heat sinks with aerodynamic cooling fins. Microelectron Reliab 44:1181–1187
11. Sathyamurthy P, Runstadler PW, Lee S (1996) Numerical and experimental evaluation of planar and staggered heat sinks. In: Proceedings of the Fifth Intersociety Conference on Thermal Phenomena in Electronic Packaging, 132–139

Design Analysis of a Blended Wing Body

Pushpal Das, Prince Roy Sharma, Rama Krishna Parida, Ayush Pradhan, Aayush Nanda, Pooja Chaubdar, and Atal Bihari Harichandan

Nomenclature

BWB	Blended Wing Body
L/D	Lift-to-Drag Ratio
NACA	National Advisory Committee for Aeronautics
α	Angle of attack

1 Introduction

A blended wing body, also called as blended body, is a fixed-wing airplane having zero unmistakable splitting line between the wings & fundamental body of the airplane. With the everyday increasing environmental reduction in fuel burn is required as well as noise emissions for the upcoming commercial air transport, at present the current operating conventional aircraft are unable to fulfil these requirements. This is the only reason behind *Blended wing body* (BWB) concept is now widely embraced across the aviation industry for its numerous benefits. The industries like Boeing, Airbus are currently working under this project. The basic goal is to make a fuel-efficient aircraft.

Analysis of the blended wing using the CFD software ANSYS was done by Tayyab Khan [1] and higher value of L/D ratio was obtained with the blended wing design. BWB design approach was designed by Peifenga LI et al. [2] with the optimization

P. Das · P. R. Sharma · R. K. Parida · A. Pradhan · A. Nanda · P. Chaubdar · A. B. Harichandan (✉)
Aerospace Engineering, School of Mechanical Engineering, KIIT Deemed to be University, Bhubaneswar 751024, India
e-mail: atal.harichandanfme@kiit.ac.in

© The Author(s), under exclusive license to Springer Nature Singapore Pte Ltd. 2023

S. Revankar et al. (eds.), *Recent Advances in Thermofluids and Manufacturing Engineering*, Lecture Notes in Mechanical Engineering,
https://doi.org/10.1007/978-981-19-4388-1_16

and inverse design approaches and methodologies being combined, according to design requirements, employing low and high fidelity aerodynamics study tools. At cruising, the BWB layout was determined to have a greater lift-to-drag ratio and pitch trim, as well as fulfil positive zero lift pitching moment and static stability design criteria. Furthermore, interior arrangement is simple to accomplish and fuel consumption is greatly lowered. A novel, robust, flexible conceptual design software was used for optimization and exploration of the design space of the BWB by Paulinus Peter [3]. The design space was then explored through a series of sensitivity tests to determine the influence of various design factors on the BWB aircraft's attributes. The results were highly consistent with the test aircraft and other research findings. Further, a review on BWB aircraft was done by Oliverio Velázquez Salazar et al. [4] in order to get an insight into the current state of art on the subject, advantages, obstacles and opportunities. A conceptual design of a BWB aircraft was proposed by Nishant [5] that can accommodate 160 passengers for a range of 9200 kms and a cruising speed of 0.77 Mach. A standard approach to aircraft design as well as a revolutionary strategy are used to develop an unorthodox configuration. Payam Dehpanaha and Amir Nejat [6] in the early stages of conceptual design, computer analysis was used to study a first scaled blended wing body airframe. In addition, based on the examination of the first aircraft, a modified airframe was built. Based on the findings of earlier ideas, a full-scaled high-capacity blended wing body design for a long-range mission was proposed.

Edwin Ordoukhanian et al. [7] conducted a review that looked at design difficulties that system architects and engineers faced during the design process, BWB was developed. The study also discusses body material, body form, selection techniques, design simulation and optimization and the benefits of the final design. The BWB aircraft structural layout design problem is solved using a two-loop optimization technique that takes into account displacement, stress, strain and buckling limitations. To improve computation efficiency, a similar finite element model is used to the aircraft structural analysis and an analytical technique is used for buckling and static analysis of the stiffened panels. [8]. A BWB aircraft baseline was designed by Roberto Merino-Martinez [9] to evaluate its aerodynamics behaviour in terms, with a focus on its difficult stability and control. The CEASIOM application for conventional aircraft design was enhanced by using DLR's CPACS as a streamlined and more extensible software framework for this project. The literature survey on the research work related to Blended Wing Body ensures enhanced aerodynamic lift and substantially higher lift-to-drag ratio that leads to greater range and fuel economy. The main purpose of the present study is to have thorough understanding of airflow patterns throughout the plane's body and pinpoint critical locations based on velocity and pressure distributions that contributed to aerodynamic forces followed by design improvement and obtaining the best L/D ratio possible.

2 Design Methodology of BWB

2.1 *Final Mixed Wing Body Geometry Design*

In this research, the design methodology from various journal articles was followed for validation of the design. Our methodology describes multiple airfoil sections at the location of the span as a percentage of the total span, using the loft, loft function together in Solidworks 2019. A similar method is again applied when defining the airfoils to the positions under the agreement. For simplicity and to do this more efficiently, the positions per chord have been defined as a percentage of the middle chord.

2.2 *Design Geometry of Centre Body of BWB*

A symmetrical NACA 0012 profile was selected as it is very much helpful in designing for the subsonic speeds, additionally, it consists of a high stall angle α. the stall process will not be attempted and allow vigorous testing to take place over an interval of α or as long as possible and also because the pilot is able to perform high α stall manoeuvres in emergency. If we choose a convex aerofoil like the NACA series, a limitation of the angle of attack between $\alpha = 5^\circ$ $\alpha = 10^\circ$, this leaves a minimum tendency for pilot errors and emergencies, increasing the danger of falls catastrophic in rotation in case BWB stand. Figure 1 shows that the centre body is formed by three aerofoil sections located at 0%, 15% and 30% span, the data is the nose wing profile of the centre body (0% span wise and 0% chord wise).

2.3 *Outer Wing*

The outer wing is also generated using NACA 0012 aerofoil section for designing simplicity. The NACA 0012 aerofoil essentially permits a scale 10 model of sufficient thickness to include internal pressure taps for testing. The outer wing is made up of three aerofoils, the first with a span of 45%, the second with a span of 65% and the third with a span of 100%. as shown in Fig. 2 (Fig. 3).

2.4 *Numerical Modelling*

The conservation of mass, momentum and energy are considered as the fluid transport equations for the present research problem. The Navier–Stokes equations have been considered as the governing equations. The Spalart–Allmaras turbulence model

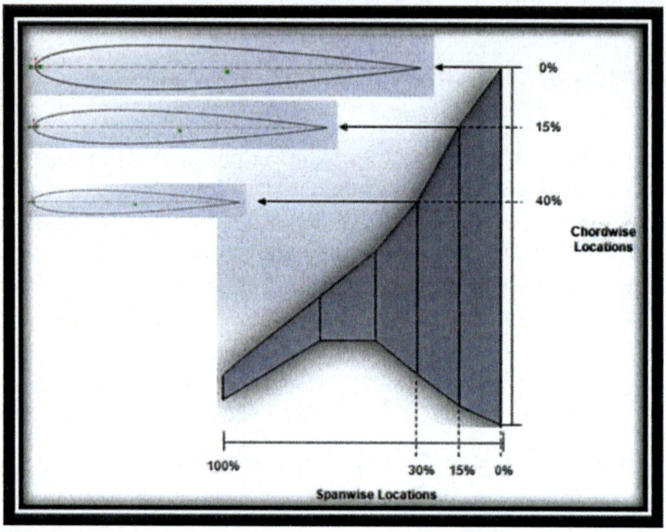

Fig. 1 Centre body

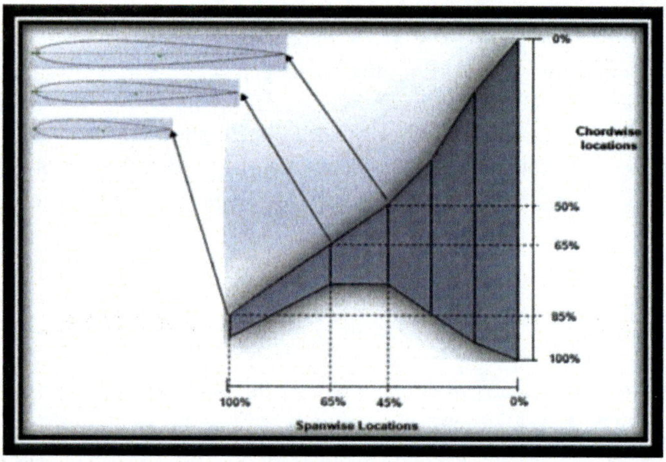

Fig. 2 Outboard wing

being used to capture the fluctuations in flow properties that solve for the kinematic eddy viscosity. Cell-centre approach has been used to solve for the variables of the present problem in the given computational mesh. In total, six variables are solved: 3 components of velocity, pressure, temperature and the kinematic eddy viscosity. The pressure far-field condition is used as the inlet boundary condition. No-slip condition is used on the body surface whereas symmetry boundary condition has been implemented for the side walls of the rectangular domain. The numerical simulation was

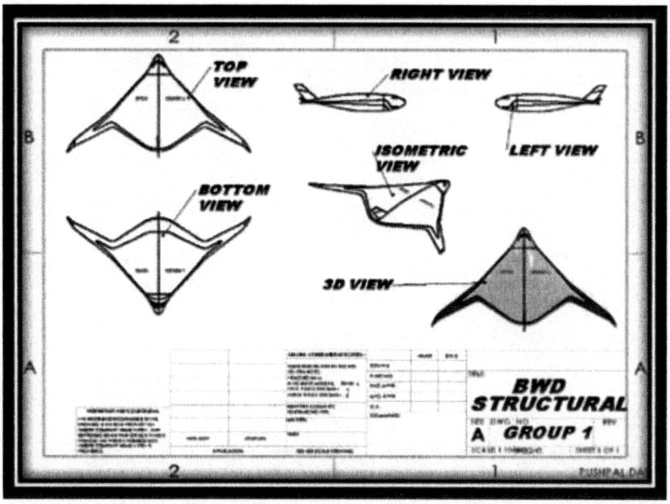

Fig. 3 Complete configuration of BWB aircraft

Table 1 Inlet reference values

Parameter	Values
Material type	Fluid
Density (kg/m^3)	1.225
Specific heat (C_P)	1006.43
Thermal conductivity	0.0242
Viscosity (kg/m-s)	1.81e-05
Area (m^2)	0.001093
Temperature (K)	288.16
Velocity (m/s)	99.999997
Ratio of specific heat	1.4

performed using CFD through ANSYS Fluent 16.0. The inlet parameters and solution methods are given for the simulation are shown in Tables 1 and 2, respectively. Mesh for the BWB geometry is shown in Figs. 4 and 5.

3 Results and Discussion

The results of the simulation were analyzed using Fluent 16.0. Figures 6, 7 and 8 show the nature of velocity streamlines, velocity contour and velocity vectors, respectively for the complete BWB aircraft for the case of subsonic flow. The velocity vectors

Table 2 Solution methods

Solution type	Method
Numerical scheme	Density based (coupled)
Gradient	Least square cell based
Pressure	Second order
Momentum	Second order
Modified turbulent viscosity	Second order
Energy	Second order
Number of iterations	500
Number of nodes	326,831
Number of elements	1,835,294

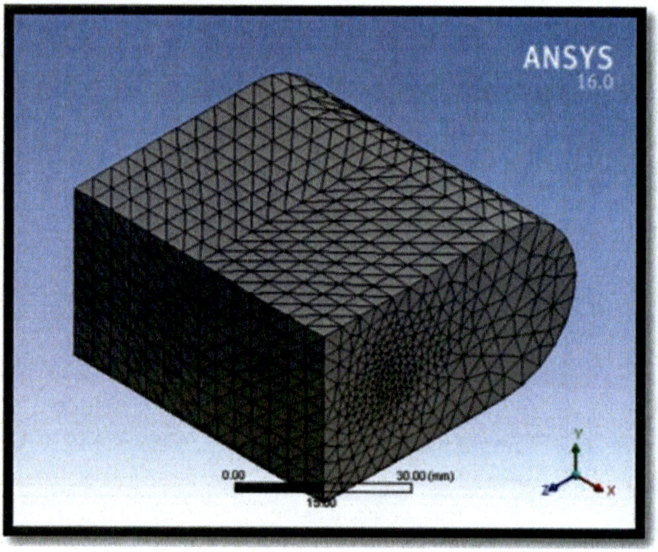

Fig. 4 Face meshing of flow domain at $\alpha = 5^{\circ}$

show how flow behaves with the parts of BWB as vector density is high on the corners of BWB, i.e. nose part, leading and trailing edge of wing and tail part.

Figure 9 represents the behaviour of pressure over whole BWB surface and the stagnation point at the nose has been noticed as the pressure attains maximum value (6.01 Pa) and velocity is minimum at the nose. It has also been observed that the pressure act higher on the sharp corners of wing and is uniformly distributed over the surface of wing which is helpful for generating more lift throughout the flight time.

The aerodynamic forces acting on the BWB surface are estimated for the present problem. The variation of lift coefficient and drag coefficient at different values of angle of attack are represented in Figs. 10 and 11, respectively. The lift coefficient

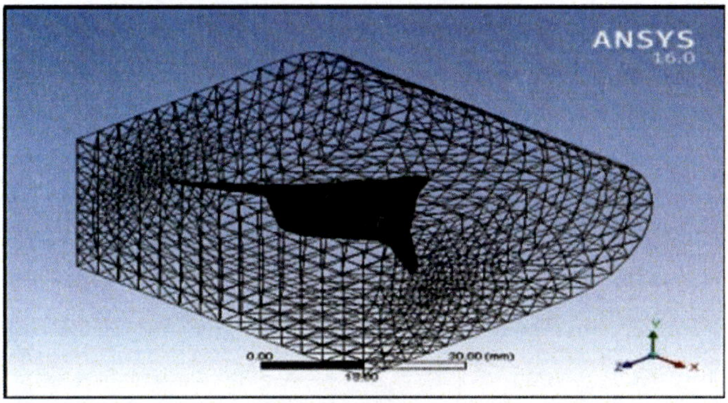

Fig. 5 Wireframe meshing at $\alpha = 5^{\circ}$

Fig. 6 Velocity streamlines

values are observed to increase with angle of attack whereas the drag coefficient reduces with angle of attack. Thus, very high value of lift-to-drag ratio for flow past BWB surface is observed. The numerical values of lift and drag coefficients are presented in Table 3.

The interior design is quiet and simple to use, whereas on the other side, it offers a significant reduction in fuel consumption. This work eventually concentrated on the aircraft's aerodynamic properties at cruising speed, which will most likely affect the BWB aircraft's take-off and landing performance and propulsion-related issues are also interesting hence will be addressed in future study.

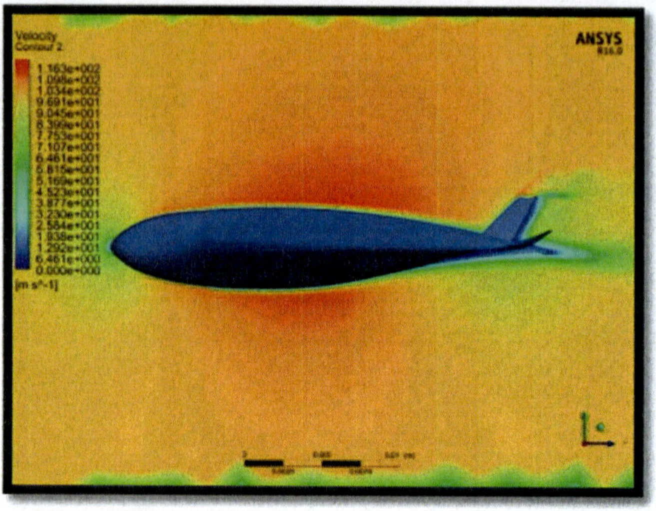

Fig. 7 Velocity contour

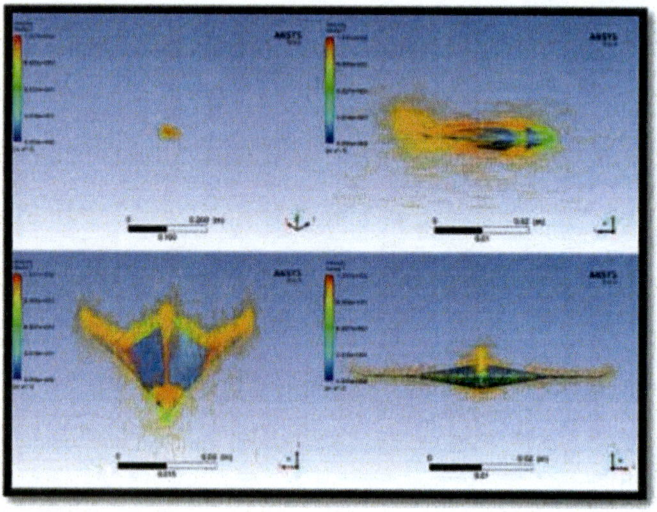

Fig. 8 Velocity vectors

4 Conclusion

In summary, the basic objective to design an eco-friendly aircraft that can produce higher lift, the *Blended Wing Body (BWB)* aircraft idea has really been encouraged to be developed because it's aerodynamically efficient and can transport a large number

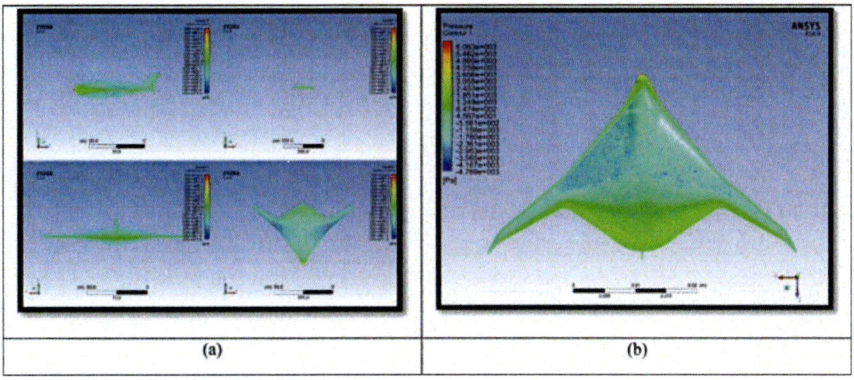

Fig. 9 Pressure distribution over BWB surface

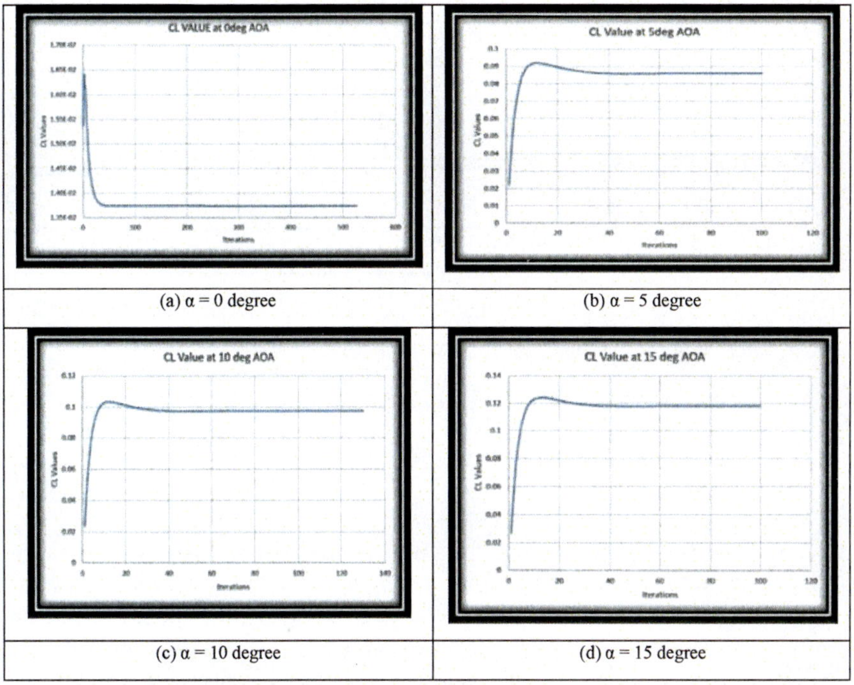

Fig. 10 Variation of lift coefficient with various angle of attacks

of people over great distances for a cheap direct operating cost. The benefits of having BWB aircraft are mostly arise due to distributed structural and aerodynamic loads, which results in better aerodynamic performance as well as provides lighter structural weight.

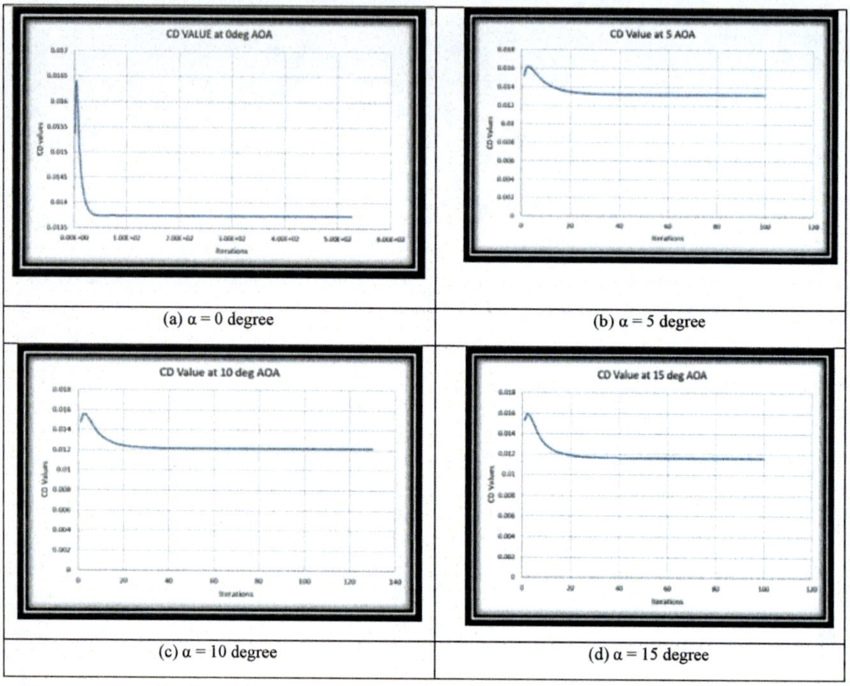

(a) α = 0 degree

(b) α = 5 degree

(c) α = 10 degree

(d) α = 15 degree

Fig. 11 Variation of drag coefficient with various angle of attacks

Table 3 Variation of aerodynamic forces on BWB surface

α	0°	5°	10°	15°
C_l	1.64e-02	0.090749	0.108511	0.123464
C_d	0.0165	0.016	0.0155	0.015

The present research work aims to build an aerodynamic design technique for BWB configuration aircraft with the goal of designing BWB passenger configurations utilizing this methodology. In the present research work; cruise of points, maximum lift-to-drag ratio and pitch attitude was originally articulated as design concepts. The BWB design methodology is established in accordance with the design requirements by combining the optimization and reverse hods of the design meet plan methodology, as well as the use of low and high aerodynamic analysis tools. The BWB design layout delivers a high lift-to-drag ratio and pitch attitude under cruising conditions as compared to the original BWB arrangement, as well as meeting positive zero lift pitch moment design criteria.

References

1. Khan T (2019) Design and CFD analysis of a Blended Wing UAV (a conceptual design). J Aerospace Engineering Mech 3(1):156–160
2. Peifeng L, Zhang B, Chen Y, Yuan C, Lin Y (2016) Aerodynamic design methodology for Blended Wing Body transport. Chinese J Aeronautics 25(4):508–516, August
3. Okonkwo PPC (2016) Conceptual design methodology for Blended Wing Body aircraft. Cranfield University, Department of Aerospace Engineering
4. Salazar OV, Weiss J, Morency F (2015) Development of blended wing body aircraft design. In: CASI 62nd Aeronautics Conference and AGM At: Montréal, January
5. Patel N (2018) Design of medium size blended wing body subsonic transport aircraft. San José State University, May
6. Dehpanaha P, Nejat A (2015) The aerodynamic design evaluation of a blended-wing-body configuration. Aerosp Sci Technol 43:96–110
7. Ordoukhanian E, Madni AM (2014) Blended wing body architecting and design: Current status and future prospects. Procedia Computer Sci 28:619–625
8. Zhu W, Yu X, Wang Y (2019) Layout optimization for Blended Wing Body aircraft structure. Int J Aeronautical Space Sci
9. Merino-Martinez R (2014) Design and analysis of the control and stability of a Blended Wing Body aircraft. Delft University of Technology, June

Development of Rayleigh Taylor Instability at Various Atwood Numbers—A Review

Ayush Boral, Souvik Dutta, Ankit Kumar, Pooja Chaubdar, and A. B. Harichandan

Nomenclature

λ	Spatial period
h	Amplitude
A_t	Atwood number
VD	Variable density
RT	Rayleigh-Taylor
RM	Richtmyer–Meshkov
NSA	Nonlinear saturation amplitudes
PAWCM	Parallel adaptive wavelet collocation method
WENO	Weighted essentially non-oscillatory
Rep	Perturbation Reynolds number

1 Introduction

Variable density (VD) flows are classified as flows in which the microscopic densities of the fluids fluctuate when mixed, as opposed to Boussinesq approximation, which tells us the densities of the fluids do not fluctuate and hence the density changes are ignored. Even though the fluids in the flow are incompressible, the velocity field in VD flow is no longer solenoidal. VD mixing can be observed in atmospheric and

A. Boral · S. Dutta · P. Chaubdar · A. B. Harichandan (✉)
Aerospace Engineering, School of Mechanical Engineering, KIIT Deemed to be University, Bhubaneswar 751024, India
e-mail: atal.harichandanfme@kiit.ac.in

A. Kumar
Department of Aeronautical Engineering, National Formosa University, Huwei, Taiwan

© The Author(s), under exclusive license to Springer Nature Singapore Pte Ltd. 2023

S. Revankar et al. (eds.), *Recent Advances in Thermofluids and Manufacturing Engineering*, Lecture Notes in Mechanical Engineering,
https://doi.org/10.1007/978-981-19-4388-1_17

oceanic flows, as well as combustion, and chemical engineering flows. In nature flows are observed in fluids with a finite difference in density caused by some kind of acceleration which results in a differential acceleration in the fluids. Gravitational force is one such force in astrophysical and geophysical flows. A fluid instability occurs when the acceleration is constant, turbulence emerges as a result of minor perturbations in the initial contact surface between the two fluids which grows with time and interacts nonlinearly.

The importance of the Rayleigh Taylor (RT) instability can be observed in a variety of applications such as the interaction of fluid with different densities, atmospheres, oceans and even in supernovae explosions happening millions and billions of light-years away from earth. RT instability develops when a disturbance with a wavelength longer than the cutoff occurs owing to surface tension (for immiscible situations) or mass diffusion (for miscible ones). Smaller and larger eddies form as the disturbance intensifies, and the flow eventually becomes turbulent due to nonlinear interactions. Group theory has also helped in the advancement of the numerical simulations and experiments [1]. It has shown that at later times, the dynamics of RT and RM flow shows certain universal features as bubbles in both cases conserves isotropy in the plane normal to direction of the acceleration. This theory obeys principles of conservation and is applicable for a broader range of nonlinear problems related to Rayleigh Taylor instability. Not a lot of reviews have been performed for numerical and experimental investigations. The review also discusses the effect of shock wave interactions with the Rayleigh Taylor instability. The review could help the researchers to get an idea of the type of research that have been attempted and the authors also have tried to discuss the possible future research prospects.

1.1 Evolution of Rayleigh Taylor Instability

The dynamic study of two fluids of varying densities has been a topic of study for many researchers [2–6] theoretically and experimentally. When two fluids of varying densities and negligible viscosities are superposed and are accelerated in perpendicular direction to interface, the stability of the surface depends upon the fluid from which the acceleration has been directed. This phenomenon has been studied by researchers [4] and has been verified experimentally by researchers [5] which shows that the stability of the surface for small deviations depends upon the acceleration whether it has been directed initially from fluid having higher density to the fluid having lower density or vice versa. If the upper fluid in a column is lighter than the lower, the frequency of the harmonic oscillations increases with diminishing wave-length indicating stability [2]. When the upper fluid is heavier, there is unstable equilibrium which increases when the wavelength is smaller. In the nonlinear regime, coherent shape of spikes and bubbles appears as the fluid with higher density penetrates the fluid with lower density forming a mushroom shaped spike. The dynamics is governed by two length scales, spatial period (λ) in normal plane and amplitude (h) in direction of gravity. Table 1 has a brief summary of some of

Table 1 List of R–T experimental investigations arranged in groups of, fluid combination and techniques of diagnosis

S. no	Refs	Fluids	At	Diagnosing Technique
1	[5]	Air/Benzene, Air-Glycerin, Air–Water	0.99	High speed shadow photography
2	[7]	Carbon dioxide-SF_6, Air-SF_6, Helium-Carbon dioxide, Helium-SF_6	0.4–0.94	Imaging, PLIF
3	[8]	Carbon tetra chloride-Air, Methanol-Air	0.107–0.997	Imaging
4	[9]	Water-Pentane, Sodium iodide-Pentane, Ethyl Alcohol-Air	0.231–0.997	Imaging
5	[10]	Brine-Water	0.048	Imaging
6	[11]	Air–Water	0.99	Imaging
7	[12–14]	Brine-Water, Brine-Water + Propan-2-ol	10^{-4}–0.05	Imaging, LIF
8	[15]	Magnetic Fluids-Air or Water	0.029–1	Imaging

the experimental researches performed for investigating Rayleigh Taylor instability. Atwood number, abbreviated as A_t, a dimensionless number, which describes density diff. between two various adjacent fluids with common interface. It is important parameter in study of RTI and Richtmyer–Meshkov instability. In RTI, the penetration distance of the heavier fluid bubbles into the lighter fluid is function of acceleration time scale.

2 Numerical Investigations on Rayleigh Taylor Instability

Over the last century, lot of research has been done for understanding the initial stages of the evolution of Rayleigh Taylor instability. Rayleigh [2] first studied the stability of an inviscid incompressible fluid layer with pressure gradient pointed against density gradient. Researchers proposed an empirical model of the saturation process and the process was thought to occur due to interaction of higher order modes and nonlinear enhancement [16]. He reported that the exponential growth of amplitude of perturbation in linear regime of the Rayleigh Taylor Instability is superseded by power-law time dependency in weak nonlinear regime.

The growth of the interface of two fluids of varying densities has been studied and is found to be asymmetrical [17]. The fluids have been considered to be incompressible, non-viscous but the effect of surface tension has been taken into consideration. An approximation of 3rd order for sinusoidal waveform initial disturbance has an amplitude of 10% of wavelength. Three values of the growth of the disturbance has been considered for three consecutive times for an initial sinusoidal disturbance. The development of the interface is asymmetrical which has been shown by spike (crest is narrowed) and bubble formation (trough is broadened) of the wave.

The asymmetric growth of the surface disturbance is much earlier for the initial disturbances of higher amplitudes than the smaller amplitudes. The analysis further determined that the systems consisting same air–liquid combination subjected to equal acceleration, those who have higher initial wave number tends more to be stable. A new numerical technique is used for the solution of time dependent Navier–Stokes equation to simulate the Rayleigh Taylor instability which has not been much used before for calculating this problem [18]. The small variations in the growth of the instability has been studied which results in changes in properties of the fluid. The authors have examined the range of density ratios from 1.1 to 10.0 and Reynolds numbers. The flow has been considered in two cases, inviscid and viscous. The results of the simulation for viscous flow have been compared to [19] who predicted variation of linear growth rate with Reynolds number and the results are found to be in agreement. For inviscid, the rate of growth is smaller than the predicted by [4] and [19] because of drag and the authors have included a modification of first order. The correlation during the early stages of growth with the modified theory is good. The fluid resistance is found to play a major role in the nonlinear phase of the unstable growth.

The NSA (Nonlinear Saturation Amplitudes) of first two harmonics of the classical RTI for inviscid, incompressible and irrational fluids were explored analytically with discontinuous profile at various arbitrary values using the assumption of nonlinear corrections up to tenth order [20]. The NSA of the fundamental mode is completely exponential to linear growth amplitude of the fundamental mode at the time of saturation when growth of the fundamental mode (1st harmonic) is decreased by 10% when compared to its corresponding linear growth. The NSA of the second harmonic could be defined in similar way.

The NSA of the fundamental mode diminishes as the Atwood number is increased, whereas the effect of HOC drastically enhances the NSA for arbitrary Atwood numbers. The NSA of second harmonic rapidly decreases, attains a minimum and then subsequently increases as the Atwood number rises. Fundamental mode NSAs with fifth, seventh and ninth-order corrections are greater than those with third-order corrections. On the other hand, second harmonic NSAs with sixth, eighth, and tenth order corrections are greater than those with fourth-order corrections at $A_t \leq 0.3$, but smaller than those with fourth-order corrections for $A_t \leq 0.3$. Effects of density gradient, Atwood number, and compressibility on NSA in classical Rayleigh Taylor Instability have been explored in present work [21]. A 2D DNS has been introduced for exploring the NSA at continuing density profiles as the investigation effects of the density gradients through analytic methods.

The WN model including Atwood number show that $k\eta_s$ is dependent on the Atwood number. The authors have observed that the RTI is destabilized by the effects of compressibility. The current research in DNS, the effect of density gradient plays a more dominant role among Atwood number, compressibility and density gradient effects. The continuous gentle density profile with low equivalent (effective) Atwood numbers have been generated in DD or ID ablative Rayleigh Taylor experiments in ICF [22–28]. The stabilizing mechanism has two impacts on evolution of ablative RTI. It decreases the linear growth [26–31] on one hand and it increases its NSA on the other hand [30]. The increase of NSA in the ablative Rayleigh Taylor Instability is quite dangerous for ignition of ICF, as a lengthier spike could break implosion shell during acceleration and destroys central ignition hot spot during the deceleration in ICF. This research is beneficial for understanding enhancement of NSA in the ablative RTI experiments [31–35]. The influence of vertical magnetic field on equilibrium of inviscid compressible fluid of varying density has been investigated stratified in vertical keeping the velocity of sound variable [36]. Although, the magnetic field stabilizes the completely but, no finite amount of rotation could do the same. The viscous effects in Rayleigh Taylor effects have also been a topic of research as the viscosity is very decisive when studying the instability. An approximate solution has been derived for the instability of the interface of the immiscible fluids [37]. The authors have first considered non viscous fluid and for second case, viscosity has been considered. Air and water has been taken as the two fluids and the interface between them with a negligible dynamic effect has been considered. The authors have extended the treatment of interfacial waves to one fluid bounded by rigid wall or free surface. The patterns clearly explain the Rayleigh Taylor instability. A simple dispersion relation of the Rayleigh Taylor has been represented through the character of growth rate of normal modes for RTI of the viscous, incompressible superposed fluids [38]. The scaled growth rates are found to be insensitive to viscosity and density of the fluid. The physical growth within small percent primarily depends upon the density-weighted avg. viscosity, surface tension and effective acceleration. The authors have also tried to recast standard dispersion relation to a form in order to prove it more amenable to the analysis and for establishing the location of its physical roots in the complex plane.

Perturbation of single mode on the interface of two fluids superposed over one other of varying densities has been investigated multiple times and the linear growth has been investigated a lot of times. The authors have attempted to employ the standard velocity potential method for expanding the velocity potential perturbation to the third order of fundamental mode, 2nd and 3rd harmonics for arbitrary Atwood numbers [39]. Nonlinear saturation amplitudes at arbitrary Atwood numbers have been discussed earlier in [20] where analytical expression of amplitudes up to the harmonics have been demonstrated with corrections up to tenth order and in [21] which discusses similar method as in [39] for generating higher harmonics by expanding the perturbation velocity. The results show that the mode coupling coefficients and the ratio of perturbation wavelength and the nonlinear saturation amplitude depends upon the Atwood number. The WNL solution follows linear theory very closely and continues till a point of the maximum amplitude after that a NL solution decreases rapidly. A function of cosine mode perturbation at interface of two fluids in which heavier fluid overlaps lighter fluid is considered, the interface is developed with time according to linear theory. Growth of varying amplitude at the interface is because of appearance of the harmonics of higher order. For a higher Atwood number, spike travels faster than bubble where the heavier fluid is bubble and the lighter fluid is spike. The equations of the evolution at the interface have been solved in [40] and solutions which satisfy initial velocity. The research conditions has raised the linearized solution which has been neglected earlier i.e. the role related to the inhibiting terms resulting a large difference in numerical simulations and the third theory of weakly nonlinearity. A third order theory of weakly nonlinearity based upon single mode Rayleigh Taylor Instability is reported.

3 Experimental Investigations on Rayleigh Taylor Instability

A theoretical and experimental quantitative investigation of the instabilities on the surface in a linear pinched discharge has been performed and reported in literature [41]. The photographs show that the instabilities have been mostly found between first and the second ounce, while there is a rapid acceleration in the discharge column. The plasma boundary which shows an unstable boundary between the immiscible fluids of varying densities shown by the authors have a general similarity to the case reported in literature [5]. Evolution of turbulence of miscible RT mixing layer is investigated experimentally for the range of A_t. nos., 0.03–0.6 for long duration reported in literature [42]. Air and helium has been considered for the gas channel experiment for studying the development of RT instability. Air is placed at the top and helium-air mixture is at the bottom. Hot wire anemometry and digital image analysis have employed for diagnosing the experiment. The gases meet at end of the plate which leads to generation of highly unstable surface and buoyancy-driven Rayleigh Taylor mixing layer. The maximum flow velocity available is 2 m/s and is controlled through dampers opening connected to blowers' suction port. Air and helium streams have been mixed uniformly prior to reaching inlet. Flow straighteners and screens are fitted at top and the bottom inlet sections for easing dissipation of free stream turbulence and also to decrease the development of boundary layer on walls and splitter plate. Density profile at Atwood numbers 0.04, 0.26, 0.47 and 0.6 has been reported by the authors across the mixing layer. Fluctuations in density and the root mean square velocity at the mixing layer for high Atwood numbers are found to be identical to behaviors at low Atwood numbers when normalized properly (Fig. 1).

New statistics is generated by conditional averaging method which separates the light fluid (bubble) and the heavier fluid (spike) dynamics. The conditional sampling showed the variations in vertical velocity fluctuations and the turbulent mass flux for spikes and bubbles. Exit splitter is used for the experiment for eliminating the gravity currents at $A_t > 0.1$. Density fluctuations and the values of root mean square velocity at centerline layer of mixing for high A_t case shows a higher similarity to the lower A_t when the behaviors are normalized properly.

The compression of the deuterium plasma through axial magnetic field generated by wide range single turn coil has been performed and reported in literature [43]. The magneto hydrodynamic RTI is generated through inwards acceleration of the

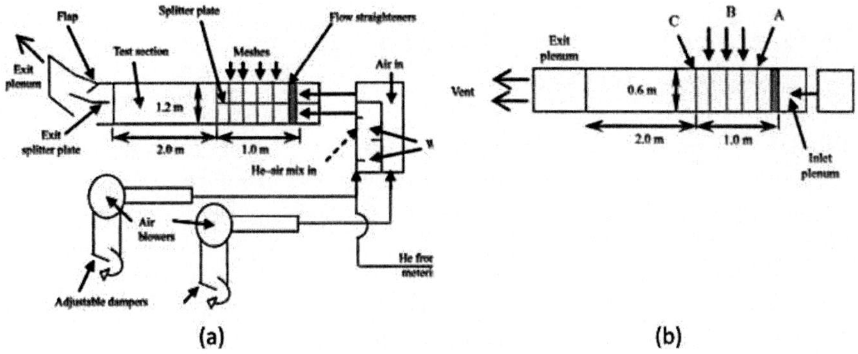

Fig. 1 Air–Helium gas channel ($0 < A_t < 0.75$) **a** Side view; **b** top view. [42]

interface which separates plasma and the magnetic field. High speed photography of instabilities have been captured which have been generated by rapidly rising the uniform axial magnetic field. The observed rate of growth and the wavelength are found to be in agreement with the simple theory. In this case, the growth of instabilities is limited by higher impulsive acceleration. At the densities and temperature in this experiment, resistivity is mainly responsible for the damping and wavelength of the "maximum instability" is calculated on this assumption which is in agreement with observed value. RTI in plasma has also been developed and investigated by researchers experimentally with two arrangements, with and without the barrier using magnetic field for projecting the plasma [44]. Flute-type instabilities of plasma have been developed. The growth of the instabilities happens for 1–4 μs after the deceleration of the surface. The instabilities develop only on forward edge of decelerated plasmoid when there is no barrier used. The wavelength of the instabilities and the maximum radius of main plasma surface are found to vary with the strength of magnetic field in the arrangement with barrier. The authors have verified that the theoretical predictions are in agreement with the Rayleigh Taylor instabilities existing in plasma. The Rayleigh Taylor Instability has to be stabilized dynamically to achieve a dynamic equilibrium at the fluid interface [45]. Oil as the lighter fluid and an aqueous solution of Potassium iodide has been considered as heavier liquid. Initially, the boundaries between the fluids are not horizontal thus, having no equilibrium. A harmonic oscillation has been enforced in the direction vertical to the system consisting two fluids creating a periodic acceleration which is perpendicular to horizontal boundary layer. The power required increases with increase in wavelength of perturbations for stability. Parametric resonances have been suppressed but can be observed using the oil having lower viscosity.

3.1 Investigations at Mach Numbers

A 2D single mode Rayleigh Taylor instability is numerically through PAWCM (Parallel Adaptive Wavelet Collocation Method) which maintained the symmetry in a single-mode Rayleigh Taylor Instability system to extremely later times [46]. The simulation is done at varying Mach and Atwood numbers for effects of variable density and wide range of flow compressibility. The results show that the flow compressibility reduces the rate of growth of RTI at lower Atwood numbers as predicted in linear stability analysis but as the Atwood numbers are increased, in intensity if the inhibition decreases. The authors could not make a conclusion statement for Atwood number = 0.3 and for higher difference in density, i.e., Atwood number > 0.3, where reverse in effect of compressibility on the growth of a single-mode RTI has been observed. The effect of compressibility was studied further for varying initial states at low Atwood number reported in literature [47] of 0.04 for the isothermal stratification at various Reynolds numbers and Mach numbers numerically [48]. A vorticity equation was derived and then the authors analyzed the compressible Rayleigh Taylor Instability through vortices dynamics.

The influence of compressibility on the 2D single-mode RTI reported in literature [49] was studied at five different Atwood numbers 0.1–0.9. The authors used a higher order central compact finite difference (FD) scheme to model the late time development of isothermal background stratification at varying isothermal Atwood and Mach nos. For varied stratification intensities, corresponding to changing isothermal Mach nos. between 0.1 and 1.0, simulations at all Atwood numbers mentioned have been provided. Except for case $A_t = 0.9$, the perturbation Re is Rep = 5000. The impact of compressibility on RTI varies depending on the Atwood number. This is because the compressibility prevents the formation of RTI at low A_t and the initial density stratification causes a stabilizing effect. As the Mach number increases, the Rayleigh Taylor instability development is reduced. For any Atwood number, the bubble velocities are comparative to those obtained from incompressible potential flow solutions at low Mach numbers. When the Mach numbers surpass A_t. 0.3, the bubble and spike velocities were suppressed.

However, the spike velocity has been modified but very slightly, and it reduces at a later stage of the reacceleration when A_t equals 0.5. As the Atwood number increases, the density stratification diminishes, and the density differential between the sides of the contact rises. When the A_t is high enough, impact of destabilization of expansion-compression motion takes over, and compressibility helps the Rayleigh Taylor instability to develop.

For $A_t = 0.9$, and $M = 0.1$, 0.3, 0.5, and 0.7, Fig. 2 illustrates the evolution of spike and bubble velocities. The dotted line plotted in Fig. 2 shows the potential flow model's asymptotic bubble velocity. There is very little effect of compressibility on the evolution of the spikes at A_t equals 0.9. The density stratification of lighter fluids in lower layer is invisible at varying Mach numbers. It has a minor stabilizing impact on the spike. Moreover, the spikes are least affected by the instability of expansion compression effect. When the upper fluid is considered to be narrow and a high-density column falls on a fluid of much less density, density stratification and the compressibility of lighter fluid has a minimal impact on its decent. When there is a significant difference in density of fluids, the A_t approach one and spike could be considered free-fall with a quadratic time-height correlation reported in literature [50] (Fig. 3).

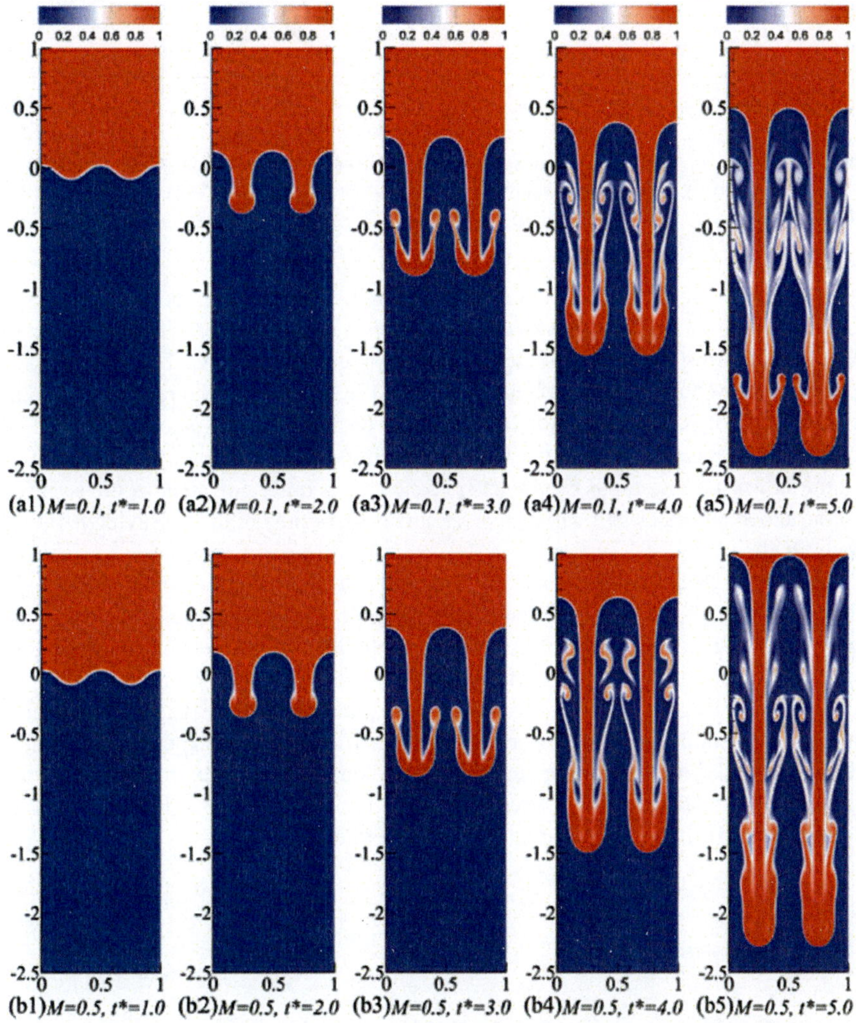

Fig. 2 In single-mode RTI, the impacts of Mach number on the concentration contours for $A_t = 0.9$. The heavy fluid is colored red ($c = 1.0$), whereas the lighter fluid is depicted by blue color ($c = 0.0$). **a** $M = 0.1$ and **b** $M = 0.5$, from top to bottom. The non-dimensional time is 1.0, 2.0, 3.0, 4.0, and 5.0, from left to right [49]

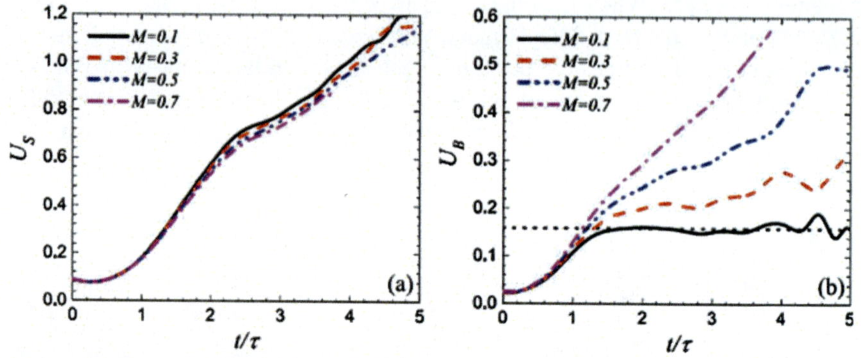

Fig. 3 For $A_t = 0.9$, the evolution of spike velocity (US) and bubble velocity (UB) are shown in **a** and **b** respectively. The asymptotic bubble velocity of potential flow model is represented by the dotted line [49]

The effect of the shock waves on generation of RTI has been investigated numerically by a 2D Navier–Stokes Equation using higher order WENO (Weighted Essentially Non-Oscillatory) finite difference shock scheme [51]. The authors have performed investigations on Mach number 6, 12, 18 and 24. The results show that the location of Rayleigh Taylor Interface moves downward as shock hits head of interface. The shocks interacting with the interface helps in speeding up transition to the full instability quite significantly [52]. This is quite prevalent as the strength of the shock increases and whether shock hits head or the tail of the Rayleigh Taylor Interface.

On the basis of results reported on linear and nonlinear theories on Rayleigh Taylor Instability, studies might be needed to explore more in turbulent mixing. The group theory has helped researchers to study the local properties of nonlinear RTI like classification of interaction, formation of pattern and asymptotic dynamics of bubble front. The initial conditions for the experiments have to be diagnosed properly and be well established for comparison with numerical models. Experimental studies on Rayleigh Taylor Instabilities involving coupled physics which might impact the growth of the instability could be carried out with more precision as RTI experiments are extremely difficult to run, diagnose and operate.

4 Conclusion

In the last few decades, considerable studied has been done on Rayleigh Taylor instability numerically and experimentally over wide range of Atwood numbers. At low Atwood number not a lot of studies have been done few decades earlier. Investigations have been done at low Atwood number (Atwood number < 0.1) numerically and experimentally. Progress in both technologies have enabled researchers to do lot of experiments. The results of the investigation of Rayleigh Taylor instability is used for exploring fundamental physics of the buoyancy-driven turbulence and model development. The theoretical investigations of the instability in linear and the nonlinear stages show that the unstable Rayleigh Taylor flow has length scale with fastest growth mode.

References

1. Abarzhi SI (2008) Review on nonlinear coherent dynamics of unstable fluid interface: conservation laws and group theory. Phys Scr T132:297681
2. Rayleigh (1882) Investigation of the character of the equilibrium of an incompressible heavy fluid of variable density. Proceedings of the London Mathematical Society, s1–14(1):170–177 https://doi.org/10.1112/plms/s1-14.1.170
3. Lamb H (1932) Hydrodynamics. Dover Publications, Inc., New York, Chap. IX
4. Taylor GI (1950) The instability of liquid surfaces when accelerated in a direction perpendicular to their planes. Royal Society of London A201(1065):19–196
5. Lewis DJ (1950) The instability of liquid surfaces when accelerated in a direction perpendicular to their planes. Royal Society of London A202(1068):81–96
6. Chang CT (1959) Dynamic Instability of Accelerated Fluids. Phy Fluids 2(6):656–663. https://doi.org/10.1063/1.1705969
7. Morgan RV, Likhachev OA, Jacobs JW (2016) Rarefaction-driven Rayleigh-Taylor instability—Part 1: Diffuse-interface linear stability measurements and theory. J Fluid Mech 791:34–60
8. Emmons HW, Chang CT, Watson BC (1960) Taylor instability of finite surface waves. J Fluid Mech 7(2):177–193
9. Read KI (1984) Experimental investigation of turbulent mixing by Rayleigh-Taylor instability. Phys D: Nonlinear Phenomena 12(1–3):45–58
10. Andrews MJ, Spalding DB (1990) A simple experiment to investigate two-dimensional mixing by Rayleigh-Taylor instability. Phy Fluids A 2(6):922–927
11. Waddell JT, Niederhaus CE, Jacobs JW (2001) Experimental study of Rayleigh-Taylor instability: Low Atwood number liquid systems with single-mode initial perturbations. Phy Fluids 13(5):1263–1273
12. Lawrie AGW, Dalziel SB (2011) Rayleigh–Taylor mixing in an otherwise stable stratification. J Fluid Mech 688:507–527
13. Banerjee A, Mutnuri LAR (2012) Passive and reactive scalar measurements in a transient high Schmidt-number Rayleigh-Taylor mixing layer. Exp Fluids 53(3):717–729
14. Dalziel SB, Linden PF, Youngs DL (1999) Self-similarity and internal structure of turbulence induced by Rayleigh-Taylor instability. J Fluid Mech 399:1–48
15. White J, Oakley J, Anderson M, Bonazza R (2010) Experimental measurements of the nonlinear Rayleigh-Taylor instability using a magnetorheological fluid. Phys Rev E 81(2):026303
16. Haan SW (1991) Weakly nonlinear hydrodynamic instabilities in inertial fusion. Phys Fluids B 3:2349–2355
17. Bart JD (1967) Numerical study of two fluid Rayleigh Taylor instability. The Physics of Fluids 10(297–307):1967. https://doi.org/10.1063/1.1762109
18. Chandrasekhar S (1962) Hydrodynamic and hydromagnetic stability. Oxford Press, London, p 438
19. Rayleigh L (1894) Theory of sound, 2nd edn, vol 2. Dover Publications Inc., New York
20. Liu WH, Wang LF, Ye WH, He XT (2012) Nonlinear saturation amplitudes in classical Rayleigh-Taylor instability at arbitrary Atwood numbers. Phys Plasmas 19(4):042705. https://doi.org/10.1063/1.3702063

21. Wang LF, Ye WH, Fan ZF, Li YJ (2010) Nonlinear saturation amplitude in the Rayleigh-Taylor instability at arbitrary Atwood numbers with continuous profiles. EPL (Europhysics Lett). https://doi.org/10.1209/0295-5075/90/15001
22. Budil KS, Lasinski B, Edwards MJ, Wan AS, Remington BA, Weber SV, Glendinning SG, Suter L, Stry PE (2001) The ablation-front Rayleigh-Taylor dispersion curve in indirect drive. Phy Plasmas 8:2344
23. Fujioka S, Shiraga H, Nishikino M, Shigemori K, Sunahara A, Nakai M, Azechi H, Nishihara K, Yamanaka T (2003) First observation of density profile in directly laser-driven polystyrene targets for ablative Rayleigh-Taylor instability research. Phys Plasmas 10:4784
24. Otani K, Shigemori K, Sakaiya T, Fujioka S, Sunahara A, Nakai M, Shiraga H, Azechi H, Mima K (2007) Reduction of the Rayleigh-Taylor instability growth with cocktail color irradiation. Phys Plasmas 14:122702
25. Mikaelian KO, Lindl JD (1984) Density gradients to reduce fluid instabilities in multishell inertial-confinement-fusion targets. Phys Rev A 29:290
26. Glendinning SG, Dixit SN, Hammel BA, Kalantar DH, Key MH, Kilkenny JD, Knauer JP, Pennington DM, Remington BA, Wallace RJ, Weber SV (1997) Measurement of a dispersion curve for linear-regime Rayleigh-Taylor growth rates in laser-driven planar targets. Phys Rev Lett 78:3318
27. Shigemori K, Azechi H, Nakai M, Honda M, Meguro K, Miyanaga N, Takabe H, Mima K (1997) Measurements of Rayleigh-Taylor growth rate of planar targets irradiated directly by partially coherent light. Phys Rev Lett 78:250
28. Sakaiya T, Azechi H, Matsuoka M, Izumi N, Nakai M, Shigemori K, Shiraga H, Sunahara A, Takabe H, Yamanaka T (2002) Ablative Rayleigh-Taylor instability at short wavelengths observed with Moiré interferometry. Phys Rev Lett 88:145003
29. Garnier J, Raviart P-A, Cherfils-Clérouin, Masse L (2003) Weakly nonlinear theory for the ablative Rayleigh-Taylor instability. Phys Rev Lett 90:185003
30. Sanz J, Ramırez J, Ramis R, Betti R, Town RPJ (2002) Nonlinear theory of the ablative Rayleigh-Taylor instability. Phys Rev Lett 89:195002
31. Ikegawa T, Nishihara K (2002) Ablation effects on weakly nonlinear Rayleigh-Taylor instability with a finite bandwidth. Phys Rev Lett 89:115001
32. Bradley DK, Braun DG, Glendinning SG, Edwards MJ, Milovich JL, Sorce CM, Collins GW, Haan SW, Page RH, Wallace RJ (2007) Very-high-growth-factor planar ablative Rayleigh-Taylor experiments. Phys Plasmas 14:056313
33. Remington BA, Haan SW, Glendinning SG, Kilkenny JD, Munro DH, Wallace RJ (1992) Large growth, planar Rayleigh-Taylor experiments on Nova. Phys Fluids B 4:967
34. Remington BA, Drake RP, Takabe H, Arnett D (2000) A review of astrophysics experiments on intense lasers. Phys Plasmas 7:1641
35. Mason RJ, Hollowell DE, Schappert GT, Batha SH (2001) Long term instability growth of radiatively driven thin planar shells. Phys Plasmas 8:2338
36. Ariel PD (1971) Rayleigh-Taylor instability of compressible fluids in the presence of a vertical magnetic field. Appl Sci Res 24(1):294–304
37. Plesset MS, Whipple CG (1974) Viscous effects in Rayleigh Taylor instability, The Physics of fluids, 17-1
38. Menikoff R, Mjolsness RC, Sharp DH, Zemach C (1977) Unstable normal mode for Rayleigh–Taylor instability in viscous fluids. Physics Fluids 20
39. Wang LF, Ye WH, Li YJ (2010) Two-dimensional Rayleigh-Taylor instability in incompressible fluids at arbitrary Atwood numbers, physics of gases, plasmas, and electric discharges. Chinese Physics Lett 27(2):025203
40. Liu W, Wang X, Liu X, Yu C, Fang M, Ye W (2020) Pure single-mode Rayleigh-Taylor instability for arbitrary Atwood numbers, scientific reports. Nature 10:4201
41. Curzon FL, Folkierski A, Latham R, Nation JA (1960) Experiments on the growth rate of surface instabilities in a linear pinched discharge. Proceedings of the Royal Society, Mathematical Physical Engineering Sciences 257:386–401

42. Arindam B, Wayne NK, Malcom JA (2010) Detailed measurements of a statistically steady Rayleigh-Taylor mixing layer from small to high Atwood numbers. J Fluid Mech 659:127–190
43. Green TS, Niblett GBF (1960) Rayleigh-Taylor instabilities of a magnetically accelerated plasma Nuclear Fusion 1:42–46
44. Dickinson H, Bostick WH, Di Marco JN, Koslov S (1962) Experimental study of Rayleigh Taylor instability in plasma. Phys Fluids 5:1048
45. Wolf GH (1969) The dynamic stabilization of the Rayleigh-Taylor instability and the corresponding dynamic equilibrium. Zeitschrift Für Physik A Hadrons and Nuclei 227(3):291–300
46. Reckinger SJ, Livescu D, Vasilyev OV (2016) Comprehensive numerical methodology for direct numerical simulations of compressible Rayleigh-Taylor instability. J Comput Phys 313:181–208
47. Wieland SA, Reckinger SJ, Hamlington PE, Livescu D (2017) Effects of background stratification on the compressible Rayleigh Taylor instability. In: 47th AIAA Fluid Dynamics Conference (American Institute of Aeronautics and Astronautics, 2017), p 3974
48. Wieland SA, Hamlington PE, Reckinger SJ, Livescu D (2019) Effects of isothermal stratification strength on vorticity dynamics for single-mode compressible Rayleigh-Taylor instability. Phys Rev Fluids 4:093905
49. Luo T, Wang J, Xie C, Wan M, Chen S (2020) 2020, Effects of compressibility and Atwood number on the single mode Rayleigh-Taylor instability. Phys Fluids 32:012110
50. Ramaprabhu P, Dimonte G, Woodward P, Fryer C, Rockefeller G, Muthuraman K, Lin PH, Jayaraj J (2012) The late-time dynamics of the single-mode Rayleigh-Taylor instability. Phys Fluids 24(7):074107
51. Zhang TT, Shu CW, Zhou Y (2006) Effects of Shock waves on Rayleigh-Taylor Instability. Phys Plasmas 13:062705
52. Herrmann M, Moin P, Abarzhi SI (2008) Nonlinear evolution of the Richtmyer-Meshkov instability. J Fluid Mech 612:311–338

Unstart Phenomenon in a Scramjet Engine Isolator

P. Arjun and S. R. Nagaraja

1 Introduction

The isolator is a duct-like device that connects the inlet and the combustion chamber of a scramjet engine. The maximum static pressure rise that the isolator can produce is an important performance parameter. Isolator also reduces the sensitivity of pressure variations inside the combustor. There will be a strong pre-combustion shock system in the dual-mode engine for lower flight Mach numbers, which reduces the flow to subsonic speed at the combustor inlet. Excessive heat addition may cause the flow to choke thermally, increasing pressure. If the pressure perturbations are large, the shock system may come out of the isolator inlet, forming a detached bow shock. This leads to a loss in engine thrust and hence the system's failure.

Wagner et al. [1] have observed a rapid increase in boundary layer separation and pressure at the isolator exit just before the unstart. Wagner et al. [2] have improved their work by introducing the particle image velocimetry technique to precisely capture the events. The PIV and Schlieren images were in close agreement. They proposed that it may be possible to delay the unstart by energizing the boundary layer. Lee and Kang [3] have successfully simulated Wagner's experiments using OpenFOAM. The k-ω SST model was incorporated to account for the turbulence effect. Jang et al. [4] have reproduced the results of Wagner et al. using RANS-based solver in both 2D and 3D. They observed a reasonable agreement between Experimental and LES simulation results. The 2D simulation was inaccurate in predicting critical flap angle correctly; employing 3D simulations has given a much closer approximation. Tan et al. [5] have used a movable plug to initiate the unstart. They observed that as the throttling ratio reduces, pressure oscillations increase. Once the backpressure

P. Arjun · S. R. Nagaraja (✉)
Department of Mechanical Engineering, Amrita School of Engineering, Amrita Vishwa Vidyapeetham, Bengaluru 560035, India
e-mail: sr_nagaraja@blr.amrita.edu

© The Author(s), under exclusive license to Springer Nature Singapore Pte Ltd. 2023
S. Revankar et al. (eds.), *Recent Advances in Thermofluids and Manufacturing Engineering*, Lecture Notes in Mechanical Engineering,
https://doi.org/10.1007/978-981-19-4388-1_18

increases, they observed the vanishing of pressure oscillation in the combustor and an increase in mean pressure to a higher value. This implies that the shock train has moved to an upstream position. Deng et al. [6] numerically simulated the entire isolator model at Mach number 4. They observed the upstream movement of the shock train as the throttling ratio decreased. Geertz et al. [7] studied the shock boundary layer interaction in a rectangular channel. They tested the model in a cold flow wind tunnel with a multi-plane shadowgraph method and oil flow visualization to capture the flow regimes. They concluded that single-axis measurements are not satisfactory to explore the physics behind the complex shock boundary layer interaction inside rectangular ducts because the sidewall interactions also have a significant role. Since flight-worthy isolator prefers a high aspect ratio, this is vital. Meng et al. [8] have studied the effect of panel vibration by attributing some user-defined vibration. They observed that the presence of vibration would make the shock train move upstream, contributing to unstart. They have also observed that a low wavelength vibration will cause the shock train and the separation region to move downstream with an increase in length. Su et al. [9] have studied the ability of air injection to stabilize the boundary layer. They have concluded that adopting air injection, low oscillation mode, and corresponding separation region can be suppressed by this method. Valdivia et al. [10] have implemented active and passive vortex generators to suppress the unstart phenomenon. The control system was found effective to delay the unstart. Unstart detection technique and response time of actuators are the limiting factors. Sethuraman et al. [11] have found that when the boundary layer flow is bypassed, the shock train slightly moves to a downstream location which is desirable. Raja Sekar et al. [12, 13] have studied separation bubble formation due to shock impingement in a scramjet engine. Vishnu Hariharan et al. [14] have found that in a supersonic combustor of 10° wedge angle, the oblique shock moves downstream for increased Mach numbers. Recently Shweta Laxman Choraghe and Nagaraja [15] have reported the effect of ramp angle and Mach number on the number of shock reflections and peak pressure inside an isolator. They have concluded that the number of reflections increases with a decrease in ramp angle, and the peak pressure increases with an increase in ramp angle.

2 Computational Methodology

An isolator with different inlet ramp angles (6°, 8°, 10°, and 12°) is used to study the shock trains and unstart time inside a dual-mode scramjet isolator. The simulations are done for Mach numbers 3, 4, 5, and 6. The boundary conditions used are pressure 5.4 kPa and temperature 55 K. These values are the same as Wagner et al. [1]. The walls were considered rigid and adiabatic. The geometry of the model is shown in Fig. 1. The model has a height (h) of 25 mm and a length of 374 mm, of which 269 mm has a constant sectional area.

The simulations are done using open-source software OpenFOAM. Density-based transient solver rhoCentralFoam in an OpenFoam environment is used. It is one of

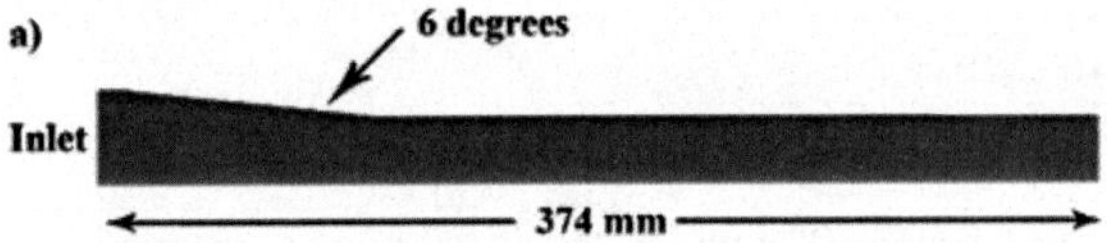

Fig. 1 Geometry of isolator model with an inlet ramp angle of 6 degrees

the most stable solvers for compressible flows. It uses central TVD of KT and KNP schemes for the advective fluxes. In this solver, three conservative variables are solved in the Eulerian frame of reference. These properties can be specified as density (ρ), Momentum density (ρu), and Total energy density (ρE).

$$\partial\rho/\partial \mathrm{t} + \nabla\cdot(\rho u) = 0 \tag{1}$$

$$\partial(\rho u)/\partial t + \nabla\cdot(\rho uu) = -\nabla p + \nabla\cdot \mathrm{T} \tag{2}$$

$$\partial(\rho E)/\partial t + \nabla\cdot(\rho uE) + \nabla\cdot(pu) = \nabla\cdot(T\cdot u) + \nabla\cdot J \tag{3}$$

Here, $E \equiv E(T) + 1/2u\cdot u$, where $E(T)$ represents the internal energy, T stands for the temperature, and p is the thermodynamic pressure related to the temperature and density through the ideal gas equation, which is given by $p = \rho \mathrm{RT}$. Here R is the gas constant. T, which can be depicted as $\tau = 2\mu *[0.5*(\nabla u + \nabla Tu) - (1/3)*(\nabla\cdot u)I]$, which is the viscous stress tensor under the assumption of Newtonian fluid with dynamic viscosity μ while I is the unit tensor and, $\mathrm{J} = -\lambda\Delta T$ which is the heat flux vector, and λ is the thermal conductivity.

Air is used as the working fluid. C_p is considered a constant. The Sutherland model is used for viscosity calculations. The variation of density is made according to the perfect gas equation. SST K Omega model is used for turbulence modeling. To calculate time derivatives first-order Euler implicit scheme is used. The central-upwind KNP (Karaganov, Noelle, and Petrova) scheme and the van Albada limiter are used to evaluate the flux at cell faces. The limited Gauss linear schemes are used as divergent schemes and corrected Gauss linear for Laplacian schemes. The mesh wave method is used for wallDist. To initiate the back pressure after establishing the fully started flow, the outlet of the model is closed partially to increase the back pressure.

Mesh independent studies were performed by considering four mesh sizes 1, 0.6, 0.5, and 0.4 mm. The pressure ratio (P_{max}/P_{inf}) is used for comparison, where P_{max} is the maximum static pressure along the bottom wall and P_{inf} is free stream static pressure. The results are shown in Table 1. The peak pressure value was same for 0.6 and 0.5 mm sized meshes, and the difference for 0.4 mm and 0.5 mm is only about 0.5%. Hence 0.5 mm mesh size is used for the analysis.

Table 1 Mesh independence test

Mesh size	Maximum pressure ratio
0.4 mm	3.83
0.5 mm	3.85
0.6 mm	3.85
1.0 mm	3.96

3 Results and Discussions

The Validation of the case setup has been done by comparing present results with experimental results [1]. The static pressure distribution along the bottom wall is shown in Fig. 2.

The maximum and minimum static pressure is found to be in close agreement. The trend of pressure distribution is also found to be similar. As shown in Fig. 3, the density contours closely match the contours obtained experimentally [1], and the number of shock reflections is also same.

Figure 4 shows the variation of maximum pressure ratio with Mach number and ramp angle. The maximum pressure obtained along the centerline for a 6-degree ramp angle varies from 2.5 to 3.07 times the freestream pressure for Mach number 3 to 6. The pressure ratio increases with Mach number and ramp angle. Also, the oscillatory nature of the pressure curve, which is a measure of the tendency for unstart increases with an increase in Mach number.

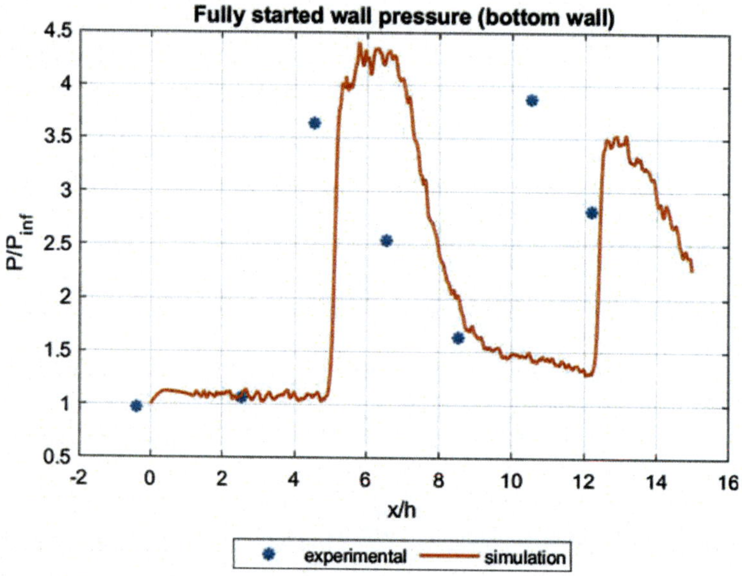

Fig. 2 Comparison of fully started wall pressure along the bottom wall

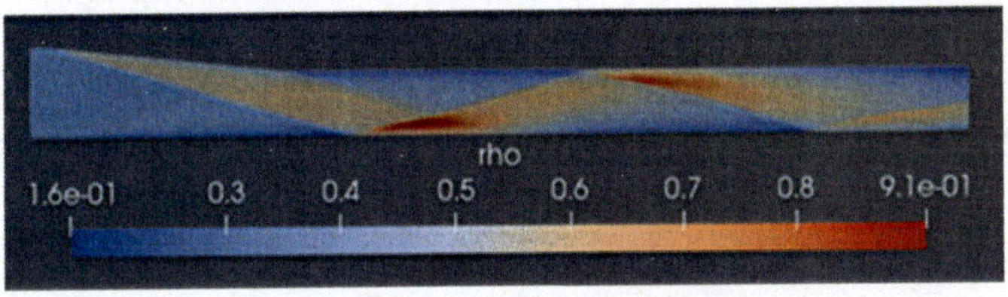

Fig. 3 Density Contour of the fully started flow from the simulation

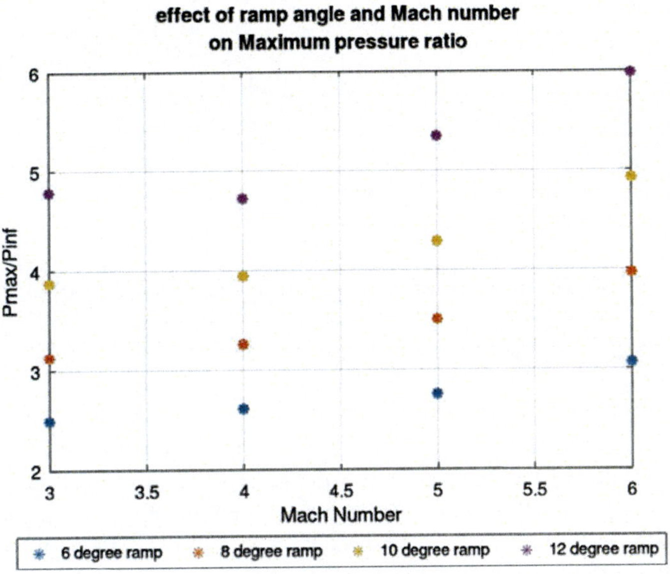

Fig. 4 Variation of pressure ratio with Mach number and ramp angle

In Wagner's experiment, the combustion pressure oscillations were replaced by a physical flow choking mechanism. A similar approach has been used to initiate the unstart process in this work. Each simulation is carried out in two phases. The first stage corresponds to initializing a fully started flow, and the second phase was linked with the propagation of an unstart shock system. The entire outlet was split into two regions a pseudo-outlet at the bottom and an outlet at the top. During the first phase of the simulation, both of these sections in the outlet were set as patches, and zero gradient conditions were mentioned. The simulations were run up to reaching a fully started condition. Then the type of bottom pseudo-outlet part was changed to wall, and no-slip condition was imposed. All the other boundary conditions were kept same. So, from that moment onwards, there will be a sudden decrease in the flow area at the outlet, and hence the back pressure will be induced. Terminal scripts were used to combine these stages to modify the "controlDict" file and "Boundary

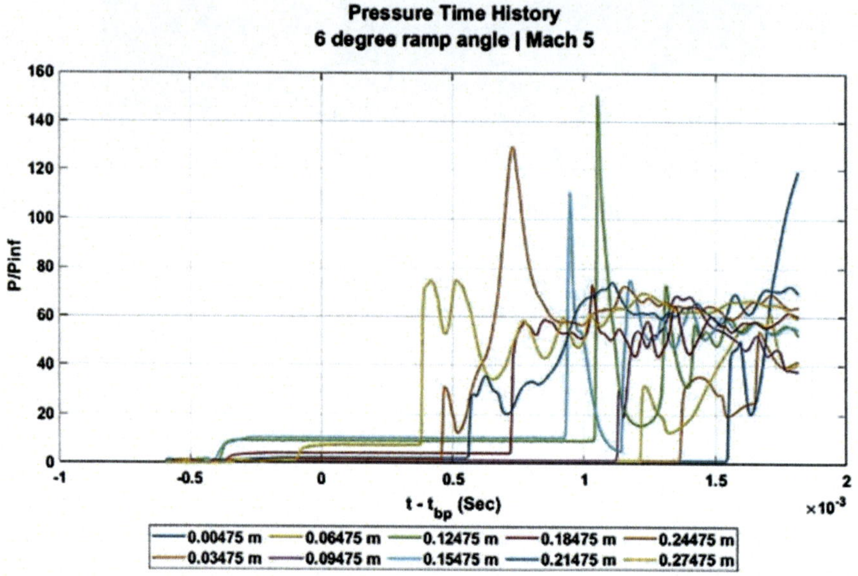

Fig. 5 Pressure Time history at various points in a 6-degree ramped isolator under Mach 5 inlet condition

file" as prescribed in all run file. In short, the first stage of simulation is used just to initialize the fully started flow field inside the isolator for the second stage.

The pressure along the bottom wall at ten different points for a ramp angle of 6 degrees and an inlet Mach number of 5 is shown in Fig. 5. The locations of probes were 0.00475, 0.03475, 0.06475, 0.09475, 0.12475, 0.15475, 0.18475, 0.21475, 0.24475 and 0.27475 m from the inlet. The distance between each probe is 30 mm. The time taken for the shock wave to cover this distance is noted, and using this information, the unstart propagation velocity and average unstart velocities are calculated. The pressure fluctuations are large near the end of the isolator.

Figure 6 shows the variation in pressure time history of three selected points, one near the inlet, one closer to the outlet, and one at the middle (0.03475 m, 0.15475 m, and 0.27475 m from inlet) with respect to Mach number. Here the quantity on x-axis is $(t–t_b)$, where t is the time and t_b is the time at which the back pressure has been introduced. It can be observed that the pressure changes depend on the position in the isolator. The amplitude of oscillation of static pressure after the passage of the unstart shock front at a point, the frequency of oscillation, and its nature depend on the location. As the Mach number increases, the peak pressure also increases, and the high-frequency low amplitude oscillations gradually change to low-frequency high amplitude oscillations.

Figure 7 shows the variation in pressure time history at the same locations as in fig for different ramp angles at Mach 5. No significant change has been observed in the peak pressure ratio. The variation in the nature of oscillations with respect to ramp

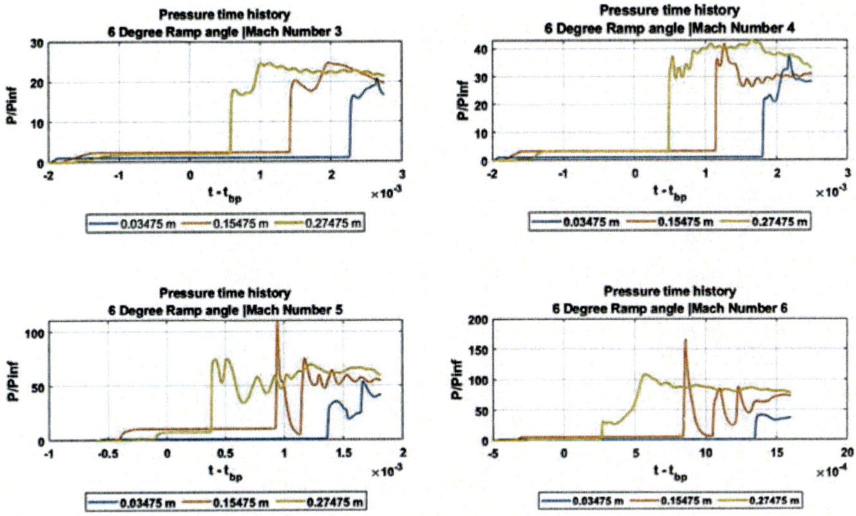

Fig. 6 Pressure time history at different locations for a 6-degree ramp angle model for different inlet Mach numbers

angle is less sensitive than variations with respect to Mach number. The variation is found to increase when the location is moved downstream.

Figure 8 compares experimental [1] and simulation values of unstart velocities. The simulations overpredict the unstart velocities. This may be due to analyzing a

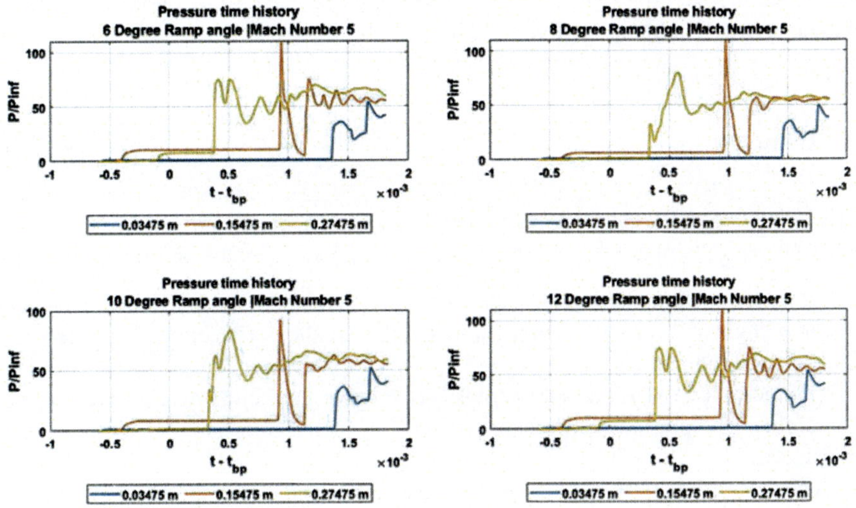

Fig. 7 Pressure time history at different points under Mach 5 inlet conditions for different ramp angles

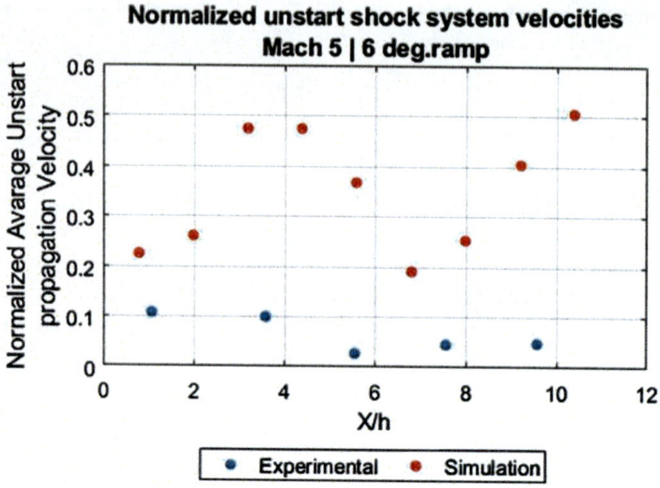

Fig. 8 Unstart velocity—comparison of results

complex 3D phenomenon in a 2D framework and the locations used for calculating unstart velocities.

Figure 9 shows average unstart velocities calculated based on two extreme points. One of them was at a distance of 0.00475 m from the inlet (Very close to the inlet), and the other one was at a distance of 0.27475 m from the inlet (point close to the outlet). The average unstart velocities were found to be decreasing with increases in Mach number, and for a given Mach number, the average unstart velocity increases with an increase in ramp angle.

4 Conclusions

The Validation of the case has been done, and a reasonable agreement was observed with the experimental results. Some of the conclusions from the present study are.

- As the Mach number increases, the number of shock reflections reduce for a given ramp angle and for a particular Mach number of shock reflections increase with increased ramp angle.
- The increase in wall pressure is more sensitive to the ramp angle than the Mach number. The pressure ratio increases with an increase in ramp angle and Mach number.
- A change in the shape of reflecting shock was observed with respect to the distance, and it is due to the increased viscous blockage because of the Shock wave boundary layer interaction.
- It is observed that the normalized unstart velocity increased with increasing ramp angle and decreased with increasing Mach number.

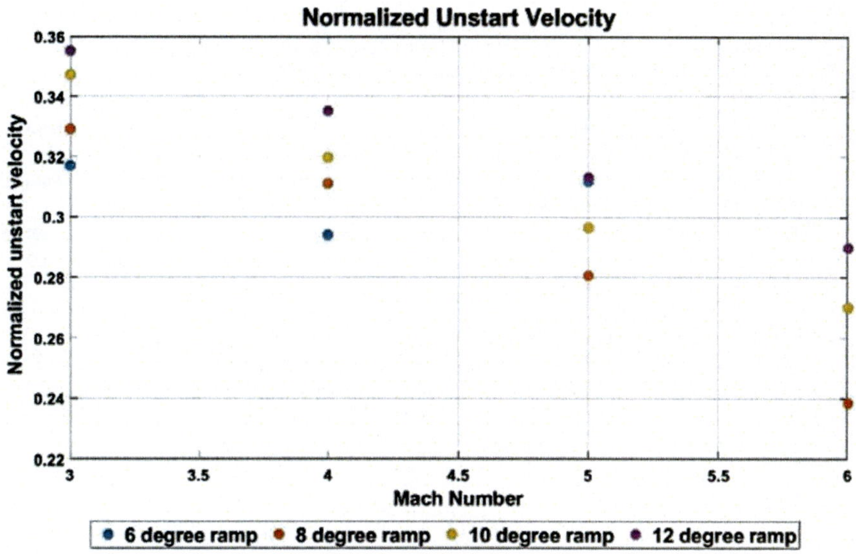

Fig. 9 Average unstart velocity calculated based on extreme points

References

1. Wagner JL, Yuceil KB, Valdivia A, Clemens NT, Dolling DS (2009) Experimental investigation of unstart in an inlet/isolator model in Mach 5 flow. AIAA J 47(6):1528–1542. https://doi.org/10.2514/1.40966
2. Wagner JL, Yuceil KB, Clemens NT (2010) Velocimetry measurements of unstart of an inlet-isolator model in Mach 5 flow. AIAA J 48(9):1875–1888. https://arc.aiaa.org/doi/abs/10.2514/1.J050037
3. Lee J, Kang SH (2019) Numerical study on the start and unstart phenomena in a scramjet inlet-isolator model. PLoS ONE 14(11):e0224994. https://doi.org/10.1371/journal.pone.0224994
4. Jang I, Pecnik R, Moin P (2010) A numerical study of the unstart event in an inlet/isolator model. Cent Turbul Res Annu Res Briefs, 93–103
5. Tan H, Li L, Wen Y, Zhang Q (2011) Experimental investigation of the unstart process of a generic hypersonic inlet. AIAA J 49(2):279–288. https://doi.org/10.2514/1.J050200
6. Deng R, Jin Y, Kim HD (2017) Numerical simulation of the unstart process of dual-mode scramjet. Int J Heat Mass Transf 105:394–400. https://doi.org/10.1016/j.ijheatmasstransfer.2016.10.004
7. Geerts JS, Yu KH (2016) Shock train/boundary-layer interaction in rectangular isolators. AIAA J 54(11):3450–3464. https://doi.org/10.2514/1.J054917
8. Meng X, Ye Z, Hong Z, Ye K (2020) Impacts of panel vibration on shock train structures and performance of two-dimensional isolators. Aerosp Sci Technol 104:105978 [Online]. https://doi.org/10.1016/j.ast.2020.105978
9. Su W-Y, Chen Y, Zhang F-R, Tang P-P, Control of pseudo-shock oscillation in scramjet inlet-isolator using periodical excitation. Acta Astronaut 143:147–154. https://www.sciencedirect.com/science/article/abs/pii/S0094576517311748
10. Valdivia A, Yuceil KB, Wagner JL, Clemens NT, Dolling DS (2014) Control of supersonic inlet-isolator unstart using active and passive vortex generators. AIAA J 52(6):1207–1218. https://arc.aiaa.org/doi/abs/10.2514/1.J052214

11. Sethuraman VRP, Kim TH, Kim HD (2021) Effects of back pressure perturbation on shock train oscillations in a rectangular duct. Acta Astronaut 179:525–535
12. Raja Sekar K, Jegadheeswaran S, Kannan R, Vadivelu P, Prasad G (2019) Numerical investigation of single ramp scramjet inlet characteristics at Mach number 5.96 due to shock wave–boundary layer interaction. In: Innovative design, analysis and development practices in aerospace and automotive engineering (I-DAD 2018). Lecture Notes in Mechanical Engineering. Springer. https://doi.org/10.1007/978-981-13-2697-4_20
13. Raja Sekar K, Jegadheeswaran S, Kannan R, Manigandan P (2019) Perforated wall in controlling the separation bubble due to shock wave–boundary layer interaction. Int J Turbo & Jet-Engines, eISSN 2191-0332, ISSN 0334-0082. https://doi.org/10.1515/tjj-2018-0048
14. Hariharan V, Velamati RK, Prathap C (2016) Investigation on supersonic combustion of hydrogen with variation of combustor inlet conditions. Int J Hydrogen Energy xxx:1–9
15. Choraghe SL, Nagaraja SR (2022) Analysis of shock trains in an isolator model. In: Govindan K et al. (eds) Advances in mechanical and materials technology. Lecture Notes in Mechanical Engineering. Springer. https://doi.org/10.1007/978-981-16-2794-1_117

Heat Transfer

Efficacy of Pulsating Heat Pipe for Laptop Cooling Applications: An Experimental Investigation

R. K. Sarangi, M. V. Rane, S. P. Kar, A. Swain, and P. C. Sekhar

1 Introduction

The adverse effect of improper thermal management is the main concern of electronic industry beginning from electronic chip to high power magnetron. The increased electrical performance of the computer industry demands power dissipation of 100 W/cm^2. The maximum junction temperature is limited to 105 °C [1]. The maximum power dissipation is in the range of 100 W over a chip size of 16 mm length and 16 mm width. There are different solutions available for the thermal management of these electronic components. In low power applications, natural convection is preferred, and in high power applications, forced convection cooling is preferred. Reduced volume requirements and lower temperature difference led to liquid cooling. The cooling is achieved in single-phase or two-phase flow. Transfer of thermal energy via heat pipes allows much of the benefits of two-phase liquid cooling while still being considered a passive device. Though heat pipe is a passive equipment, wick adds an additional thermal resistance to heat flow path. Kipping all those point in consideration oscillating heat pipe or pulsating heat pipe without wick inside is the best option. It has a very simple structure and a fast thermal response. Pulsating heat pipe (PHP) was firstly proposed by Akachi [2] and was considered for high heat flux electronic cooling [3].

The PHP consists of long capillary tube of small inner diameter bent into number of U-turns and consists of evaporator, condenser and adiabatic sections. The inner diameter of PHP should be small enough to satisfy the Bond number criteria as given by Eq. 1. Heat is transported from evaporator to condenser by the pulsating

R. K. Sarangi (✉) · S. P. Kar · A. Swain · P. C. Sekhar
School of Mechanical Engineering, KIIT Deemed to be University, Bhubaneswar, India
e-mail: sarangi_radhakanta@yahoo.co.in

M. V. Rane
Department of Mechanical Engineering, IIT Bombay, Mumbai, India

© The Author(s), under exclusive license to Springer Nature Singapore Pte Ltd. 2023
S. Revankar et al. (eds.), *Recent Advances in Thermofluids and Manufacturing Engineering*, Lecture Notes in Mechanical Engineering,
https://doi.org/10.1007/978-981-19-4388-1_19

action of liquid plugs and vapor bubbles. The creation of vapor bubbles in evaporator section and collapse in condenser section creates a pressure imbalance and causes to-and-fro motion of liquid plugs. This is called pulsation. The pulsation is measured by its frequency and amplitude which can be measured by observing temperature fluctuation mostly in adiabatic section. The PHP may be classified as looped and un-looped, and looped type has smaller thermal resistance due to better oscillation [4].

$$Bo = d_i \sqrt{\frac{g(\rho_l \rho_v)}{\sigma_l}} \tag{1}$$

Gi [5] has reported the performance of "O"-shaped flat PHP with R141b as a working fluid. The PHP was made of copper capillary tube of 2 mm ID and six number of turns with a fill ratio of 60%. The fan had a blowing rate of 2 m/s. The thermal resistance of the PHP was 1.1 °C/W for a heat load of 50 W and a chip temperature of 80 °C. Zhuang [6] and Kwang [7] have experimentally studied the PHP as a cooling device for desktop PC and have reported the thermal performance which is better than the heat sink. Gi and Meazawa [8] have studied the performance of flat type PHP for cooling the CPU of notebook PC and found it better. Rittidech et al. [9] have reported the thermal performance of PHP, which was used as a cooling device for CPU in desktop PC. Two sets of capillary tubes with an inner diameter of 2 mm, an evaporator length of 50 mm and condenser length of 160 mm, were used in their experimental setup. Each set had six numbers of turns. The working fluid was R134a, and fill ratio was 50%. In this experiment, the thermal resistance of PHP was found to be 0.83 °C/W, whereas the conventional heat sink had a thermal resistance of 0.99 °C/W. The heat dissipation rates of heat sink and PHP were 55 W and 54 W, respectively. Maydanik et al. [10] have investigated a compact cooler which consists of a PHP with ID of 1.2 mm and OD of 2 mm in the form of a 3D spiral containing 17 turns. The total length was 5.6 m. Three working fluids, R141b, methanol and water, were used. The fill ratio was 65% for methanol and R141b, whereas in case of water, the fill ratio was 60%. The minimum thermal resistance of the cooler was 0.32 °C/W, in case of water and methanol as working fluids, for a uniform load of 250 W. But the minimum thermal resistance increased to 0.62 °C/W in case of concentrated heat load of 125 W, over an area of 1 cm^2.

Mostly, the PHP has been investigated as a thermal management solution for desktop computer. There are no much investigations of PHP for laptop cooling applications. The aim of the present experiment is to study the thermal performance of closed loop pulsating heat pipe in laptop cooling application, for a maximum die temperature of 105 °C. The tests were conducted for the conventional PHP of different dimensions and various fill ratios. The working fluids were water and ethanol. The inner diameters are 0.66, 1.0 and 1.3 mm. Maximum number of turns is limited to four and the total length of evaporator, adiabatic and condenser is limited to 150 mm, due to the space limitation inside the laptop.

2 Experimental Setup and Testing

In the laptop cooling application, the available space for the cooling device is the main constraint. As per the requirement, the distance between the evaporator and condenser section should be between 100 and 200 mm, the height of the cooling device including copper plate should not be more than 10 mm, and the space for the cooling device is limited to an area of $40 \times 40\ mm^2$. The detail assembly and experimental setup of 1-turn PHP are given in Fig. 1a and b, respectively. The PHP is supported by a copper backing plate of 3 mm thickness and $52 \times 52\ mm^2$ area.

The PHP is brazed to this copper backing plate at onside, and other side of the backing plate is pressed to the die by four spring nuts and support plate at the back side of the PCB. Thermal Interface Material (TIM) of appropriate thickness makes the thermal contact between the backing plate and the die. The condenser section of the PHP is brazed to the heat exchanger. The complete assembly is given in the Fig. 1a. In the Fig. 1a, each turn of the evaporator and condenser section of the PHP is brazed to the support plate and heat exchanger, respectively.

The experimental setup consists of a Printed Circuit Board (PCB) on which die and 4-wire RTD are mounted. The 4-wire RTD is powered by a mili-Amp constant current source, and a multimeter is used to measure the milivolt output of the 4-wire RTD and the condenser exit temperature. This source is accurate within $\pm 1\%$. Die is powered by a power meter of 0–50 V and 0–6 A capacity, which is connected to the main supply of 230 V. Power meter has separate displays for current and voltage. The supply power to the die is accurate within $\pm 2.5\%$. The experimental setup is shown in Fig. 1b. Due to the limited space above the heat exchanger surface, in case of two-turn PHP, each turn of the PHP is brazed to top and bottom surface of the heat exchanger. This is shown in the Fig. 2a. The die has a dimension of $10 \times 10\ mm^2$. The condenser section of the PHP consists of a 0.5 W DC cooling fan of $65 \times 65 \times 10$ mm dimension which has a flow rate of $0.92\ m^3/min$ at 5 mm of H2O, and a heat exchanger consists of a copper heat sink of $65 \times 30 \times 10$ mm dimension. The heat sink consists of 30 fins which are having 10 mm height and 30 mm length with a gap

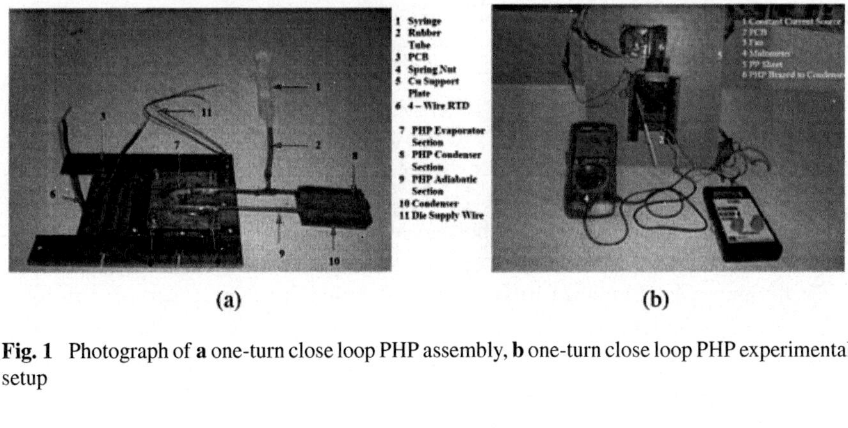

(a) (b)

Fig. 1 Photograph of **a** one-turn close loop PHP assembly, **b** one-turn close loop PHP experimental setup

(a) (b)

Fig. 2 Photograph of **a** two-turn close loop PHP setup, **b** four-turn close loop PHP setup

of 3 mm between them. More than two turns are difficult to accommodate over the copper support plate due to limited space over the copper support plate. So, a slotted Al plate is used to support more number of turns. In another set of experiment, slotted aluminum plate of 3 mm thickness and 52 × 52 mm^2 area is used as a backing plate instead of copper to support more number of turns. The slots are of 2 mm wide and 2 mm deep. So, the PHP of outer diameter 2 mm can be well pressed into these slots. Four-turn PHP assembly with slotted Al plate is shown in Fig. 2b. Before testing, the PHP is evacuated and charged with working fluid. Evacuation is done by two stage vacuum pump, whereas charging is carried out by an injection syringe attached to the PHP though a T-section.

3 Results and Discussions

The tests are conducted for horizontal as well as vertical orientation for different PHPs of 0.66, 1 and 1.3 mm ID. Water has higher heat transport capability and safe working fluid. So, in laptop cooling applications, mostly water is used as a working fluid. Ethanol may be flammable at high temperature, but in some orientation, it may give better performance [13]. So, the working fluids are chosen as water and ethanol in this project. Fill ratio is the ratio of working fluid volume and total inner volume of PHP. Fill ratios are varied to observe the dependency of thermal performance on the fill ratio. The complete experimental matrix is shown in the Table 1.

All the experiments start with 0% fill ratio and end at 100% fill ratio. If pulsation is created for intermediate fill ratios, then the thermal performance of the PHP should be better than 0 and 100% fill ratios. At 0 and 100% fill ratios, heat dissipation through the PHP is by conduction. Figure 3 shows the thermal performance of four-turn PHP. It has an inner diameter of 0.66 mm and a total length of 585 mm.

Water and ethanol both are used as working fluids. It can be observed that, for a variation of heat input from 4.7 to 12.2 W, corresponding variation in die temperature is from 58 to 98 °C. There is no change in the die temperature for ethanol and water at various fill ratios. Figure 3 also shows the die temperature variation with heat

Table 1 Experimental matrix

Fluid	ID	L_{tot}	L_e	L_a	L_c	FR	N_{turn}
#	mm	mm	mm	mm	mm	%	#
H_2O, C_2H_5OH	0.66	585	27.6	15.0	29.1	0, 20 40, 70	4
H_2O, C_2H_5OH	1.0	1120	52	35	50	0, 20, 30 40, 50	4
H_2O, C_2H_5OH	1.3	300, 600	32	55	62	0, 18 22, 35, 40	1, 2

L_{tot}: total length, L_e: evaporator length, L_a: adiabatic length, L_c: condenser length
FR: fill ratio, N_{turn}: number of turns

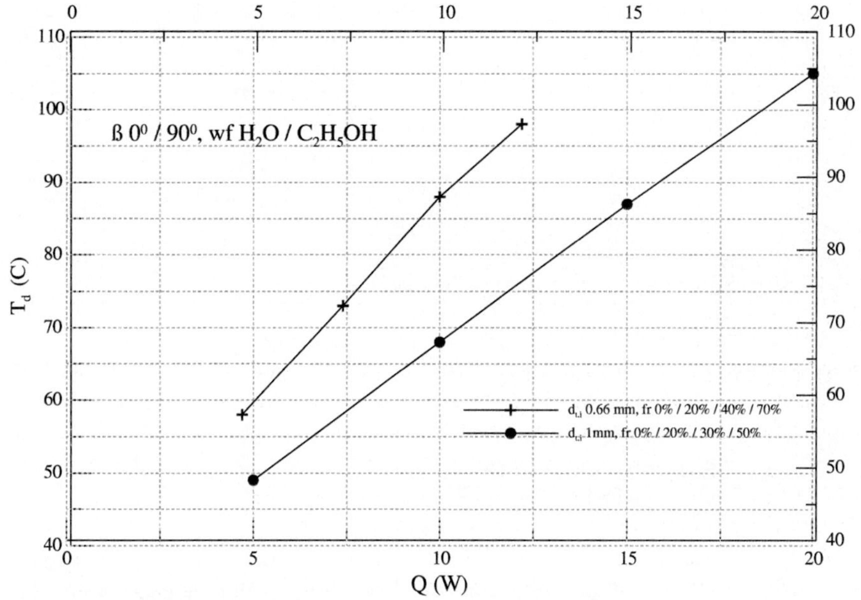

Fig. 3 Variation of die temperature (T_d) with heat input (Q) of 4-turn PHP

input for 1 mm ID and 4-turn PHP. It shows a variation in die temperature from 49 to 105 °C for a heat input variation from 5 to 20 W. There is no variation in thermal performance for water and ethanol at different fill ratios. So, it can be concluded that there is no pulsation in both the cases. In both the cases, the die temperature variation with heat input is linear, which indicates that the thermal resistance is constant. No variation in thermal resistance confirms that the heat transfer is by conduction, not by pulsation.

The variation of die temperature with the heat input for 1.3 mm ID and two-turn PHP is shown in Fig. 4. There is no pulsation for 35% fill ratio as the die temperature variation in this case is similar to that of 0% fill ratio. This is because heat transfer

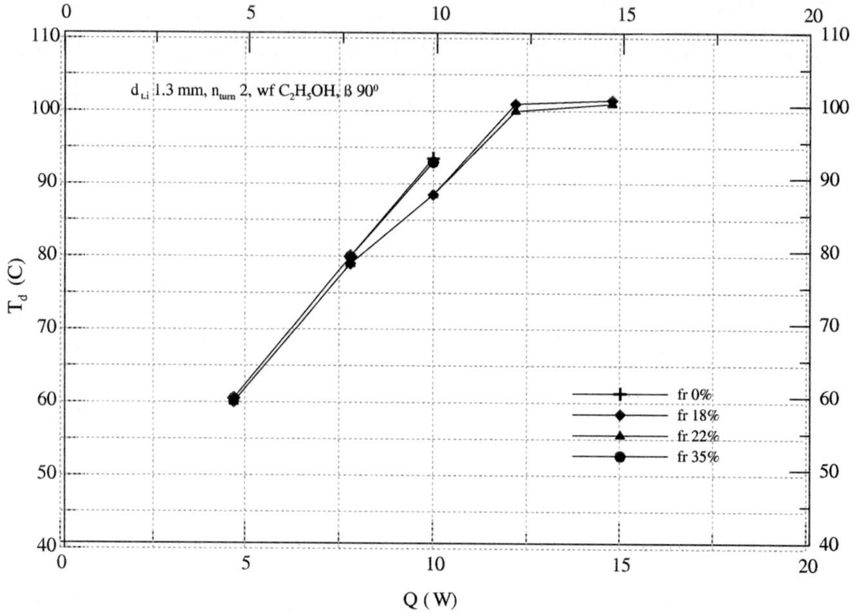

Fig. 4 Variation of die temperature (T_d) with heat input (Q) of 2-turn PHP

is by conduction. As thermal conductivity of working fluid present inside the PHP is insignificant compared to the thermal conductivity of the wall of PHP which is copper, there is no significant variation of die temperature between 0 and 35% fill ratios. But there is pulsation for 18 and 22% fill ratios. At 10 W, the die temperatures are observed as 93.5 and 93 °C for 0 and 35% fill ratios, whereas for same die power, the die temperatures are 88.5 °C for 18 and 22% fill ratios.

At 12.2 W, the die temperature was increasing continuously and was not stable at even 105 °C for 0 and 35% fill ratios. So, that reading is not included in the plot. But the die temperatures are 101 and 100 °C at same heat input of 12.2 W for fill ratios of 18 and 22%, respectively. The PHP was tested for a maximum heat input of 14.8 W due to limitation of die temperature which should not exceed 105 °C. At this maximum heat input, the die temperatures are 101.5 and 101 °C for the fill ratios of 18 and 22%. This PHP was also tested for horizontal operation, but no pulsation was found. Even no pulsation was found for water as a working fluid. As indicated in Table 1, the tests were also performed for 1.3 mm ID and 1-turn PHP but there were no pulsation in these cases. So, these cases are not presented.

It can be found from the test results that the inner diameter and number of turns play a vital role for creating the pulsation. In fact, there is also a working range of fill ratio and heat input in addition to proper combination of inner diameter and number of turns. No quantitative relationship between the number of turns and pulsation was found in the literature. But this can be explained qualitatively. The PHP turns in form of U-bend actually accumulate liquid at its lowest point [11]. This results in the liquid plug formation inside the bend in the evaporator section, which blocks the vapor bubble movement. This restriction builds up the pressure of the vapor bubble in upstream direction. When this pressure is sufficient, the liquid plugs are pushed out of the bends to the evaporator section. This sudden motion of the liquid plug is associated with random pressure fluctuation and instability in the PHP, which facilitates the working of the PHP. Similarly in the condenser section, liquid lobes are formed either on outer bend tube or inner bend tube of a U-turn, depending on the motion of the vapor bubbles, which is either in anticlockwise direction or in clockwise direction [12]. These lobes break the vapor bubble, which result in the pressure fluctuation and instability.

So, the number of turns creates instability and facilitates the working of PHP. The lobe formation in the condenser section and liquid slug formation in the bends are more favorable in vertical orientation than horizontal orientation for same number of turns. So, the horizontal operation requires more number of turns than vertical operation to start the pulsation. This fact is also supported in the experimental result, which shows pulsation for 1.3 mm ID and 2-turn PHP in vertical orientation, but not in horizontal orientation. The tendencies of lobe formation and liquid plug formation are more in case of higher inner diameter than lower inner diameter tube. So, if inner diameter decreases, number of turn increases for the satisfactory operation of the PHP. So, in the present experiment 0.66 mm ID and 4-turn PHP and 1 mm ID and 4-turn PHP do not show pulsation, whereas 1.3 mm ID and 2-turn PHP show pulsation. In addition, number of turn and inner diameter type of working fluid are also important for creating the pulsation. Some fluids are better to startup and some fluids are better to take up more loads at normal operation of the PHP.

As regard to working fluids, dp/dt that is ratio between change in saturation pressure to change in saturation temperature, surface tension, viscosity and latent heat of vaporization play a major role on the performance of the PHP. The ratio between saturation pressure and temperature should be more for better performance, because of more driving force for the same evaporator and condenser temperature. Surface tension and viscosity should be less for better pulsation, as these parameters are responsible for dampening the oscillation. Latent heat should be less for easy startup and more for more heat load taking capacity at normal operation. Water has higher latent heat, viscosity and surface tension than that of ethanol for the same working range of temperatures. So, water is a better working fluid for higher inner diameter and ethanol is a better fluid for lower inner diameter. This explanation is also supported by the experimental result of Charoensawan et al. [13], where he has found out that water gave a poor performance compared to ethanol for 1 mm ID and 16-turn PHP in spite of its higher latent heat, but water gave a better performance than ethanol for 2 mm ID and 16-turn PHP. This fact is further supported in the present experiment, where water does not give pulsation in case of 1.3 mm ID and 2-turn PHP, whereas ethanol gave pulsation.

It has been observed in the previous experiments that more number of turns are required for the better performance of the PHP. But due to the space limitation inside the laptop, more than four numbers of turns cannot be accommodated in case of conventional PHP. So conventional U-turn PHPs may not be suitable for laptop cooling applications. So, it is required to design and develop some new configured PHP of more number of turns which can be fitted into the limited available space inside the laptop. This may give better performance and meet the future challenges.

4 Conclusions

The experiments were carried out for PHP of different inner diameters that is 0.66, 1 and 1.3 mm consisting of 1, 2 and 4 numbers of turns. It has also tested for different fill ratio starting from 0 to 100%. Results were presented. Pulsations were observed only in 1.3 mm ID and 2-turn PHP, in vertical orientation and ethanol as a working fluid. The maximum heat load in this case is 14.8 W at a fill ratio of 22%. The same PHP with more number of turns could have been given better performance. But due to the space limitation inside the laptop, it was not possible to accommodate more number of conventional U-turns. It was also recommended that some new configured PHP of more number of turns can be designed and developed for better performance which may meet the future challenges of laptop cooling applications.

Acknowledgements The authors are thankful to Intel India Pvt. Ltd. for their partial funding of the research work under the sponsorship grant 08 Intel 001.

References

1. Azar K (2006) Managing power requirements in the electronics industry. Electron Cooling 6(4):6–10
2. Akachi H (1990) Structure of heat pipe, U S Patent, 4921041
3. Han X, Wang X, Zheng H, Xu X, Chen G (2016) Review of the development of pulsating heat pipe for heat dissipation. Renew Sustain Energy Rev 59:692–709
4. Jo J, Kim J, Kim SJ (2019) Experimental investigations of heat transfer mechanisms of a pulsating heat pipe. Energy Convers Manage 331–341. https://doi.org/10.1016/j.enconman.2018.12.027
5. Gi K (1999) CPU cooling of notebook PCs, proceedings of the 11th IHPS. Tokyo, pp 467–472
6. Zhuang J (2002) Research on CPU heat pipe coolers. In: 12th international heat pipe conference, Sect; F1-F9, Moscow, Russia
7. Kwang SK, Won HM, Kim JW (2002) Heat pipe cooling technology for desktop PC CPU. In: 12th international heat pipe conference, Sect; F1–F9, Russia
8. Gi K, Meazawa S (2002) CPU Cooling of notebook by oscillating heat pipe. In: Proceedings 6th international heat pipe symposium. Chiang Mai, Thailand, pp 166–169
9. Rittidech S, Boonyaem A, Tipnet P (2005) CPU Cooling of desktop PC by closed end oscillating heat Pipe. Am J Appl Sci 2:1574–1577
10. Maydanik YF, Dmitrin VI (2009) Pastukhov: compact cooler for electronics on the basis of pulsating heat pipe. Appl Therm Eng 29:3511–3517
11. Tong BY, Wong TN, Ooi KT (2001) Closed loop pulsating heat pipe. Appl Therm Eng 21:1845–1862

12. Xu JL, Li YX, Wong TN (2005) High speed flow visualization of a closed loop pulsating heat pipe. Int J Heat Mass Transf 48:3338–3351
13. Charoensawan P, Khandekar S, Groll M, Terdtoon P (2003) Closed loop pulsating heat pipes Part A: parametric experimental investigations. Appl Therm Eng 23:2009–2020

Influence of Surface Unevenness on Enhancement of Heat Transfer During Saturated Pool Boiling of Acetone

Rajiva Lochan Mohanty, Bibhu Bhusan Sha, and Mihir Kumar Das

1 Introduction

Heat transmission via phase change processes, notably nucleate boiling, is unique of the furthermost efficient heat transfer system seen in various applications such as heat-exchanger systems, power plants, cooling systems of electronic device and refrigeration systems. Due to the high rate of heat transfer, this process takes a substantial quantity of heat energy from a heated surface to maintain a comparatively lower surface temperature. The nucleate boiling heat transfer (NBHT) is significant in the boiling process for a surface where production, development on, bubble and detachment from a heated surface occurs often, resulting in a lower surface temperature. The most important point to remember is that the boiling limit is determined by the physical burnout of materials on the heated surface. The heated surface achieves a maximum value of heat flux, known as critical heat flux, at this state (CHF). The phenomenon of CHF leads to a considerable reduction in the heat transfer coefficient, resulting in a rapid increase in surface temperature. As a result, the CHF should not be exceeded in order to guarantee system safety.

Many researchers have developed various surfaces to theoretically explain the CHF enhancement mechanism. The very well-known theoretical concept of CHF mechanism for pool boiling phenomena was developed by Kutateladza [1] and Zuber [2] and using this theory many studies on CHF observation from coating surfaces have been carried out [3]. The experimental observations [4, 5] on microporous

R. L. Mohanty (✉)
School of Mechanical Engineering, KIIT Deemed to be University, Bhubaneswar, Odisha 751024, India
e-mail: rajiva.mohantyfme@kiit.ac.in; rajivamohanty@gmail.com

R. L. Mohanty · B. B. Sha · M. K. Das
School of Mechanical Sciences, Indian Institute of Technology Bhubaneswar, Bhubaneswar, Odisha 752050, India

© The Author(s), under exclusive license to Springer Nature Singapore Pte Ltd. 2023

S. Revankar et al. (eds.), *Recent Advances in Thermofluids and Manufacturing Engineering*, Lecture Notes in Mechanical Engineering,
https://doi.org/10.1007/978-981-19-4388-1_20

coating surfaces have been conducted to show CHF enhancement by 100% over a plain surface. In pool boiling experiments on modified porous-layer coated surfaces, Liter and Kaviany [6] detected an increase in CHF. Wu et al. [7] also discovered CHF on a hydrophilic titanium dioxide (TiO_2) nanoparticle-modified surface. The TiO_2 coated surface raised CHF by 38.2% for FC-72, according to their findings.

By anodizing a Zircaloy-4 surface with distilled water, Ahn et al. [8] created a micro/nano-structured surface. To further understand the boiling mechanism, Furberg and Palm [9] have done experiments with R134a and FC-72 on a dendritic and micro-porous surface during pool boiling. Misale et al. [10] investigated the pool boiling heat transfer and CHF of saturated HFE-7100 on a restricted smooth copper surface at atmospheric pressure. Byon et al. [11] attempted to determine the impacts of particle size, cluster size and coating thickness on the pool boiling CHF in order to explore the critical heat flux of bi-porous sintered copper coatings. Researchers [12] found that the CHF of the coated surface increased due to the unlike morphologies of the surface and interconnected pore structures. Kirby and Westwater [13] first established a relationship between CHF and dry spots found on the heated surface and observed that the occurence of CHF was due to the presence of dry spots Semeria and Martinet [14] developed mathematical formulae based on actual observations and temperature conditions found from dry region ("calefaction spot"). The dry spot spreading and its dynamics were originally noticed experimentally by Van Ouwerkerk [15]. Using the whole reflection approach, Chung and No [16] investigated the dynamic behaviour of dry patches on a sapphire plate. According to Theofanous and Dinh [17], the instability and dynamics of liquid microlayer found on heated surface are responsible to control the value of CHF. Using a high-speed infrared thermometry approach, Kim et al. [3] published experimental data of nucleate boiling events under diverse surface heat flow circumstances (known as DEPICT). The growth, merging and its spreading behaviour of dry spots observed on a heated surface, and its frequency and density grew dramatically as the value of imposed heat flux raised. Higher surface roughness levels allow improved heat transmission, according to Umesh et al. [18].

From the literature, it is observed that nucleate boiling has been studied over the different surface fluid combinations. As per the authors knowledge, there are very few studies on boiling heat transfer over stainless steel surface taking acetone as working fluid under different roughness. Based on the foregoing information, the current motivation is to investigate the boiling heat transfer of acetone over flat stainless steel surfaces with varying roughness values under saturated and atmospheric conditions.

2 Experimentation

Figure 1 illustrates the current investigation's pool boiling experimental setup with various components.

The test vessel consists of side Teflon covers, upper Teflon cover, toughened glasses, and a Teflon base that contains the arrangement for thermocouple, test section, and DC heater. At the bottom part of the test vessel a Teflon base of 260 mm

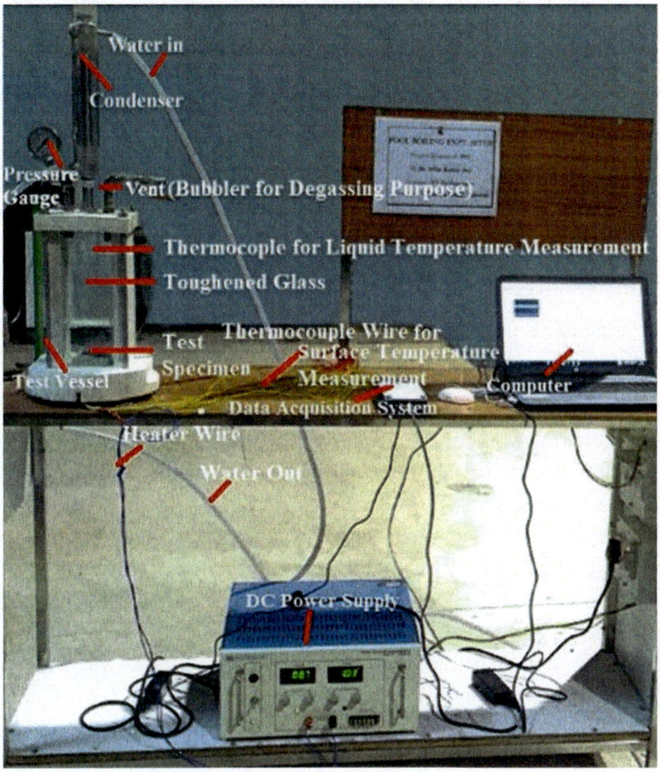

Fig. 1 Pool boiling experimental setup

in diameter and 65 mm in thickness is having rectangular slot to put the side Teflon cover of dimensions of 120 mm × 295 mm × 30 mm. The toughened glasses are inserted into the cavity made inside the side Teflon. The arrangements for bubbler and pressure gauge were fitted at the upper Teflon cover. At atmospheric pressure, the test vessel housed a 150 mm-deep pool of FC-72 working fluid. During pool boiling, the saturated condition in the vessel is monitored with a thermocouple probe form the top part of the test vessel and a pressure gauge is mounted on it to ensure the atmospheric condition. A DC flat heater with dimensions of 40 mm × 30 mm × 12 mm is placed at the bottom part of the flat test section and is operated by DC power supply. A double pipe heat exchanger is mounted on the top part of the experimental setup. It condenses the vapour created during experimentation and maintains a steady liquid level.

The rectangular flat test section of dimensions (52 mm × 42 mm × 8 mm) is made with 304 grade stainless steel using CNC milling available in Central Tool Room and Training Centre, Bhubaneswar as displayed in Fig. 2.

The test section's surface was first polished with 1500 grade emery paper, then with 2000 grade emery paper. It was polished with emery paper and diamond paste

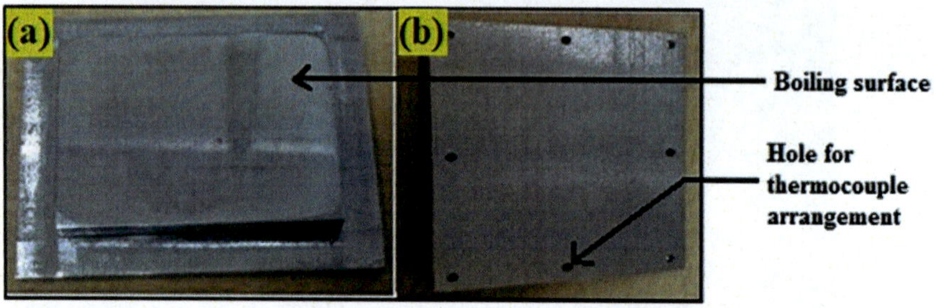

Fig. 2 **a** Test section upper part. **b** The bottom surface of the test section with holes for thermocouple arrangement

once more to get a near-mirror sheen. In this manner, the desired surface roughness values of 0.3–0.5, and 0.8–1 μm are maintained over the test section. First, acetone was used to clean the surface, followed by 1% HCL. Finally, distilled water was used to rinse the surface. The holes were made on the test section bottom part to measure the surface temperature of the surface assuming 1-D steady state heat conduction equation. Before assembling the experimental setup, the DC heater of dimensions (40 mm × 30 mm) is placed inside the cavity made in the Teflon base. To lower contact resistance, conductive paste is put at the interface between the heater and the test area. The test section with thermocouples arrangement is kept above the DC heater; then, the experimental setup is assembled to check thoroughly the leakage from the test vessel using distilled water. For the surface temperature measurement in this investigation, K-type thermocouples were employed.

With the aim of check accurate and equal measurement of temperature distribution at each thermocouple position, the computational study is to be required. Therefore, the 2D model considering the test section, DC heater and the cylindrical Teflon sheet is created in COMSOL Multi-Physics Software 5.0. The computational study using 1D steady state heat conduction equation has been used to calculate the temperature distribution of the test section when it is energized. As demonstrated in Figs. 2 and 3, the temperature at whatever height from the interface of the test section and the heater is the same along the plane parallel to the heater surface. Before starting the experimentation, the test vessel and test specimen are thoroughly cleaned by using acetone. The leakage test of the setup was done with compressed air and then with distilled water. Heater holes and thermocouple holes are also cleaned using acetone. The test specimen is push fitted into the cavity made at centre of the base Teflon. The heaters and thermocouples are inserted into their respective holes in the test specimen. In order to negate the effect of thermal contact resistance, thermal conductive paste is applied on the peripheral area of heaters and to the tip of thermocouples. The temperature measurements are obtained by connecting the thermocouples to the data gathering equipment. To energize the test specimen during testing, the heaters are linked to a DC power supply. For safety reason, the maximum current that can be drawn by DC heaters is made fixed in the DC power supply. This is calculated from heater rating and maximum voltage supply from DC power source. An initial power supply is given to heaters to provide the heat flux around 11.16 kW/m^2. With the set heat flux the test fluid takes nearly one hour to reach the saturation temperature i.e. 56 °C. Then the heat flux from the surface is gradually increased in steps of 10 kW/m^2 by increasing the voltage in DC supply. The heat flux is varied from 11.16 to 52.53 kW/m^2. The readings of surface temperature are taken under steady state. The pressure gauge readings are checked in periodic intervals to ensure that the experiment is running at atmospheric pressure. At each heat flux, interim video clips are recorded and numbers of images are taken to study the vapour bubble generation using high speed camera.

The DC heater was operated to achieve thermal stabilization for 24 h before conducting experiments. The liquid pool inside the test vessel is degasified to remove the dissolved gases through bubbler. Thermocouples are calibrated to get accurate experimental temperature distribution on heating surface. Thermal conductive paste is provided to diminish the contact resistance at surface interface of heater and test section. The thermo-physical properties of acetone are listed in Table 1.

The thermocouples are put into the test specimen's wall thickness to detect the surface temperature during the experiment. However, this is not actual surface temperature. So a drop of temperature (δT_w) in between thermocouple and boiling surface is evaluated through the wall thickness using the linear heat conduction equation as given below.

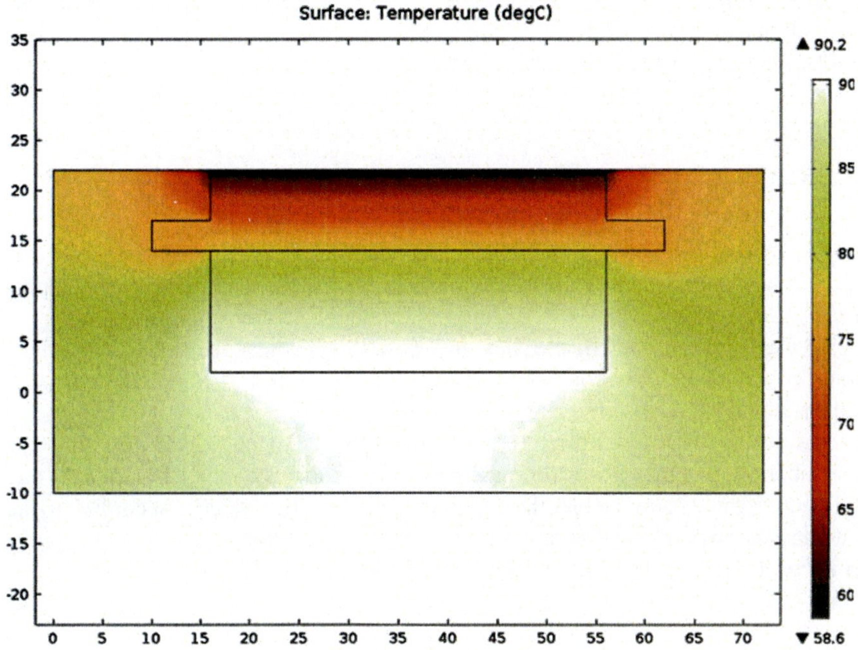

Fig. 3 Computational study of temperature distribution of the test section

$$\delta T_w = q\left(\frac{\delta x}{k}\right) \tag{1}$$

The BHTC (h) is determined using Eq. 2 as follows [19–22]:

$$h = \frac{q}{T_{\text{avg}} - T_{\text{sat},l}} \tag{2}$$

where

$$T_{\text{avg}} = \frac{\begin{array}{c}(T_{s1} - \delta T_w) + (T_{s2} - \delta T_w) + (T_{s3} - \delta T_w) + (T_{s4} - \delta T_w) + \\ (T_{s5} - \delta T_w) + (T_{s6} - \delta T_w) + (T_{s7} - \delta T_w) + (T_{s8} - \delta T_w)\end{array}}{8}$$

T_{avg} = Average surface temperature of the test specimen at the surface liquid interface.

$T_{\text{sat},l}$ = Saturation temperature of liquid (FC-72) inside the test vessel.

$T_{s1}, T_{s2}, T_{s3}, T_{s4}, T_{s5}, T_{s6}, T_{s7}$ and T_{s8} are the individual surface temperature measured by thermocouples at eight various locations on the wall thickness of the test specimen.

An uncertainty analysis is conducted for the current investigation using the error propagation approach [23], with maximum uncertainties of 0.3% for surface temperature, 1% for heat flux and 5.2% for BHTC.

3 Results and Discussions

The current part expands the findings of the investigation such as the influence of heat flux on wall superheat and its effect on BHTC at various test specimen roughness levels. It also includes the existing author's validation of the current work.

Figure 4 presents the effect of heat flux on wall superheat and coefficient of heat transfer with the surface roughness of a Ra (0.3–0.5 μm ± 0.005 μm) along with experimental data of existing author.

The results of current experiment are well matched with the available existing data [18] obtained at a lower heat flow, as shown in Fig. 4a. The fluctuation in heat flux follows the same pattern under the higher range of heat flux values. According to the findings, our experimental setup is capable of accurately predicting the boiling curve at atmospheric pressure. The change of the BHTC with wall superheat for saturated boiling of acetone on a plain stainless steel flat surface at atmospheric pressure, as illustrated in Fig. 4b, yields similar results.

Figure 5a, b demonstrate the influence of surface roughness on wall superheat and BHTC for different roughness values with various heat fluxes. The increased value of heat flux leads to raise the value of wall superheat as illustrated in Fig. 5a.

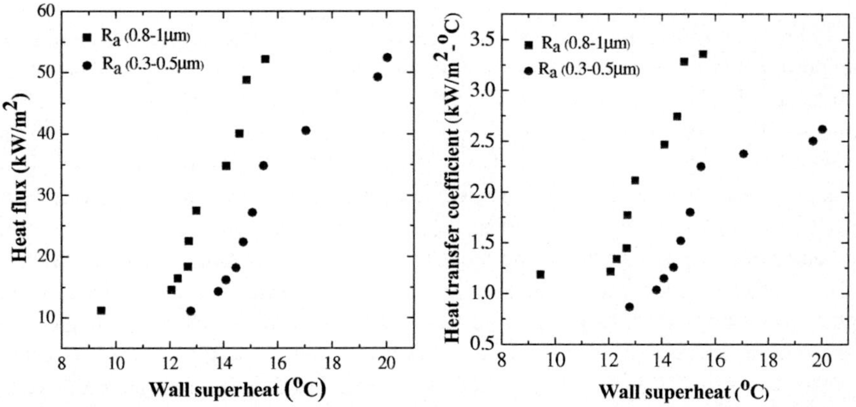

Fig. 4 **a** Influence of heat flux on observed value of wall superheat, **b** influence of wall superheat on BHTC

This is due to its natural qualities and material features. When the degree of surface roughness is raised, however, the value of wall superheat drops. The boiling curve changed to the left as a result. For a given heat flow, a surface with a roughness value of 0.8–1 has less wall superheat than 0.3–0.5 μm unevenness value. This occurs as the heated surface is subjected to more activation of a number of tiny cavities. As a result, the heat transmission from the heated surface to the boiling fluid improves. Consequently, the heat is not absorbed by the surface, resulting in low wall superheat. For the above-mentioned surface unevenness, the observed wall superheats are 12.71 and 14.73 °C at 22.33 kW/m^2 value of heat flux.

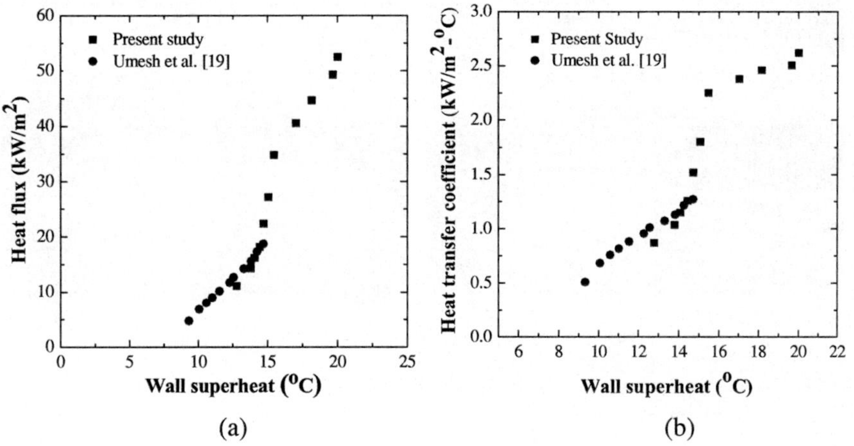

Fig. 5 For various roughness values **a** influence of heat flux on wall superheat, **b** influence of wall superheat on BHTC

Figure 5b represents the influence of surface unevenness on the BHTC during experimentation. As wall superheat increases, its effect is to increase the magnitude of BHTC. In addition, for a given magnitude of wall superheat, the BHTC rises when the surface unevenness increases. The value of BHTC becomes maximum for the higher surface unevenness value (0.8–1 μm), than the lower value of 0.3–0.5 μm. As previously stated, a greater surface unevenness value promotes the activation of tiny voids presented on the heated surface. This event draws out a larger population of active nucleation sites throughout the heated surface, resulting in a higher BHTC. The number of active nucleation sites detected on the heated surface increases when the surface texture value is increased, as can be observed in high-speed images like the one shown in Fig. 6. In contrast, other researchers [24–27] have reported that surface unevenness improves the NBHT for a variety of fluids and surfaces. It's worth noting that as surface unevenness rises, so does the amount of superheat on the wall and the BHTC. In this investigation, the most substantial increase in the BHTC for stainless steel surfaces with increasing surface unevenness is observed to be 15%.

As illustrated in Fig. 6, high-speed photographs of boiling processes have also been explored at various surface roughnesses. The photos show that for a uniform heat flux value of 34.38 kW/m^2, the bubble density rises as the nucleate site density increases at increased surface unevenness. The bubble population is clearly reduced at low surface roughness in Fig. 6a, but the bubble population grows at high roughness, as seen in Fig. 6b. The coefficient of heat transmission rises when the heated surface's roughness value rises as a result of this phenomenon.

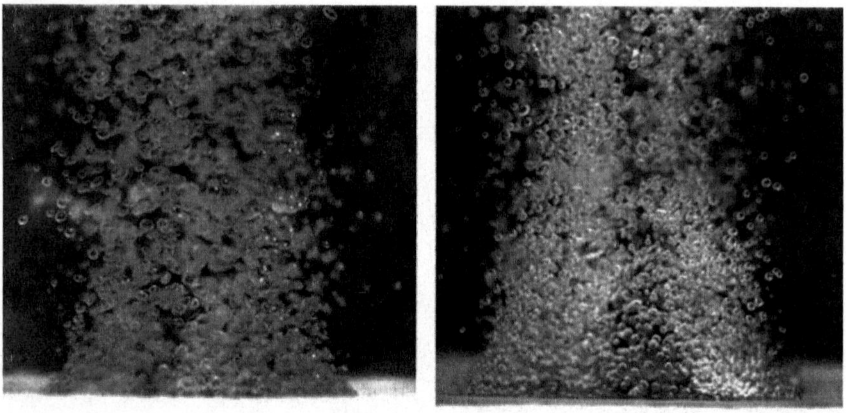

(a) (b)

Fig. 6 Photographic image of boiling phenomena with heat flux of 34.38 kW/m^2 **a** R_a = 0.3–0.5 μm, **b** R_a = 0.8–1 μm

Table 1 Thermo-physical properties of acetone

Parameters	Values	Units
Boiling point (1 atm)	56.08	°C
Critical temperature (T_{crit})	235	K
Critical pressure (P_{crit})	46.9×10^5	Pa
Vapour pressure (P_v)	30.8×10^3	Pa
Latent heat of vapourization (h_{lv})	534,000	J/kg
Liquid density (ρ_l)	784.5	kg/m^3
Kinematic viscosity (υ_l)	3.9×10^{-7}	m^2/s
Absolute viscosity (μ_l)	308.94×10^{-7}	Pa s
Liquid specific heat $\left(c_{pl}\right)$	2140	J/kg K
Liquid thermal conductivity (k_l)	0.179	W/m K
Surface tension (σ)	0.023	N/m
Coefficient of expansion (β_l)	0.00143	°C^{-1}
Molecular weight (M)	58.08×10^{-3}	kg/mole

4 Conclusion

The present investigation aims to study the influence of surface unevenness on the heat transfer for various heat fluxes under saturated condition at atmospheric pressure during boiling over stainless steel surfaces. The experimental results are well matched with the available existing experimental data in the literature. The BHTC rises as wall superheat increases, independent of surface roughness levels. When the unevenness of the surface rises, the minimum value of wall superheat is obtained. On the other hand, for a given het flux value, the BHTC upturns with a rise in the value of surface unevenness. Comparing the lower surface roughness value, the higher value gives the maximum BHTC with the lowest wall superheat. An augmentation of about 15% is found in the BHTC when the roughness is increased two fold. The findings of present investigation will be helpful where this surface fluid combination is to use in different aspects.

Acknowledgements Department of Science and Technology (DST) funded this research investigation under DST-RFBR project grant named as INT/RUS/RFBR/P-211. The authors acknowledge DST for the financial support to accomplish this research work.

References

1. Zuber N (1959) AEC Rep., AECU-4439
2. Kutateladze SS (1948) Kotloturbostroenie No. 3:10
3. Kim DE et al (2015) Exp Thermal Fluid Sci 66:173–196
4. Chang JY et al (1997) International Journal of Heat Mass Transfer 40(18):4437–4447

5. Chang JY et al (1997) Int J Heat Mass Transfer 40(18):4449–4460
6. Liter SG et al (2001) Int J Heat Mass Transfer 44(22):4287–4311
7. Wu W et al (2010) Int J Heat Mass Transfer 53:1773–1777
8. Ahn HS et al (2010) Nuclear Eng Design 240:3350–3360
9. Furberg R et al (2011) Appl Therm Eng 31:3595–3603
10. Misale M et al (2011) Exp Thermal Fluid Sci 35:772–779
11. Byon C et al (2013) Int J Heat Mass Transfer 65:655–661
12. Seo H et al (2015) Int J Heat Mass Transfer 82:490–502
13. Kirby DB et al (1965) Chem Eng Prog Symp Ser 57:238
14. Semeria R et al (1965) Symposium on boiling heat transfer in steam generating equipment and heat exchangers, Manchester
15. Ouwerkerk HJV (1972)Int J Heat Mass Transfer 15:25–34
16. Chung HJ et al (2003) Int J Heat Mass Transfer 46:2239–2251
17. Theofanous TG et al (2006) Multiph Sci Technol 18(3):251–276
18. Umesh V et al (2016) Experimental Heat Transfer 29:414–425
19. Swain A et al (2017) Heat Mass Transf 53(8):2487–2495
20. Mohanty RL et al (2020) Heat Mass Transf 56(7):2079–2092
21. Mohanty RL et al (2020) Int J Thermal Sci 156:106433
22. Mohanty RL et al (2021) Proceedings of the institution of mechanical engineers. Part C J Mech Eng Sci. https://doi.org/10.1177/09544062211057471
23. Kline SJ et al (1953) Mech Eng 75:3–8
24. Mohanty RL et al (2017) Renew Sustain Energy Rev 78:466–494
25. Jabardo JMS et al (2009) Exp Thermal Fluid Sci 33:579–590
26. Benjamin RJ et al (1997) Exp Thermal Fluid Sci 15:32–42
27. Mohanty RL et al (2018) International heat transfer conference digital library, pp 1439–1448

Flow Characteristics of Electromagnetohydrodynamics of a Third-Grade Fluid Through Large Parallel Plates with Moving Upper Plate

Sumanta Chaudhuri, Rajiva Lochan Mohanty, Paromita Chakraborty, and Vijay Kumar Mishra

1 Introduction

Magnetohydrodynamic (MHD) refers to the study of the interaction of a magnetic field with electrically conductive fluids, which is important in a variety of natural and man-made phenomena. MHD flow may be seen in the movement of fluids in the Earth's core. Magnetic fields are widely used in industry for fluid transfer and heating. Over the last decade, flow and heat transfer in micro and nanoscale heat exchangers has become a hot topic of research. The flow of fluid is operated in these microscale geometries by applying a magnetic field, an electric field, or both. As a result, researchers are interested in EMHD flow and EMHD convection. Microscale devices employ a combination of conventional pressure gradient-driven flow actuation mechanisms and electric and magnetic fields, according to Chakraborty et al. [1].

Jang and Lee [2] were the first to propose the concept of an EMHD micro pump. Wang et al. [3] conducted a numerical investigation of laminar, fully developed flow. Ho [4] investigated the flow via an EMHD micro pump's rectangular duct. In many situations, the channels are not smooth and have roughness from the fabrication process on the wall surface. Buren et al. [5] and Buren and Jian [6] used the standard perturbation approach to analyse EMHD flow via micro channels with corrugated

S. Chaudhuri · R. L. Mohanty (✉) · V. K. Mishra
School of Mechanical Engineering, Kalinga Institute Industrial Technology, Bhubaneswar, Odisha 751024, India
e-mail: rajivamohanty@gmail.com

S. Chaudhuri
e-mail: sumanta.chaudhurifme@kiit.ac.in

P. Chakraborty
School of Civil Engineering, Kalinga Institute Industrial Technology, Bhubaneswar, Odisha 751024, India

© The Author(s), under exclusive license to Springer Nature Singapore Pte Ltd. 2023
S. Revankar et al. (eds.), *Recent Advances in Thermofluids and Manufacturing Engineering*, Lecture Notes in Mechanical Engineering,
https://doi.org/10.1007/978-981-19-4388-1_21

walls. Approximating the corrugated top and lower walls as small-amplitude wavy surfaces yields the solution. The movement of Newtonian fluids was addressed in all of the studies mentioned above. However, non-Newtonian fluids are widely utilised in industry. The influence of a magnetic field on several electrically conducting non-Newtonian fluids were explored. As indicated by Nayak et al. [7], electrically conducting polymeric liquids can play an essential role in polymer technology, as mentioned by Nayak et al. [6]. In an investigation of wire coating, Nayak et al. [7] looked at MHD flow and heat transfer of a 3rd grade non-Newtonian fluid. MHD slip flows of non-Newtonian second-grade fluid and Walter's liquid B were investigated by Turkyilmazoglu [7]. Ellahi et al. [8] investigated the influence of the slip factor on the movement of third-grade fluids between concentric cylinders analytically. On the velocity, the impact of pressure gradients, non-Newtonian parameters and so on have been explored. Researchers looked on issues of practical importance regarding third-grade fluids. Wang et al. [9] explored the EMHD flow and heat transmission of a third-grade fluid through two plates that are parallel to each other. They solved the velocity as well as temperature distributions analytically, considering the viscous dissipation and joule heating effects. They expressed the results in terms of third-grade fluid parameters, Hartmann number and Brinkman number.

From the aforementioned observations, an EMHD flow of a third-grade fluid has been performed between two plates of significant size that are parallel to each other. One of the varieties of non-Newtonian fluid is third-grade fluid. Lubricating oils, polymers and slurry are all studied using this model [9]. Whether a fluid is second grade or third grade is determined by the number of terms in the Rivlin-Erickson tensor. Researchers have already considered EMHD flow with heat transmission in a fluid (3rd grade) [9]. In the investigation, the pressure gradient term was zero. The current study is mostly concerned with the influence of the top plate in motion. Furthermore, the study focuses on the impact of the pressure gradient term. The LSM (Least Square Method) has been used to solve the non-linear governing equations [10, 11]. In LSM, the presence of a tiny parameter in the issue is not essential. As a result, the current research is applicable to a variety of issues involving various third-grade fluid parameter values. The influence of pressure terms dominates the effect of the moving plate (top).

2 Problem Formulation

Figure 1 illustrates the physical problem under consideration. The flow conditions are considered to be laminar, stable and hydro-dynamically fully developed. The plates' widths are believed to be extremely large. The study's premise is that properties are constant and temperature insensitive. The coordinate system is used with the x, y and z axes in the flow, vertical and breadth directions, respectively. The pressure gradient is imposed perpendicular to the flow direction. Along the lateral (-z) axis, an electric field is applied and a uniform magnetic field with B as the flux density

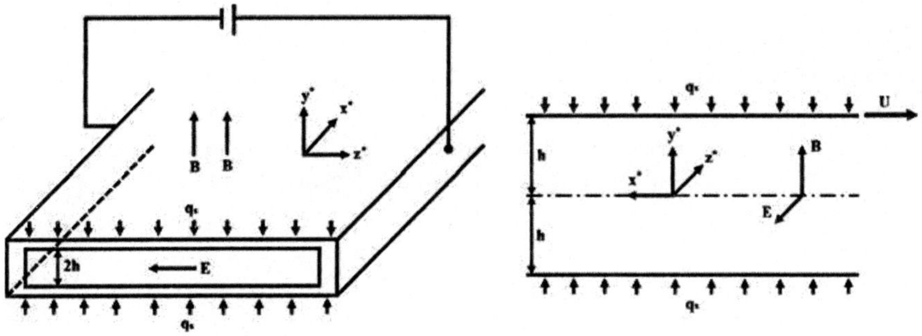

Fig. 1 Line diagram of EMHD flow between two parallel plates with the upper plate moving

is applied vertically (positive y-axis). Both the upper and bottom walls get constant and uniform heat fluxes.

3 Governing Equations

The fluid is propelled by the electromagnetic force created by the pressure gradient. The electromagnetic force was created when the applied magnetic field and electric field collided. The continuity equation for an incompressible fluid is expressed as below:

$$\nabla.\vec{V} = \vec{0} \tag{1}$$

where V denotes the velocity vector. The Lorentz force is created by the interaction of the magnetic and electric fields and it adds to the body force term. The following is the momentum conservation equation:

$$\rho\frac{D\vec{V}}{Dt} = \nabla.\underset{\sim}{\tau} + \vec{f} \tag{2}$$

where, f, τ and ρ are the body force per unit volume, stress and density, respectively.

$$\vec{f} = \vec{J} \times \vec{B} \tag{3}$$

$\vec{f}$, $\vec{J}\vec{B}$where, are the body force, the current density and the applied magnetic field. For steady-state, the left-hand side of Eq. (2) is zero. E and B are linked to current density in the presence of electric and magnetic fields as follows:

$$\vec{J} = \sigma(\vec{E} + \vec{V} \times \vec{B}) \tag{4}$$

where the electrical conductivity of the fluid is σ. The stress component in the momentum equation is made up of two parts. The first is linear, whereas the latter is non-linear. The non-linear component, also known as the deviatoric stress tensor, gives a non-linear connection between stress and strain rate for various non-Newtonian fluids. For third-grade fluids, the additional stress tensor is expressed as:

$$\underset{\sim}{\tau} = -p\underset{\sim}{I} + \mu\underset{\sim 1}{A} + \alpha_1\underset{\sim 2}{A} + \alpha_2\underset{\sim 1}{A}^2 + \beta_1\underset{\sim 3}{A} + \beta_2\underset{\sim 1}{A}\underset{\sim 2}{A} + \underset{\sim 2}{A}\underset{\sim 1}{A}) + \beta_3(tr\,A^2)\underset{\sim 1}{A} \tag{5}$$

where p is the static pressure, μ, α_1, α_2, β_1, β_2, β_3 are the material constants of the fluid and A_1, A_2 and A_3 represent the first, second and third Rivlin-Erickson tensors, respectively. The following relations can be used to get these tensors:

$$\underset{\sim 1}{A} = (\text{grad}\vec{V}) + (\text{grad}\vec{V})^{\text{Transpose}} \tag{6}$$

$$\underset{\sim nn}{A} = \frac{d\underset{\sim nn-1}{A}}{dt} + \underset{\sim nn-1}{A}(\text{grad}\vec{V}) + (\text{grad}\vec{V})^{\text{Transpose}}\underset{\sim nn-1}{A},\ nn = 1, 2, 3 \tag{7}$$

The velocity gradient matrix grad$V$* is used for calculating the fluid element motion and shear strain rate. Equations (5)–(7) uses a velocity gradient vector to explain the link of shear stress with shear strain rate. In Eq. (5), only up to A_1 is used for modelling Newtonian fluids, simplifying the constitutive equation to linear. Other factors, such as A_2 and A_3, must be included for non-Newtonian fluid models in order to capture the fluid's non-linear behaviour.

The energy conservation equation is as follows:

The velocity component u exists and is dependent on y, as per the assumptions. According to the assumptions, all other velocity components become zero. As a result, the expression of velocity vector is obtained as follow:

$$V = [u(y), 0, 0] \tag{8}$$

Upon putting of the expression of velocity from Eq. (8) into Eq. (2), using Eqs. (3)–(7), the conservation of momentum equations in the three directions are presented as below:

x Momentum Conservation Equation:

$$\frac{\partial p}{\partial x} = \mu\frac{d^2u}{dy^2} + 2(\beta_2 + \beta_3)\frac{d}{dy}\left(\frac{du}{dy}\right)^3 - \sigma B^2 u + \sigma BE \tag{9}$$

y momentum conservation equation:

$$\frac{\partial p}{\partial y} = (2\alpha_1 + \alpha_2)\frac{\mathrm{d}}{\mathrm{d}y}\left(\frac{\mathrm{d}u}{\mathrm{d}y}\right)^2 \tag{10}$$

z momentum conservation equation:

$$\frac{\partial p}{\partial z} = 0 \tag{11}$$

It is obvious from Eqs. (9), (10) and (11) that the pressure gradient remains constant in the flow direction and its.

3.1 Boundary Conditions

In order to solve the second-order momentum conservation equation, two boundary conditions needed. These are given as:

$$u(-H) = 0, u(H) = U_p \tag{12}$$

3.2 Non-dimensionalisation

The following reference velocity and dimensionless variables are introduced for reducing the momentum conservation equations into their dimensionless forms:

$$u^* = \frac{u}{U_p}, y^* = \frac{y}{H} \tag{13}$$

where, y^* and u^* are the non-dimensional coordinate in the y-direction and non-dimensional form of velocity. U_p is the velocity of the upper plate and $N1$ indicates a non-dimensional pressure gradient. The dimensionless momentum is calculated as follows using the dimensionless variables specified by Eq. (13):

$$\frac{\mathrm{d}^2u^*}{\mathrm{d}y^*2} + 6A\left(\frac{\mathrm{d}u^*}{\mathrm{d}y^*}\right)^2\frac{\mathrm{d}^2u^*}{\mathrm{d}y^*2} - \mathrm{Ha}^2u^* + H_1 = N_1 \tag{14}$$

$$A = \frac{\beta_2 + \beta_3}{\mu}\left(\frac{Up}{H}\right)^2, \mathrm{Ha} = BH\sqrt{\frac{\sigma}{\mu}},$$
$$H_1 = \frac{\sigma BEH^2}{\mu U_p} \tag{15}$$

where A refers the intensity of the non-Newtonian effect, Ha denotes the strength of the magnetic field and H_1 is the strength of the electric field. The following are the boundary conditions in dimensionless form that should be utilised to solve the momentum conservation equation

$$u^*(-1) = 0, \text{ and } u^*(1) = 1 \tag{16}$$

The asterisk (*) is removed for convenience and the dimensionless variables are indicated without the asterisks.

Equation (14) is difficult to acquire correct analytical answers because of its non-linear character. Using diverse methodologies, however, semi-analytical or approximation analytical answers are available. LSM is one such method. In order to solve the equations in this study, LSM is being used. The existence of any minor parameters is not required by LSM. Because of some simplifying assumptions, the issue is not made linear. Readers can refer to the researchers' studies [10, 11] for further information about LSM.

4 Solution

A semi-analytical technique called the Least Square Method is utilised to solve the set governing equation according to the problem formulation. The capacity of LSM to create accurate non-linear equation solutions is exploited. The first step is to select a trial function that meets the boundary requirements. The trial functions might be of the kind $(1 - y^{n1})$ (where n^1 is even). As a result, the velocity may be expressed as:

$$u = k\left(\frac{1+y}{2}\right) + c_1\left(1 - y^2\right) + c_1\left(1 - y^4\right) \tag{17}$$

Now, substituting Eq. (17) into Eq. (14), an error function is obtained. This error can be expressed as a polynomial of the following form:

$$R(y) = a_0 + a_1 y + a_2 y^2 + a_3 y^3 + a_4 y^4 + a_5 y^5 + a_6 y^6 + a_7 y^7 + a_8 y^8 \tag{18}$$

where,

$$a_0 = \left(-2c_1 - 3Ak^2 c_1 - 0.5kHa^2 - Ha^2 c_1 - Ha^2 c_2 + H - N\right) \tag{19}$$

$$a_1 = 12Akc_1^2 - 0.5Ha^2, a_2 = -12c_2 - 48Ac_1^3 - 18Ak^2 c_2 + Ha^2 c_1$$

$$a_3 = 120Akc_1 c_2, a_4 = -480Ac_1^2 c_2 + Ha^2 c_2, a_5 = 128Akc_2^2, a_6 = -1344Ac_1 c_2^2$$

$$a_7 = 0, a_8 = -1152Ac_2^3$$

The error $R(y)$ is minimised with respect to the unknown coefficients c_1 and c_2. Differentiating $R(y)$ with respect to c_1, and c_2 we get:

$$\int_{-1}^{1} R\frac{\partial R}{\partial c_1}\mathrm{d}y = 0 \tag{20}$$

$$\int_{-1}^{1} R\frac{\partial R}{\partial c_2}\mathrm{d}y = 0 \tag{21}$$

where,

$$\frac{\partial R}{\partial c_1} = b_0 + b_1y + b_2y^2 + b_3y^3 + b_4y^4 + b_5y^5 + b_6y^6 + b_7y^7 + b_8y^8 \tag{22}$$

$$\frac{\partial R}{\partial c_2} = d_0 + d_1y + d_2y^2 + d_3y^3 + +d_4y^4 + d_5y^5 + d_6y^6 + d_7y^7 + d_8y^8 \tag{23}$$

$$b_0 = -2 - 3Ak^2 - Ha^2, b_1 = 24Akc_1, b_2 = -144Ac_1^2 + Ha^2, b_3 = 120Akc_2 \tag{24}$$

$$b_4 = -960Ac_1c_2, b_5 = 0, b_6 = -1344Ac_2^2, b_7 = 0, b_8 = 0$$

$$d_0 = -Ha^2, d_1 = 0, d_2 = -12 - 18Ak^2, d_3 = 120Akc_1, d_4 = -480Ac_1^2,$$
$$d_5 = 256Akc_2, d_6 = -2688Ac_1c_2, d_7 = 0, d_8 = -3456Ac_2^2$$

Equations (20) and (21) are non-linear, coupled, algebraic equations. These are solved by symbolic computational tool in MATLAB for Ha.

5 Results and Discussions

To begin, the current study's findings were cross-checked against the previously published results of Wang et al. [9]. In the current method, $k = 0$ is substituted for validation reasons, which simplifies the problem of fixed lower and upper plates. The results of Wang et al. [9] for both fixed plates are obtained when they are exchanged. The results of Wang et al. [9] are compared to those of the current investigation in Fig. 2. The results are well-matched, proving the usefulness of LSM for non-linear situations, as seen in the figure. The impact of various factors on the dimensionless velocity distributions is furthermore addressed.

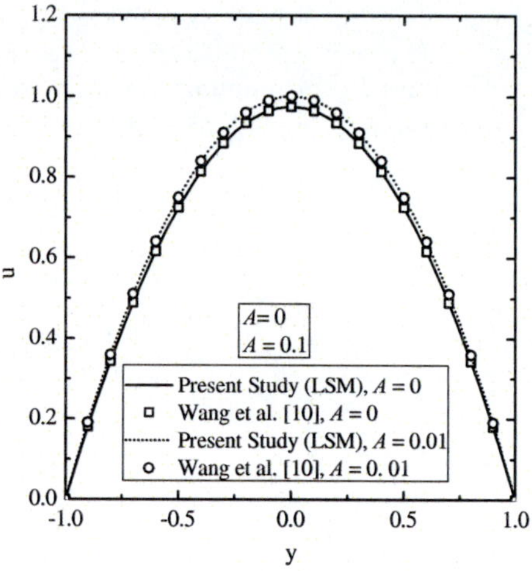

Fig. 2 Validation of the present work with Wang et al. [9]

Figure 3 demonstrates non-dimensional flow velocity distributions for several values of A.

The graphic shows that as the third-grade fluid parameters are increased, velocity drops. The fluid motion in this problem is caused by both the action of the top plate's motion and the pressure gradient. The fluid velocity will rise as the viscosity increases due to the sliding top plate. This is due to the larger drag force value. However, the resistance to motion increases when the effective viscosity increases as a result of the pressure gradient, and velocity drops. The influence of the pressure gradient prevails in the result shown in Fig. 3, and the velocity falls as A increases.

Figure 4 depicts the influence of the magnetic force on velocity. The graphic shows that when the value of Ha increases, the velocity drops. Ha stands for magnetic force

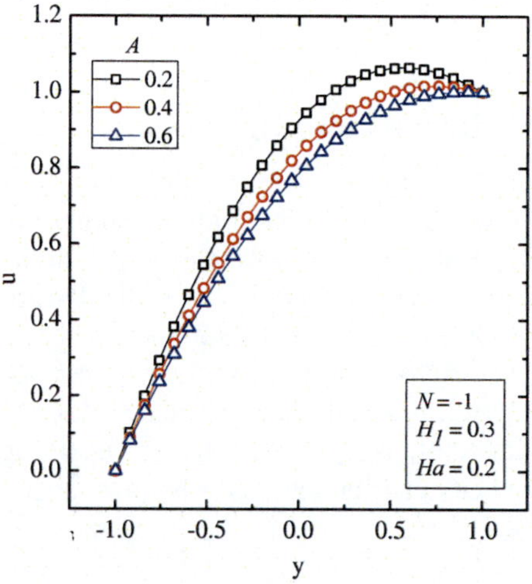

Fig. 3 Dimensionless velocity profile according to various A, for $N = -1$, $H_1 = 0.3$, and $\mathrm{Ha} = 0.5$

strength, and a greater value of Ha corresponds to a stronger magnetic force, and vice versa. As a result, a high Ha value causes a significant force on the fluid, which causes higher resistance to motion. As a result, fluid velocity will be lowered. When the value of A is increased from 0.2 to 0.6, the maximum velocity drops from 1.02 to 0.94. The observations show that increasing Ha from 0.5 to 1 causes a considerable drop in velocity.

Figure 5 shows non-dimensional velocity distributions for various H_1. The electric field strength employed in the problem is denoted by H_1. The force experienced by the charged particles of the fluid rises as the electric field increases. As a result, as H_1 rises, velocity rises as well.

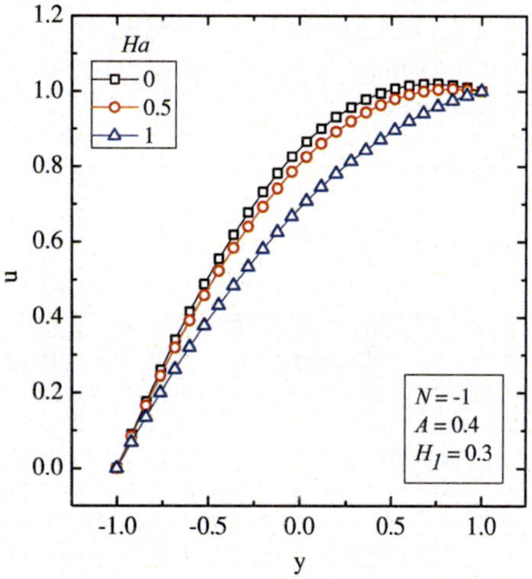

Fig. 4 Dimensionless velocity profile for different Ha, when $N = -1$, $A = 0.4$ and $H_1 = 0.3$

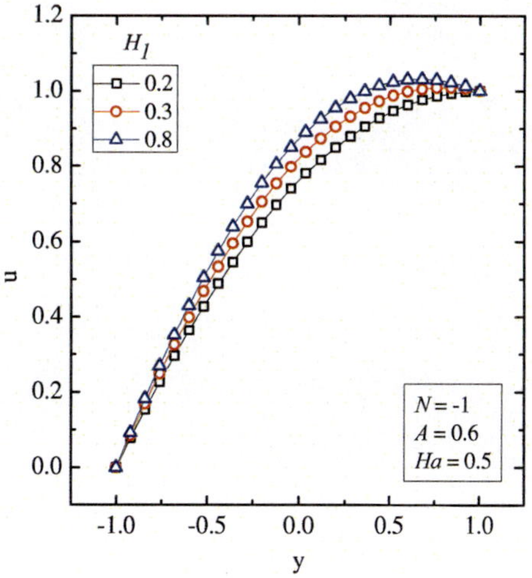

Fig. 5 Dimensionless velocity profile for different H_1, when $N = -1$, $A = 0.6$ and Ha $= 0.5$

6 Conclusion

EMHD flow of a fluid (3rd grade) flowing between parallel plates is investigated. The top plate is thought to be travelling at a constant speed. In the range of parameters investigated, the results show that the influence of the pressure gradient dominates the velocity. The velocity decreases as A increases. Higher A values cause the moving plate to exert more drag on the fluid (upper). The dominance of pressure causes a drop in velocity under the conditions utilised in the study. For higher Ha, the influence of Ha on velocity is substantial. The findings of this study can help researchers better understand the impact of altering the upper plate in EMHD flow and term dominance. The velocity field acquired in this study may be applied to conduct a heat transfer analysis in an EMHD flow for further research. Furthermore, qualities that are temperature sensitive can be employed to determine their impact on heat transmission.

References

1. Chakraborty R et al (2013) Int J Heat Mass Transf 67:1151–1162
2. Jang J et al (2000) Sensors Actuators A Phys 80:84–89
3. Wang PJ et al (2004) Biosens 20:115–121
4. Ho JE (2007) J Mar Sci Technol 15:315–321
5. Buren, M., et al., Journal of Physics D: Applied Physics 47 (2014).
6. Buren M et al (2015) Electrophoresis 36:1539–1548
7. Nayak MK et al (2014) Int J Heat Mass Transf 79:1087–1095
8. Turkyilmazoghu M et al (2012) Comput Fluids 70:53–58
9. Ellahi R et al (2010) Non-linear analysis-real world applications 10:1139–1146
10. Wang L et al (2016) Colloids Surf A 494:87–94
11. Chaudhuri S et al (2019) J Thermal Sci Eng Appl 11:024504
12. Chaudhuri S et al (2021) Heat Transfer 50(7):6737–6757

Scaled Conjugate Gradient Algorithm Trained ANN for Parameter Retrieval in Combined Heat Transfer in Porous Ceramic Matrix

Swagatika Acharya, Vijay Kumar Mishra, Jitendra Kumar Patel, Gaurav Gupta, Sumanta Chaudhuri, and Niraj Kumar Mishra

Nomenclature

A	Surface area of ceramic matrix (per unit volume basis), $^1/_{\mathrm{m}}$
c	Specific heat of air at constant pressure, J/kg K
G	Emissive power, $^W/_{\mathrm{m}^2}$
h	Heat transfer coefficient, $^W/_{\mathrm{m}^2 \cdot K}$
i	Radiation intensity, $^W/_{\mathrm{m}^2 \cdot sr}$
k	Thermal conductivity, $^W/_{\mathrm{m} \cdot K}$
L_x	Ceramic length in $x-$ direction, m
L_y	Ceramic length in $y-$ direction,
q_R	Radiative heat flux, $^W/_{\mathrm{m}^2}$
$\dot{Q}$	Volumetric heat generation source term, $^W/_{\mathrm{m}^3}$
S_{av}	Average source term, $^W/_{\mathrm{m}^2}$
T	Temperature, K
u	Velocity, $^m/_s$

S. Acharya · V. K. Mishra (✉) · J. K. Patel · S. Chaudhuri
School of Mechanical Engineering, KIIT Deemed to be University, Bhubaneswar 751024, India
e-mail: mishra.vdm@gmail.com

G. Gupta
Vellore Institute of Technology, Vellore, India

N. K. Mishra
Department of Mechanical Engineering, NIT Uttarakhand, Sumari, India

© The Author(s), under exclusive license to Springer Nature Singapore Pte Ltd. 2023
S. Revankar et al. (eds.), *Recent Advances in Thermofluids and Manufacturing Engineering*, Lecture Notes in Mechanical Engineering,
https://doi.org/10.1007/978-981-19-4388-1_22

Greek Symbols

β	Extinction coefficient of ceramic, $^{1}/_{\mathrm{m}}$
δ	Unit step function
ε	Emissivity of ceramic
η	Coordinate (in dimensionless form)
θ	Temperature (in dimensionless form)
ρ	Air density, $^{kg}/_{\mathrm{m}^3}$
σ	Stefan-Boltzmann constant, $5.67 \times 10^{-8}\, ^{W}/_{m^2 \cdot K^4}$
φ	Porosity of porous ceramic matrix
Ψ_{Rad}	Non-dimensional radiative heat flux
ω	Scattering albedo

Subscripts and Superscripts

E	East direction
W	West direction
N	North direction
S	South direction
e	Exit side
g	Gas phase
i	Inlet side
s	Solid phase
*	Non-dimensional

1 Introduction

Any thermal problem involves computation of velocity/temperature/heat flux, etc. (dependent variables) by solving the applicable governing equations subjected to knowledge of physical/geometric/flow properties (independent properties) together with boundary conditions. Such problems come under direct problems. In various practical problems, dependent variables are known but independent variables are required to be computed. These are called inverse problems and are very useful in situations such as: inaccessible location of measurement site, chances of disturbances in flow due to probe intrusion in narrow channels, etc. Inverse problems play a vital role in design and performance enhancement of any thermal device [1–3]. Inverse problems are difficult to solve due to their ill-posed nature.

Porous matrix made of ceramics such as silicon oxide (SiO_2), alumina (AL_2O_3), zirconia (ZrO), etc. have various unique properties such as: high surface area per unit volume, high strength in high temperature, heat transfer enhancement through conduction and radiation. These qualities of porous ceramic matrix (PCM) make it very popular for applications like burners, filters, bioceramics, etc. Conventional materials like metals, polymers, wood, glass, etc. are not suitable for such applications [4]. Burners based on PCM offer many advantages as compared to conventional free flame burners. PCM-based burners are able to modify the heat transport in the burner by conduction and radiation. Some of the advantages of PCM-based burners are: high heat generation rate leading to compact design, low emissions due to modified temperature distribution, extended flammability limits, etc. [5]. Large volume of work is available in literature to understand the PCM-based burner, but use of inverse analysis and optimization methods to study PCM-based burner is quite rare [6].

Artificial neural network (ANN) comes very handy when it comes to solving problems where the relationship between independent variables and dependent variables is either not explicitly known or very complicated [7–9]. ANN offers a fast and non-iterative tool for high-quality results [10, 11]. Vast application of inverse analysis and search for simple and accurate methods for design and improvement of PCM-based burner leads to ANN approach. Scarcity of ANN-based approaches for solving PCM-based burner problems, in the open literature, motivated the present work [12–15]. Scaled conjugate gradient (SCG) algorithm is very popular training algorithm due to its speed and good accuracy. Thus, SCG algorithm is used in the present ANN network.

Present work involves a burner with PCM, and the applicable governing equations are solved by finite volume method (FVM). The temperature profile of gas and the solid phase are obtained by FVM, for 49 different values of heat transfer coefficient (HTC). This data is used in the training of ANN model by employing SCG algorithm for training of neurons. Once the ANN model is trained, it is analyzed for its robustness, i.e., whether it can give highly accurate result consistently. After successfully analyzing the ANN model, it is fed with an unknown pair of gas and solid-phase temperature profile. The ANN gives back the corresponding HTC as output with good accuracy. The accuracy of the ANN model trained with SCG algorithm is good. This approach is very useful and simple for design of a PCM-based burner.

2 Problem Formulation

A 2D rectangular PCM is considered and air is allowed to flow from left to right (Fig. 1). The axis system together with the dimensions of the PCM is depicted in Fig. 1. Effect of combustion on heat flow is considered by placing a heat generation zone (HGZ) at the center of the PCM. Since the gas and solid may have different temperatures due to thermal local non-equilibrium, different energy equations are considered for each of the two phases. Because of conduction and radiation in PCM, heat is redistributed in the burner. The gas-phase takes heat from the solid phase in the entrance region, while at the exit region heat transfer from gas to solid phase takes place. Also, due to conduction and radiation in the solid phase, heat flows from HGZ to upstream and downstream directions. Thus, this technology of heat generation embedded in a PCM, helps us to dissipate heat from a location to throughout the PCM. All the properties of the gas and PCM are assumed to be free from temperature change. Under steady conditions, the two energy equations governing the problem are:

$$\text{Gas phase: } \phi\rho_g c_g u\frac{\partial T_g}{\partial x} + (1-\phi)Ah(T_g - T_s) = \phi\,\dot{Q}\,\delta(x) + \phi k_g\left(\frac{\partial^2 T_g}{\partial x^2} + \frac{\partial^2 T_g}{\partial y^2}\right) \tag{1}$$

$$\text{Solid phase: } (1-\phi)Ah(T_g - T_s) + (1-\phi)k_s\left(\frac{\partial^2 T_s}{\partial x^2} + \frac{\partial^2 T_s}{\partial y^2}\right) - \nabla\cdot q_{\text{Rad}} = 0 \tag{2}$$

Heat generation is considered in $0.45L_x < x \leq 0.55L_x$, with the help of Dirac-delta function $\delta(x)$ which is one in heat generation zone and zero elsewhere. In Eq. (2), the volumetric radiative term $\nabla\cdot q_{\text{Rad}}$ is computed as:

$$\nabla\cdot q_{\text{Rad}} = (1-\omega)\beta\left(4\sigma T^4 - G\right) \tag{3}$$

Further, in Eq. (3) the incident radiation G is calculated by following expression:

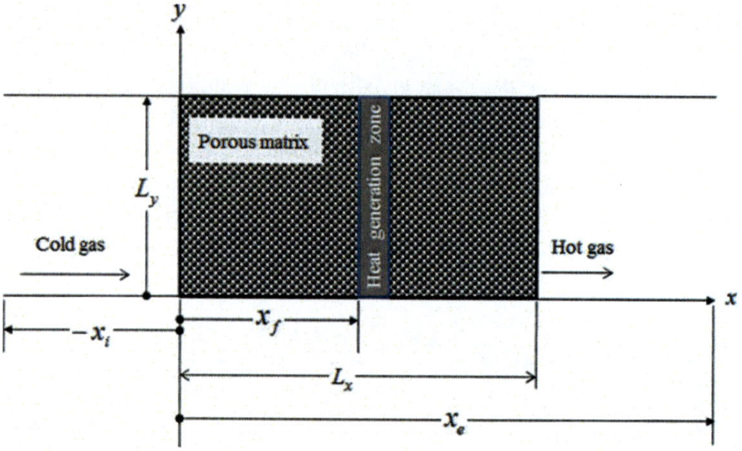

Fig. 1 Schematic of the conduction radiation problem in PCM

$$G = \int_{\Omega=0}^{4\pi} I(\Omega)\mathrm{d}\Omega = \int_{\mu=0}^{2\pi} \int_{\lambda=0}^{\pi} I(\lambda, \mu) \sin\lambda \, \mathrm{d}\lambda \, \mathrm{d}\mu$$
$$\approx \sum_{k=1}^{M_\mu} \sum_{l=1}^{M_\lambda} I^m(\lambda_l^m, \mu_k^m) 2\sin\lambda_l^m \sin\left(\frac{\Delta\lambda_l^m}{2}\right) \Delta\mu_k^m \tag{4}$$

Here M_λ and M_μ are the number of discrete points taken in polar and azimuthal directions, respectively. Therefore, $M_\lambda \times M_\mu$ indicates the number of directions in which intensities are calculated and can be expressed as:

$$I_P^m = \frac{\frac{|D_x^m| A_{EW}}{\gamma_x} I_W^m + \frac{|D_y^m| A_{NS}}{\gamma_y} I_S^m + (V\,\Delta\Omega^m) S_P^m}{\frac{|D_x^m| A_E}{\gamma_x} + \frac{|D_y^m| A_N}{\gamma_y} + \beta V\,\Delta\Omega^m} \tag{5}$$

Here, the cross-sectional areas of the 2-D control volume are represented by A_{EW} and A_{NS}, in x and y directions, respectively. For any cell, average value of intensity I_P^m is assigned at the center and calculated as:

$$I_P^m = \gamma_x I_E^m + (1-\gamma_x) I_W^m = \gamma_y I_N^m + (1-\gamma_y) I_S^m \tag{6}$$

Here γ is taken as 0.5, and used to give different weights to the intensities at different faces of the cell. In Eq. (5), D_x^m and D_y^m are computed as:

$$D_x^m = \int_{\Delta\Omega^m} \sin\lambda \cos\mu \, \mathrm{d}\Omega = \int_{\mu^m - \frac{\Delta\mu^m}{2}}^{\mu^m + \frac{\Delta\mu^m}{2}} \int_{\lambda^m - \frac{\Delta\lambda^m}{2}}^{\lambda^m + \frac{\Delta\lambda^m}{2}} \cos\mu \sin^2\lambda \, \mathrm{d}\lambda \, \mathrm{d}\mu$$
$$= \cos\mu^m \sin\left(\frac{\Delta\mu^m}{2}\right) [\Delta\lambda^m - cos\, 2\lambda^m \sin(\Delta\lambda^m)] \tag{7a}$$

$$D_y^m = \int_{\Delta\Omega^m} \sin\lambda \sin\mu \, \mathrm{d}\Omega = \int_{\mu^m - \frac{\Delta\mu^m}{2}}^{\mu^m + \frac{\Delta\mu^m}{2}} \int_{\lambda^m - \frac{\Delta\lambda^m}{2}}^{\lambda^m + \frac{\Delta\lambda^m}{2}} \sin\mu \sin^2\lambda \, \mathrm{d}\lambda \, \mathrm{d}\mu$$
$$= \sin\mu^m \sin\left(\frac{\Delta\mu^m}{2}\right) [\Delta\lambda^m - cos\, 2\lambda^m \sin(\Delta\lambda^m)] \tag{7b}$$

here, $\Delta\Omega^m$ is elemental solid angle and it is expressed as:

$$\Delta\Omega^m = \int_{\Delta\Omega^m} d\Omega = \int_{\mu^m - \frac{\Delta\mu^m}{2}}^{\mu^m + \frac{\Delta\mu^m}{2}} \int_{\lambda^m - \frac{\Delta\lambda^m}{2}}^{\lambda^m + \frac{\Delta\lambda^m}{2}} \sin\lambda \, d\lambda \, d\mu$$
$$= 2\sin\lambda^m \sin\left(\frac{\Delta\lambda^m}{2}\right)\Delta\mu^m \tag{8}$$

For any radiatively participating medium, the source term required in Eq. (5) is given by:

$$S = (1-\omega)\beta\frac{\sigma T^4}{\pi} + \left(\frac{\omega\beta}{4\pi}\right)G \tag{9}$$

For solving the two energy equations simultaneously, volumetric radiative information $\nabla \cdot q_{\text{Rad}}$ is required and can be calculated by using Eq. (3). The analysis is made more generalized by non-dimensionalizing the governing equations with the help of following terms:

$$\begin{aligned} \theta &= \frac{T - T_i}{T_i}, \eta_x = \frac{x}{L_x}, \eta_y = \frac{y}{L_x} \\ \Psi_{\text{Rad}} &= \frac{q_{\text{Rad}}}{\dot{Q}L_x}, P_1 = \frac{\rho_g c_g u T_i}{\dot{Q}L_x}, P_2 = \frac{hAT_i}{\dot{Q}} \\ P_3 &= \frac{k_g T_i}{\dot{Q}L_x^2}, P_4 = \frac{k_s T_i}{\dot{Q}L_x^2}, P_5 = \frac{hL_x}{k} \end{aligned} \tag{10}$$

here, θ is temperature of gas and solid phases depending upon suffixes, η_x is distance measured in x direction (direction of flow), η_y is distance measured in y direction, Ψ_{Rad} is radiative heat flux, P_1 is enthalpy associated with the advection, P_2 is heat transfer coefficient, P_3 is gas-phase thermal conductivity, P_4 is PCM thermal conductivity, P_5 is Biot number. All the terms in Eq. (10) are non-dimensional and lead to simplification of the two energy equations as:

$$\text{Gas phase:} \phi P_1\frac{\partial\theta_g}{\partial\eta_x} + (1-\phi)P_2(\theta_g - \theta_s) = \phi\delta(\eta_x) + \phi P_3\left(\frac{\partial^2\theta_g}{\partial\eta_x^2} + \frac{\partial^2\theta_g}{\partial\eta_y^2}\right) \tag{11}$$

$$\text{Solid phase: } (1-\phi)P_2(\theta_g - \theta_s) + (1-\phi)P_4\left(\frac{\partial^2\theta_s}{\partial\eta_x^2} + \frac{\partial^2\theta_s}{\partial\eta_y^2}\right) - \nabla\cdot\Psi_{\text{Rad}} = 0 \tag{12}$$

Gas-phase boundary conditions:

$$\theta_g = 0 \text{ at } \eta_x = \eta_{x,i}, \frac{\partial \theta_g}{\partial \eta_x} = 0 \text{ at } \eta_x = \eta_{x,e}$$
$$\frac{\partial \theta_g}{\partial \eta_y} = 0 \text{ at } \eta_y = 0, \frac{\partial \theta_g}{\partial \eta_y} = 0 \text{ at } \eta_y = \eta_{Ly} \tag{13}$$

Solid-phase boundary conditions:

$$-\frac{\partial \theta_s}{\partial \eta_x} = P_5(\theta_g - \theta_s) \text{ at } \eta_x = 0, \frac{\partial \theta_s}{\partial \eta_x} = P_5(\theta_g - \theta_s) \text{ at } \eta_x = 1$$
$$\frac{\partial \theta_s}{\partial \eta_y} = 0 \text{ at } \eta_y = 0, \frac{\partial \theta_s}{\partial \eta_y} = 0 \text{ at } \eta_y = \eta_{Ly} \tag{14}$$

The volumetric radiative term $\nabla \cdot \Psi_{\text{Rad}}$ in non-dimensional form can be expressed as:

$$\nabla \cdot \Psi_{\text{Rad}} = \beta L_x (1 - \omega)\left[4\pi\, \Phi(1 + \theta_s)^4 - G^*\right] \tag{15}$$

In Eq. (15), the non-dimensional emissive power at inlet is represented by $\Phi = \frac{\sigma T_i^4}{Q L_x}$, and non-dimensional incident radiation is represented by $G^* = \frac{G}{Q L_x}$. The source term given in Eq. (9) can be expressed in non-dimensional form as:

$$S^*(\tau) = (1 - \omega)\frac{\Phi(1 + \theta)^4}{\pi} + \left(\frac{\omega}{4\pi}\right) G^* \tag{16}$$

The intensities at the boundary is calculated as:

$$I_b^* = \varepsilon_b \frac{\Phi(1 + \theta_b)^4}{\pi} + \frac{(1 - \varepsilon_b)}{\pi} \sum_{k=1}^{M_\mu} \sum_{l=1}^{M_\lambda/2} I^{*,m}\left(\lambda_l^m, \mu_k^m\right) \sin \lambda_l^m \cos \lambda_l^m \sin \Delta\lambda_l^m \Delta\mu_k^m \tag{17}$$

Here $I^* = \frac{I}{Q L_x}$ is the non-dimensional intensity.

The energy equation of solid and gas phases are solved simultaneously. The temperature for both the gas and the solid phase are assigned some guess values, and values of different parameter used for solution is given in Table 1. By using Gauss-Siedel method, both the energy equations are solved simultaneously, and the values of temperature profile is updated. First of all, the temperature values are assigned with some guessed values. After some iteration, the radiative information is updated by solving the radiative transfer equation. While computing radiative information, boundary intensities are computed initially. Then, the intensity is computed for the downstream direction along that direction. Such intensity computation is repeated

Table 1 Non-dimensional values of relevant parameters

Parameters	Values
P_1	0.01
P_3	2.5×10^{-4}
P_4	0.02
P_5	5
ω	0.5
ε	1.0
$\Phi_i = \Phi_e$	2.98×10^{-5}

for all the four quarters of polar angle. Again with the help of new radiative information, the energy equations are solved to update the gas and the solid-phase temperature profiles. This alternating method of solving energy equations and the radiative transfer equation continues till convergence is reached. Convergence criteria used for development of the FVM model is: change in gas- and solid-phase temperature is less than 10^{-6}.

3 Results and Discussion

By employing FVM, radiative transfer equation and the two energy equations are solved to generate temperature profile for gas and the solid phase. The result of FVM solution is validated with the numerical results available in literature [16], and shown in Fig. 2. The temperature profiles of both the phases (solid and gas) are merged due to large HTC, and the agreement is good. Present problem is solved for high aspect ratio $\left({L_y}/{L_x} = 10\right)$ to reduce the problem to 1D, and then validated. The generation of temperature profile for gas and the solid phase by employment of FVM and DTM method is called direct modeling. Further, use of the above-computed temperature profiles for predicting the corresponding HTC, inverse modeling is employed. Present inverse model employs ANN, trained with SCG algorithm. The details of input variable numbers, hidden number of neurons, and output variable numbers, used in the ANN model are is shown in Fig. 3. In the input layer, 300 nodes for gas temperature and 300 nodes for solid temperature are taken. Number of neurons considered in the hidden layer is 10. Whereas, on the output side, one node is taken for HTC. Number of input variable numbers and output variable numbers decides the nodes for input layer and output layer, respectively. For hidden layer, number of neurons is selected based on trial and error. Low values of neurons in hidden layer lead to poor training of ANN and high error in retrieved values. While too much neurons in hidden layer will lead to overtraining of the ANN and reduces robustness of the ANN model. Such overtrained ANN models will not be able to give accurate results for new cases. For training the ANN, large amount of data is required. Temperature profiles of both the phases corresponding to 49 different values of HTC are computed by

FVM and used for training of ANN. For training of the ANN model, SCG algorithm is employed. Once ANN is trained, detailed analysis is done to check its robustness, and is described below.

Figure 4 shows the histogram of the SCG-trained ANN model. It is clearly observed that the ANN model with SCG algorithm is good. Since sufficient data falls near the zero error line (ZEL), and some data falls away from the ZEL. Large data accumulation near the ZEL ensures that the ANN model is well acostum to the cases used for the ANN model. Few data away from the ZEL helps the ANN model to accommodate new cases. Such distribution of data across the ZEL is true for training, validation, and testing stage. Thus, such ANN models are robust and

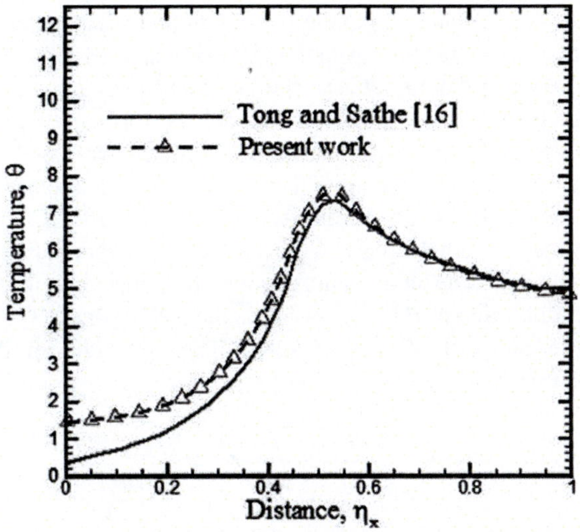

Fig. 2 Validation of direct model of present work with the work of Tong and Sathe [16] for $P_2 = 500$ and $\beta = 1.0$

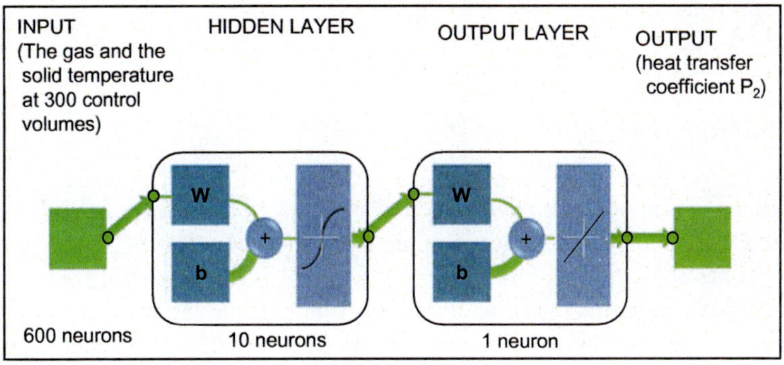

Fig. 3 Details of the used ANN model for inverse analysis

versatile to handle the cases similar to the data used for training, and new cases as well. Performance curves of the ANN model with SCG algorithm is presented in Fig. 5. It is found that the convergence of the validation curve is achieved at 100th iteration, and the validation and the testing curves are not diverging from each other. Error reduces with the increase of iteration, for all the curves (training, validation, and testing). Reduction of error in the training stage ensures that the ANN model is able to correlate the output from the input data. During validation stage, reduction of the error indicates that the ANN model finetunes its weights and biases to get the output with higher accuracy. Testing stage checks that the ANN model is able to give the correct output for the input data which is new to the ANN model. Apart from performance curves and histogram, regression analysis provides further insight into the ability of the ANN model to handle a problem and is presented in Fig. 6. It is observed that the value of regression coefficient R is near one. Thus, the ANN model is able find a good correlation between the input and the output. All the tools for prediction of the performance of the ANN model (Performance curves, histogram, and regression coefficient) indicate that the ANN model will be able to give good results for the cases similar to the input data, and can handle the new cases as well. Table 2 shows the results obtained from ANN model with the SCG algorithm. The temperature profiles of both the phases (gas and solid) is fed to the ANN model, was not a part the input data. The temperature profiles are corresponding to the high HTC ($P_2 = 500$), but this value of HTC is not fed to the ANN model. The inverse model of ANN gives the value of HTC to be 472.6883. The error in retrieved value of HTC from ANN model is computed as:

$$E = \frac{|P_{2,\,\mathrm{act}} - P_{2,\,\mathrm{est}}|}{P_{2,\,\mathrm{act}}} \times 100 \tag{18}$$

The developed ANN model is analyzed for the robustness with the help of performance curves, histogram, and regression coefficient. The accuracy of results obtained from the ANN model is found to be good. However, the present approach is suitable for steady-state treatment only. For transient problems, different networks with time-dependent data need to be used.

4 Conclusions

An artificial neural network (ANN) model is employed to perform inverse analysis involving 2-D rectangular porous ceramic matrix (PCM). The PCM was used for burner application. The governing equations were solved by employing finite volume method (FVM), and this constituted the direct part. Then, temperature profiles for gas and the solid phase were computed for various values of heat transfer coefficient (HTC), and were used in training of ANN. Scaled conjugate gradient (SCG) algorithm was used to train the ANN. After training the ANN, unknown pair of gas and solid-phase temperature profiles was fed into the ANN model as input. The ANN gives

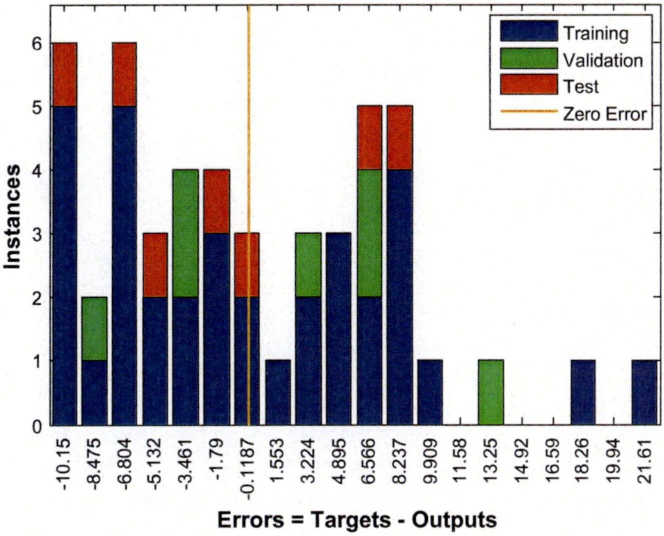

Fig. 4 Analyzing the SCG-trained ANN model by histogram

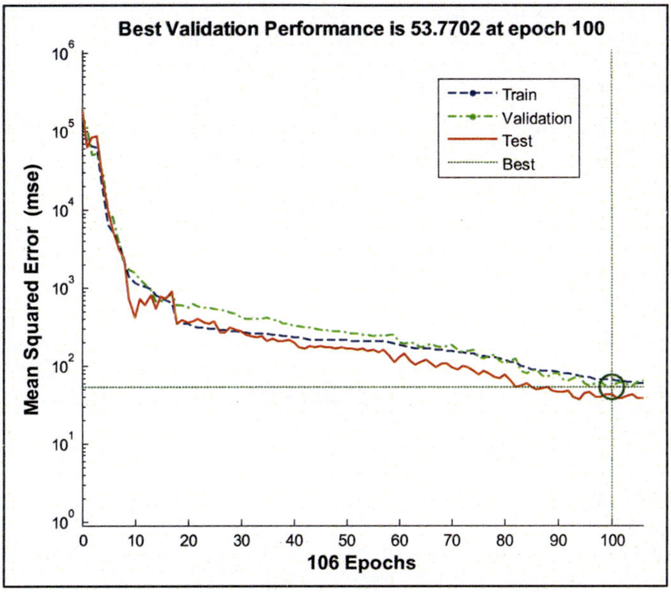

Fig. 5 Analysis of the SCG-trained ANN model by performance curves

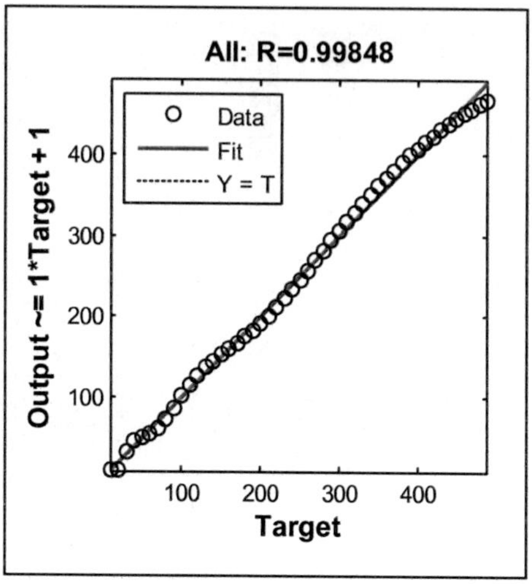

Fig. 6 Analysis of SCG-trained ANN model by regression coefficient

Table 2 Analysis of SCG algorithm in training of neuron in ANN model

S. no.	Algorithm	$P_{2,act}$	$P_{2,\,ret}$	% Absolute error in retrieval
1	SCG	500	472.6883	5.46

the corresponding value of HTC as output. The accuracy of the result from ANN is 5.4%, and the ANN model has high regression coefficient of 0.998.

References

1. Backus G (1988) Hard and soft prior bounds in geophysical inverse problems. Geophys J 94:249–261
2. Özisik MN, Orlande HRB (2000) Inverse heat transfer fundamentals and applications. Taylor & Francis, New York
3. Osman AM, Beck JV (1990) Investigation of transient heat transfer coefficients in quenching experiments. ASME J Heat Transfer 112:843–848
4. Saggio-Woyanski J, Scott C (1992) Processing of porous ceramics. Am Ceram Soc Bull 71:1674–1682
5. Kakati S, Mahanta P, Kakoty S (2007) Performance analysis of pressurized kerosene stove with porous medium inserts. J Sci Ind Res 66:565–569
6. Song FQ, Wen Z, Dong ZY, Wang EY, Liu XL Numerical study and optimization of a porous burner with annular heat recirculation. Appl Thermal Eng. https://doi.org/10.1016/j.applthermaleng.2019.113741
7. Jambunathan K, Hartle S, Ashforth-Frost S, Fontama VN (1996) Evaluating convective heat transfer coefficients using neural networks. Int J Heat Mass Transf 39:2329–2332
8. Sablani SS (2001) A neural network approach for non-iterative calculation of heat transfer coefficient in fluid-particle systems. Chem Eng Process 40:363–369
9. Khan MZ, Nawaz I, Tiwari GN, Meraj M (2021) Effect of top cover cooling on the performance of hemispherical solar still. Mat Today Proc 38(1):384–390

10. Mishra VK, Chaudhuri S (2021) Implementation of stochastic optimization method-assisted radial basis neural network for transport phenomenon in non-Newtonian third-grade fluids: assessment of five optimization tools. Arab J Sci Eng 46:11797–11818
11. Mishra VK, Chaudhuri S (2021) Genetic algorithm-assisted artificial neural network for retrieval of a parameter in a third grade fluid flow through two parallel and heated plates. Heat Transfer 50:2090–2128
12. Mishra VK, Dasgupta U, Patra S, Pal R, Anand K (2022) A dynamic two-level artificial neural network for estimation of parameters in combined mode conduction-radiation heat transfer in porous medium: an application to handle huge dataset with noise. Heat Transfer 51:1306–1335
13. Anand K, Bhardwaj A, Chaudhuri S, Mishra VK (2022) Self-organizing map network for the decision making in combined mode conduction-radiation heat transfer in porous medium. Arab J Sci Eng. https://doi.org/10.1007/s13369-021-06489-4
14. Horsman AP, Daun KJ (2011) Design optimization of a two-stage porous radiant burner through response surface modeling. Numer Heat Transf (A) 60(9):727–745
15. Mishra NK, Muthukumar P (2018) Development and testing of energy efficient and environment friendly porous radiant burner operating on liquefied petroleum gas. Appl Therm Eng 129:482–489
16. Tong T, Sathe S (1991) Heat transfer characteristics of porous radiant burners. J Heat Transfer 113:423–428

Design Optimization for Heat Transfer Through a Circular Pipe

Souvik Dutta, Ayush Boral, Anwesha Das, Pooja Chaubdar, and A. B. Harichandan

Abstract Pipe flow has been a subject of intensive study because of the wide application, it finds in every domain. The objective of this current research work is focussed on the enhancement of heat transfer characteristics of fluid flow in a pipe by changing the internal geometry of pipe wall. Four different configurations were used including the circular pipe considered as a baseline. The flow inside the pipes with heated walls was simulated at varying Reynolds number using the k-omega turbulence model. Water was taken as the working medium. The authors found that a 104% increase in Nusselt number was observed at low-Reynolds number (Re = 10,000) for the pipe with sine-wave wall. A 49% increase in Nusselt number was observed at high-Reynolds number (Re = 500,000) for the pipe with scallop pattern. The increase in Nusselt number ensures significant increase in the efficiency and reduction in size of current heat exchangers without any significant increase in material costs. The performance of pipe with scallop pattern establishes the fact that the scallop pattern which is frequently observed in natural structures plays a vital role in increasing the rate of heat transfer.

Keywords Heat transfer · Circular pipe · Reynolds number · Nusselt number · Turbulence model · Heat exchanger

1 Introduction

Efforts towards improving the efficiencies of power generation and storage systems are of paramount importance to fulfil the ever-increasing demand of energy. Most of the power generation systems rely on the effective transfer of heat from the source to the working fluid for energy generation. By increasing the efficiency of heat exchangers, the overall efficiency of a plant can be effectively improved. Concerted efforts have been made in the recent past to improve the efficiency of heat transfer

S. Dutta · A. Boral · A. Das · P. Chaubdar · A. B. Harichandan (✉)
Aerospace Engineering, School of Mechanical Engineering, KIIT Deemed to be University, Bhubaneswar 751024, India
e-mail: atal.harichandanfme@kiit.ac.in

© The Author(s), under exclusive license to Springer Nature Singapore Pte Ltd. 2023

S. Revankar et al. (eds.), *Recent Advances in Thermofluids and Manufacturing Engineering*, Lecture Notes in Mechanical Engineering,
https://doi.org/10.1007/978-981-19-4388-1_23

machinery by increasing the heat transfer coefficients keeping in mind the economic feasibility and material saving. Passive and active techniques have been considered to increase the heat transfer rates in heat exchangers. Passive techniques do not require any additional power. These include rough or treated surfaces, extended surfaces, swirl frow devices, coil tubes, etc. Whereas active techniques rely on external power source in order to improve the thermal characteristics such as fluid or surface vibration by mechanical aids, jet impingement, etc. Ahmadvand et al. [1] used a swirl generator to improve the heat transfer characteristics of a pipe and analysed it numerically as well as experimentally. Compared to a pipe flow, they achieved a heat transfer enhancement between 50 and 110% using the swirl mechanism depending on vane angle. They found that the heat transfer enhancement is strongly dependant on the blade angle which is independent of Reynolds number, but Reynolds number plays a significant role in the selection of vane angle. Kurnia et al. [2] used twisted tape insert in helical pipe and found that increasing the twisting ratio on helical pipe enhances the heat transfer rate at the penalty of higher friction pressure drop in pipe walls. They also concluded that helical pipe with twisted tape insert is best at higher Reynolds number and lower twisting ratio when the working fluid is water.

Smithberg et al. [3] studied the effect of twisted tape turbulence generators on the heat transfer coefficients both analytically and experimentally. Algifri et al. [4] studied the swirl flow in a pipe and proposed a calculation procedure to calculate swirl intensity, axial, and tangential velocity profiles by using the inlet conditions prevailing at the pipe inlet. Eiamsa-ard et al. [5] carried out an experimental study to investigate the heat transfer enhancement by means of helical tape inserts for a concentric tube heat exchanger. They further determined the optimal space ratio required to get the maximum heat transfer with the minimum possible pressure loss. They observed that the mean Nusselt number can be increased by 160% using a full-length helical tape with centered rod, 150% using a full-length helical tape without rod, and 145% using a regularly spaced helical tape, with spacing ratio 0.5, in comparison with plain tube. Hay [6] studied the swirl flow in a pipe produced by a single tangential slot at different angles. They proposed that this method of heat transfer enhancement can be applied for turbine blade cooling. Kurnia et al. [7] carried out experiment on conical, spiral, and helical non-cylindrical duct and found that the heat transfer rate is higher in coiled ducts but there is higher pressure drag penalty imposed lowering its merit. The researchers also mentioned that these coiled ducts can be useful when the space available is limited. Sasmito et al. [8] conducted research on six different cross section in spiral duct and found rectangular and triangular cross section has higher performance amongst in pane spiral ducts.

All these researches depended on the generation of swirl flow for heat enhancement either by employing a twisted tape, or other swirling mechanisms. Swirl flow mechanisms have been used in solar water heaters for increasing their performance. The major disadvantage of employing swirl flow heat transfer enhancement techniques is the increased pressure loss which leads to the requirement of higher pumping power. In the present research, an effort has been made to improve the heat transfer in pipe flow by altering the internal geometry of the pipe (passive control). The aim is to induce turbulence in the flow to enhance the heat transfer characteristics. Since, altering the internal structure of the conduit is both economic and material saving, hence, this research can prove useful in the pursuit of achieving the maximum efficiency in an energy plant. Moreover, since no swirl mechanisms are used, so the heat transfer rates can be augmented without significantly increasing the pressure drop and pumping power. This research may also find its use in the field of automobile, turbo-machinery, and even refrigeration systems. Numerical simulations have been carried in ANSYS Fluent which is a proven numerical tool preferred by most researchers. The geometries have been designed using Autodesk Fusion 360.

2 Computational Model

In the present discussion, a cylindrical domain is taken where fluid flow takes place internally. The walls are specified with a constant temperature. The surface Nusselt number gives us the effectiveness of the heat transfer from the wall which we intend to optimise. Four different arrangements were studied as shown in Fig. 1.

For all the given arrangements, the internal diameter is 10 mm, the external diameter is 12 mm, and the length of the pipe is 15 mm. The internal structure of the pipe with scallop pattern is of particular interest since similar geometries have been found in natural underwater caves and glaciers. The authors believe that these structures intimately alter the heat transfer characteristics and propose to investigate the effect of the particular geometry on the effectiveness of heat transfer. For each of the following pipe cross-sections, the inlet was specified as velocity inlet, and outlet was specified with the outflow boundary conditions. The internal walls of the pipe were specified with a constant temperature. To study the effectiveness of the heat transfer from the pipe to the fluid flowing inside, the surface Nusselt number was measured which is a non-dimensional number reflecting the ratio of convective to conductive heat transfer. For the current research, unstructured mesh was used in order to accommodate the complex internal structure of the conduit. The k-omega turbulence model was used to analyse the internal pipe flow with 2nd order accuracy of convective, diffusive, and temporal terms. In the current study, the working medium is taken to be water at 293 K. The temperature of the pipe wall was taken to be 333 K.

3 Results

The first objective of the authors was to validate the computational model for the cylindrical pipe case with the existing open literature. The results for the cylindrical pipe case for different Reynolds number have been presented in Fig. 2. These results are validated by the research work published independently by Gnielinski [9] and Gül [10]. It can be clearly seen from the comparison that the results obtained by the

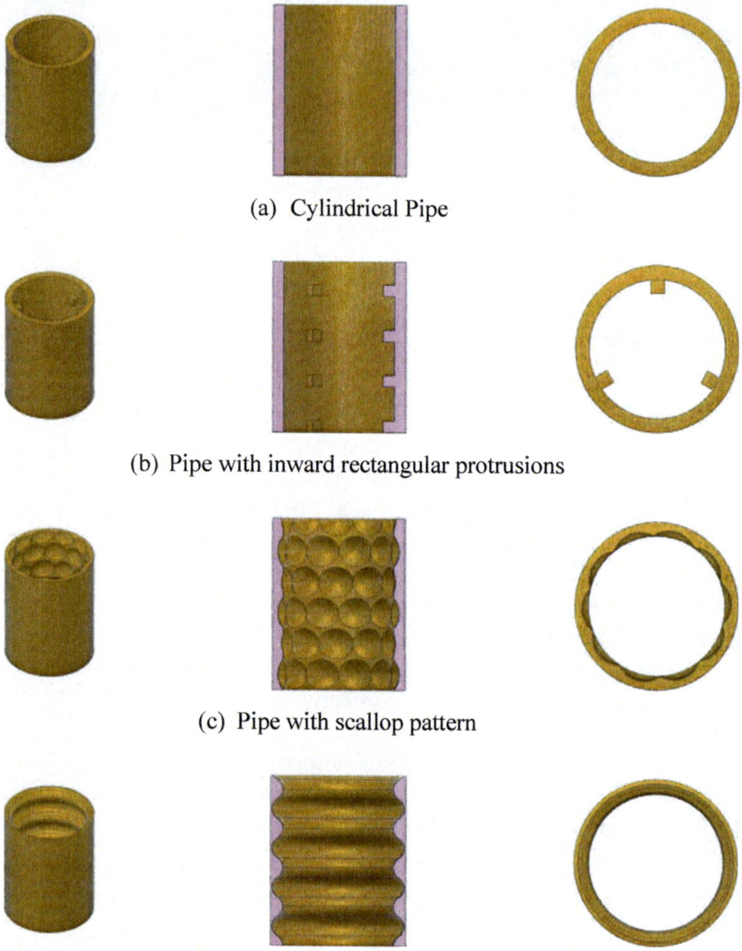

(a) Cylindrical Pipe

(b) Pipe with inward rectangular protrusions

(c) Pipe with scallop pattern

(d) Pipe with sine-wave wall

Fig. 1 Schematic of different circular pipe configurations

current numerical model are accurate. Once the results were validated, the authors moved forward towards the simulation of other geometries as discussed in Fig. 1.

After the validation, numerical simulation was carried out on the three different geometries for the Reynolds number range of Re = 10,000 to Re = 500,000. The results obtained are presented in Fig. 3. The curve shows the variation of Nusselt number (Nu) with varying Reynolds number. It is a general trend that the rate of heat transfer is proportional to the Reynolds number of the flow, hence there is a close relationship between Nusselt number and Reynolds number which can be clearly seen from the plot. Further, when we compare the Nusselt number for the cylindrical pipe case and the other three geometries, we can conclude that for the same Reynolds

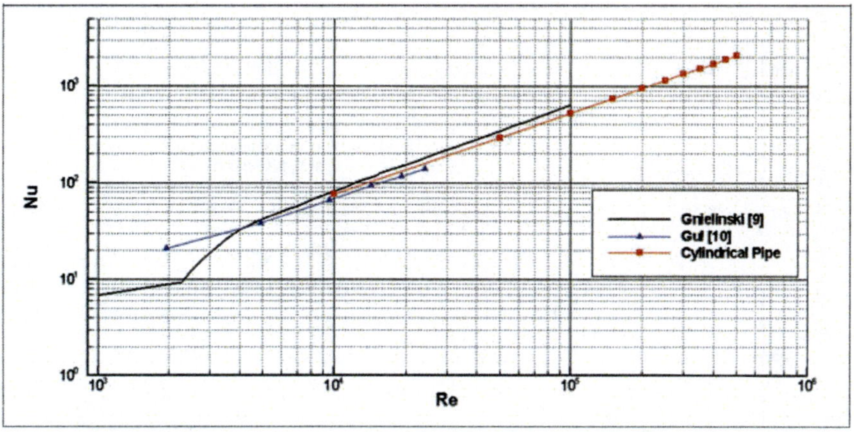

Fig. 2 Validation of the model

number, all of the following geometries effectively enhance the rate of heat transfer thus resulting in higher Nusselt number values. Moreover, for low-Reynolds number (Re = 10,000), the pipe with sine-wave wall is the best choice for optimal heat transfer.

The Nu versus Re curve shows the % increase in Nusselt number compared to the cylindrical pipe case as presented in Fig. 4. It can be clearly seen that for low-Reynolds number case, the % increase in Nu for pipe with sine-wave wall is greater thus better thermal performance. But with increasing Re, the pipe with scallop pattern is the optimal choice. This point further establishes the authors believe that the scallop pattern which is frequently observed in natural structures plays a vital role in increasing the rate of heat transfer.

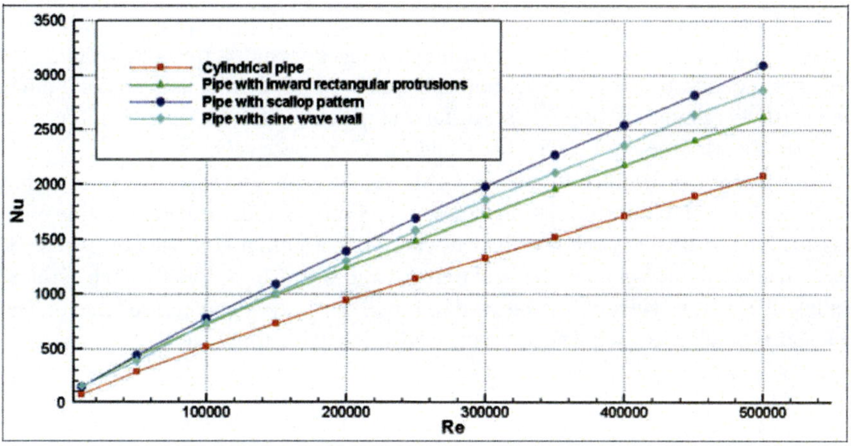

Fig. 3 Nu versus Re for different geometries

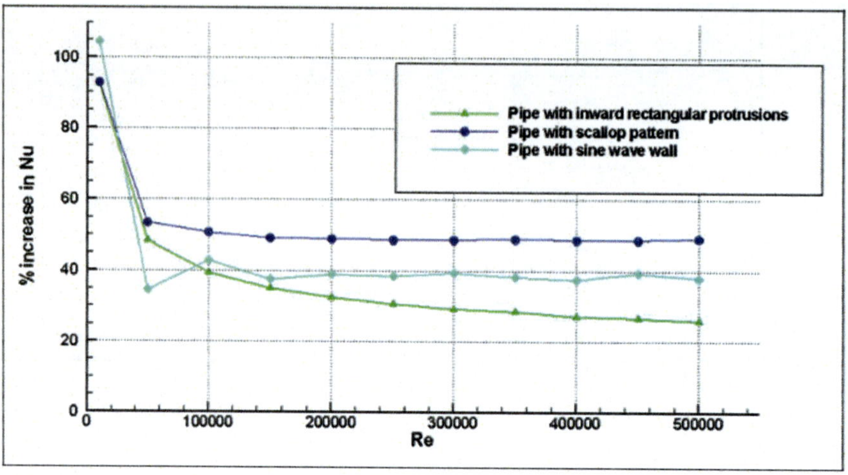

Fig. 4 % increase in Nu for different geometries compared to cylindrical pipe

Bushuk et al. [11] found that due to scallops, the form drag increases because of adverse pressure gradient in the scallop trough. This increases the total turbulent kinetic energy and melt rates in ice. The pipe with the sine-wave wall, even though it has a better performance in the low-Reynolds number regime, is not a good choice in the opinion of the authors. This is because it shows a drastic variation in the range of Re < 100,000, and a fluctuation is observed between Re = 50,000 to Re = 150,000. This drastic variations in heat transfer can lead to inherent dynamic instabilities due to a large margin of change in heat transfer coefficients. So, we can conclude that the pipe with scallop pattern is the best overall choice which can increase the heat transfer rate up to 105%. Moreover, the % increase in Nu almost remains constant for Re > 100,000. So, the pipe with a scallop pattern can provide an optimised thermal performance economically. Also, for increasing the overall performance of an energy plant, no substantial investment is required, but as concluded from the results, just the cross-section of the coupling elements needs to be altered. The velocity angle and vorticity magnitude contours for different pipe configurations at varied Reynolds number are presented in Fig. 5, Fig. 6, and Fig. 7, respectively.

As it is evident from the figures given above, the pipe with scallop formation has vortices attached to the scallops, and the vorticity magnitude is greater than the other cases. In the case of a cylinder with sine-wave wall, a similar kind of vorticity can be seen attached to the wall, but the heat transfer rate seems to be less than the cylinder with scallops. It is a clear observation that for all the cases, the increased rate of heat transfer is a direct consequence of increased vorticity near the wall.

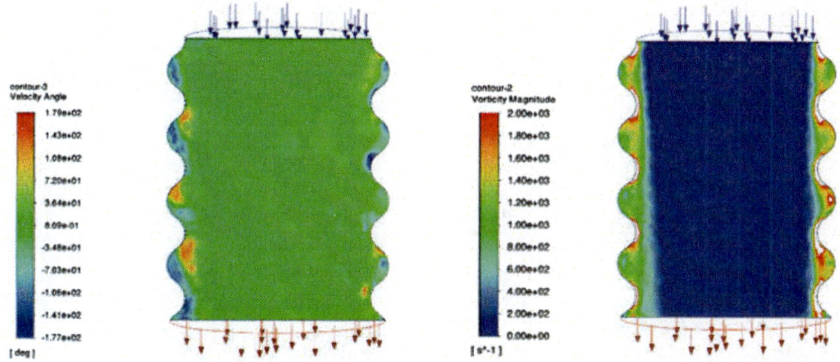

(a) Pipe with sine-wave wall

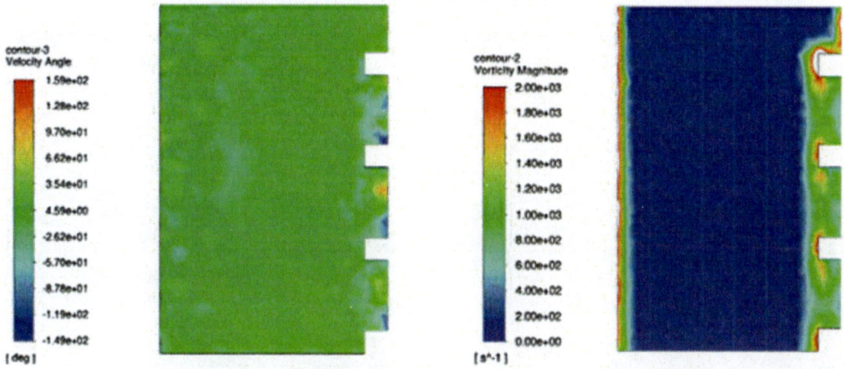

(b) Pipe with inward rectangular protrusions

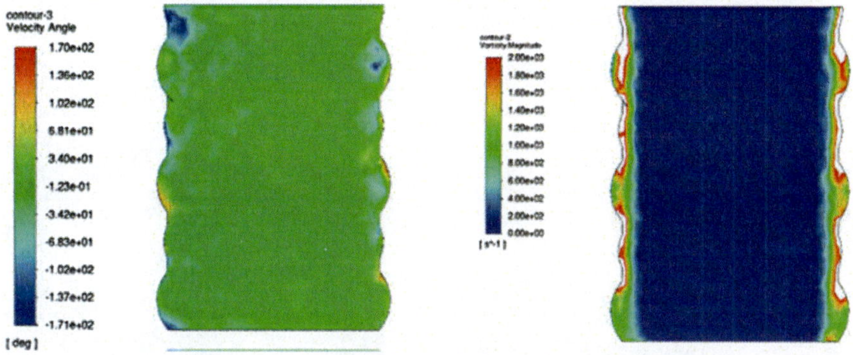

(c) Pipe with scallop pattern

Fig. 5 Velocity angle and vorticity magnitude contours at Re = 10,000

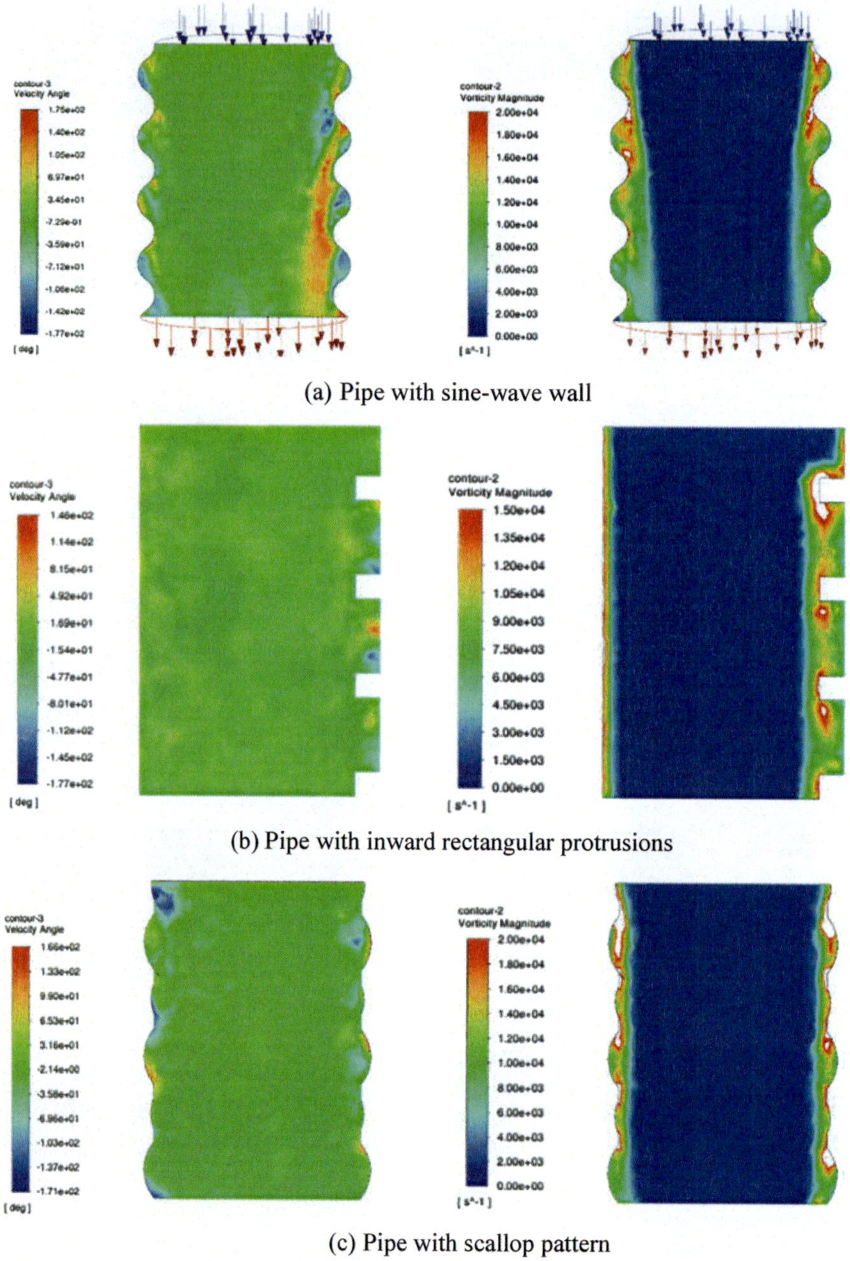

Fig. 6 Velocity angle and vorticity magnitude contours at Re = 100,000

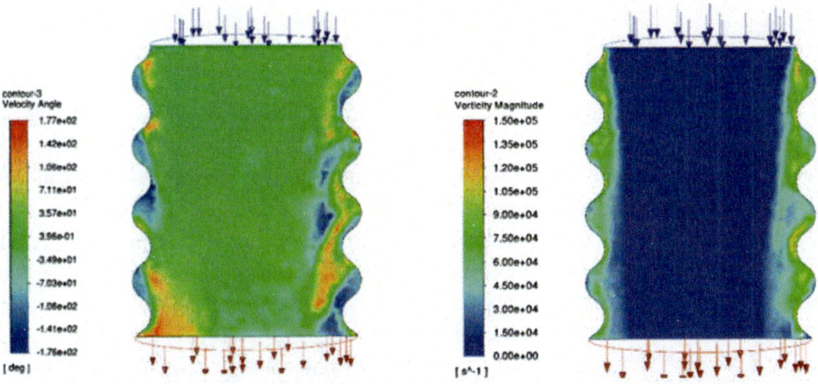

(a) Pipe with sine-wave wall

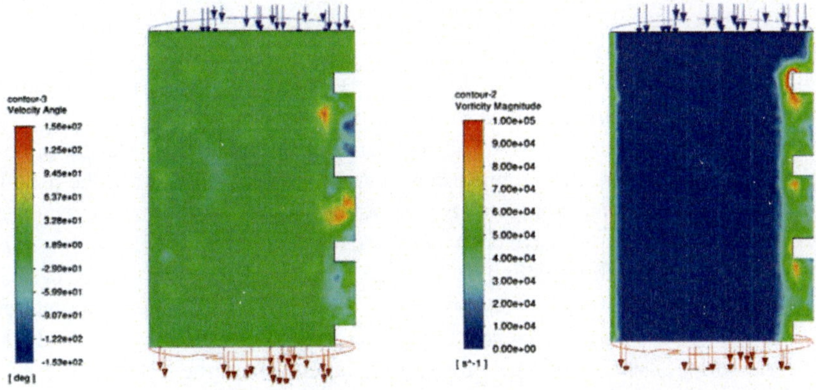

(b) Pipe with inward rectangular protrusions

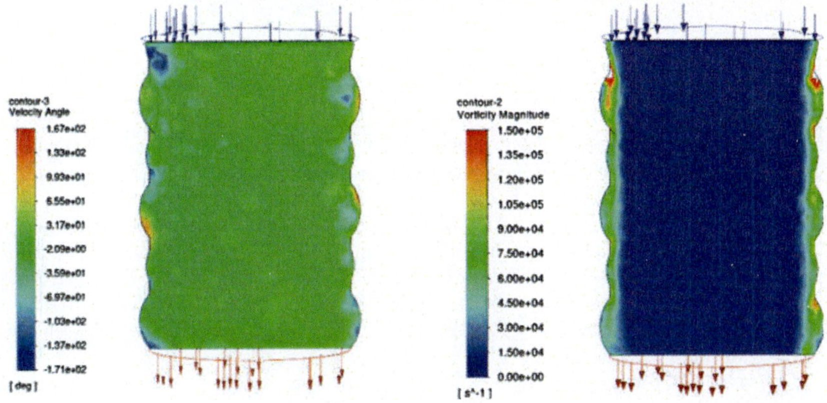

(c) Pipe with scallop pattern

Fig. 7 Velocity angle and vorticity magnitude contours at Re = 500,000

4 Conclusion

All the shapes mentioned in the current research show an improved heat transfer rate compared to the cylindrical pipe case. For low-Reynolds number, the pipe with sine-wave wall had a greater vorticity region near the wall and was better at heat transfer than all the models compared above. A 104% increase in Nusselt number was observed for this case in comparison with the cylindrical pipe case at low-Reynolds number (Re = 10,000). The pipe with scallop pattern had the highest heat transfer at high-Reynolds number in spite of having a lower vorticity region near the wall compared to the pipe with sine-wave wall. A 49% increase in Nusselt number was observed in comparison with the cylindrical pipe case at high-Reynolds number (Re = 500,000). The observed enhancement in heat transfer rate (Nusselt number) was due to increased vorticity near the wall surface and increased surface area in comparison with the cylindrical pipe case. From the current investigation, the authors concluded that the pipe with scallop formation is best fit for use in the flow regime ranging from Re = 10,000 to Re = 500,000 due to its consistent performance. However, for low-Reynolds number flows, the pipe with sine-wave wall is preferred. This research can find its use to enhance the performance and reduce the size of heat exchangers in aircrafts, ships, submarines, automobiles, etc. It can also be used in the energy production and storage plants including solar farms, thermal, and nuclear power plants. Many studies have been carried to improve turbine blade cooling. The authors believe that putting scallop pattern in cooling passages of turbine blades can cool the blades more efficiently. This will allow for higher turbine inlet temperatures, thus improving propulsive efficiency. Further research needs to be done to study the effect of swirlers used in the pipe with scallop pattern for further increment in heat transfer rates.

References

1. Ahmadvand M, Najafi AF, Shahidinejad S (2010) An experimental study and CFD analysis towards heat transfer and fluid flow characteristics of decaying swirl pipe flow generated by axial vanes. Meccanica 45(1):111–129
2. Kurnia JC, Chaedir BA, Sasmito AP (2020) Laminar convective heat transfer in helical tube with twisted tape insert. Int J Heat Mass Transfer 150:119309
3. Smithberg E, Landis F (1964) Friction and forced convection heat-transfer characteristics in tubes with twisted tape swirl generators. ASME J Heat Transfer 86(1):39–48
4. Algifri AH, Bhardwaj RK, Rao YVN (1987) Prediction of the decay process in turbulent swirl flow. J Mech Eng Sci 201(4):279–283
5. Eiamsa-ard S, Promvonge P (2005) Enhancement of heat transfer in a tube with regularly-spaced helical tape swirl generators. Sol Energy 78:483–494
6. Hay N, West PD (1975) Heat transfer in free swirling flow in a pipe. ASME J Heat Transfer 97(3):411–416
7. Kurnia JC, Sasmito AP, Mujumdar AS (2011) Evaluation of the heat transfer performance of helical coils of non-circular tubes. J Zhejiang Univ Sci A 12:63–70

8. Sasmito AP, Kurnia JC, Wang W, Jangam SV, Mujumdar AS (2012) Numerical analysis of laminar heat transfer performance of in-plane spiral ducts with various cross-sections at fixed cross-section area. Int J Heat Mass Transfer 55:5882–5890
9. Gnielinski V (2013) On heat transfer in tubes. Int J Heat Mass Transfer 63:134–140
10. Gül H (2006) Enhancement of heat transfer in a circular tube with tangential swirl generators. Exp Heat Transfer 19(2):81–93
11. Bushuk M, Holland DM, Stanton TP, Stern A, Gray C (2019) Ice scallops: a laboratory investigation of the ice-water interface. J Fluid Mech 873:942–976

Improvement of Heat Transfer by Natural Convection in a Three-Dimensional Square Cavity with a Heated Floor

El Bachir Lahmer, Youssef Admi, Mohammed Amine Moussaoui, and Ahmed Mezrhab

Nomenclature

C	Lattice speed
C_s	Speed of sound (m s^{-1})
e_i	Discrete velocities
f	Flow Distribution function
F	Body force
g	Thermal Distribution function
Kf	Thermal conductivity of the solid
Ks	Thermal conductivity ratio
Kr	Characteristic length (m)
Lc	19 × 19 Transformation matrix of the fluid flow
M	Moment space for fluid flow
m	Equilibrium moment for fluid flow
m^{eq}	7 × 7 Transformation matrix of the thermal process
n	Moment space for temperature field
n^{eq}	Equilibrium moment for temperature field
Pr	Prandtl number
Q	Relaxation matrix for thermal field
Ra	Rayleigh number
S	Relaxation matrix for fluid flow
T_h	Dimensionless hot temperature
T_c	Dimensionless cold temperature
U	Velocity component of x-direction (m s^{-1})

E. B. Lahmer (✉) · Y. Admi · M. A. Moussaoui · A. Mezrhab
Mechanics & Energy Laboratory, Faculty of Sciences Oujda, Mohammed First University, Oujda, Morocco
e-mail: lahmerelbachir@gmail.com; lahmer_elbachir1718@ump.ac.ma

© The Author(s), under exclusive license to Springer Nature Singapore Pte Ltd. 2023
S. Revankar et al. (eds.), *Recent Advances in Thermofluids and Manufacturing Engineering*, Lecture Notes in Mechanical Engineering,
https://doi.org/10.1007/978-981-19-4388-1_24

u	Lattice velocity
V	Velocity component of y-direction (m s^{-1})
W	Velocity component of z-direction (m s^{-1})
X	Dimensionless direction of x
Y	Dimensionless direction of y
Z	Dimensionless direction of z

Greek Symbols

δ	Dimensionless position of the heat source
ε	Solid thickness (m)
ρ	Density of the fluid (Kg. m^{-3})
θ_c	Cold temperature (°K)
θ_h	Hot temperature (°K)

Subscripts

c	Cold
h	Hot
exp	Experimental
MRT	Multi-relaxation time
LBM	Lattice Boltzmann method

1 Introduction

Natural convection in a cubic cavity within a heat source is widely employed to study fluid flow and heat transfer enhancement in various fields. This type of system is considered as one of the focused subjects of many researchers because of its preferential applications in many industrial and engineering applications related to the thermal insulation of buildings, cooling of electronic chips, and so on [1–3]. The research of the three-dimensional configuration natural convection heat transfer represents one of the challenges in figuring out the mechanism and behavior of the fluid flow and heat transfer by numerical simulations linked to real physical problems. The time and cost of experimental studies are some of the main problems that have guided most researchers to find alternative solutions to reduce time and the lack of experimental equipment [4]. Numerical simulation allows solving physical problems in a masterful and efficient way, where it offers valuable solutions by providing a clear view of complex systems. In this context, the Boltzmann Lattice Method (LBM), as a numerical simulation, allows describing the behavior of fluids and heat transfer using different computational models [5–7]. Wang et al. [8] led a numerical simulation of a three-dimensional heated cubical cavity for a high Rayleigh number, and the main purpose was to study the evolution of the heat transfer and fluid flow behavior inside a differential heated cavity. The numerical results showed that the thermal and velocity contour formed close to the adiabatic walls, and the thickness of the contour layers narrowness near these walls for high values of Rayleigh number. Chavez-Modena et al. [4] executed the under-resolved turbulent flow simulation using the D3Q19 model with multi-relaxation time with the central moment. They conclude that the model used to obtain a satisfactory result compared with BGK and MRT-CM model results for the under-resolved systems. Liu et al. [9] examined 3D convection heat transfer in porous media at the REV scale. They checked if the model employed for studying the porous channel is appropriate to investigate the heat transfer and the fluid flow. Moussaoui et al. [3] performed a wide study on the effect of the Rayleigh number of two-dimensional heated obstacles within a square enclosure by using the double MRT-LBM. They found that the increase of obstacle dimension improves the heat transfer by increasing the surface exchange. The recirculation zone loses its symmetric when the position of the heated obstacle varied, and the isotherm field also disturbed because of this variation in the local heating [3].

The present research conducted a comprehensive analysis of the conjugate heat transfer inside a square enclosure with underfloor heating under the effects of Rayleigh number, Prandtl number, and thermal conductivity ratio.

2 Structural System and Numerical Estimation

2.1 Configuration Problem

The structural system studied is a cubical cavity consisting of two vertical cold walls and a heated floor as shown in Fig. 1. The cavity base undergoes a dimensionless temperature $T_h = 1$ according to different values of the thermal conductivity (Ks). While the cold surfaces kept fixed cold dimensionless temperatures $Tc = 0$. The front, rear, and top cavity surfaces are insulated. The thickness of the solid cavity base ε is characterized by a different thermal conductivity (Ks) compared to the thermal conductivity of the airflow (Kf). The continuity of the temperature and heat flux is taken into account for the interface solid/fluid (see Fig. 1).

2.2 Numerical Method

The exploitation of numerical models for the simulation of fluid flow and heat transfer is considered an alternative method that allows to acquire time and facilitate the work

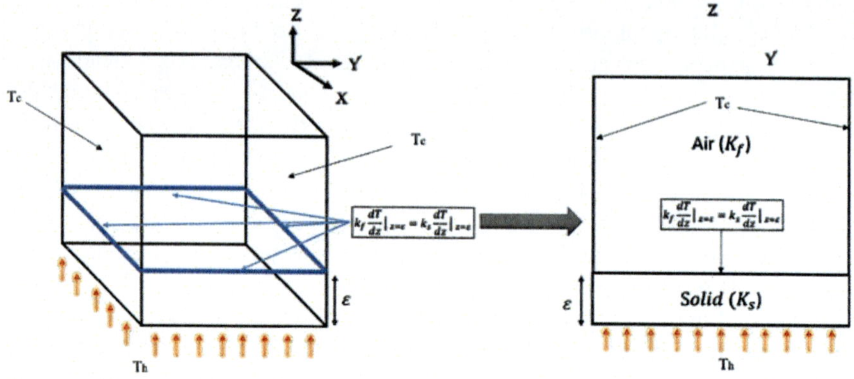

Fig. 1 3D and 2D configurations under investigation

tasks in front of the experimental methods that require high cost for their realization. The numerical method based on the Boltzmann equation has a powerful resolution to model physical phenomena in different fields related to fluid engineering and heat transfer quality [10].

The lattice Boltzmann equation (LBE) is represented by two main terms being the evolution of advection and collision of the fluid particle, the space–time discretization of this equation allows to solve the fluid behavior and the heat exchange by the numerical calculation of the distribution function with taking into consideration the D3Q19 and D3Q7 model to describe the fluid and temperature field, respectively as illustrated in Fig. 2 [4, 8, 9].

The nineteen discrete speeds for the D3Q19 model $\langle e_i | i = 0, \ldots, 18 \rangle$ was employed to define the velocity direction of the fluid flow, in other hands, the seven discrete temperatures $\langle e_i | i = 0, \ldots, 6 \rangle$ was proposed for the D3Q7 scheme with taking into account the accordance between the temperature direction to seven speed

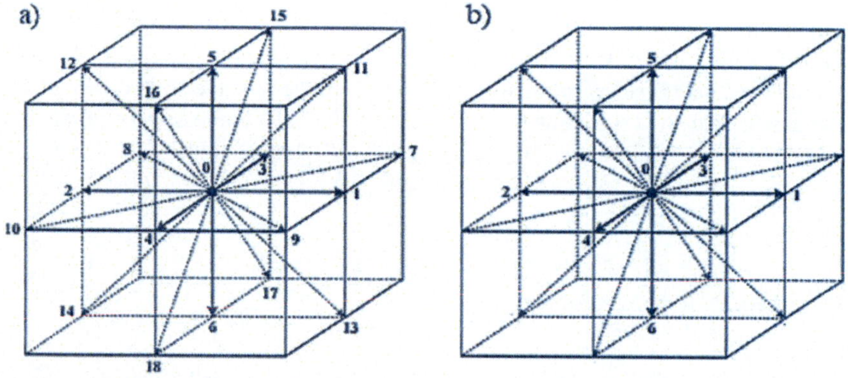

Fig. 2 **a** D3Q19 model for fluid behavior, **b** D3Q7 model for thermal field

direction of the D3Q19 scheme [4, 9],

$$e_i = c\begin{bmatrix} 0110000111111110000 \\ 0001100111100001111 \\ 0000011000011111111 \end{bmatrix} \quad (1)$$

where $\boldsymbol{c}$ considers the lattice speed of LBM equation which is corresponding to the speed of sound as shown in the expression below.

$$c^2 = 3 \times c_s^2 \quad (2)$$

2.2.1 Lattice Boltzmann Method for Thermal and Fluid Flow

The equation of the lattice Boltzmann associated with the multiple relaxation times can be expressed by the following equation related to the D3Q19 model for the streaming field is expressed as follows.

$$f(x + ei \times \Delta t, t) - f(x, t) = -M^{-1} \times S \times (m(x, t) - m^{\mathrm{eq}}(x, t)) + F \quad (3)$$

The quantities M and S denote the 19 × 19 transformation matrix and the relaxation matrix, respectively, which is related to the moment $\boldsymbol{m}$ and the equilibrium moment m^{eq}. The quantity F is the body force related to the gravity aspect [3].

The macroscopic quantities can be determined after the computational process of the Boltzmann equation which is the density and momentum [7, 10].

$$\rho = \sum_{i=0}^{18} f_i$$
$$\rho u = \sum_{i=0}^{18} f_i e_i \quad (4)$$

To simulate the temperature field using the D3Q7 model, the thermal LBE-MRT equation can be formulated, as indicated in the following expression.

$$g(x + ei \times \Delta t, t) - g(x, t) = -N^{-1} \times Q \times (n(x, t) - n^{\mathrm{eq}}(x, t)) \quad (5)$$

The 7 × 7 transformation matrix and the relaxation matrix are defined by the quantities N and Q, respectively. The data generated from the thermal Boltzmann equation can be used to represent the temperature field at each position of the calculation domain. So, after the numerical computation, the final treatment is converting obtained data to the macroscopic quantity which is temperature [1, 5, 6, 11].

$$T = \sum_{i=0}^{6} g_i \tag{6}$$

All other parameters utilized in the LBM-MRT equation for the velocity and temperature field that we did not specify are detailed in the reference [4, 8, 9].

2.3 Boundary Conditions

2.3.1 Flow Boundary Conditions

The implementation of boundary conditions in the numerical calculation is one of the necessary requirements to confirm the stability of the code during the numerical simulation [12]. The "bounce-back" technique was used to characterize the cavity walls in terms of fluid flow, as indicated in the formula below [7],

$$f_i(x, y, z) = f_{\tilde{i}}(x, y, z) i = 0, \ldots, 18 \tag{7}$$

where i indicates the discrete velocity direction and $\tilde{i}$ denotes the opposing speed direction. The x, y, and z coordinates indicate the position of each cubical cavity wall.

2.3.2 Thermal Boundary Conditions

The boundary conditions used to determine the cold and hot walls in the thermal issue are stated by the equation linked to the macroscopic quantity. Therefore, the boundary conditions for the hot bottom wall and cold vertical walls may be written using the following equations.

$$\begin{cases} Z = 0 \rightarrow g_5(x, y) = T_{\text{hot}} - \sum_{\substack{i=0 \\ i \neq 5}}^{6} g_i \\ Y = 0 \rightarrow g_1(x, z) = T_{\text{cold}} - \sum_{\substack{i=0 \\ i \neq 1}}^{6} g_i \\ Y = 0 \rightarrow g_2(x, z) = T_{\text{cold}} - \sum_{\substack{i=0 \\ i \neq 2}}^{6} g_i \end{cases} \tag{8}$$

The Neumann Boundary Condition was applied to the adiabatic walls through using the following expressions.

$$\begin{cases} z = H \rightarrow g_6(x, y) = g_5(x, y) \\ x = 0 \rightarrow g_4(y, z) = g_3(y, z) \\ x = H \rightarrow g_3(y, z) = g_4(y, z) \end{cases} \tag{9}$$

3 Numerical Results and Discussion

We conducted an intensive investigation to establish the consistency and effectiveness of our simulation code in order to determine whether the present code is better appropriate for evaluating fluid flow and convective heat transfer in diverse configurations.

3.1 Validation of Numerical Code

The results of the generated code are evaluated to those found in the literature [13]. The first comparison was conducted using Corvaro and Paroncini's experimental data [13] for the isotherm and velocity contour. For Rayleigh numbers equal to $Ra = 2.02 \times 10^5$ and the position of the heat source equal to $\delta = 0.5$, the thermal and velocity field results are consistent with the experimental and numerical results illustrated in Fig. 3 and Fig. 4.

The consistency of our simulation method was ensured by the calculation of the Nusselt number for various Rayleigh number values where $\delta = 0.5$. The Nusselt

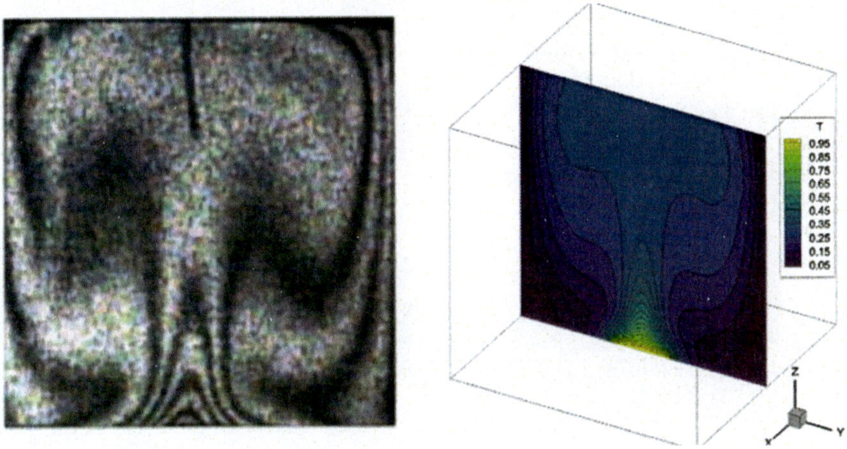

Fig. 3 Comparison of the experimental measurement utilizing the double-exposure interferogram [13] with the 3D and 2D projections of the LBM-MRT simulation results for $Ra = 2.02 \times 10^5$, $\delta = 0.5$

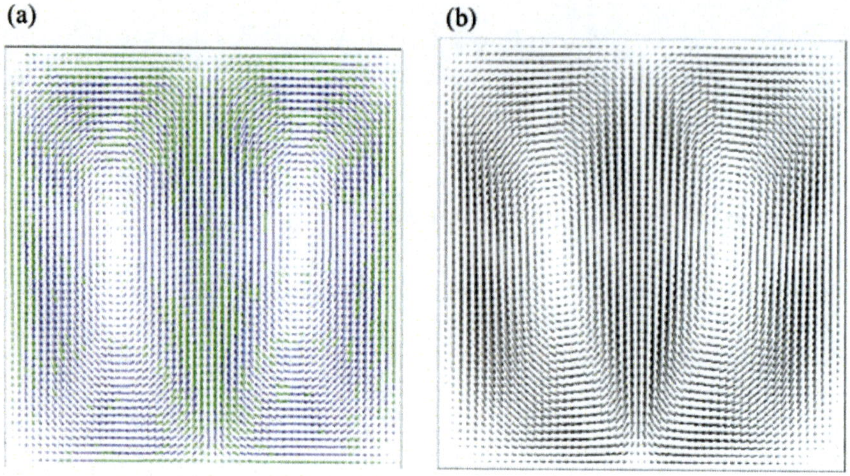

Fig. 4 Comparison study between the experimental data of the vector map (**a**), and the 2D projection of velocity trace of LBM-DMRT simulation results (**b**) for Ra = 2×10^5, $\delta = 0.5$

number, as given in the formula below, enables us to describe the nature of heat transmission between the hot cavity base and the two vertical cold walls [14],

$$\mathrm{Nu} = -\frac{L_c}{T_h - T_c} \times \frac{\mathrm{d}T}{\mathrm{d}z}|_z \tag{10}$$

where L_c represents the characteristic length of the heated section of the cubical cavity and T denotes dimensionless temperature such as $T = \frac{\theta - \theta_c}{\theta_h - \theta_c}$.

As shown in Table 1, the experimental and numerical results obtained by the literature [13] converge toward the results acquired by the numerical simulation performed using the double MRT-LBM.

One can note that the average Nusselt number for different values of the Rayleigh number has a maximum deviation of less than 2%. As a result, the code employed in the current work is capable of performing numerical simulations of the fluid flow and heat transfer problems. The 55 × 55 × 55 mesh is adopted for all the numerical

Table 1 Assessment of the average Nusselt number between the experimental and numerical results for $\delta = 0.5$

	Reference	Our results	Standard deviation
Ra	Nu (exp)	Nu (LBM)	$\|\Delta\|[\%]$
1.38E+05	5.859	5.962	1.760%
1.71E+05	6.300	6.250	0.800%
1.98E+05	6.450	6.458	0.130%
2.32E+05	6.650	6.695	0.670%
2.50E+05	6.810	6.810	0.001%

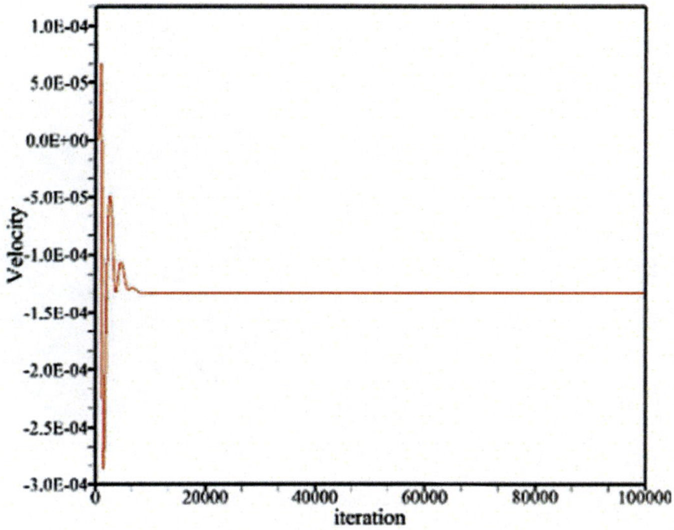

Fig. 5 Time series for U-velocity at the center of the cavity for 55 × 55 × 55 lattice

results obtained during this research, where the stability is approved very well as shown in Fig. 5.

3.2 *Results and Discussion*

This section interprets the numerical results of isotherm contours, stream traces, and average Nusselt number for varying parameters of the Rayleigh number, thermal conductivity ratio ($K_r = K_s/K_f$), and nature of fluid flow (Prandtl number). The purpose of this study is to analyze the heat transfer development by cooling a heated cubical box base using fluid flow.

Figures 6 and 8 illustrate the 2D and 3D stream traces of the fluid flow inside a cubical cavity for a fixed thermal conductivity ratio. The mounting of lateral cold walls in front of the heated base provokes an apparition of two recirculation zone clockwise and anticlockwise because of the natural convection involvement inside the cavity. The shape of this vortex changes when the Rayleigh number varying, where the strength of the recirculation increases when the Ra increases.

The thermal field was also discussed in this stage to figure out the evolution of the convective heat transfer for $Kr = 10$. Figures 7 and 9 show the 2D projection of temperature field and 3D iso-surfaces.

The results indicate that the thermal energy climbs significantly upward of the cavity mid-length, where the temperature of the fluid increases in accordance with the augmentation of Ra. Consequently, the heat transfer enhances in the vicinity of

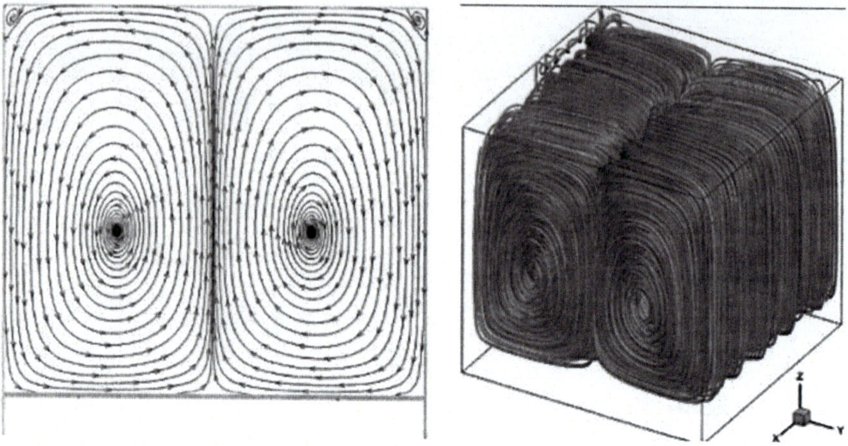

Fig. 6 2D and 3D presentation of the stream traces at Ra = 5.10^3, $Kr = 10$

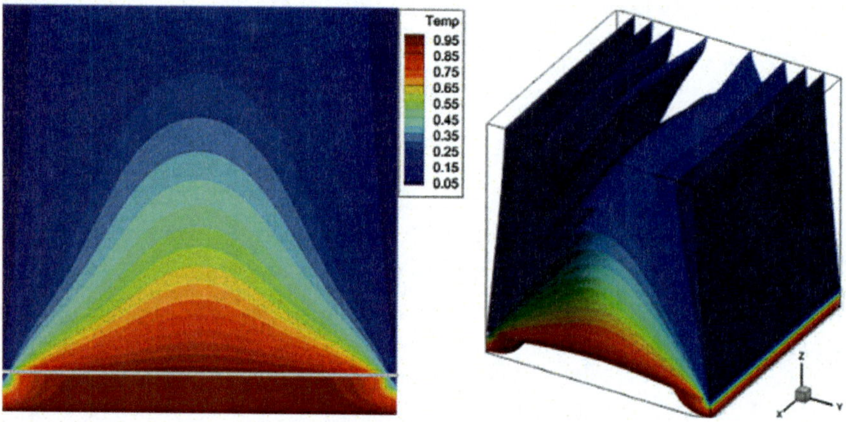

Fig. 7 2D Temperature distribution and 3D iso-surfaces Ra = 5.10^3, $Kr = 10$

the solid bottom cavity wall and particularly near the bottom of the left and right corners.

The assessment of the Prandtl number is also highlighted in this research in order to pursue the evolution of the thermal field. As shown in Figs. 10a and 11a, the fluid heats up due to direct contact between the fluid and the heated cavity base so that the heated fluid flows to the top of the mid-length cavity by natural convection, and the fluid becomes typically heated. It can be interpreted that the temperature of the fluid increases rapidly due to the high value of the thermal conductivity ratio compared to the previous case where $Kr = 10$ (see Fig. 9).

Regarding the fluid flow behavior, one can remark that the strength of propagation of fluid decreases when the number of Prandtl increases (see Figs 10a and 11a),

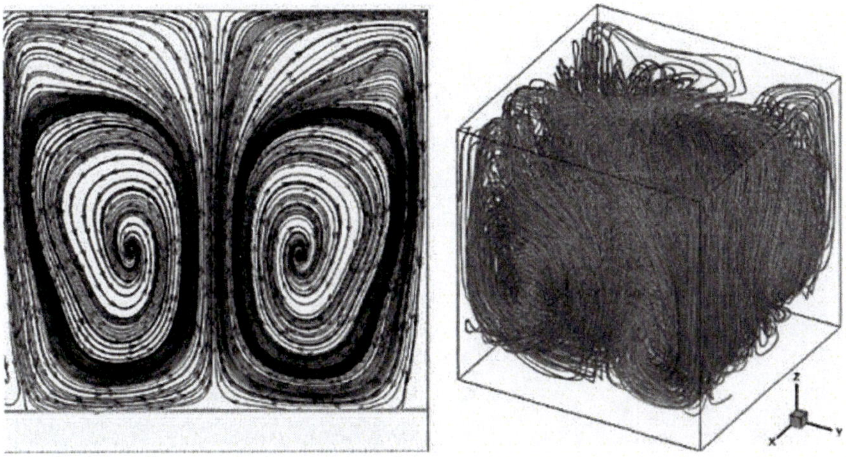

Fig. 8 2D and 3D presentation of the stream traces at $\mathrm{Ra} = 1.10^6$, $Kr = 10$

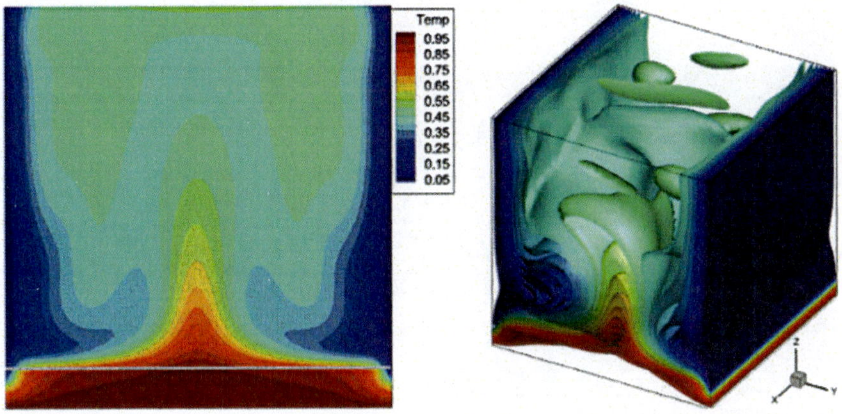

Fig. 9 2D temperature distribution and 3D iso-surfaces $\mathrm{Ra} = 1.10^6$, $Kr = 10$

in other words, the movement of water ($Pr = 7.01$) under the effect of natural convection is weak because of the high value of their density and viscosity in front of the propagation of the air ($Pr = 0.71$).

To gain a better understanding of the heat transfer process involved in this study, the local Nusselt number evolution of the heated solid wall was analyzed for different values of Rayleigh number, thermal conductivity ratio, and Prandtl number with two different ways: perimeter ($X = 1/2$, $Z = 0.1$) and surface area of the solid wall (see Fig. 12). For fixed Kr and Pr number (see Fig. 12a and b), the average Nusselt number increases when the Rayleigh number increases, therefore, the heat transfer increases due to the rapidity of the fluid propagation by the natural convection.

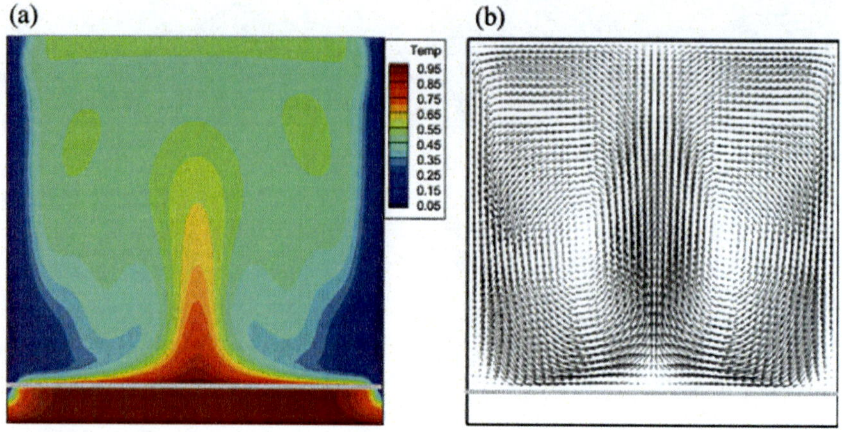

Fig. 10 2D projection ($X = 1/2$, $Z = 0.1$) at Ra = 1.10^6 and $Kr = 1000$, $Pr = 0.71$ (airflow): **a** isotherms, **b** velocity vector

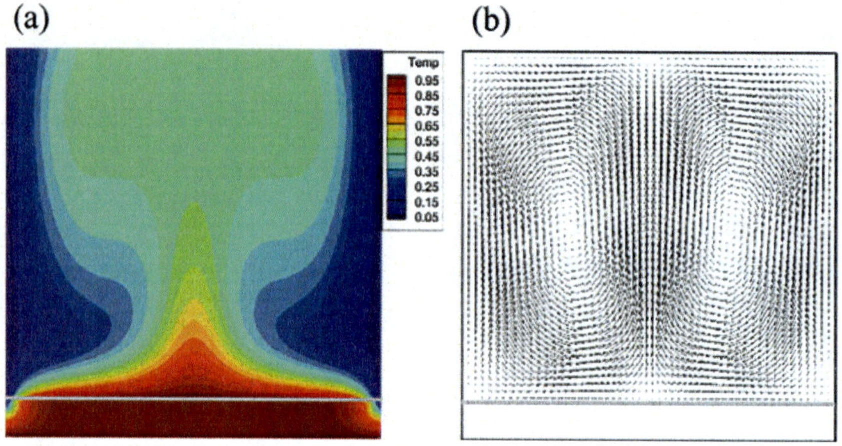

Fig. 11 2D projection ($X = 1/2$, $Z = 0.1$) at $Ra = 1.10^6$ and $Kr = 1000$, $Pr = 7.01$ (waterflow): **a** isotherms, **b** velocity vector

Regarding the variation of the thermal conductivity ratio for fixed *Ra* and *Pr*, the heat transfer decreases due to the fact that the heated solid of the cavity base rapidly transfers the energy via the surrounding fluid, resulting in a significant increase in the temperature of the fluid.

Concerning the variation of the Prandtl number for $Kr = 1000$, and Ra $= 10^6$ (see Fig. 12c, d), the evolution of the average Nusselt number remains relatively the same value when the properties of fluid changed (water flow instead of the airflow). It can be concluded that the heat exchange between the fluid and the heated solid is very weak in the case of water utilization because of the influence of its viscosity

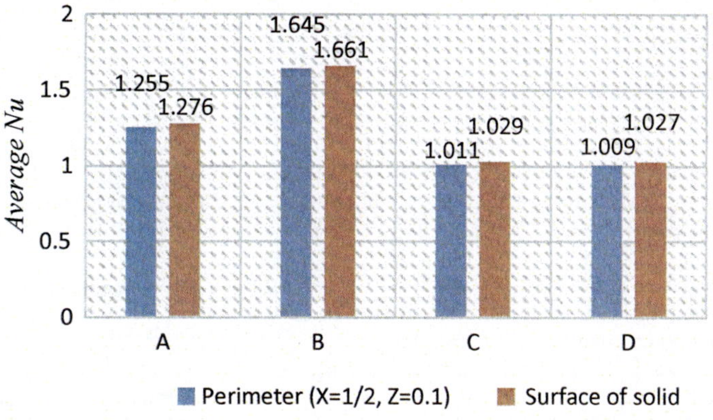

Fig. 12 Evaluation of the average Nusselt number for different cases: **a** Ra = 5.10^3, $Kr = 10$, $Pr = 0.71$, **b** Ra = 1.10^6, $Kr = 10$, $Pr = 0.71$, **c** Ra = 1.10^6, $Kr = 1000$, $Pr = 0.71$, **d** Ra = 1.10^6, $Kr = 1000$, $Pr = 7.01$

[11]. Therefore, to increase the heat transfer it is necessary to increase the Rayleigh number in order to increase the buoyancy of the water.

4 Conclusions

In the present work, three-dimensional natural convection was successfully simulated using the lattice Boltzmann method that describing the fluid flow and heat transfer within a cubical cavity contain a heat source that comes from the cavity slab. The obtained results by the utilization of the double MRT-LBM approach agreed well with the experimental research, lending credence to the present simulation code. The main purpose of the current study is to understand and predict the fluid flow behavior and the convective heat transfer in a cubical cavity under the effect of several parameters that can either increase or decrease the heat transfer between the solid wall and the studied fluid. The results show that the heat transfer quality increases when the Rayleigh number increases by around 31% for $Kr = 10$ and $Pr = 0.71$. While the heat exchange decreases when the thermal conductivity increases by around 38% for Ra = 10^6 and $Pr = 0.71$. In this situation, it is necessary to reduce the thermal conductivity of the solid part of the cavity slab. The thermal transfer relatively changes when the water is used instead of air inside the cavity in the conditions where $Kr = 1000$ and Ra = 10^6, so the increase of Rayleigh number is necessary in order to increase the water buoyancy which allows increasing the thermal transfer quality.

References

1. Moussaoui MA et al (2011) A computation of flow and heat transfer past three heated cylinders in a vee shape by a double distribution MRT thermal lattice Boltzmann model. Int J Therm Sci. https://doi.org/10.1016/j.ijthermalsci.2011.03.011
2. Mezrhab A et al (2010) Double MRT thermal lattice Boltzmann method for simulating convective flows. Phys Lett Sect A Gen At Solid State Phys. https://doi.org/10.1016/j.physleta.2010.06.059
3. Moussaoui MA et al (2019) Natural convection heat transfer in a square enclosure with an inside hot block. In: 2019 International conference wireless technology embedded intelligent systems WITS 2019, pp 1–6. https://doi.org/10.1109/WITS.2019.8723863
4. Chávez-Modena M et al (2020) Optimizing free parameters in the D3Q19 Multiple-Relaxation lattice Boltzmann methods to simulate under-resolved turbulent flows. J Comput Sci 45. https://doi.org/10.1016/j.jocs.2020.101170
5. Moussaoui MA et al (2009) Convective heat transfer over two blocks arbitrary located in a 2D plane channel using a hybrid lattice Boltzmann-finite difference method. Heat Mass Transf und Stoffuebertragung. https://doi.org/10.1007/s00231-009-0514-9
6. Moussaoui MA et al (2017) Study of heat transfer in an enclosure with a square cylinder using Lattice Boltzmann method. EPJ Appl Phys. https://doi.org/10.1051/epjap/2017170080
7. Lahmer EB et al (2019) Investigation of laminar flow and convective heat transfer in a constricted channel based on double MRT-LBM. In: 2019 international conference on wireless technologies, embedded and intelligent systems, WITS 2019. https://doi.org/10.1109/WITS.2019.8723820
8. Wang P et al (2017) Numerical study of three-dimensional natural convection in a cubical cavity at high Rayleigh numbers. Int J Heat Mass Transf 113:217–228. https://doi.org/10.1016/j.ijheatmasstransfer.2017.05.057
9. Liu Q et al (2019) Three-dimensional multiple-relaxation-time lattice Boltzmann models for single-phase and solid-liquid phase-change heat transfer in porous media at the REV scale. Appl Therm Eng. https://doi.org/10.1016/j.applthermaleng.2019.02.057
10. Lahmer EB et al (2019) Double constricted channel for laminar flow and heat transfer based on double MRT-LBM. In: Double constricted channel for laminar flow and heat transfer based on double MRT-LBM, p 16 24ème Congrès Français de Mécanique, Brest
11. Benhamou J et al (2020) Numerical study of natural convection and acoustic waves using the lattice Boltzmann method. Heat Transf. https://doi.org/10.1002/htj.21800
12. Moussaoui MA et al (2021) Numerical investigation of convective heat transfer in fluid flow past a tandem of triangular and square cylinders in channel. IOP Conf Ser Mater Sci Eng 1091, 1, 012058. https://doi.org/10.1088/1757-899x/1091/1/012058
13. Mohamad AA (2011). Lattice Boltzmann method: fundamentals and engineering applications with computer codes. https://doi.org/10.2514/1.J051744
14. Corvaro F, Paroncini M (2008) A numerical and experimental analysis on the natural convective heat transfer of a small heating strip located on the floor of a square cavity. Appl Therm Eng. https://doi.org/10.1016/j.applthermaleng.2007.03.018

Retrieval of Parameter in Combined Mode Conduction–Radiation Problem in Porous Ceramic Matrix by Artificial Neural Network

Swagatika Acharya, Vijay Kumar Mishra, Jitendra Kumar Patel, Gaurav Gupta, Mrityunjay K. Sah, and Pinky Shah

Nomenclature

A	Porous matrix surface area per unit, $^1/_{\mathrm{m}}$
η	Dimensionless coordinate
c	Specific heat at constant pressure, $^{\mathrm{J}}/_{\mathrm{kg}\cdot\mathrm{K}}$
θ	Dimensionless temperature
G	Emissive power, $^{\mathrm{W}}/_{\mathrm{m}^2}$
ρ	Density, $^{\mathrm{kg}}/_{\mathrm{m}^3}$
h	Heat transfer coefficient, $^{\mathrm{W}}/_{\mathrm{m}^2\cdot\mathrm{K}}$
σ	Stefan–Boltzmann constant, $5.67 \times 10^{-8}\,^{\mathrm{W}}/_{\mathrm{m}^2\cdot\mathrm{K}^4}$
i	Intensity of radiation, $^{\mathrm{W}}/_{\mathrm{m}^2\cdot\mathrm{sr}}$
φ	Porosity of porous ceramic matrix
k	Thermal conductivity, $^{\mathrm{W}}/_{\mathrm{m}\cdot\mathrm{K}}$
Ψ_{Rad}	Non-dimensional radiative heat flux
L_x	Length in x-direction, m
ω	Scattering albedo of solid
L_y	Length in y-direction, m

S. Acharya · V. K. Mishra (✉) · J. K. Patel · M. K. Sah · P. Shah
School of Mechanical Engineering, KIIT Deemed to be University, Bhubaneswar 751024, India
e-mail: mishra.vdm@gmail.com

G. Gupta
Vellore Institute of Technology, Vellore, India

© The Author(s), under exclusive license to Springer Nature Singapore Pte Ltd. 2023
S. Revankar et al. (eds.), *Recent Advances in Thermofluids and Manufacturing Engineering*, Lecture Notes in Mechanical Engineering,
https://doi.org/10.1007/978-981-19-4388-1_25

Subscripts and Superscripts

q_R	Radiative heat flux, W/m^2
E	East direction
$\dot{Q}$	Heat generation rate per unit volume, W/m^3
W	West direction
S_{av}	Average source term, W/m^2
N	North direction
T	Temperature, K
S	South direction
μ	Velocity, m/s
e	Exit of PCM

Greek Symbols

g	Gas phase
β	Extinction coefficient, $1/m$
i	Inlet of PCM
δ	Unit step function
s	Solid phase
ε	Emissivity
*	Non-dimensional

1 Introduction

In the development of any thermal device, temperature profile and/or heat flux, etc., is assumed to be known, and other properties are required for proper functioning of the device. This type of problems, where dependent variables are known and independent variables are required to be calculated, comes under inverse type of problem. Inverse problem is very useful for problems where direct measurement of properties is difficult due to variety of reasons such as: inaccessible locations, chances of disturbance in flow due to introduction of probe, complicated properties or high associated costs [1–3].

Porous ceramic matrix (PCM) is very useful material, especially where convention materials like metals and polymers are not suitable, such as: thermal insulations, burners, filters, and bioceramics [4]. In gas burners, PCM helps in redistribution of heat and hence enhances efficiency and reduces emissions [5–9]. In conventional burner-free blame-based type, heat because of combustion is generated in air. But, in PCM base burner, heat is generated within the cavities of the PCM. Then, the heat transfer through conduction and radiation takes place toward upstream and downstream direction. This redistribution of heat helps in preheating of the fresh fuel air mixture, thus helping to burn leaner fuel air mixture also. Also, the temperature in the PCM-based burner helps in reducing emissions such as NOx, CO, and unburned hydrocarbons (UHC). In gas burners, PCM is used by various researchers, but use of inverse analysis and optimization techniques is quite rare [10–12].

Artificial neural network (ANN) is very useful tool in problems where either the governing equations are unknown or are too complicated to solve. ANN is non-iterative approach, and once trained, it is able to give result of a problem very quickly [13–16].

In the present work, PCM is used in gas burner application. Finite volume method (FVM) is employed to numerically solve the governing equations together with prescribed boundary conditions. The solution gives the gas- and the solid-phase temperature profile corresponding to 49 different heat transfer coefficients (HTCs) values, leading to development of direct model. In the second part, ANN model is developed by using the two temperature profile corresponding to different HTCs. The number of neurons in the hidden layer considered in the ANN model is 10. Once the ANN model is developed, it is analyzed for suitability to handle new types of cases. After satisfaction of the robustness of the ANN model, an unknown pair of gas and the solid temperature profiles is fed into ANN as input; the ANN returns back the related HTC as output. LM is employed to train the neurons in neural network. The retrieved value of HTC by LM algorithm is very accurate.

In the present case, even though the problem could be solved by direct method, complexity of the direct method and demand for high computational resources makes it necessary to search for alternative approach. ANN-based method could be very handy for development of PCM-based technology for different application, considering its simplicity and fastness. The proposed method of ANN is very novel for PCM-based burners.

2 Formulation

Gas stream is allowed to flow through the PCM, and a heat source is situated at the center of the PCM to mimic combustion and represented schematically in Fig. 1. The PCM is conducting and radiating, whereas the gas is assumed to be transparent to the thermal radiation. Local thermal non-equilibrium is assumed and represented by considering two distinct energy equations for each phase, i.e., the gas and the solid phase. PCM helps to redistribute the heat from heat generation zone to upstream and downstream direction by conduction and radiation. Constant values for all the properties of solid and gas are assumed. The governing equations for steady-state scenario are as follows:

$$\text{Gas phase: } \phi \rho_g c_g u \frac{\partial T_g}{\partial x} + (1 - \phi) A h (T_g - T_s)$$

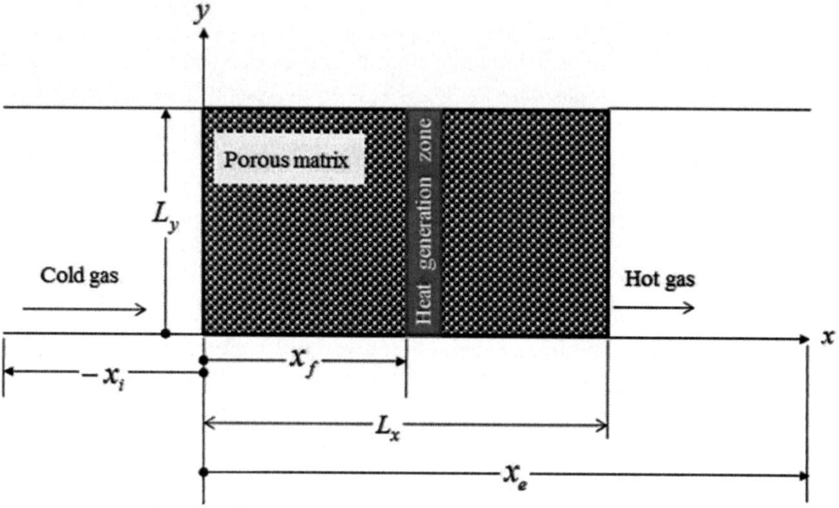

Fig. 1 Schematic of the conduction–radiation problem in PCM

$$= \phi\, \dot{Q}\, \delta(x) + \phi k_g \left(\frac{\partial^2 T_g}{\partial x^2} + \frac{\partial^2 T_g}{\partial y^2} \right) \tag{1}$$

$$\text{Solid phase: } (1-\phi)Ah(T_g - T_s) + (1-\phi)k_s \left(\frac{\partial^2 T_s}{\partial x^2} + \frac{\partial^2 T_s}{\partial y^2} \right) - \nabla \cdot q_{\text{Rad}} = 0 \tag{2}$$

where δ(x) is the Dirac-delta function and is zero except in heat generation region $0.45L_x < x \le 0.55L_x$. The volumetric radiative term $\nabla \cdot q_{\text{Rad}}$ in Eq. (2) is computed as follows:

$$\nabla \cdot q_{\text{Rad}} = (1-\omega)\beta\left(4\sigma T^4 - G\right) \tag{3}$$

The incident radiation G in Eq. (3) is numerically calculated as:

$$G = \int_{\Omega=0}^{4\pi} I(\Omega) d\Omega = \int_{\mu=0}^{2\pi} \int_{\lambda=0}^{\pi} I(\lambda, \mu) \sin\lambda \, d\lambda \, d\mu$$

$$\approx \sum_{k=1}^{M_\mu} \sum_{l=1}^{M_\lambda} I^m\left(\lambda_l^m, \mu_k^m\right) 2 \sin\lambda_l^m \sin\left(\frac{\Delta\lambda_l^m}{2}\right) \Delta\mu_k^m \tag{4}$$

Here, M_λ and M_μ represent the number of discrete points taken in polar and azimuthal directions, respectively. Therefore, $M_\lambda \times M_\mu$ represents the discrete directions in which intensities are calculated.

I_{P}^{m} in Eq. (4) represents intensity at any cell center is obtained by:

$$I_{\mathrm{P}}^{m} = \frac{\frac{|D_{\mathrm{x}}^{m}|A_{\mathrm{EW}}}{\gamma_{\mathrm{x}}} I_{\mathrm{W}}^{m} + \frac{|D_{\mathrm{y}}^{m}|A_{\mathrm{NS}}}{\gamma_{\mathrm{y}}} I_{\mathrm{S}}^{m} + (V\,\Delta\Omega^{m}) S_{\mathrm{P}}^{m}}{\frac{|D_{\mathrm{x}}^{m}|A_{\mathrm{E}}}{\gamma_{\mathrm{x}}} + \frac{|D_{\mathrm{y}}^{m}|A_{\mathrm{N}}}{\gamma_{\mathrm{y}}} + \beta V\,\Delta\Omega^{m}} \tag{5}$$

where A_{EW} represents area in x-face and A_{NS} represents area in y-face, of the 2D control volume. This helps in calculating intensities at all the cell centers. Within a control volume, intensity is computed for all quadrants, starting form the boundaries. Average intensity I_{P}^{m} for any cell is computed by:

$$I_{P}^{m} = \gamma_x I_{E}^{m} + (1-\gamma_x) I_{W}^{m} = \gamma_y I_{N}^{m} + (1-\gamma_y) I_{S}^{m} \tag{6}$$

Here γ is the finite difference weighting factor (=0.5), and it helps to compute the intensities at a point for a given direction by taking average value of intensities from different directions. The values for D_{x}^{m} and D_{y}^{m} are given by:

$$D_{x}^{m} = \int_{\Delta\Omega^{m}} \sin\lambda \cos\mu \, d\Omega = \int_{\mu^{m}-\frac{\Delta\mu^{m}}{2}}^{\mu^{m}+\frac{\Delta\mu^{m}}{2}} \int_{\lambda^{m}-\frac{\Delta\lambda^{m}}{2}}^{\lambda^{m}+\frac{\Delta\lambda^{m}}{2}} \cos\mu \sin^{2}\lambda \, d\lambda \, d\mu$$
$$= \cos\mu^{m} \sin\left(\frac{\Delta\mu^{m}}{2}\right) [\Delta\lambda^{m} - cos\, 2\lambda^{m} \sin(\Delta\lambda^{m})] \tag{7a}$$

$$D_{y}^{m} = \int_{\Delta\Omega^{m}} \sin\lambda \sin\mu \, d\Omega = \int_{\mu^{m}-\frac{\Delta\mu^{m}}{2}}^{\mu^{m}+\frac{\Delta\mu^{m}}{2}} \int_{\lambda^{m}-\frac{\Delta\lambda^{m}}{2}}^{\lambda^{m}+\frac{\Delta\lambda^{m}}{2}} \sin\mu \sin^{2}\lambda \, d\lambda \, d\mu$$
$$= \sin\mu^{m} \sin\left(\frac{\Delta\mu^{m}}{2}\right) [\Delta\lambda^{m} - cos\, 2\lambda^{m} \sin(\Delta\lambda^{m})] \tag{7b}$$

The value for elemental solid angle $\Delta\Omega^{m}$ is calculated as:

$$\Delta\Omega^{m} = \int_{\Delta\Omega^{m}} \mathrm{d}\Omega = \int_{\mu^{m}-\frac{\Delta\mu^{m}}{2}}^{\mu^{m}+\frac{\Delta\mu^{m}}{2}} \int_{\lambda^{m}-\frac{\Delta\lambda^{m}}{2}}^{\lambda^{m}+\frac{\Delta\lambda^{m}}{2}} \sin\lambda \, \mathrm{d}\lambda \, \mathrm{d}\mu$$
$$= 2\sin\lambda^{m} \sin\left(\frac{\Delta\lambda^{m}}{2}\right) \Delta\mu^{m} \tag{8}$$

For e medium which is radiatively participating, the source term at any point is computed as:

$$S = (1 - \omega)\beta\frac{\sigma T^4}{\pi} + \left(\frac{\omega\beta}{4\pi}\right)G \tag{9}$$

Volumetric radiative term $\nabla \cdot q_{\text{Rad}}$ is obtained from Eq. (3); then, the two energy equations (for the gas as well as the solid phase) are solved simultaneously. To obtain generalized solution, the conservation equations are non-dimensionalized by using the following terms.

$$\begin{aligned} &\theta = \frac{T - T_i}{T_i}, \quad \eta_\text{x} = \frac{x}{L_\text{x}}, \quad \eta_\text{y} = \frac{y}{L_\text{x}} \\ &\Psi_\text{Rad} = \frac{q_\text{Rad}}{\dot{Q}L_\text{x}}, \quad P_1 = \frac{\rho_\text{g} c_\text{g} u T_i}{\dot{Q}L_\text{x}}, \quad P_2 = \frac{hAT_i}{\dot{Q}} \\ &P_3 = \frac{k_\text{g} T_i}{\dot{Q}L_\text{x}^2}, \quad P_4 = \frac{k_\text{s} T_i}{\dot{Q}L_\text{x}^2}, \quad P_5 = \frac{hL_\text{x}}{k} \end{aligned} \tag{10}$$

In Eq. (10), θ is temperature; η_x is distance in x-direction (direction of flow); η_y is distance in y-direction; Ψ_Rad is radiative heat flux in porous medium; P_1 is flow enthalpy in gas; P_2 is heat transfer coefficient (between solid and the gas phase); P_3 is conductivity of gas phase; P_4 is conductivity of solid phase; P_5 is biot number, and all the terms are non-dimensional. Thus, the energy equations can be non-dimensionalized as:

$$\begin{aligned} \text{Gas phase:}\ &\phi P_1 \frac{\partial \theta_\text{g}}{\partial \eta_\text{x}} + (1 - \phi)P_2(\theta_\text{g} - \theta_\text{s}) \\ &= \phi\delta(\eta_\text{x}) + \phi P_3\left(\frac{\partial^2\theta_\text{g}}{\partial \eta_\text{x}^2} + \frac{\partial^2\theta_\text{g}}{\partial \eta_\text{y}^2}\right) \end{aligned} \tag{11}$$

$$\begin{aligned} \text{Solid phase: }&(1 - \phi)P_2(\theta_\text{g} - \theta_\text{s}) + (1 - \phi)P_4\left(\frac{\partial^2\theta_\text{s}}{\partial \eta_\text{x}^2} + \frac{\partial^2\theta_\text{s}}{\partial \eta_\text{y}^2}\right) \\ &- \nabla \cdot \Psi_\text{Rad} = 0 \end{aligned} \tag{12}$$

Boundary conditions for the energy equations are given in Eqs. (13) and (14).
Gas phase:

$$\begin{aligned} &\theta_\text{g} = 0 \text{ at } \eta_\text{x} = \eta_{\text{x},i}, \quad \frac{\partial \theta_\text{g}}{\partial \eta_\text{x}} = 0 \text{ at } \eta_\text{x} = \eta_\text{x,e} \\ &\frac{\partial \theta_\text{g}}{\partial \eta_\text{y}} = 0 \text{ at } \eta_\text{y} = 0, \quad \frac{\partial \theta_\text{g}}{\partial \eta_\text{y}} = 0 \text{ at } \eta_\text{y} = \eta_\text{Ly} \end{aligned} \tag{13}$$

Solid phase:

$$-\frac{\partial \theta_s}{\partial \eta_x} = P_5(\theta_g - \theta_s) \text{ at } \eta_x = 0, \quad \frac{\partial \theta_s}{\partial \eta_x} = P_5(\theta_g - \theta_s) \text{ at } \eta_x = 1$$
$$\frac{\partial \theta_s}{\partial \eta_y} = 0 \text{ at } \eta_y = 0, \quad \frac{\partial \theta_s}{\partial \eta_y} = 0 \text{ at } \eta_y = \eta_{Ly} \tag{14}$$

In non-dimensional form, the volumetric radiative term $\nabla \cdot \Psi_{Rad}$ becomes:

$$\nabla \cdot \Psi_{Rad} = \beta L_x(1-\omega)\left[4\pi\,\Phi(1+\theta_s)^4 - G^*\right] \tag{15}$$

Here, $\Phi = \frac{\sigma T_i^4}{\dot{Q}L_x}$ is the radiative emissive power at the entry of PCM $\eta_{x,i}$ in non-dimensional form, and $G^* = \frac{G}{\dot{Q}L_x}$ is the incident radiation in non-dimensional form. The source term dimensionless form becomes:

$$S^*(\tau) = (1-\omega)\frac{\Phi(1+\theta)^4}{\pi} + \left(\frac{\omega}{4\pi}\right)G^* \tag{16}$$

The boundary intensity is calculated as:

$$I_b^* = \varepsilon_b \frac{\Phi(1+\theta_b)^4}{\pi} + \frac{(1-\varepsilon_b)}{\pi}\sum_{k=1}^{M_\mu}\sum_{l=1}^{M_\lambda/2} I^{*,m}\left(\lambda_l^m, \mu_k^m\right)\sin\lambda_l^m \cos\lambda_l^m \sin\Delta\lambda_l^m \Delta\mu_k^m \tag{17}$$

Here, $I^* = \frac{I}{\dot{Q}L_x}$ is the non-dimensional intensity.

First of all, both phase temperatures are given some guess values. Values of different parameters utilized in the analysis are presented in Table 1. Then, the values of temperature fields are updated by solving the energy equations simultaneously by Gauss–Seidel method. After solving for few iterations, the radiative information is updated by solving the radiative transfer equation. Again, the energy equation is solved; this process is repeated until convergence is achieved.

Table 1 Non-dimensional values of relevant parameters

Parameters	Values
P_1	0.01
P_3	2.5×10^{-4}
P_4	0.02
P_5	5
ω	0.5
ε	1.0
$\Phi_i = \Phi_e$	2.98×10^{-5}

3 Results and Discussion

The two energy equations of gas and solid phase together with the radiative transfer equations (RTE) are solved by employing FVM. The gas and the solid temperature profiles obtained by FVM are compared with the numerical results available in the literature [17] and are shown in Fig. 2. For validation purpose, the aspect ration in the present problem is taken to be large $\left({L_y}/{L_x} = 10 \right)$ value to reduce the problem to a 1D one. High value of HTC (P2) leads to high rate of heat transfer between the two phase, collapse of solid and gas temperature curves to a single one.

After getting good agreement with the literature, temperature distributions for solid and the gas phase are computed for 49 different HTCs. This constitutes the first part of the present work. In the next part, the gas and the solid temperature distributions along with the corresponding HTCs are used as the input data for training the ANN model. The architecture of network is shown in Fig. 3, with one hidden layer having 10 number of neurons. Best way to select suitable number of hidden neurons is to start with low value. Keep an eye on different performance metrics (histogram, performance curves, etc.) and continue to increase the number of neurons, till desired accuracy is obtained. Proper number of neurons in the hidden layer is very necessary for robust network. Too low neurons may lead to under fitting, leading to improper training, and poor accuracy. Whereas, too much hidden neurons lead to over training, and the network will not give accurate results for new cases. In the input layer, 600 neurons were taken to take the gas and the solid phase temperatures corresponding to 300 control volumes. The output layer is made up of 1 neuron, for HTC. In the ANN model, neurons are trained by employing LM algorithm. The ANN model is then checked for robustness with the help of different performance curves and histogram.

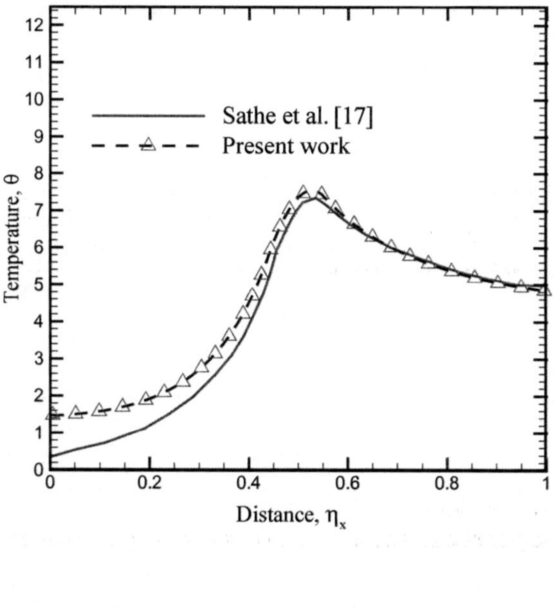

Fig. 2 Comparison of results of present direct numerical model with the results of Tong and Sathe [17] for $P_2 = 500$ and $\beta = 1.0$

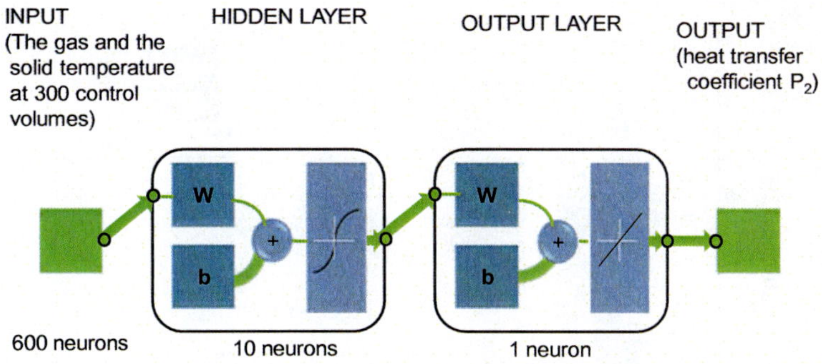

Fig. 3 Architecture of ANN model with one hidden layer

Then, a new pair of solid and the gas temperature is fed into the ANN model as input, and the ANN returns back the corresponding HTC as output.

Figure 4 shows the histogram of the network trained by LM algorithm. Sufficiently, large amount of data falling on zero error line (ZEL) shows that the network is properly trained for the supplied cases. Data falling away from ZEL show that the network is able to nicely deal with new types of casesl, for which the ANN model is not supplied with data as input. Distribution of validation and testing data, across the ZEL, is also a good sign of a network. In a good ANN model, the data should be normally distributed around ZEL. The maximum of the normal distribution should be near to ZEL. Figure 5 shows the performance curves of the network trained by using LM algorithm. In ANN, the total input data are segmented into three components, namely: training data, validation data, and testing data. With the rise of iteration, contineous decrease in mean square error (MSE) is the sign of proper traing of the network. During the validation stage, convergence is achieved at 35th iteration; thus, the parameters in ANN such as weights and biases are modified. The slope of testing curve is also decreasing, indicating that the ANN model is able to give result for new case with high accuracy. This gives the indication that the network with LM algorithm is neither over-trained nor under-trained. Over-trained ANN model gives good result for the trained data but miserably fails to handle new scenarios. Whereas, under-trained ANN model predicts wrong result even corresponding to the input cases. Regression analysis of the network with LM algorithm is presented in Fig. 6. It is clear from the regression analysis that there exist a linear relation between the input (the gas and the solid temperature profiles) and the output (HTCs) of the ANN model. High value of regression coefficient ($R = 1$) and the solid line for fitness is also diagonal, which means the model is able to establish linear relationship between the input and output data. Thus, present ANN model is robust and could give correct result.

Table 2 shows the retrieved values obtained from the network, trained by employing the LM algorithm. The gas and the solid temperature profiles corresponding to HTC P2, act = 500 (obtained from direct part), were fed into the ANN model, and this data were not viewed by the ANN model. The ANN model developed

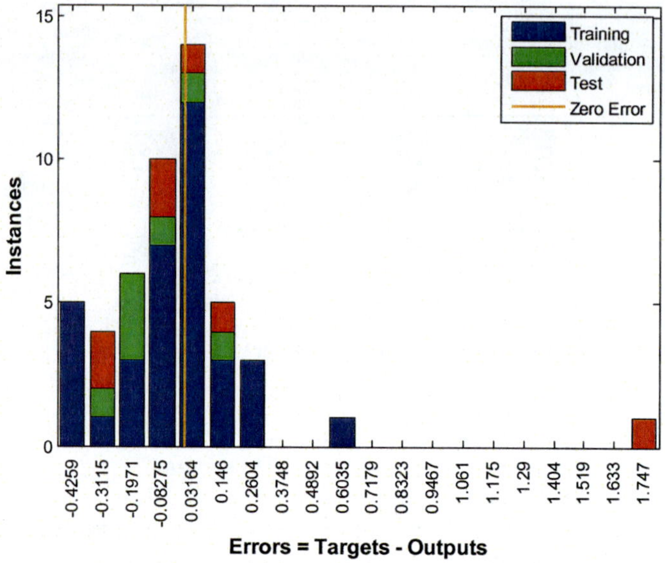

Fig. 4 Performance evaluation of the ANN model by histogram

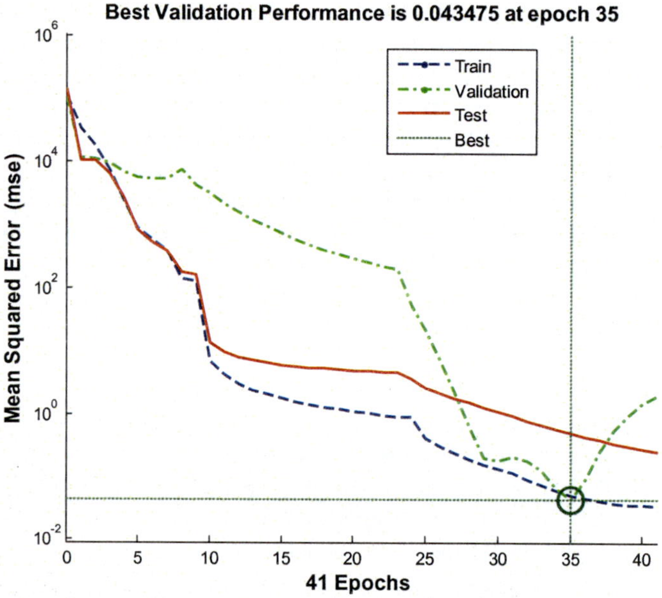

Fig. 5 Performance evaluation of the ANN model trained with LM algorithm by curves for training, validation, and testing stage

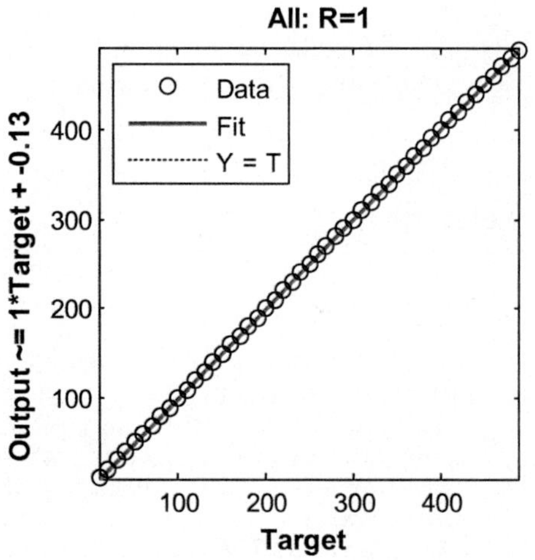

Fig. 6 Regression analysis of ANN model with LM algorithm

Table 2 Comparison of LM and SCG algorithm in training of neuron in ANN model

S. No.	Algorithm	$P_{2,act}$	$P_{2,ret}$	Absolute error in retrieval (%)
1	LM	500	499.1728	0.16

for the inverse analysis takes the new data (not viewed during training stage) and returns the retrieved value of HTC (P2) to be 499.17. This retrieved value of HTC is well within 0.16% accuracy. The error in retrieved value is computed by:

$$E = \frac{|P_{2,act} - P_{2,est}|}{P_{2,act}} \times 100 \tag{18}$$

The accuracy of retrieved HTC value is very high with the ANN model trained by LM algorithm.

4 Conclusions

An artificial neural network (ANN) model is developed by using Levenberg–Marquardt (LM) as training algorithm. The ANN model is trained by using the data from a 2D conduction–radiation problem in porous ceramic matrix (PCM). The governing equations of the problem are solved by using finite volume method (FVM). The temperature distribution of the gas and the solid phase corresponding to various heat transfer coefficients (HTCs) is utilized for training the ANN model. After development of the ANN model, an unknown temperature profiles of gas and solid phase are used in the ANN model as input, and the ANN returns the related value of HTC as output. Performance of LM in training of neurons is analyzed. LM algorithm is found to be appropriate for such problems, as the retrieved value of HTC is accurate within 0.16% absolute error.

References

1. Backus G (1988) Hard and soft prior bounds in geophysical inverse problems. Geophys J 94:249–261
2. Özisik MN, Orlande HRB (2000) Inverse Heat Transfer Fundamentals and Applications. Taylor & Francis, New York
3. Osman AM, Beck JV (1990) Investigation of transient heat transfer coefficients in quenching experiments. ASME J Heat Transf 112:843–848
4. Saggio-Woyanski J, Scott C (1992) Processing of porous ceramics. Am Ceram Soc Bull 71:1674–1682
5. Kakati S, Mahanta P, Kakoty S (2007) Performance analysis of pressurized kerosene stove with porous medium inserts. J Sci Ind Res 66:565–569
6. Prosuntsov PV, Barinov DY (2021) Analysis of combined radiation-conductive heat transfer during destruction of porous carbon-ceramic matrix composite material of thermal protection. In: AIP conference proceedings, Vol 2318, 020024
7. Mishra VK, Mishra SC, Basu DN (2015) Combined mode conduction and radiation heat transfer in a porous medium and estimation of optical properties of the porous matrix. Numerical Heat Transfer (A) 67(10):1119–1135
8. Chopade RP, Agnihotri E, Singh AK, Kumar A, Uppaluri R, Mishra SC, Mahanta P (2011) Application of a particle swarm algorithm for parameter retrieval in a transient conduction-radiation problem. Numerical Heat Transfer (A) 59(9):672–692
9. Sans M, Schick V, Parent G, Farges O (2020) Experimental characterization of the coupled conductive and radiative heat transfer in ceramic foams with a flash method at high temperature. Int J Heat and Mass Transf 148:119077
10. Anand K, Bhardwaj A, Chaudhuri S, Mishra VK (2022) Self-organizing map network for the decision making in combined mode conduction-radiation heat transfer in porous medium. Arab J Sci Eng
11. Mishra VK, Dasgupta U, Patra S, Pal R, Anand K (2022) A dynamic two-level artificial neural network for estimation of parameters in combined mode conduction-radiation heat transfer in porous medium: an application to handle huge dataset with noise. Heat Transfer 51:1306–1335
12. Song FQ, Wen ZD, Wang ZY, Liu EYXL (2019) Numerical study and optimization of a porous burner with annular heat recirculation. Appl Therm Eng. https://doi.org/10.1016/j.applthermaleng.2019.113741
13. Jambunathan K, Hartle S, Ashforth-Frost S, Fontama VN (1996) Evaluating convective heat transfer coefficients using neural networks. Int J Heat Mass Transf 39:2329–2332
14. Sablani SS (2001) A neural network approach for non-iterative calculation of heat transfer coefficient in fluid-particle systems. Chem Eng Process 40:363–369
15. Mishra VK, Chaudhuri S (2021) Implementation of stochastic optimization method-assisted radial basis neural network for transport phenomenon in non-newtonian third-grade fluids: assessment of five optimization tools. Arab J Sci Eng 46:11797–11818
16. Mishra VK, Chaudhuri S (2021) Genetic algorithm-assisted artificial neural network for retrieval of a parameter in a third grade fluid flow through two parallel and heated plates. Heat Transfer 50:2090–2128
17. Tong T, Sathe S (1991) Heat transfer characteristics of porous radiant burners. J Heat Transf 113:423–428

Combustion Analysis

Effect of Compression Ratio on Performance, Combustion and Emission Characteristics of a Diesel Engine Fueled with Rubber Seed Biodiesel–Diesel Blends

Chinmoy Jit Sarma, Bhaskor J. Bora, Dilip K. Bora, and Bhaskar J. Medhi

Nomenclature

BSEC	Brake-specific energy consumption
BTE	Brake thermal efficiency
CO	Carbon monoxide
CO_2	Carbon dioxide
CR	Compression ratio
EGT	Exhaust gas temperature
HC	Hydrocarbon
IC	Internal combustion
NHRR	Net heat release rate
NOx	Nitrogen oxides
PCP	Peak cylinder pressure
RBD	Rubber seed biodiesel
VCR	Variable compression ratio

C. J. Sarma · B. J. Bora (✉) · B. J. Medhi
Assam Energy Institute, Sivasagar, Center of Rajiv Gandhi Institute of Petroleum Technology, Sivasagar, Assam 785697, India
e-mail: bjbora@rgipt.ac.in

Energy Institute, Bengaluru, Center of Rajiv Gandhi Institute of Petroleum Technology, Bengaluru, Karnataka 561003, India

D. K. Bora
Department of Mechanical Engineering, Assam Engineering College, Jalukbari, Guwahati, Assam 781013, India

© The Author(s), under exclusive license to Springer Nature Singapore Pte Ltd. 2023

S. Revankar et al. (eds.), *Recent Advances in Thermofluids and Manufacturing Engineering*, Lecture Notes in Mechanical Engineering,
https://doi.org/10.1007/978-981-19-4388-1_26

1 Introduction

Internal combustion (IC) engines can be considered as one of the panaceas for power production. IC engines have got wide industrial applications and in transportation sector. Biodiesel can be utilized as an alternate fuel in any diesel engine without making any engine design change and is also eco-friendly [1]. India experimented the first biodiesel run flight from Delhi to Dehradun on 27th of August 2018 [2].

Biomass can be considered as solar energy stockpiled in the form of organic matter like plants or animal remnants having a net positive worth as a chemical reserve [3]. Biodiesel is produced from oils or fats through chemical conversion process, and it resembles diesel in terms of fuel properties. Biodiesel is identified as a potential candidate to substitute petroleum diesel, and therefore, it is drawing worldwide attention in research and development. One of the benefits of using biodiesel is that it does away with the intolerable odor and black smoke of diesel fuel. Biodiesel promotes burning of the hydrocarbons in a blend as it contains about 11% oxygen. Biodiesel run diesel engines emit less carbon monoxide (CO) and thus reduce air pollution and hence are less harmful in comparison to conventional diesel fuel. Spillage of biodiesel results in less environmental sabotage as it is biodegradable in equivalence to conventional diesel.

Keeping in mind the fact of depletion of petroleum reserves and surge in environmental pollution, biofuels can substitute to diesel as is renewable and environmentally friendly. Among different types of biofuels available, rubber seed biodiesel seems to offer a tremendous potential. India produces about 5000 tons of rubber seed oil per annum [4]. The chemical and physical characteristics of rubber seed biodiesel resemble to that of diesel fuel.

Vegetable oils have comparable intensity of energy use, cetane number, and latent heat of vaporization with that of diesel. They emit less hazardous gases and are environmentally benign. But, due to their high viscosity, fuel pulverization is inadequate which results in partial burning. The thickness of the biofuels can be curtailed by mixing it with diesel. The thickness of biofuels can be curtailed by directly mixing with diesel, preheating, and transesterification. Senthil et al. [5] curtailed the thickness of rubber seed oil by mixing it with diesel in various percentages, tested its thickness at different temperatures and run a diesel engine with it. The resultant brake thermal efficiencies (BTEs) were 28.64, 27.74, 26.82, and 26.2% for B25, B50, B75, and B100 and that obtained for diesel mode at full load is 29.92%. Due to slow burning of vegetable oils, exhaust gas temperature (EGT) exhibited is greater with all mixtures of rubber seed oil (RSO) at 100% load in correlation to diesel oil. For RSO and its blend, carbon monoxide (CO) and hydro carbon (HC) emissions are maximum at full load, and for B25 biodiesel blend, CO, NO, and HC emissions are optimal. Geo et al. [6] evaluated the performance of RSO, rubber seed oil methyl ester (RSOME) run diesel engine run with diesel as a primary fuel, and hydrogen as an added fuel at 25, 50, 75, and 100% load conditions. The preblend burning rate increases due to addition of hydrogen which raises the BTE. At 100% load, the BTE of RSO and RSOME raises from 26.56 to 28.12% and 27.89

to 29.26% with hydrogen addition. HC and CO emission decreases for all loads because of addition of hydrogen; however, NOx emission raises due to the rise in burning temperature because of the high pre-blend burning. Geo et al. [7] studied the performance of rubber seed oil with diethyl ether (DEE) addition run diesel engine at various injection rates. Due to addition of DEE in RSO, the BTE of the engine raises, and maximum BTE obtained is 28.5% at 200 g/h injection rate of DEE. HC and CO emission decreases due to addition of DEE; however, NO_x emission raises due to raise in pre-blend heat release rate. Patil et al. [8] evaluated the performance of RSME diesel blend run diesel engine with fractional addition of butanol. BTE, exhaust gas temperature, carbon dioxide (CO_2), and NO_x emission rises with rise in engine load. Krishna et al. [9] evaluated the performance and emission characteristics of rubber seed biodiesel with diethyl ether (DEE) addition run semi-adiabatic diesel engine (SADE) with EGR (exhaust gas recirculation). Low values of BTE with raise in brake-specific energy consumption (BSEC), EGT and NOx emissions were obtained when the blend A15B85 (additive DEE 15% + rubber seed biodiesel 85%) is run in an ordinary diesel engine (ODE). When the blend A15B85 is run in a SADE, BTE increased, & BSEC, EGT, cylinder pressure values decreased with advancement in start of injection (SOI) timing; however, NOx emission continues to rise due to raise in pre-blend heat release rate. Xuan et al. [10] evaluated the performance, combustion, and emission parameters by adding 10% dimethylfuran (DMF) with petrol in spark-ignition engine or with diesel in compression ignition engine and found that the results obtained were better than pure petrol or diesel. Swarup et al. [11] evaluated the emission parameters of producer gas produced from waste babul wood pieces-blended Jatropa biodiesel in single as well as dual fuel mode diesel engines at gas flow rate of 21.69 kg/h. CO_2, CO, and hydrocarbon emission level rises with rise in load in dual fuel mode, however, NOx and smoke emission level drops with rise in engine load. Swarup et al. [12] evaluated the emission parameters of Jatropa oil methyl ester-blended biodiesel in a diesel engine. CO, HC, and NO emission level for the blend was lesser than that of conventional diesel, whereas smoke darkness was higher for the biodiesel blend because of high thickness, low volatility, and low calorific value of the blend.

The exhaustive literature highlights that no literatures have been reported on the effect of the operating parameter such as compression ratio on the performance and emission characteristic of a rubber seed biodiesel run diesel engine. This forms the basis of the present study.

2 Materials and Methods

A 3.5 kW, single cylinder, four stroke, water cooled, direct injection, stroke length 110 mm, bore 87.5 mm, 661 cc variable compression ratio (VCR) diesel engine (Compression ratio ranging from 12 to 18) operating at 1500 rpm and having injection timing variation option from 0° to 27° BTDC has been considered as test setup as

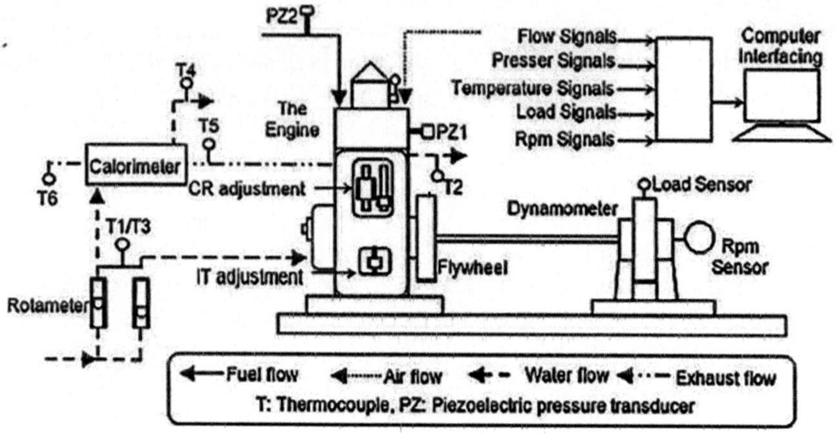

Fig. 1 Schematic diagram of the VCR engine setup

shown in Fig. 1. The description of the test setup is already discussed in the previous study [13].

Rubber seed biodiesel (RBD) is considered for the study as it is non-edible oil. The emission investigation is performed by a flue gas analyzer made of Testo. The flue gas analyzer specifications are given in Table 2. The working of the gas analyzer is discussed in the previous work [13]. The characteristic properties like density, calorific value, cetane number, flash point, and fire point of RBD are 875 kg/m^3, 39.73 MJ/kg, 52.6, 420, and 440 K, respectively, while that for conventional diesel are 840 kg/m^3, 42 MJ/kg, 45–55, 346, and 355 K. The performance, combustion, and emission parameters are evaluated for three different compression ratios (CRs) of 17, 17.5, and 18 at standard injection timing of 23° BTDC for 100% RBD by varying the engine load from 20 to 100%, and the outcomes have been matched with the outcomes attained for 100% diesel at standard CR of 17.5 for the same loads (Tables 1, 2, 3 and 4).

The relative errors of independent variables as well as the uncertainty linked with various performance parameters are given in Tables 5 and 6.

Table 1 Engine specifications

Engine	Make Kirloskar, single cylinder, 4 stroke, water cooled, stroke 110 mm, bore 87.5 mm, 661 cc
	Diesel mode: 3.5 KW@ 1500 rpm, CR range 12-18. Injection variation:0–27^0 BTDC

Table 2 Gas analyzer specifications

Sensors	Sensor type/range/repeatability/resolution
CO	Electrochemical /0–10000 ppm/' ± 5% of reading for < 200 to 2000 ppm, ± 10% of reading up to > 2000 ppm, ± 10 ppm for 0 to 199 ppm/ ± 1 ppm
CO_2	INFRARED /0–20% v/v/ ± 1%/ ± 0.1% v/v
NO	Electrochemical/0–4000 ppm/' ± 5% of reading up to < 2000 ppm, ± 10% of reading up to > 2000 ppm/ ± 1 ppm
HC	Electrochemical/methane: 100 to 40,000 ppm, propane: 100 to 21,000 ppm, butane: 100 to 18,000 ppm/ ± 2%/10 ppm

Table 3 Fuel properties

Properties	Diesel	RBD
Chemical composition	C12H26	C18.05 H34.7 O_2*
Density (kg/m3) at 298 K and 1 atm	840	875
Lower calorific value (MJ/kg)	42	39.73
Cetane number	45–55	52.6
Flash point (K)	346	420
Fire point (K)	355	440

Table 4 Experimental matrix

Mode	Fuel used	CR	IT	Loading condition
Diesel	100%	17.5	23°	20% (BP = 0.7 kW),
	Diesel			40% (BP = 1.4 kW),
Biodiesel	100% RBD	18	BTDC	60% (BP = 2.1 kW),
		17.5		80% (BP = 3.8 kW),
		17		100% (BP = 3.52 kW)

Table 5 Relative errors of independent variables

Independent variable	Relative error (%)
Engine speed	0.5
Engine load	0.5
Liquid fuel flow rate	1
LHV of liquid fuel	1
CO, CO_2, NO_X, and HC emissions	3

Table 6 Uncertainty linked to various performance parameters

Parameters	Uncertainty (%)
BTE	1.5
BSFC	1.5

Fig. 2 BTE at different load for CR 17, 17.5,18

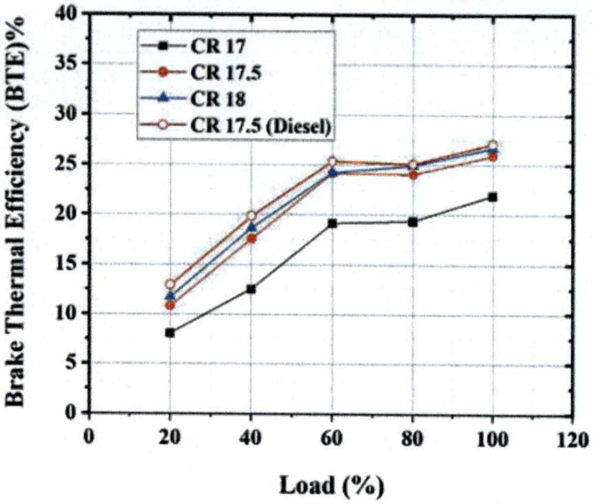

3 Results and Discussion

The current study has evaluated the performance, combustion, and emission characteristics of a rubber seed biodiesel run diesel engine at CR of 17, 17.5, and 18, respectively.

3.1 *Performance Analysis*

From Fig. 2, it is evident that for RBD brake thermal efficiency (BTE) increases for CR 17, 17.5, 18 as load increases from 20 to 100%. Whereas for diesel at standard CR of 17.5, BTE increases with increase in load from 20 to 60%, from 60 to 80% load increase BTE decreases, and again, BTE increases from 80 to 100% load increase. At 100% load, the BTEs are found to be to be 26.71, 25.97, and 21.93% for CRs of 18, 17.5, and 17, respectively, for biodiesel mode in contrast to 27.11% under diesel mode. From Fig. 3, it is evident that for RBD exhaust gas temperature (EGT) increases for CR 17, 17.5, 18 as load increases from 20 to 100%. Whereas for diesel at standard CR of 17.5, EGT increases with increase in load from 20 to 80% and from 80 to 100% load increase EGT decreases.

3.2 *Combustion Analysis*

From Fig. 6, it is evident that when the engine is running with RBD, less net heat is released than in diesel mode because of low heating value of RBD than diesel. The

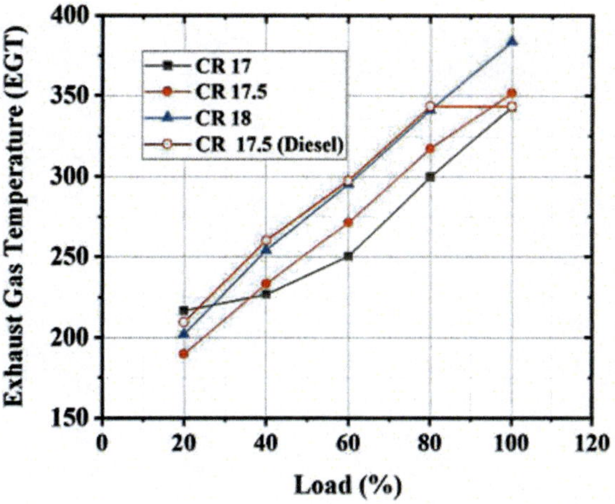

Fig. 3 EGT at different load for CR 17, 17.5, 18

net heat release rate (NHRR) at 100% load for RBD at CR of 17, 17.5, and 18 is 25.26, 41.95, and 47.6 J/deg.CA and that for diesel is 51.43 J/deg.CA at CR of 17.5. It can thus be concluded that as CR is raised NHRR also raises. From Fig. 4, it is evident that when the engine is running with RBD, peak cylinder pressure (PCP) is less than when the engine is running with diesel for all the CRs. It is also evident form Fig. 5 that the PCP approaches toward TDC with rise in CR when the engine is running with RBD. When the engine is running with RBD, as CR raises PCP also raises. The PCP rises with the rise of CR in RBD mode. It has been found that when the engine is running with RBD, PCP falls by 24.64, 22.67, and 20.44% for CRs of 17, 17.5, and 18. Similar findings on the variation of BTE, EGT NHRR, and PCP with the change of compression ratio have been reported by previous study [13] (Fig. 6).

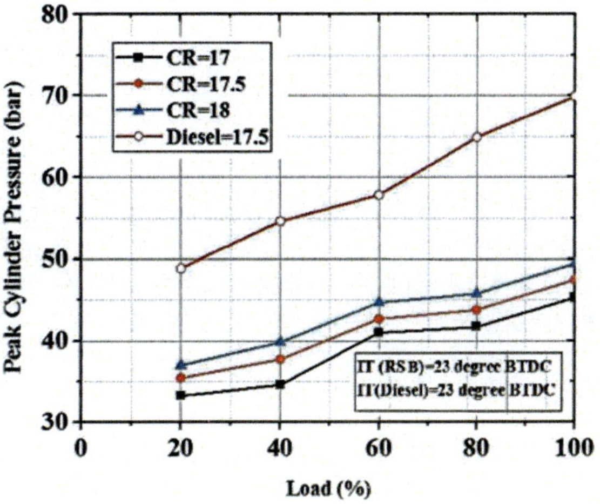

Fig. 4 Peak cylinder pressure at various loads for different CR

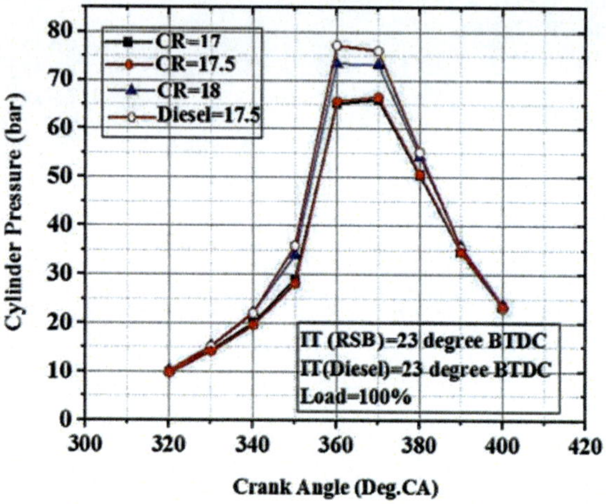

Fig. 5 Cylinder Pressure at different crank angles for CR 17, 17.5, 18

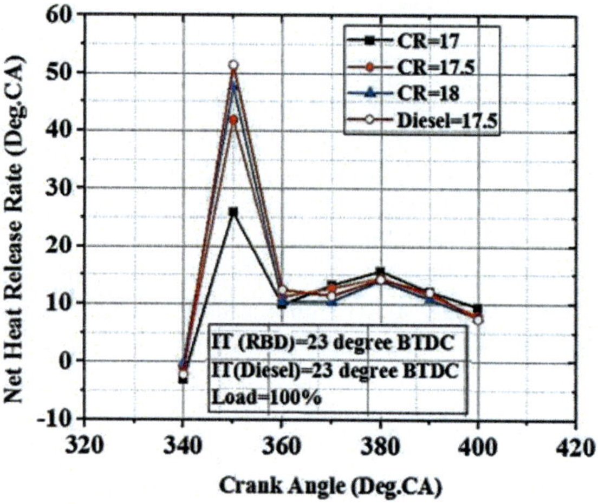

Fig. 6 Neat heat release rate at different crank angles for CR 17, 17.5, 18

3.3 Emission Analysis

In Fig. 7, it is evident that for RBD at CR 17.5, CO emission decreases when load increases from 20 to 80%, and when the load is further increased from 80 to 100%, CO emission increases. Whereas, for diesel, CO emission exponentially increases as load increases from 20 to 100%, but its values are much lower than RBD. HC emission for both RBD and diesel decreases with increase in load from 20 to 80% and from 80 to 100% load increase HC emission increases. The nitrogen oxides (NO_x) emission for RBD increases with increase in load from 20 to 60%, from 60 to 100% load increase NO_x emission remains constant. Whereas, for diesel, NO_x emission slightly decreases as load increases from 20 to 40% and beyond that NO_x emission exponentially increases as load increases from 40 to 100%. CO_2 emission remains almost constant and low as load is raised for both RBD and diesel. From Fig. 8, this is evident for RBD at CR 18, CO emission increases when load increases from 20 to 80%, and when the load is further increased from 80 to 100%, CO emission decreases. Whereas, for diesel, CO emission exponentially increases as load increases from 20 to 100%. HC emission for both RBD and diesel decreases with raise in engine load from 20 to 80%, and from 80 to 100% load increase, HC emission increases. For both RBD and diesel, NO_x emission first rises with increase in load from 20 to 40%; NO_x emission level falls with increase in load from 40 to 80%, and NO_x emission again rises with rise in load from 80 to 100%. CO_2 emission remains almost constant, and low as load is raised for both RBD and diesel. From Fig. 9, this is evident for RBD at CR 17, CO emission levels are higher than diesel mode from 20 to 60% load increase, and beyond 60% load increase, CO emission levels are much lower than diesel. HC emission for RBD as well as for diesel drops with rise in engine load from 20 to 80% and beyond 80% load increase HC level slightly surges up; however, the HC emissions for RBD are slightly higher than diesel at all loads. NO_x emissions for RBD rise with rise in engine load from 20 to 60%, and beyond 60% load increase, NO_x emission remains constant, whereas for diesel NO_x emission level drops with rise in engine load up to 40%, and beyond 40% load increase, NO_x emission level rises. However, NO_x emission level for RBD at all loads is higher than diesel. CO_2 emission remains almost constant & low when load is raised for RBD & diesel. Similar trend on the variation of CO, NO_x, and HC emission with the change of compression ratio has been reported by previous study [13].

In summary, BTE for RBD mode raises with rise in engine load for CR 17, 17.5, and 18, respectively. BTE for diesel mode raises with rise in engine load from 20 to 60%; further, increase in engine load from 60 to 80% shows a fall in BTE; again, further increase in engine load 80–100% shows rise in BTE. At 100% load, the BTEs are found to be to be 26.71, 25.97, and 21.93% for CRs of 18, 17.5, and 17 under biodiesel mode in contrast to 27.11% under diesel mode.

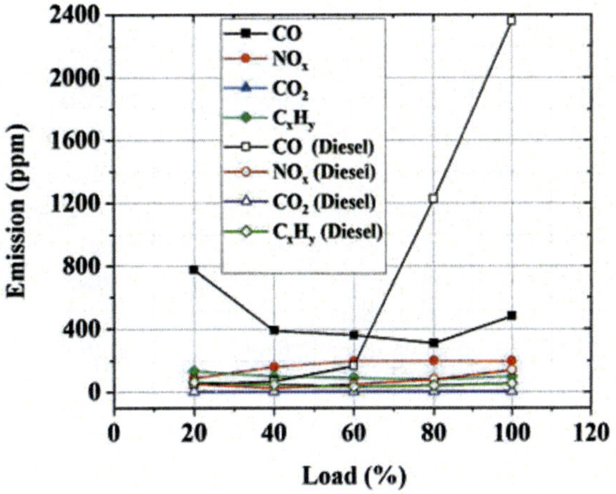

Fig. 7 Emission at different load for CR 17.5

Fig. 8 Emission at different load for CR 18

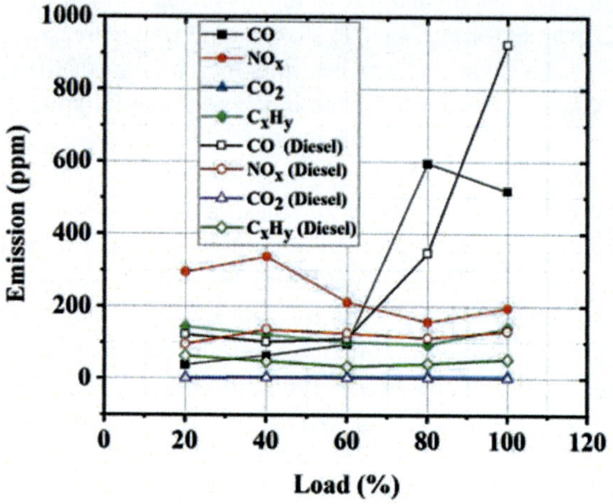

Fig. 9 Emission at different load for CR 17

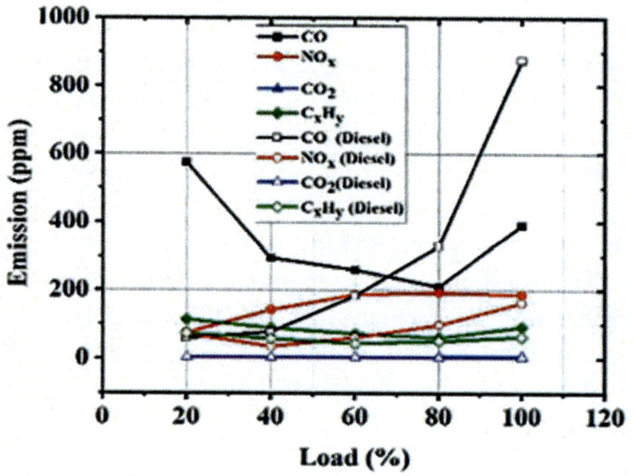

For RBD, exhaust gas temperature (EGT) increases for CR 17, 17.5, and 18, respectively, as load increases from 20 to 100%. Whereas, for diesel at standard CR of 17.5, EGT increases with increase in load from 20 to 60%, from 60 to 80% load increase EGT decreases, and again, EGT increases with 80–100% load increase. When the engine is running with RBD, less net heat is released than in diesel mode because of low heating value of RBD than diesel. NHRR at full load for RBD at CR of 17, 17.5, and 18, respectively, is 25.26, 41.95, and 47.6 J/deg.CA and that for diesel is 51.43 J/deg.CA at CR of 17.5. It can thus be interpreted that as CR is raised NHRR also raises for RBD mode.

It is evident that when the engine is running with RBD, PCP is less than when the engine is running with diesel for all the CRs. It is also evident that the PCP approaches toward TDC with rise in CR when the engine is running with RBD. When the engine is running with RBD, as CR raises, PCP also raises. The PCP rises with the rise of

CR for RBD mode. It has been found that when the engine is running with RBD, PCP falls by 24.64%, 22.67%, and 20.44% for CRs of 17, 17.5, and 18, respectively.

For RBD at CR of 17 and 17.5, with raise in engine load from 20 to 80%, CO emission rate drops, and when the load is further raised from 80 to 100%, CO emission rate increases. For RBD at CR of 17 and17.5, HC emission increases with increase in load from 20 to 60%; from 60 to 80% load increase, HC emission decreases, & again, HC emission increases from 80 to 100% load increase. For RBD at CR of 17, NOx emission rate raises with raise in engine load from 20 to 80%; from 80 to 100% load increase, NOx emission rate decreases. For RBD at CR of 17.5, NOx emission rate raises with raise in engine load from 20 to 60%; from 60 to 100% load increase, NOx emission remains constant.

4 Conclusions

The present study investigates the effect of compression ratio on the performance, combustion, and emission characteristic of a rubber seed biodiesel run diesel engine. Rubber seed biodiesel can be utilized in a diesel engine as an alternative fuel as the results in laboratory is interesting. The efficiency rises with the use of high compression ratio 18:1 for rubber seed biodiesel. At CR 18, CO emission slightly increases when load increases from 20 to 80%, but when the load is further increased from 80 to 100%, CO emission decreases. Emission rate of HC marginally rises with raise in engine load. With raise in engine load from 20 to 40%, NOx emission rate increases; with further rise in engine load from 40 to 80%, NOx emission rate actually drops, therefore using this fuel at high load capacity (say ~80%) seems beneficial.

Acknowledgements The authors are indebted to the Department of Energy, Assam Science and Technology University (ASTU), Guwahati, Assam for giving permission to conduct experiments in Internal Combustion Engine Laboratory.

References

1. Chintala V, Subramanian KA (2016) Experimental investigation of hydrogen energy share improvement in a compression ignition engine using water injection and compression ratio reduction. Energy Convers Manage 108:106–119
2. Sai Bharadwaj AVSL, Subramaniapillai N, Begum MK, Sheriffa M, Narayanan A (2021) Effect of rubber seed oil biodiesel on engine performance and emission analysis. Fuel 296:120708
3. Nayak SK, Mishra PC, Noorb MM, Hagosb FY, Kadirgamab K, Mamatb R (2019) The performance of turbocharged diesel engine with injected calophyllum inophyllum methyl ester blends and inducted babul wood gaseous fuels. Fuel 257(1):116060
4. Kole C, Joshi CP, Shonnard DR (2012) Handbook of bioenergy crop plants, 1st Edn. CRC Press

5. Senthil Kumar S, Purushothaman K, Devan PK (2015) Performance and emission characteristics of a diesel engine using rubber seed oil and its diesel blends. Int J Chem Tech Res 8(2):669–675. CODEN (USA): IJCRGG ISSN: 0974-4290
6. Edwin Geo V, Nagarajana G, Nagalingamb B (2008) Studies on dual fuel operation of rubber seed oil and its bio-diesel with hydrogen as the inducted fuel. Int J Hydrog Energy 33:6357–6367
7. Edwin Geo V, Nagarajana G, Nagalingamb B (2010) Studies on improving the performance of rubber seed oil fuel for diesel engine with DEE port injection. Fuel 89:3559–3567
8. Patil VV, Patil RS (2018) Effects of partial addition of butanol in a rubber seed oil methyl ester diesel blend. Mater Today Proc 5:23119–23127
9. Vamsi Krishna K, Sastry GRK, Murali Krishna MVS, Deb Barma J (2018) Investigation on performance and emission characteristics of EGR coupled semi adiabatic diesel engine fuelled by DEE blended rubber seed biodiesel. Eng Sci Technol Int J
10. Nguyena PX, Hoang TA, Olçer AI, Engel D, Phama VV, Nayak SK, Biomass-derived 2,5-dimethylfuran as a promising alternative fuel
11. Nayak SK, Mishra PC (2016) Emission from utilization of producer gas and mixes of Jatropa biodiesel. Energy Sour Part A: Recov Utilization Environmental Effects 38(14):1993–2200
12. Nayak SK, Behera GR, Mishra PC, Kumar A, Functional characteristics of Jatropa biodiesel as a promising feedstock for engine application. Energy Sour Part A: Recovery, Utilization, Environmental Effects
13. Bora BJ, Saha UK (2016) Experimental evaluation of a rice bran biodiesel- biogas run dual fuel diesel engine at varying compression ratios. Renew Energy 87:782–490
14. Ramadhas AS, Jayaraj S, Muraleedharan C (2008) Dual fuel mode operation in diesel engines using renewable fuels: Rubber seed oil and coir-pith producer gas. Renew Energy 33:2077–2083
15. Choudhary KD, Nayyar A, Dasgupta MS (2018) Effect of compression ratio on combustion and emission characteristics of C.I. Engine operated with acetylene in conjunction with diesel fuel. Fuel 214:489–496
16. Debnath BK, Bora BJ, Sahoo N, Saha UK (2014) Influence of emulsified palm biodiesel as pilot fuel in a biogas run dual fuel diesel engine. J Energy Eng 140(3):A4014005–1–A4014005–9
17. Shay EG (1993) Diesel fuel from vegetable oils: status and opportunities. Biomas Bioener 4:227–242
18. Chavanne G (1938) Procedure for the transformation of vegetable oils for their uses as fuel (Belgian Patent No: 422877)
19. Konwer D, Baruah K (1985) Petroleum-like hydrocarbons from Mesua ferrea L seeds. Chem Ind (London), 447–448
20. Konwer D, Taylor SE, Gordon BE, Otvos JW, Calvin M (1989) Liquid fuels from Mesua ferrea L seed oil. J Am Oil Chem Soc 66:223–226
21. Pioch D, Lozano P, Rasoanantoandro MC, Graille J, Geneste P, Guida A (1993) Biofuels from catalytic cracking of tropical vegetable oils. Oleagineux 48:289–291
22. Backers L (1982) Vegetable oil fuels. In: Proceedings of the international conference on plant and vegetable oils as fuels (ASAAE)
23. Vasudevan PT, Briggs M (2008) Biodiesel production- current state of the art and challenges. J Ind Micro Biotechnol 35:421–430
24. Zheng S, Kates M, Dubé MA, McLean DD (2006) Acid-catalyzed production of biodiesel from waste frying oil. Biomas Bioener 30:267–272
25. Liu K (1994) Preparation of fatty acid methyl esters for gas-chromatographic analysis of lipids in biological materials. J Am Oil Chem Soc 71:1179–1187
26. Dizge N, Keskinler B (2008) Enzymatic production of biodiesel from canola oil using immobilized lipase. Biomas Bioener 32:1274–1278
27. Uosukainen EL, Linko MY, Linko P, Leisola M (1999) Optimization of enzymatic transesterification of rapeseed oil ester using response surface and principal component methodology. Enzy Micro Technol 25:236–243

Experimental Analysis on Combustion Characteristics of a DICI Engine Utilizing Argemone Biodiesel with Diesel Blend

Manoj Kumar Parida, Mamuni Arya, and Akshya Kumar Rout

1 Introduction

The rate of energy consumption increased day by day which causes rapid environmental degradation. Petroleum fuel is having limited resources and is also economically not preferable to use for a long year according to various authors [1, 2]. There is wide application of alternative bio-fuels that promote sustainable growth [3]. Vegetable oil was not favored as an alternative biodiesel for CI engines because of its properties like high viscosity, which might clog fuel filters. The volatility and viscosity were improved by heating and mixing with diesel in vegetable oils, but polyunsaturated structures were remaining same for the oils [4].

Biodiesel extracted from Neem, Polanga, Jatropha, and Karanja, etc. were blended with diesel in different percentages in volume basis were tested for combustion and performance characteristics with varying load at constant compression ratio [5]. According to the research, the optimum fuel blend is Polanga biodiesel with maximum pressure. All of the fuels tested had a faster ignition time than diesel, as the load was increased. The ignition delay of neat Jatropha biodiesel is shorter than that of other neat biodiesels tested in the lab. The combustion parameters of Karanja biodiesel and its mix with diesel in compression ignition engine were investigated while varying engine load and maintaining a 1500 rpm engine speed [6]. The study found that Karanja oil mixed with diesel has a shorter delay period and a slower burning rate, which results in a longer combustion duration. Due to incorrect fuel atomization/mixing, a lower blend concentration changes the combustion

M. K. Parida · M. Arya (✉)
Mechanical Engineering, C.V. Raman Global University, Bhubaneswar, Odisha 752054, India
e-mail: mamuniarya23@gmail.com

A. K. Rout
School of Mechanical Engineering, KIIT Deemed to be University, Bhubaneswar, Odisha 752024, India

© The Author(s), under exclusive license to Springer Nature Singapore Pte Ltd. 2023

S. Revankar et al. (eds.), *Recent Advances in Thermofluids and Manufacturing Engineering*, Lecture Notes in Mechanical Engineering,
https://doi.org/10.1007/978-981-19-4388-1_27

process. Higher mixes are not acceptable for engines that have not been changed. However, without any engine modifications, up to 20% blended fuel was advised. In a 4-stroke DICI engine, engine characteristics using castor biodiesel blends were explored, and the output was compared to standard data [7]. In comparison to higher mixes, lower blends exhibit a delay in heat release. Castor biodiesel blends might thus be used to replace diesel in compression ignition engines. To boost efficiency and minimize emissions of the diesel engine, Jatropha biodiesel mix with the diesel fuel was used as a fuel in many laboratories [8]. Thermal efficiency of 27.93% and 28.18%, smoke density of 61.9 HSU and 81.2 HSU, and heat release rate 53.57 and 57.75 J/degree, respectively, were obtained at peak load using B20 Methyl Ester of Jatropha and diesel. MEOJ blends have comparable performance and lower emissions, making them a viable diesel substitute. The CIME biodiesel (CIME100, CIME30, and CIME60) reduced BTE by a small amount [9]. Under various loading situations, CIME100 combustion metrics such as P_{max}, ID, and NHRR are equivalent to diesel fuel.

In this experimental work, the engine combustion characteristics like maximum in-cylinder pressure, Net heat release rate, Mass fraction burned, Rate of Pressure Rise, and Ignition delay were determined using pure Argemone Mexicana biodiesel and diesel blends in 20, 40, and 60% (volume basis) with different percentage of load at constant compression ratio of 17.5:1 in DICI engine. The above characteristics were further compared with the diesel fuel using the same engine.

2 Experimental Methodology

2.1 Biodiesel Preparation

Argemone Mexicana biodiesel are the main ingredient in this study. Argemone Mexicana belongs to Papaveracae family and Mexicana prickle poppy species, those flowers are yellow in color and the seeds are black in color and round-shaped. Argemone Mexicana plants are shown in "Fig. 1". The Argemone Mexicana crude oil was extracted from the seeds by mechanical oil extraction methods, but crude oils were not fit for engine operation because of high viscosity and less volatility. First de-gumming process was carried out to remove the fatty acid of the crude oil, after that esterification and transesterification processes were occurred to reduce the acidic value of crude Argemone Mexicana oil from 19.30 mg KOH/g to 0.32 mg KOH/g [10].

The Argemone Mexicana biodiesel preparation processes are shown in "Fig. 2". The physical and chemical parameters of Argemone Mexicana biodiesel and its blends with diesel were assessed by the equipment using normal ASTM test procedures, and the comparison with standard diesel is presented in Table 1.

Fig. 1 **a** Argemone Mexicana flower. **b** Argemone Mexicana seed

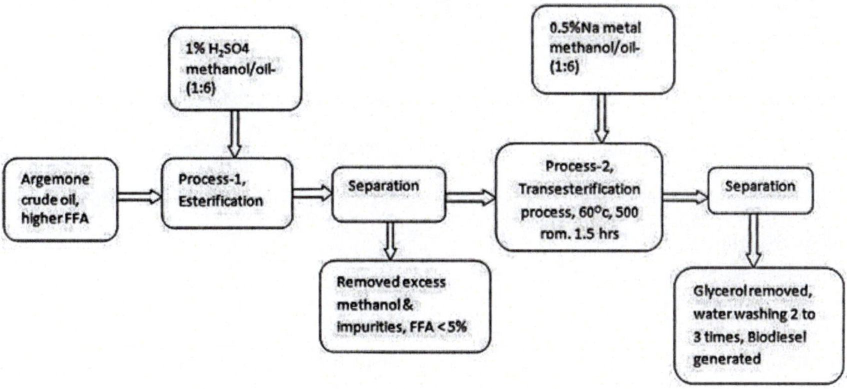

Fig. 2 Argemone Mexicana biodiesel preparation process

Table 1 Blended fuel properties

Sl. No.	Properties	D100	AMME	B20	B40	B60
1	Kinematic Viscosity, 40 °C (cst)	3.68	5.07	4.1	4.32	4.58
2	Density at 20 °C (kg/m^3)	830	868	843	849	858
3	Calorific values, (Kj/kg)	42,800	41,500	42,540	42,280	41,980
4	Flash point (°C)	65	170	86	107	125

2.2 *Performance Test*

A four-stroke, single-cylinder, completely computerized dual fuel variable compression ratio engine was used for this work (see Fig. 3). The fuel flow was controlled by the governor to keep the engine speed at approximately 1500 ± 3% rpm. At a fixed compression ratio of 17.5, the engine was evaluated for standard diesel, B20, B40, B60, and B100. Two pressure transducers measure the pressure in the

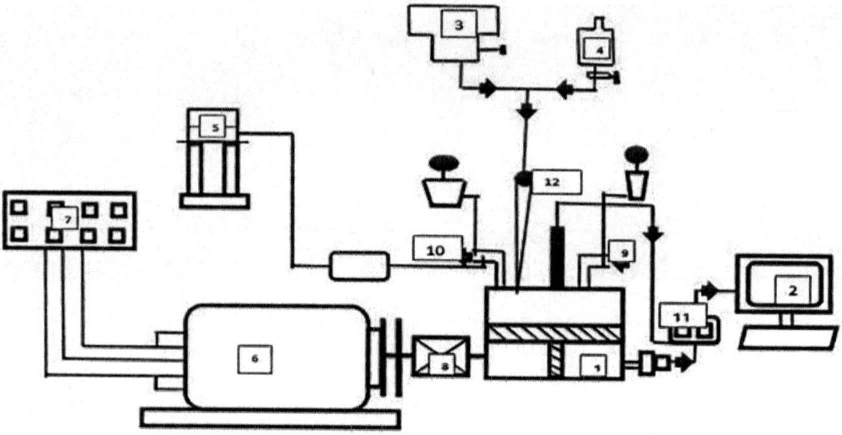

1-Engine 2-Computer 3-Fuel tank 4-Biodiesel 5-Exhaust gas analyzer
6-Eddy current dynamometer 7-Dynamometer controller 8-Coupling
9-Air intake 10-Exhaust 11-Combustion data acquisition system 12-Fuel injector

Fig. 3 Experimental set-up

cylinder and the diesel line. These signals are interfaced with a computer for pressure crank-angle graph generation. These pressure transducer signals are sent into a charge amplifier. A high precession encoder is utilized to communicate signals for TDC and crank angle. To build a *p*–*θ* diagram, the data collection system, which is connected to the computer, receives signals from the charge amplifier and the crank-angle encoder. The "Engine soft" software, which is based on LabVIEW, records numerous combustion parameters. To remove cyclic variation, pressure data from 10 consecutive engine cycles were collected, and average values were used to analyze and calculate combustion parameters such as P_{max}, NHRR, MFB, ID, and ROPR.

3 Results and Discussions

3.1 Maximum In-Cylinder Pressure

From the experiment the maximum in-cylinder pressure for B20 observed was 67.8 bar, for B40 was 66.2 bar, for B60 was 66.4 bar and for B100 was 67.6 bar, respectively at full load conditions. With improved spark timing, a smooth ascent in pressure during the compression stroke keeps bringing about a most extreme in chamber pressure. For all of the test fuels, the peak pressure rises as the applied load rises; more fuel is consumed at higher loads [11]. Probably due to its shorter ignition

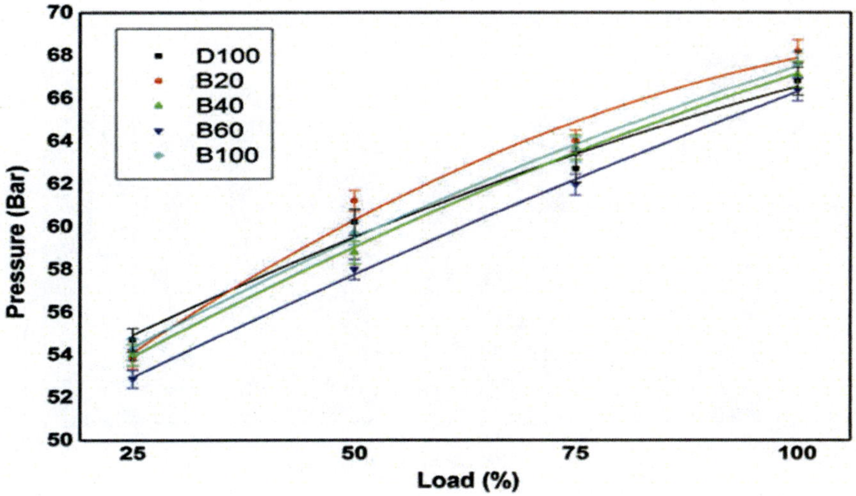

Fig. 4 Maximum in-cylinder pressure versus % load

delay than other blends, blend B20 produces the highest combustion pressure. After injection, the fuel blend absorbs more heat from the cylinder, resulting in a longer delay period [12]. As the load grows, more fuel being consumed, leading in longer combustion duration and the occurrence of maximum in-cylinder pressure (P_{max}). As the fraction of biodiesel in the mix rises, the angle of crank-angle position moves away. "Fig. 4" shows the difference in peak cylinder pressure (P_{max}) with variation of load in engine for all the test fuels.

3.2 Mass Fraction Burned

Mass fraction of fuel burned can be determined according to the rate at which the fuel burns. It describes the combustion process of chemical energy as a fraction of crank angle on a range of 0–90%. In comparison to diesel, the half and 90% mass part consumed, the crank-angle positions for those fuel mixes were hindered [13]. This is owing to the fact that combining or atomizing Argemone oil mixes takes longer. Probably the blend's low heating value necessitates more quantity of fuel, the induced time is increased as well as the time of combustion also increased. Regardless of the proportion of biodiesel with the diesel mix, the combustion curves are utmost identical and are nearer for crank angle 342°–386°. Mass fraction of fuel burned for diesel, 20, 40, 60, and 100% biodiesel mixes at maximum load were plotted against crank angle is as displayed in "Fig. 5".

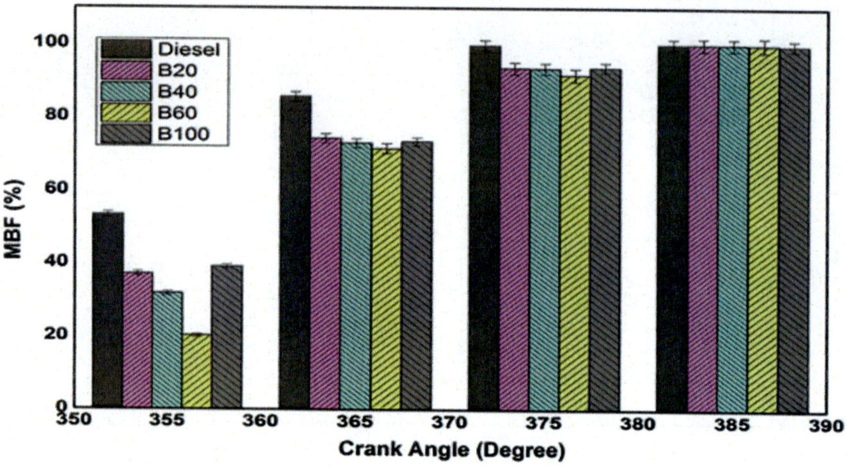

Fig. 5 MBF versus crank angle

3.3 *Net Heat Release Rate (NHRR)*

At increased engine loads, the mean gas temperature inside the cylinder rises, leaving less time for fuel vaporization and diffusion mixing, reducing the delay period and increasing the NHRR. The NHRR for diesel is higher at low and medium loads during the premixed burning stage as compared to Argemone bio-fuel blends due to further developed blending in with air and better atomization because of its low thickness and boiling range. However, because to its lower volatility, substantially and higher viscosity Argemone oil blends released more heat during the controlled combustion phase at higher loads [13]. In this experiment, the maximum net HRR value for diesel was 79.56 J/CA, where by using 20%, 40%, 60%, and 100% biodiesel the NHRR value was 84.48, 78.49, 76.7, and 73.38 J/CA, respectively at peak load condition. The NHRR increases even at lower percentages of biodiesel mix with the diesel, due to the presence of oxygen in bio-fuel, which enhances the premixed burning phase. "Fig. 6" depicts the fluctuation of NHRR with variation of load in the above said fuel blend. Because biodiesel has a lower heating value, NHRR decreases as the percentage of biodiesel increases.

3.4 *Ignition Delay*

The delay period (ID) is the first phase of the combustion process, which is controlled by a number of factors such as the A/F ratio, fuel type, fuel quality, pressure, engine speed, and intake air temperature [14]. The period between fuel injection (23° before TDC) and ignition is known as ID, and it is an important metric in combustion

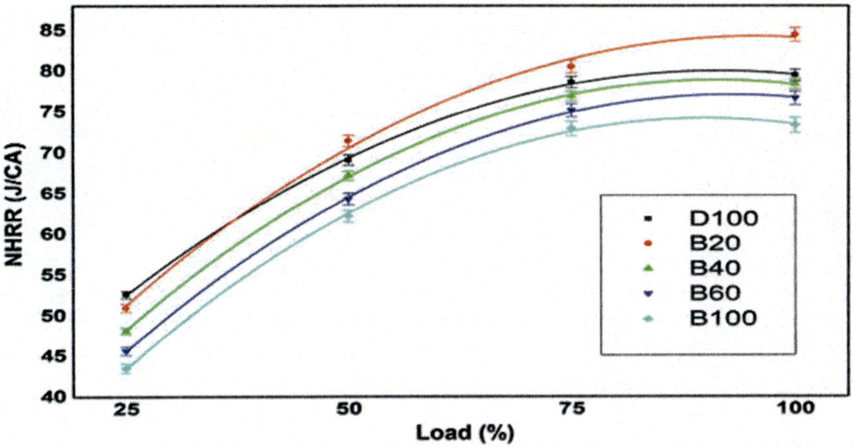

Fig. 6 NHRR versus % load

analysis. As the load inside the cylinder rises, the pressure and temperature rise, and more fuel is injected, resulting in early start of burning. For full load conditions, a delay period of 15° for diesel, 16.5° delay using 20% biodiesel, 17° delay by taking 40% biodiesel with diesel, 18° delay by using 60% of biodiesel, and 18.5° delay in crank-angle rotation was recorded in 100% biodiesel. Using B20, B40, B60, and B100 blend the ignition delay periods were found 1.5, 2, 3, and 3° crank angle, respectively longer than diesel. The reason given could be related to biodiesel blends greater Cetane Number, P_{max}, and mean gas temperature [15]. "Fig. 7" depicts the fluctuation of delay duration with load for various fuels at conventional compression ratio of 17.5:1.

3.5 Rate of Pressure Rise (ROPR)

Maximum Rate of Pressure Rise (ROPR) is an important characteristic in the whole combustion process vis-a-vis engine operation; it reflects combustion roughness. The fluctuation of ROPR $(dP/d\theta)_{max}$ with engine load for different fuel blends is as shown in "Fig. 8". For all above test fuels in this experiment, the ROPR value increases with increasing engine load. For diesel, the value of pressure rise rate was 8.72 bar/degree. By using 20% biodiesel the value of pressure rise rate was 9.14 bar/degree and by adding 40% biodiesel with diesel the value of pressure rise rate was 8.6 bar/degree. For 60% biodiesel mixing the value of pressure rise rate was 8.21 bar/degree whereas for 100% biodiesel the value of pressure rise rate was 7.92 bar/degree at 100% load condition. Higher the ROPR, the greater amount of the induced fuel could be burned in the premixed combustion stage [16]. Since biodiesel includes longer carbon-chain length atoms with a higher edge of boiling point and lower volatility, ROPR can

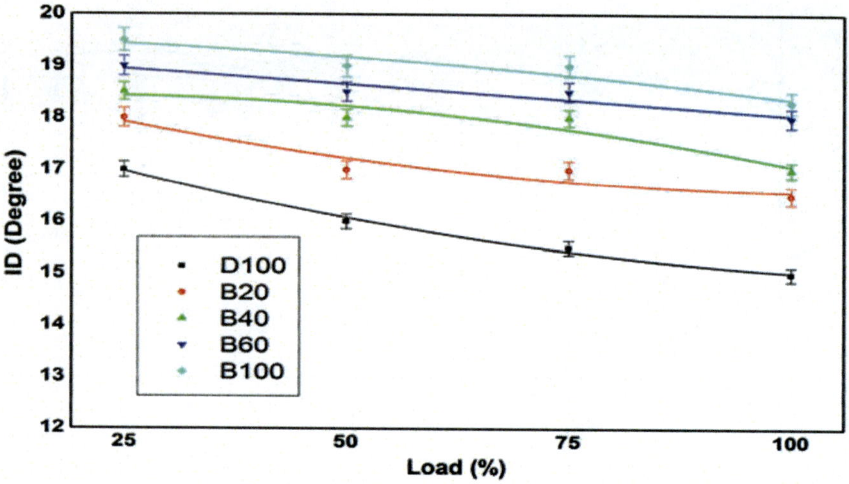

Fig. 7 Ignition delay versus % load

be higher for lower blends as well and vice versa. At greater loads, the cylinder temperature rises rapidly, allowing the low volatile parts of the fuel to dissipate and blend sufficiently with the air.

Due to longer ignition delay, in-cylinder pressure was obtained maximum for B20 blend at full load condition. The burned mass fraction with position of crank angle for the fuel blends was delayed in contrast to commercial diesel. Probably the

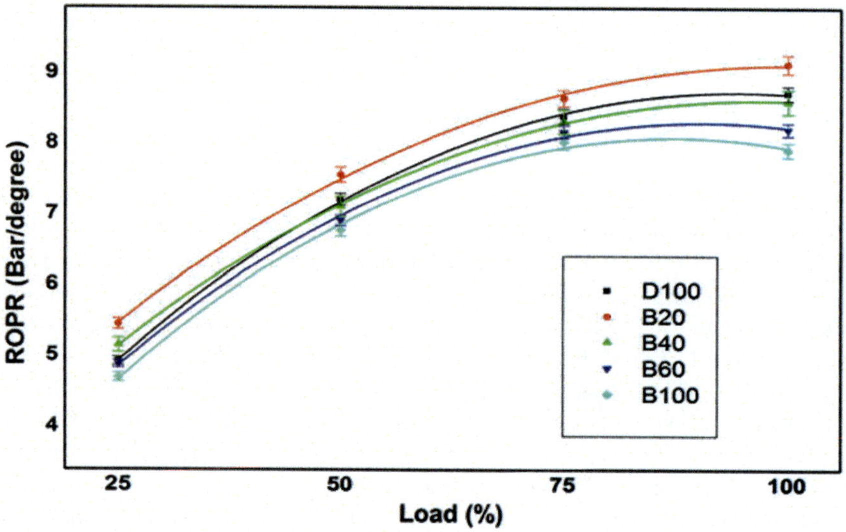

Fig. 8 ROPR versus % load

atomization of Argemone oil blends takes longer. The net heat release rate of B20 was obtained higher than diesel fuel. The rate of pressure rise was obtained for B40 fuel as 8.6 bar/degree, whereas for diesel was ~8.72 bar/degree at full load conditions.

4 Conclusions

In present work, the physical and chemical features of Argemone Mexicana biodiesel with diesel blend fuels were studied and compared to diesel fuel. The Calorific value of B60 was very close to diesel fuel. The difference of densities between B20 fuel and diesel fuel was very less. It was found that all combustion parameters were higher for B20 blend except for the smaller value of delay period as compared to other blends. As the viscosity and density of the fuel rise, the delay period increases. With a higher percentage of biodiesel, there may be a reduction in cone angle and less penetration, resulting in a longer delay. It was found that all combustion parameters were higher for B20 blend except the smaller value of delay period as compared to other blends.

Acknowledgements The authors would like to acknowledge the support extended by Biodiesel testing Laboratory, Department of Mechanical Engineering, C. V. Raman Global University, Bhubaneswar.

References

1. Sen S, Ganguly S (2017) Opportunities, barriers and issues with renewable energy development—a discussion. Renew Sustain Energy 69:1170–1181
2. Kotebavi V, Shetty D, Sahu D (2016) Performance and emission characteristics of a CI engine run on waste cooking oil-diesel blends. Pollut Res 35:159–166
3. Utlu Z, Kocak MS (2008) The effect of biodiesel fuel obtained from waste frying oil on direct injection diesel engine performance and exhaust emissions. Renew Energy 33:1936–1941
4. Bora DK (2009) Performance of single cylinder diesel engine with karabi seed biodiesel. J Sci Ind Res 68:960–963
5. Sahoo PK, Das LM (2009) Combustion analysis of Jatropha, Karanja and Polanga based biodiesel as fuel in a diesel engine. Fuel 88:994–999
6. Agarwal AK, Dhar A (2013) Expt. investigations of performance, emission and combustion characteristics of Karanja oil blends fueled DICI engine. Renew Energy 52:283–291
7. Das D, Pathak V, Yadav AS, Upadhyaya R (2017) Evaluation of performance, emission and combustion characteristics of diesel engine fueled with castor biodiesel. Biofuels 8:225–233
8. Sivaganesan S, Chandrasekaran M (2016) Performance and emission analysis of compression ignition engine with methyl ester of jatropha and diesel, Indian. J Sci Technol 9:1–6
9. Ashok B, Nanthagopal K, Vignesh DS (2018) Calophylluminophyllum methyl ester biodiesel blend as an alternate fuel for diesel engine applications. Alex Eng J 57:1239–1247
10. Pramanik P, Das P, Kim PJ (2012) Preparation of biofuel from argemone seed oil by an alternative cost-effective technique. Fuel 91:81–86
11. Rao GLN, Prasad BD, Sampath S, Rajagopal K (2007) Combustion analysis of diesel engine fueled with jatrophaoil methyl ester-diesel blends. Int J Green Energy 4:645–658

12. Devan PK, Mahalakshmi NV (2009) Performance, emission and combustion characteristics of poon oil and its diesel blends in a DI diesel engine. Fuel 88:861–867
13. Parida MK, Rout AK (2017) Combustion of biodiesel blends in a constant-volume DICI engine. Biofuels 10:537–543
14. Ozsezen AN, Canakci M, Turkcan A, Sayin C (2009) Performance and combustion characteristics of a DI diesel engine fueled with waste palm oil and canola oil methyl esters. Fuel 88(2009):629–636
15. Gumus M (2010) A comprehensive experimental investigation of combustion and heat release characteristics of a biodiesel (hazelnut kernel oil methyl ester) fueled direct injection compression ignition engine. Fuel 89:2802–2814
16. Parida MK, Routaray I (2018) RSM analysis of in-cylinder pressure in a DICI engine fuelled with Argemone Mexicana biodiesel-diesel blend. Int J Eng Technol 7:28–31

Pyrolysis Characterization of Biomass Feedstock Using Thermogravimetric Analysis

Debarshi Mallick, Mayuri Goswami, and Devasish Bhuyan

Nomenclature

DTG	Differential thermogravimetric analysis
FWO	Flynn-Wall-Ozawa method
TGA	Thermogravimetric analysis

1 Introduction

The consumption of energy has increased as a result of increased population, growth of the economy as well as and industrial expansion [1]. It has been reported that more than 19% of the global energy intake is met by renewable resources of energies wherever biomass provides ~9% of energy [2, 3]. The world now has given more emphasis on renewable energy. Sources of biomass energy can be classified as the residue of agriculture crops, energy plantations, and municipal and industrial waste. India is the 4th largest country regarding the availability of agro and forest-residues [4]. So, biomass is one of the encouraging routes for India to replace fossil fuels to extract energy from it. There are several technologies available to convert biomass to bioenergy, which can be classified as thermochemical, biochemical, and physicochemical [5]. Amongst all technologies, thermochemical conversion has been reported to be a more promising option to convert biomass to bioenergy [6, 7].

There are three common thermochemical routes of conversion of biomass to bioenergy, which are combustion, gasification, and pyrolysis [8–10]. Combustion

D. Mallick (✉) · M. Goswami · D. Bhuyan
Girijananda Chowdhury Institute of Management and Technology, Guwahati, Assam 781017, India
e-mail: debarshimallick123@gmail.com

© The Author(s), under exclusive license to Springer Nature Singapore Pte Ltd. 2023

S. Revankar et al. (eds.), *Recent Advances in Thermofluids and Manufacturing Engineering*, Lecture Notes in Mechanical Engineering,
https://doi.org/10.1007/978-981-19-4388-1_28

produces heat, gasification produces producer gas, heat, and char, and pyrolysis produces bio-oil and char. Pyrolysis or thermal decomposition is commonly used thermochemical conversion process [11]. Various research papers are available in the area of pyrolysis analysis, such as coconut shell, perennial grass, rice husk, sawdust, hazelnut husk, pine dust, and Nahor seed [11–18]. Thermogravimetric analysis has been reported to be the most common and venerable technique for analyzing the thermochemical decomposition of organic and inorganic feedstocks [19–22]. The thermochemical decomposition activities of the biomass are calculated by assessing the weight loss of the biomass as a function of time and operating temperature [10]. The 1st derivative of the weight loss graph acquired from the thermogravimetric analysis shows decomposition intensity, which is known as DTG analysis [23].

This study here focuses on the characterization and comparative analysis of three different types of lignocellulosic and organic waste biomasses, viz., sugarcane bagasse, elephant grass, and rice straw that are easily and abundantly available in India. The activation energy of the biomasses was determined using the model free-Flynn-Wall-Ozawa method.

2 Material and Methods

The three feedstocks selected for this study were elephant grass, sugarcane bagasse, and rice husk. These were collected from Guwahati, Assam located in the northeast part of India. Samples were dried in sun, ground using a blender, and then sieved to obtain the average particles size of the biomasses using standard procedure [24]. The proximate analysis of the feedstocks was conducted using a muffle furnace (optical technology) following ASTM standard (ASTM E870-82). The elemental composition of the biomasses was determined by using a CHNS analyzer (Euro Vector EA3000). The calorific value of each feedstock was calculated using a bomb calorimeter (Parr, 1341) as shown in Eq. (1).

$$\mathrm{HHV} = \frac{W \times \Delta t - e_1 - e_2 - e_3}{m} \tag{1}$$

where

HHV	Higher heating value in MJ/Kg
W	Energy equivalent in Cal/°C
Δt	Temperature difference in °C
e_1, e_2, e_3	Correction factor for the presence of HNO_3, H_2SO_4, and heat of combustion of fuse wire, respectively
m	Weight of the sample in gm.

2.1 Thermogravimetric Analysis (TGA)

The pyrolysis study of various feedstocks was conducted in a thermogravimetric analyzer (Hitachi STA 7200) to study the decomposition of biomasses occurring in the presence of an inert atmosphere. Approximately, 10 mg samples of the biomass were taken in a container without the lid and heated at the temperature range of 30 to 900 °C maintaining a standard of 1 atm pressure. To obtain better heat transfer in-between the crucible and thermocouples, platinum crucibles were selected for his study [23]. Four different heating rates (β) such as 10, 15, 20, and 30 °C/min were used for the study. High-purity N_2 gas at the flow rate of 50 ml/min was used as an inert gas during the experiment to eliminate the presence of any residual oxygen from the furnace chamber.

2.2 Kinetic Study of Biomass

The decomposition rate of the biomass sample can be expressed by Eq. (2).

$$\mathrm{d}\alpha/\mathrm{d}t = k(T)f(\alpha) \tag{2}$$

Here, α represents the fraction of biomass that degrades over time t, and k represents the Arrhenius kinetic constant. The detailed kinetic study of the biomass has been discussed in our previous research paper [16].

2.3 Flynn-Wall-Ozawa (FWO) Method

The FWO equation for the constant heating rate can be written in the following form:

$$\ln(\beta) = \text{Const.} - 1.052\left(\frac{E_a}{RT}\right) \tag{3}$$

For each α value, ln (β) versus ($1/T$) plot shows a straight line having the slope of $\left(-\frac{E_a}{R}\right)$, hence, the E_a is obtained in the conversion function [16].

3 Results and Discussion

The mean particle size of the biomasses was found to be 240 μm for elephant grass (EG), 264 μm for rice straw (RS), and 361 μm for sugarcane bagasse (SB), respectively. Smaller particle size means greater surface area. The greater surface

Table 1 Properties of the various biomass feedstocks

Analysis	Properties	Sugarcane bagasse (SB)	Elephant grass (EG)	Rice husk (RH)
The particle size of the biomass	Mean particle size (μm.)	361	240	264
Proximate analysis (wt.%), dry basis	Moisture content	9.90	9.24	9.54
	Volatiles content	74.98	73.83	63.21
	Ash content	1.94	2.74	16.02
	Fixed carbon content	13.18	14.19	11.23
Ultimate analysis (wt.%), dry basis	N	0.13	0.12	0.54
	C	43.36	43.21	39.98
	H	7.38	6.92	6.37
	O	49.14	49.75	53.11
	S	–	–	–
Calorific value	Higher heating value (MJ/kg)	16.44	16.86	14.95

area of the particle essentially represents a greater chance of collisions, thus faster the rate of reaction. The result of the mean particle size analysis is presented in Table 1. From the table, it is observed that sugarcane bagasse has more moisture content (9.90 wt.%) and volatile matter content (74.98 wt.%) compared to the other two biomasses.

3.1 *Volatile Content*

The presence of a higher amount of volatile matter in biomass is very attractive for the production of bio-oil, which makes the EG and SB a better source of energy compared to RH. A higher amount of moisture present in biomass essentially lowers the heating value as a substantial amount of heat is released in vaporizing the moisture during combustion. As such, higher moisture content is not preferred. Whereas more volatile matter content means it is easier to ignite, has better flame stability, and improved carbon burnout. As such, it is preferred. The traces of ash are exceptionally high for rice husk (16.02 wt.%) compared to sugarcane bagasse (1.94 wt.%) and rice husk (2.74 wt.%), due to the presence of silicon oxide in the ash of rice straw. Ash is the amount of inorganic incombustible material that remains after burning due to which high-ash content is not suitable. Again, fixed carbon content is maximum for elephant grass (14.19 wt.%) which is presented in Table 1.

From the ultimate analysis, it is observed that sugarcane bagasse particles contain more hydrogen (7.38 wt.%) and carbon content (43.36 wt.%) as illustrated in Table 1. As hydrogen is a non-toxic, non-metallic, and highly combustible gas, the presence of more amounts of hydrogen is good as it helps the biomass to burn easily. The more amount of carbon in sugarcane bagasse explains the more amount of ash found in it during proximate analysis. Also, the carbon when combusted forms carbon-di-oxide, which is a greenhouse gas. The highest amounts of nitrogen and oxygen are found in rice husk (0.54 wt.% and 53.11wt.%, respectively). The presence of oxygen is helpful for the complete combustion of the fuel, thus ash will be less. There is no sulphur present in the three biomasses, which is good because when sulphur reacts with oxygen, it produces sulphur dioxide (SO_2), which is an air pollutant.

As depicted in Table 1, it is seen that elephant grass has the maximum calorific value (16.86 MJ/kg), and rice husk has the minimum calorific value (14.946 MJ/kg) which is due to the presence of a greater amount of silica in rice husk particles, and it remains as the ash (non-combustible material) after combustion. The rice husk has lesser heating power than sugarcane bagasse and elephant grass.

3.2 *Effect of Pyrolysis*

Figure 1 represents the TGA curve of temperature vs. % weight loss profile of the biomasses at the heating rate of 10 °C/min. The weight residue left as char at the end of the pyrolysis process is 21.29 wt.% for sugarcane bagasse, 20.45 wt.% for elephant grass, and 32.65 wt.% for rice husk. During pyrolysis of the biomasses, sugarcane bagasse and elephant grass lost a higher amount of weight (79.71 and 79.55 wt%, respectively), in comparison with rice husk (67.35 wt%). This result can be attributed to higher volatile matter present in sugarcane and elephant grass compared to rice husk. The maximum char for rice husk is due to the presence of silicon in the sample of it, and after burning, it remains as silicon dioxide (SO_2) in the ash. After the pyrolysis, a minimum amount of char in the biomass is preferred. So elephant grass and sugarcane bagasse can be better fuels for the production of energy than rice husk.

From Fig. 2, it is observed that the pyrolysis process can be characterized by 3 distinct steps of thermal degradation. In stage 1, the water molecule and lower hydrocarbons present in the biomass sample are decomposed. The graph is approximately a straight line, which shows that it does not have much effect on weight loss. The initial deviation is due to the instrumental and environmental error present in the TGA analyzer. In stage 2, drastic weight loss occurs. Here, cellulose and hemicellulose present in the biomass sample are removed, and maximum decomposition of biomass sample occurs at this stage. In stage 3, the crystalline portion of cellulose, lignin, and aromatic hydrocarbons is removed from the biomass sample. Here, the graph is almost a straight line, and a little amount of weight loss occurs at this stage. The rate of degradation of the feedstock was found to be very slower than stage 1 and 2 which signified higher thermal stability of biomass. All the biomasses selected for this study show similar stages of decomposition in TGA analysis. A similar trend of results has been also stated by Mallick et al. [16] and Kaur et al. [25].

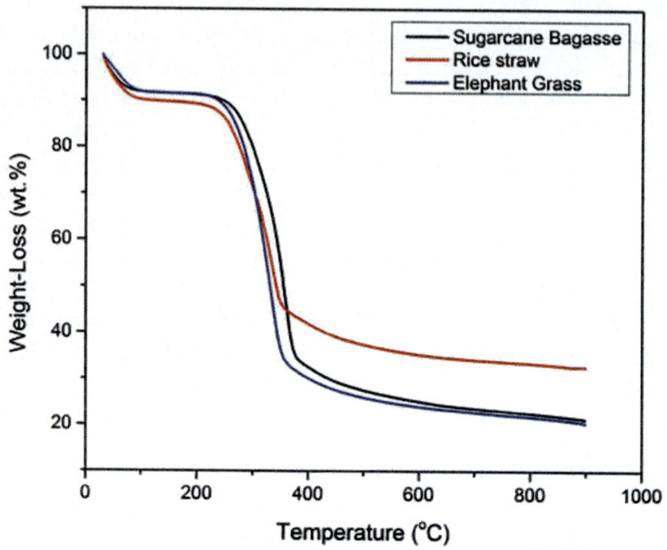

Fig. 1 TGA weight loss profile for various biomasses at $\beta = 10$ °C/min

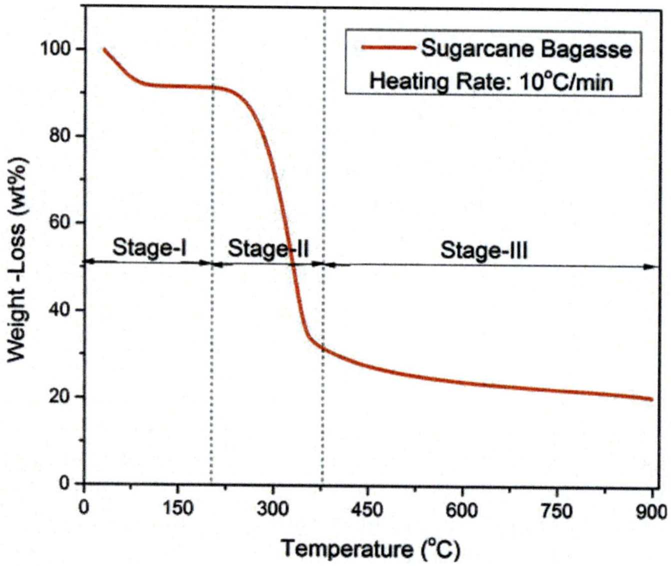

Fig. 2 Various states of pyrolysis of sugarcane bagasse at $\beta = 10$ °C/min

3.3 Thermal Stability

Figure 3 represents the derivative thermogravimetric (DTG) graph of the biomasses. From the curve, the maximum decomposition intensity is found to be 327 °C at the decomposition rate of 0.6 wt.%/°C for rice husk, 354 °C at the decomposition rate of 0.86 wt.%/°C for elephant grass, and 355 °C at the decomposition rate of 0.89 wt.%/°C for sugarcane bagasse. The peak degradation temperatures are found to be similar for SB and EG and the lowest for the rice husk. The lowest degradation temperature of the rice husk may be due to the comparatively lower amount of cellulose and higher amount of lignin present in the biomass [16].

Figure 4 depicts the discrepancy of the degree of conversion of biomasses with operating temperature. It is seen from the curve that the maximum decomposition of biomasses takes place in the range of 225–400 °C. This is owing to the thermal decomposition of cellulose and hemicellulose and the release of volatile matters from the biomasses. Beyond 400 °C, the conversion of the biomasses is observed to be very slow due to the slow breakdown of lignin content in the feedstocks.

To obtain the physical insight of the pyrolysis of feedstocks, the TGA data at different heating rates were studied using FWO methods for obtaining E_a for all biomasses in the conversion range of $\alpha = 0.1$ to 0.8 is shown in Fig. 5. The activation energy for each value of α was determined and depicted in Table 2. From the table, the average values of E_{a} using the FWO method were 209.4 kJ/mol for rice husk, 182 kJ/mol for sugarcane bagasse, and 130.4 kJ/mol for elephant grass, respectively. Similar values of E_{a} are also reported in the previous literature for biomasses with

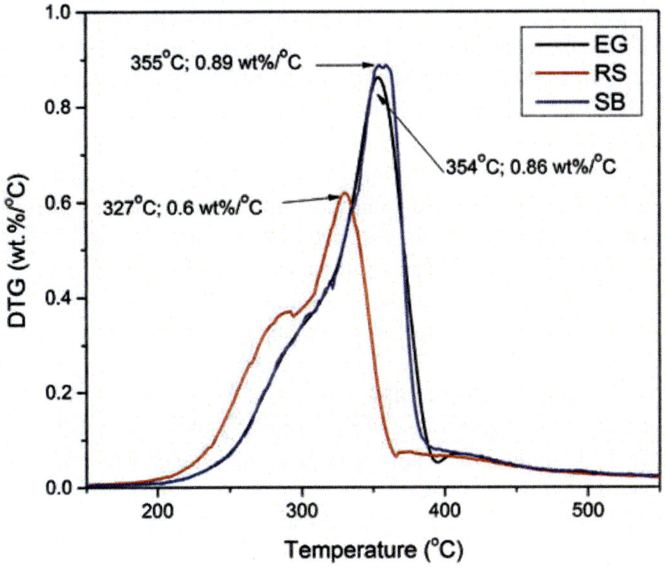

Fig. 3 DTG profile for various biomasses at $\beta = 10^{\circ}$ C / min

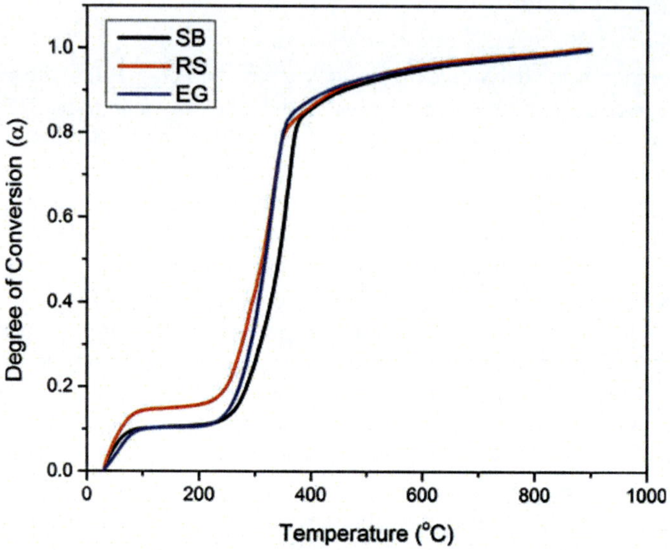

Fig. 4 Degree of conversion of various biomasses at $\beta = 10$ °C/min

analogous compositions. Varma and Mondal (2016) have reported the average value of E_a of 104.43 kJ/mol for pyrolysis of sugarcane bagasse, for $\alpha = 0.1$–0.8 using the FWO method [26]. Ceylan and Topcu (2014) had investigated the kinetic analysis of hazelnut husk and observed the average value of E_a of 131.1 kJ/mol for $\alpha = 0.2$ to 0.8 [27]. Mallick et al. (2018) had investigated the kinetic analysis of various biomasses, namely sawdust, rice husk, and bamboo dust using thermogravimetric analysis. The authors have reported the average values of E_a of 188.74, 271.63, and 189.34 kJ/mol for bamboo dust, rice husk, and sawdust, respectively, using the FWO method for $\alpha = 0.2$ to 0.8 [16].

The correlation coefficient (R^2) value for each value of conversion was found to be not less than 0.97 as depicted in Table 2. From the values of E_a values, it has been observed that the rice husk the highest average activation energy (209.4 k J/mol) compared to the other two biomasses. It is because RH requires more amount of energy to decompose the holocellulose content of the feedstock. This result can be attributed to the associated with its high-ash content that hinders the thermal as well as mass transfer limits [28].

The activation energy refers to the minimum value of energy required to start a chemical reaction. The higher value of E_a essentially indicates the slower reactions [29]. An increasing trend of activation energy up to $\alpha = 0.3$ with conversion essentially indicates the exothermic nature of reaction as presented in Fig. 6. After $\alpha = 0.3$ to 0.8, almost a constant value of activation energy is obtained for all the biomasses. The different values of E_a at different degrees of conversions essentially indicate the multistage characteristic of the thermal decomposition process of the selected biomasses. The high-average values of activation energy in the case of RH compared

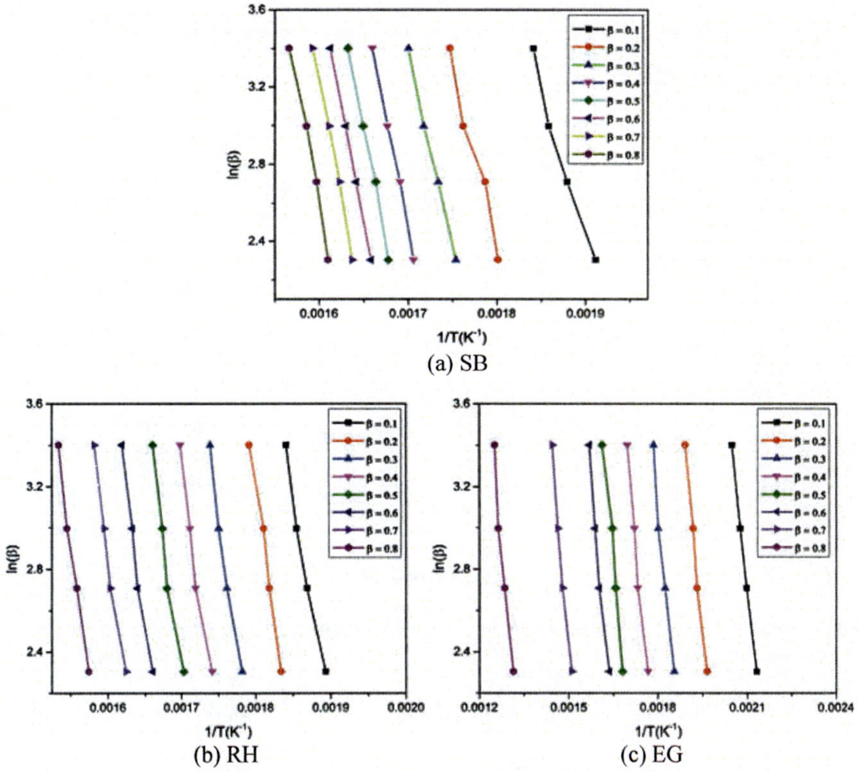

Fig. 5 Variation of ln (β) versus 1/T for **a** SB, **b** RH, and **c** EG biomass

Table 2 Activation energy of biomasses using Flynn and Wall kinetic method

Conversion (α)	Rice husk		Sugarcane bagasse		Elephant grass	
	E_a (kJ/mol)	R^2	E_a (kJ/mol)	R^2	E_a (kJ/mol)	R^2
0.1	171	0.98	126	0.98	108	0.99
0.2	216	0.99	157	0.97	129	0.99
0.3	209	0.98	167	0.99	126	0.98
0.4	210	0.98	194	0.99	133	0.98
0.5	213	0.97	199	0.99	135	0.98
0.6	219	0.98	201	0.99	135	0.98
0.7	215	0.98	205	0.99	138	0.98
0.8	222	0.99	207	0.99	139	0.98
Average	209.4		182		130.4	

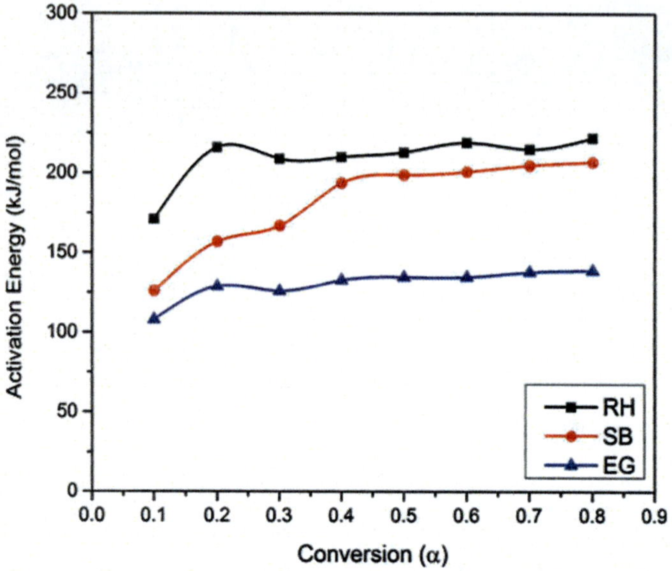

Fig. 6 Variation of E_a versus extent of conversion of biomass

to EG and SB are might be due to the greater amount of minerals (SiO_2) present in it.

Thus, the pyrolysis of biomasses is categorized into 3 discrete steps. The maximum weight loss at the end of pyrolysis is found to be in elephant grass amongst the three feedstocks at a constant heating rate of 10 °C/min. The maximum decomposition intensity was found in the case of sugarcane bagasse 0.89 wt%/o C at 355 °C and minimum in the case of rice husk 0.06 wt%/°C at 327 °C.

4 Conclusions

This study analyzes the physicochemical properties and thermal decomposition behaviour of waste lignocellulosic and organic biomasses. From the result of proximate analysis of the feedstocks, it is found that ash content is exceptionally high for rice husk than the other two feedstocks. The ultimate analysis shows that the sugarcane bagasse has the highest amount of carbon and hydrogen content. The pyrolysis study of the feedstock reveals that the maximum weight loss occurs for elephant grass (79.55 wt.%). The activation energy of the feedstocks was evaluated. The average values of the E_a were observed to be 209.4 kJ/mol for rice husk, 182 kJ/mol for sugarcane bagasse, and 130.4 kJ/mol for elephant grass, respectively. Therefore, the elephant grass and sugarcane bagasse as biofuel have almost the same characteristic values.

References

1. http://refhub.elsevier.com/S1364-0321(14)00601-7/sbref1. Accessed on 21 Sept 2021
2. Bordoloi N, Narzari R, Chutia RS, Bhaskar T, Kataki R (2014) Pyrolysis of Messua Ferrea and Pongamia glabra seed cover: characterization of bio-oil and its subfractions. Biores Technol 178:83–89
3. Xu Y, Chen B (2013) Investigation of thermodynamic parameters in the pyrolysis conversion of biomass and manure to biochars using thermogravimetric analysis. Biores Technol 146:485–493
4. Mallick D, Mahanta P, Moholkar VS (2017) Co-gasification of coal and biomass blends: chemistry and engineering. Fuel 204:106–128
5. Velden VD, Baeyens M, Boukis I (2008) Modeling CFB biomass pyrolysis reactor. Biomass Bioenerg 32:128–139
6. Bridgwater AV (2012) Review of fast pyrolysis of biomass and product upgrading. Biomass Bioenerg 38(3):68–94
7. Mallick D, Bora BJ, Barbhuiya SA, Banik R, Garg J, Sarma R, Gogoi AK (2019) Detailed study of pyrolysis kinetics of biomass using thermogravimetric analysis. In AIP conference proceedings, Vol 2091, No 1, pp 020014. AIP Publishing LLC.4–264. https://doi.org/10.1063/1.5096505
8. Ceylan S, Topcu Y (2014) Pyrolysis kinetics of hazelnut using thermogravimetric analysis. Bioresour Technol 156:182–188
9. Mallick D, Bora BJ, Kakati RK, Barbhuiya SA, Banik R, Garg J, Sarma R, Gogoi AK, Das PK (2019) Pyrolysis kinetics of water hyacinth using thermogravimetric analysis. Carbon: Sci Technol 11(2):22–28
10. Mallick D, Baruah D, Mahanta P, Moholkar VS (2018) A comprehensive kinetic analysis of bamboo waste using thermogravimetric analysis. In: 2018 2nd international conference on power, energy and environment: towards smart technology (ICEPE), pp 1–6. IEEE. https://doi.org/10.1109/EPETSG.2018.8658672
11. Siddiqi H, Kumari U, Biswas S, Mishra A, Meikap BC (2020) A synergistic study of kinetic reaction heat transfer with multi-component modeling approach for the pyrolysis of biomass waste. Energy 204:117933
12. Saikia R, Baruah B, Kalita D, Pant K, Gogoi N, Kataki R (2020) Pyrolysis and kinetic analysis of a Perennial Grass (Saccharum ravannae L.) from North-East India: optimization through response surface methodology and product characterization, bioresource technology. 17-07025
13. Xiao R, Yang W, Cong X, Dong K, Xu J, Wang D, Yang X (2020) Thermogravimetric analysis and reaction kinetics of lignocellulosic biomass pyrolysis. Energy 201:117537
14. Poletto M, Zattera AJ, Sanata RMC (2012) Thermal decomposition of wood: kinetics and degradation mechanisms. Biores Technol 126:7–12
15. Fermanelli SC, Cordoba A, Pierella BL, Saux C (2020) Pyrolysis and co-pyrolysis of three lignocellulosic biomass from residues from the agro-food-industry: a comparative study. WasteManage 102:362–370
16. Mallick D, Poddar MK, Mahanta P, Moholkar VS (2018) Discernment of synergism in pyrolysis of biomass blends using thermogravimetric analysis. Biores Technol 261:294–305
17. Braga RM, Melo DM, Aquino FM, Freitas JC, Melo MA, Barros JM, Fontes MS (2014) Characterization and comparative study of pyrolysis kinetics of the rice husk and the elephant grass. J Therm Anal Calorim 115(2):1915–1920
18. Mallick D, Mahanta P, Moholkar VS (2020) Co-gasification of biomass blends: performance evaluation in circulating fluidized bed gasifier. Energy 192:116682
19. Mallick D, Buragohain B, Mahanta P, Moholkar VS (2018) Gasification of mixed biomass: analysis using equilibrium, semi-equilibrium, and kinetic models. In: De S, Agarwal A, Moholkar V, Thallada B (Eds) Coal and biomass gasification. Energy, environment, and sustainability. Singapore: Springer, pp 223–441

20. Mallick D, Mahanta P, Moholkar VS (2018) Synergistic effects in gasification of coal/biomass blends: analysis and review. In: De S, Agarwal A, Moholkar V, Thallada B (eds) Coal and biomass gasification. Energy, environment, and sustainability. Singapore: Springer, pp 273–497
21. Mallick D, Sharma SD, Kushwaha A, Brahma HS, Nath R, Bhowmik R (2022) Emerging commercial opportunities for conversion of waste to energy: aspect of gasification technology. In Waste-to-energy approaches towards zero waste. Elsevier, pp 105–127
22. Kushwaha A, Bajgai RC, Mallick D, Singh A, Goswami L, Bhan U, Hussain CM (2022) Biohythane production from organic waste: challenges and techno-economic perspective. In Waste-to-energy approaches towards zero waste. Elsevier, pp 373–392
23. Hossain AK, Badr O (2007) Prospects of renewable energy utilization for electricity generation in Bangladesh. Renew Sustain Energy Rev 11:1617–1649
24. Kunil D, Levenspiel O (2005) Fluidization Engineering, 2nd edn. Elsevier, Washington (USA)
25. Kaur R, Gera P, Jha MK, Bhaskar T (2018) Pyrolysis kinetics and thermodynamic parameters of castor (Ricinus communis) residue using thermogravimetric analysis. Biores Technol 1(250):422–428
26. Varma AK, Mondal P (2016) Physicochemical characterization and pyrolysis kinetic study of sugarcane bagasse using thermogravimetric analysis. J Energy Res Technol 138(5):052205
27. Ceylan S, Topçu Y (2014) Pyrolysis kinetics of hazelnut husk using thermogravimetric analysis. Biores Technol 156:182–188
28. Sait HH, Hussain A, Salema AA, An FN (2012) Pyrolysis and combustion kinetics of date palm biomass using thermogravimetric analysis. Bioresour Technol 118:382–389
29. Gai C, Dong Y, Zhang T (2013) The kinetic analysis of the pyrolysis of agricultural residue under non-isothermal conditions. Biores Technol 1(127):298–305

The Effect of Varying Compression Ratio's Upon Combustion and Performance Parameters of a TDI Diesel Engine

Swarup Nayak, Prakash Ghose, Achinta Sarkar, Jitendrax K. Patel, Basanta K. Rana, and Prakash Ghose

1 Introduction

NicolousOtto constructed the very first gasoline engine, which was preceded by Rudolph Diesel's invention of diesel engine in the end of the nineteenth century [1]. Catastrophic advancements and adjustments have already been achieved in the domain of internal combustion engines and now they will serve humanity in transport and power generation [2]. In today's modern world, diesel vehicles are more widely used than SI engines. Because of the enhanced compression ratio's superior torque and efficiency, humans prefer diesel engines for a wide range of applications. Automobiles and steady-speed gas turbines are subject to increasingly stringent pollution laws, signalling significant technical breakthroughs [3]. Numerous trial conditions, including modification of injection time, compression ratio and injector pressures, have been carried out by various researchers in order to increase performance and fall in line with pollutants [4]. The ignition of a gasoline engine is considered to really be complicated and it may be divided into four phases: ignition delay, combustor phase, regulated burning sequence and post-combustion process [5]. The time lag between initiation of direct injection as well as the onset of ignition is referred to as delay period. Inside the premixed phase, fast-burning with substantial heat release is visualised. The burning pace during direct combustion is regulated by that of the pace of the burned mixture and indeed the latter ignition stage incorporates the continuance of heat release at a slower rate [6]. The efficiency, emissions and combustion efficiency of gasoline and WCO Biodiesel had been studied at varied compression ratios with 50% load and indeed the findings revealed that exhaust gas temperature rose with increasing compression ratios [7]. The mechanical efficiency steadily reduced

S. Nayak (✉) · P. Ghose · A. Sarkar · J. K. Patel · B. K. Rana · P. Ghose
School of Mechanical Engineering, KIIT Deemed to Be University, Bhubaneswar, Odisha 751024, India
e-mail: drswarupkumarnayak@gmail.com; swarup.nayakfme@kiit.ac.in

© The Author(s), under exclusive license to Springer Nature Singapore Pte Ltd. 2023

S. Revankar et al. (eds.), *Recent Advances in Thermofluids and Manufacturing Engineering*, Lecture Notes in Mechanical Engineering,
https://doi.org/10.1007/978-981-19-4388-1_29

even as compression ratio increased. As the CR raised, the thermal performance of the braking enhanced and indeed the fuel economy of the braking system decreased. With a rise in CR, the peak cylinder pressure likewise improved. The heat release rate declined because as compression ratio was increased and the heat release rate of petro-diesel had been found to be greater than that of the mixtures [8]. Experiments were conducted on an adjustable speed power train with varied compression ratios using biodiesel blends that demonstrated an increment in engine power for raising compression ratio throughout all rpm. Since this compression ratio was improved, the thermal performance of the brakes continued to rise for biodiesel and other mixtures. The ignition delay declined since this compression ratio increased, which has been seen at all speeds [9]. A substantial decrement in the delay period had been found, that might be attributed to quick ignition as compression ratios improved. With a dual-fuel engine idling upon biogas, the influence of compression ratios on performance, combustion and emission characteristics had been investigated [10]. The thermal efficiency for dual operation enhanced when compression ratios increased, that might be attributed to full incineration of biogas as well as a reduction in exhaust gas temperature as compression ratios continued to increase throughout all loads. To a lower compression ratio, overall peak cylinder pressure decreased as well as shifted relatively close to top dead centre. As the compression ratio was raised, so did the heat release. Across all loads, raising the compression ratio lowered the delay period. The shortest igniting lag happened when the compression ratio was indeed the lowest just at least load [11].

Therefore in current investigation, the engine performance attributes of petro-diesel at all loads have been assessed and compression ratios of 18, 17 and 16 have been included in the analysis. Throughout all loads, performances such as thermal efficiency, specific fuel consumption and exhaust temperature had been analysed and so were ignition characteristics including cylinder pressure, peak pressure rise and net heat release rate.

2 Materials and Methods

2.1 Experimental Setup and Experimentation

The gasoline engine experiment setup includes a Kirloskar single-cylinder, four-stroke, direct-injection engine. The engine gets air-cooled and seems to have a rated power of 5.02 kW at 2000 rpm. Figure 1 illustrates layout sketch of the diesel engine. Table 1 shows the additional comprehensive specs, including the engine displacement capacity of 594 cc. Even though the temperature gradient bands available in several components fluctuate, thermocouples including RTD, PT 100 and K-type sensors have been used in the engine to monitor engines' water temperatures and exhaust temperatures. The arrangement employs a pressure transducer including an acoustic wire in the range of 10,000 PSI to detect combustion pressure and fuel injection pressure. The operating system will have a 15 L fuel tank, an airflow container, a manometer and a fuel burette. The load is measured by a strain gauge load cell, while the speed is measured with a tachometer [2, 5]. The compression ratio may be modified while the engine is operating and a specifically engineered tilting block arrangement is employed to do so without affecting the cylinder head [6]. For all of the loads, performance metrics including thermal efficiency, specific fuel consumption had been measured, along with ignition metrics like in-cylinder pressure, net heat release, rate of pressure rise and cumulative heat release. Determining the gradient of pressure in relation towards the crank angle would be used to compute the rate of pressure increase. The preceding equation is used to calculate net heat release [12, 13].

$$\frac{\mathrm{d}q_{\mathrm{HR}}}{\mathrm{d}\theta} - \frac{\mathrm{d}q_{HT}}{\mathrm{d}\theta} = \frac{\beta}{\beta - 1} R \frac{\mathrm{d}W}{\mathrm{d}\theta} + \frac{1}{\beta - 1} W \frac{\mathrm{d}R}{\mathrm{d}\theta} \tag{1}$$

On equating the above equation can be further written as

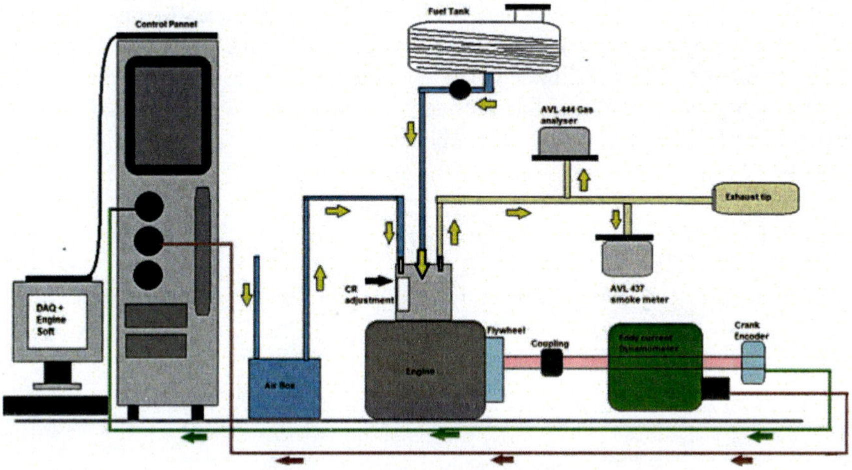

Fig. 1 Layout sketch of the experimental setup

Table 1 Diesel engine specification

Make and model type	TV1 Kirloskar eine
Bore x stroke	87.5 × 110 mm
Rated power	5.02 kW@2000 r
Injection timing	21°bTDC
Injection pressure	210 bar
Compression ratio	16.5:1
gine loading	Eddy curre dynamometer

$$\frac{dq_P}{d\theta} = \frac{\beta}{\beta - 1} R \frac{dW}{d\theta} + \frac{1}{\beta - 1} W \frac{dR}{d\theta} \tag{2}$$

$\frac{dq_P}{d\theta}$ is considered as the net heat release rate.

$\frac{dq_{HR}}{d\theta}$ is considered as the heat released amid ignition and $\frac{dq_{HT}}{d\theta}$ refers to the heat transferred within the cylinder walls. For lower compression ratios of 17 and 16, a consistent approach had been used. The ideal compression ratio was determined by releasing the allen bolts upon that cylinder as well as tightening the clamp bolt and adjustor. Subsequently, as before, the efficiency and combustion of CR 17 and 16 were recorded.

2.2 *Uncertainty Analysis*

Iition and performance characteristics had been examined using a variety of equipment made utilising diverse methodologies and technology. The precision is heavily influenced by the environment and operational circumstances. Errors can result during an issue of uncertainty, that could be either fixed or random [5, 14]. As a result, the equipment has been calibrated. The uncertainty for speed, load, temperatures, combustion pressure and fuel line pressure are determined which results in overall experimental uncertainty of ±2.81%. The study's overall percentage of uncertainty is shown in Table 2.

$$\sqrt{\text{SFC}^2 + \text{BTE}^2 + \text{BP}^2 + \text{PP}^2 + \text{Crank angle encoder}^2}$$
$$= \sqrt{1.8^2 + 1^2 + 1^2 + 0.6^2 + 1.5^2}$$
$$= \pm 2.81\%$$

Table 2 Uncertainty analysis for instruments used during experimentation

Instrument used	Measurement	Uncertainty
Load cell	Loading device	±1.2
Speed sensor	Speed	±1.0
Temperature indicator	Exhaust gas temperature	±0.4
Crank angle encoder	Crank angle	±1.5

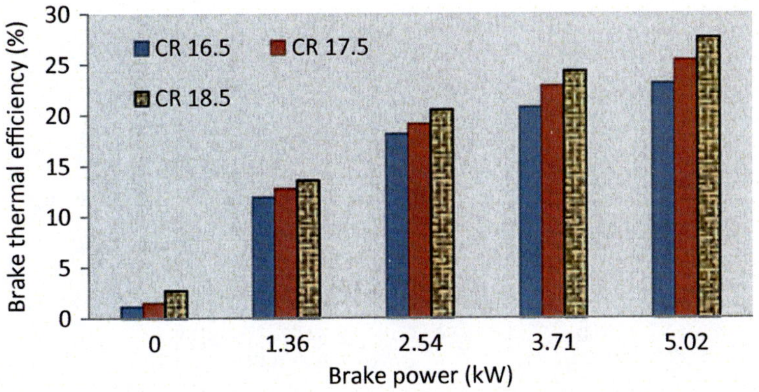

Fig. 2 Variation in brake thermal efficiency with respect to engine loading for all test fuel blends at varying compression ratio

3 Results and Discussion

3.1 Brake Thermal Efficiency

Figure 2 briefly elaborates the variation in brake thermal efficiency with respect to engine loading for all test fuel blends at varying compression ratios. From the above figure it is clearly viewed that with increase in load, BTE increases for all the prepared blends. This may be due to excess energy generated by the engine at maximum loading conditions [11, 15]. Again from Fig. 2 is also depicted that with increase in CR, BTE shows a upward trend. This might be due to complete combustion at elevated compression ratios [15]. With increment in CR from 16.5 to 18.5, BTE showed an improvement in about 14.3% at highest load. At maximum loading condition of 5.02 kW, BTE for CR16.5, CR17.5 and CR18.5 are 23.09%, 25.37% and 27.58%, respectively for single operational mode.

3.2 Brake-Specific Fuel Consumption

Figure 3 briefly elaborates the variation in brake-specific fuel consumption with respect to engine loading for all test fuel blends at varying compression ratios. From the above figure it is clearly viewed that with increase in load, BSFC declines for all the prepared blends. This may be due to diminished loss of heat at maximum loading conditions [16]. Again from Fig. 2 is also depicted that with increase in CR, BSFC shows a downward trend. This might be due to improved combustion at elevated compression ratios as well as lower loss of heat resulting in diminished BSFC [11, 16]. With increment in CR from 16.5 to 18.5, BSFC showed an reduction

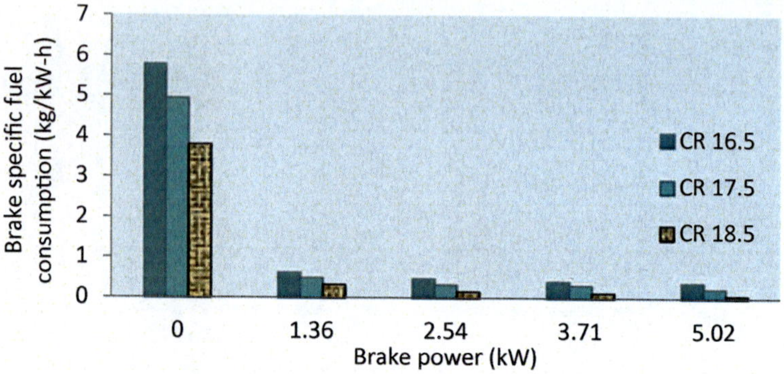

Fig. 3 Variation in brake-specific fuel consumption with respect to engine loading for all test fuel blends at varying compression ratio

in about 23.6% at highest load. At maximum loading condition of 5.02 kW, BSFC for CR16.5, CR17.5 and CR18.5 are 0.4 kg/kW-h, 0.25 kg/kW-h and 0.08 kg/kW-h respectively for single operational mode.

3.3 Exhaust Gas Temperature

Figure 4 briefly elaborates the variation in exhaust gas temperature with respect to engine loading for all test fuel blends at varying compression ratios. From the above figure it is clearly viewed that with increase in load, EGT increases for all the prepared blends. This may be due to excess fuel consumption for generating excess amount of power amid combustion [17, 18]. Again from Fig. 4 is also depicted that with increase in CR, EGT diminishes down. This might be due to complete combustion at elevated compression ratios as well as decrement in ignition delay period (IDP) [13, 15]. With increment in CR from 16.5 to 18.5, EGT showed decline of about 2.7% at peak load. At maximum loading condition of 5.02 kW, EGT for CR16.5, CR17.5 and CR18.5 are 293.84 °C, 280.81 °C and 263.72 °C, respectively for single fuel mode of operation.

3.4 Cylinder Pressure (CP)

Figures 5, 6 and 7 briefly elaborates the variation in-cylinder pressure with respect to engine crank angle for all test fuel blends at varying compression ratios of 16.5, 17.5 and 18.5, respectively for no load, part load and full loading conditions. Cylinder pressure plays a very vital role in determining the combustion characteristics for an engine [16, 18]. In case of no-load condition, in-cylinder pressure for higher CR of

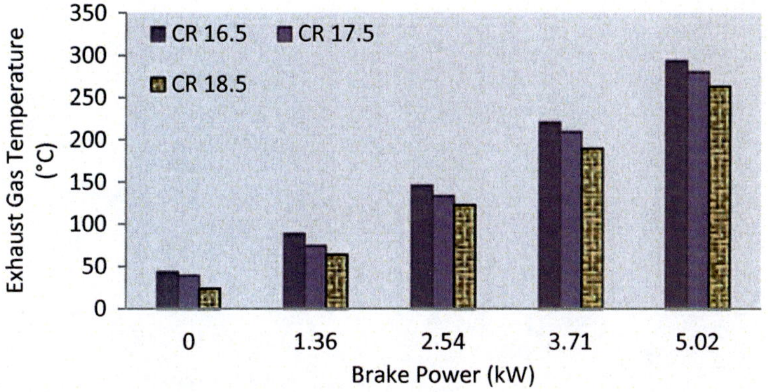

Fig. 4 Variation in exhaust gas temperature with respect to engine loading for all test fuel blends at varying compression ratio

18.5 was found to be 55.7 bar and further declined by 19.3% while the compression ratio was lowered to 16.5. Considering part-load condition, in-cylinder pressure for CR of 18.5 was found to be 66.9 bar and further declined by 16.4% while the compression ratio was lowered to 16.5. Moreover, at peak loading condition of 5.02 kW, in-cylinder pressure for higher CR of 18.5 was found to be 71.8 bar which further declined by 9.2% and 11.5% while the compression ratio was lowered to 17.5 and 16.5, respectively.

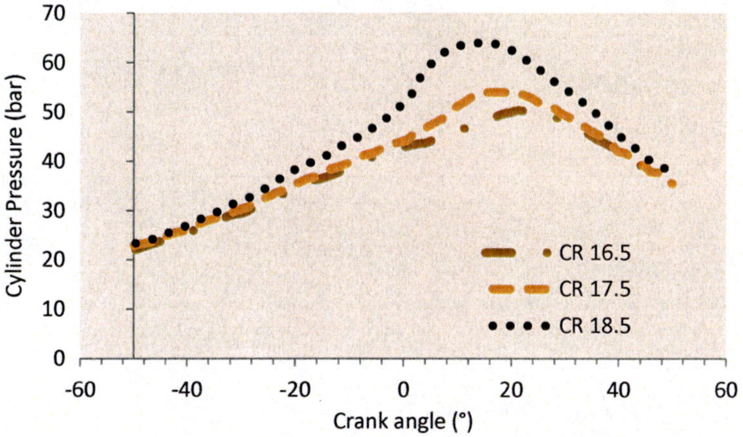

Fig. 5 Variation in-cylinder pressure with respect to engine crank angle for all test fuel blends at varying compression ratios of 16.5, 17.5 and 18.5, respectively for no-load conditions

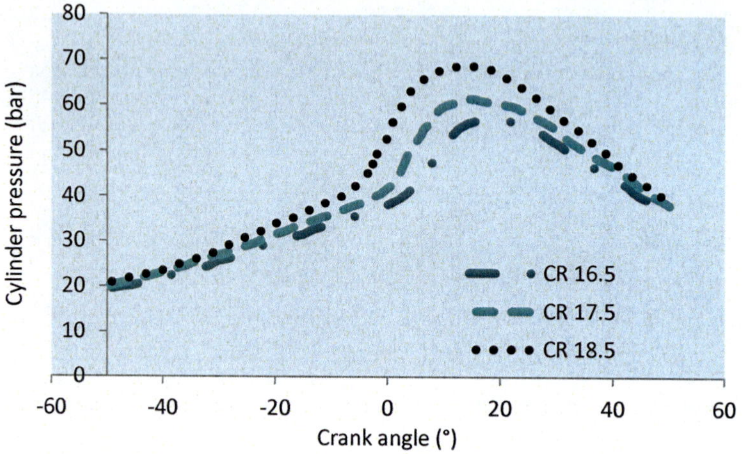

Fig. 6 Variation in-cylinder pressure with respect to engine crank angle for all test fuel blends at varying compression ratios of 16.5, 17.5 and 18.5, respectively for part-load conditions

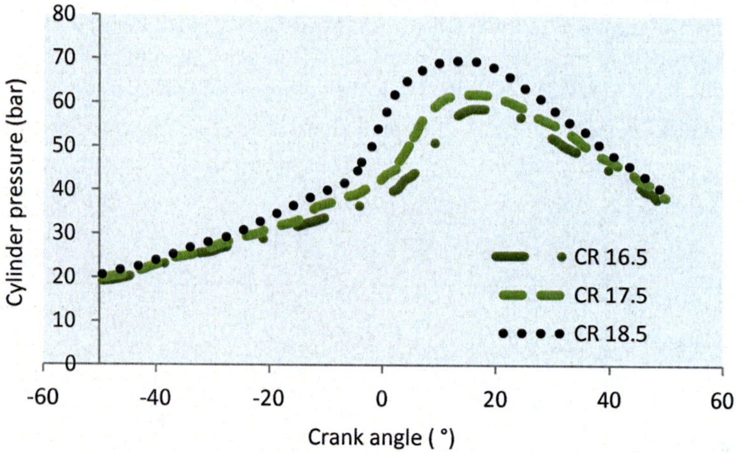

Fig. 7 Variation in-cylinder pressure with respect to engine crank angle for all test fuel blends at varying compression ratios of 16.5, 17.5 and 18.5, respectively for peak load conditions

3.5 Peak Cylinder Pressure

Figure 8 briefly elaborates the variation in peak cylinder pressure with respect to engine loading for all test fuel blends at varying compression ratios. From the above figure it is clearly viewed that with increase in load, PCP increases for all the prepared blends. This may be due to excess energy generated by the engine at maximum loading conditions [19, 20]. Again from Fig. 8 is also depicted that with increase

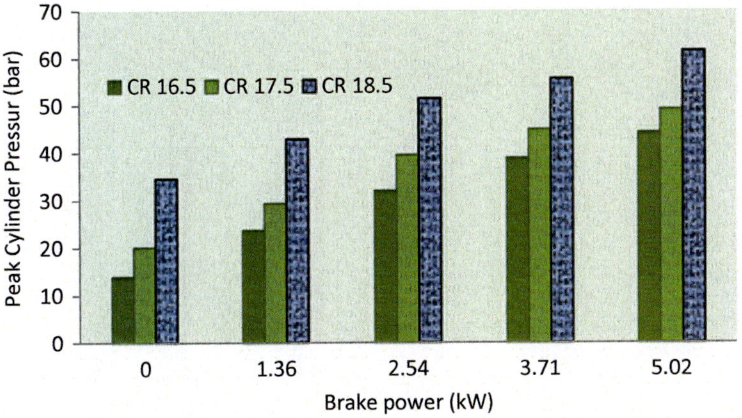

Fig. 8 Variation in peak cylinder pressure with respect to engine loading for all test fuel blends at varying compression ratios

in CR, PCP shows an upward trend. This might be due to complete combustion at elevated compression ratios [5, 6]. At maximum loading condition of 5.02 kW, PCP for CR16.5, CR17.5 and CR18.5 are 44.25 bar, 49.15 bar and 61.47 bar, respectively for single fuel mode of operation.

3.6 Ignition Delay Period (IDP)

Figure 9 briefly elaborates the variation in ignition delay period (IDP) with respect to engine loading for all test fuel blends at varying compression ratios. From the above figure it is clearly viewed that with increase in load, IDP decreases for all the prepared blends. This is because the residual air, as well as sidewall temperatures, raised when the load was elevated, resulting in a greater charge temperature amid injection leads to shorter IDP [4, 20]. Again from Fig. 9 is also depicted that with increase in CR, IDP shows a downward trend. This might be due to complete combustion at elevated compression ratios [6, 11]. With increment in CR from 18.5 to 16.5, IDP showed declination of about 9.4%. At maximum loading condition of 5.02 kW, IDP for CR16.5, CR17.5 and CR18.5 are 11.46°CA, 8.27°CA and 5.01°CA, respectively for single fuel mode of operation.

3.7 Net Heat Release Rate (NHRR)

Figures 10, 11 and 12 briefly elaborate the variation in net heat release rate (NHRR) with respect to engine crank angle for all test fuel blends at varying compression

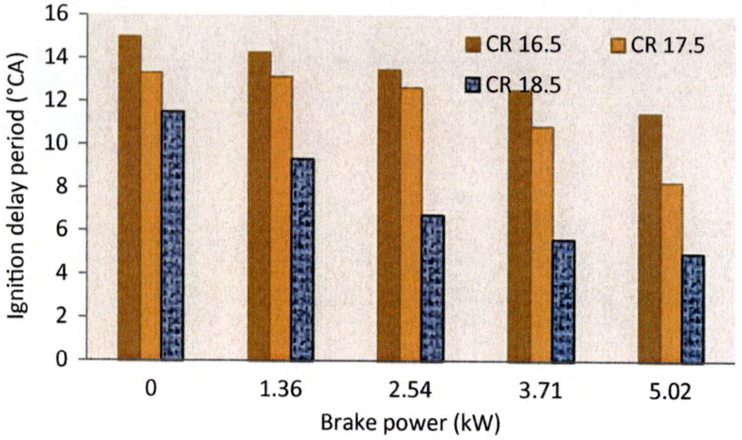

Fig. 9 Variation in ignition delay period (IDP) with respect to engine loading for all test fuel blends at varying compression ratios

ratios of 16.5, 17.5 and 18.5, respectively for no load, part load and full loading conditions. HRR plays a very vital role in determining the start of combustion, fuel fraction ignited in the pre-mix phase of combustion stage [20, 21]. In case of no-load condition, NHRR for higher CR of 18.5 was found to be 24.8 J/°CA and further declined by 19.35% for CR 17.5 and 23.4% for CR 16.5, respectively. Considering part-load condition, NHRR for CR of 18.5 was found to be 41.7 J/°CA and further declined by 5.07% and 8.63% while the compression ratio was lowered to CR 17.5 and CR 16.5. Moreover, at peak loading condition of 5.02 kW, NHRR for higher CR of 18.5 was found to be 56.4 J/°CA which further declined by 9.57% and 12.2% while the compression ratio was lowered to 17.5 and 16.5, respectively.

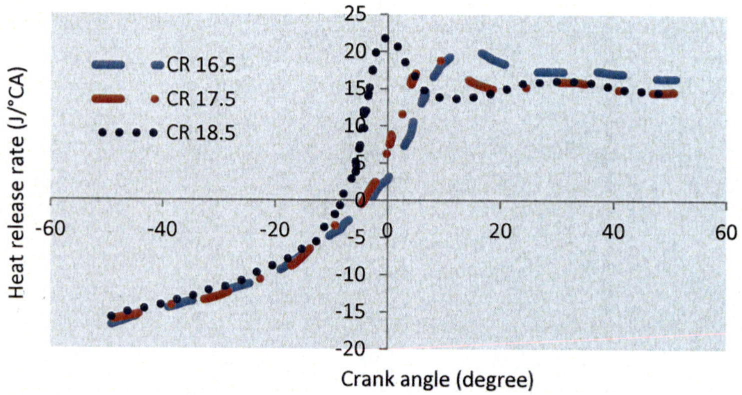

Fig. 10 Variation in net heat release rate (NHRR) with respect to engine crank angle for all test fuel blends at varying compression ratios of 16.5, 17.5 and 18.5, respectively for no-load conditions

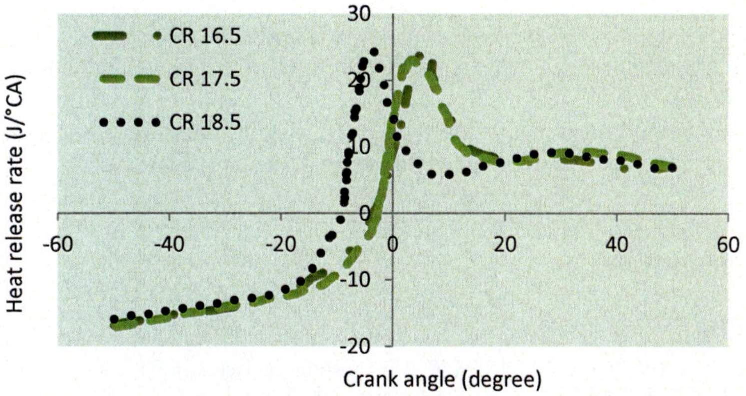

Fig. 11 Variation in net heat release rate (NHRR) with respect to engine crank angle for all test fuel blends at varying compression ratios of 16.5, 17.5 and 18.5, respectively for part load conditions

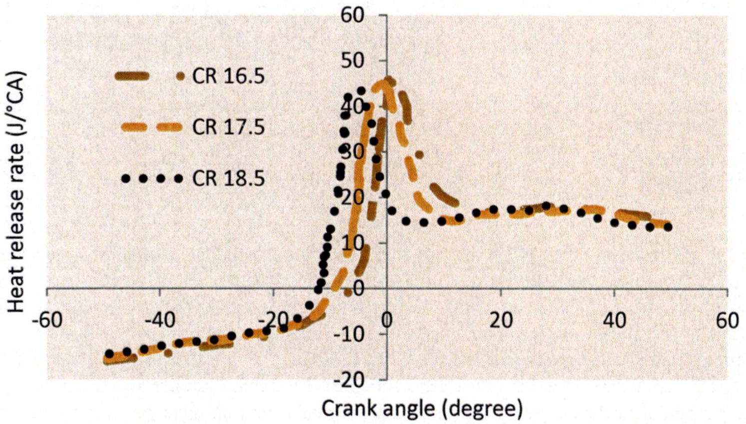

Fig. 12 Variation in net heat release rate (NHRR) with respect to engine crank angle for all test fuel blends at varying compression ratios of 16.5, 17.5 and 18.5, respectively for full load conditions

Therefore the research outcomes (from above discussion) were made are as follows:

(a) With increment in CR value from 16.5 to 18.5, leads to an increment in BTE of about 12.9% while a drastic diminishment in both EGT and BSFC of about 2.7% and 23.6% for engine operated in single fuel mode at peak loading condition.
(b) Considering the engine combustion behaviour, cylinder pressure showed major declination of about 19.3%, 16.4% and 11.5%, respectively for no load, part load and peak loading conditions.

(c) The least time delay can be seen at peak load with maximum CR of 18.5:1, with an ultimate measure of 9.4% reduction in IDP was reported with an elevated CR of 18.5:1.
(d) With advancement in CR from 16.5 to 18.5, net heat release rate seemed to be near to top dead centre and the HRR was significant at peak loading conditions.

4 Conclusion

Diesel engines are becoming increasingly important for human applications. The engine operating parameters are also important in tweaking the engine to meet improved performance and pollution criteria. Throughout this experiment, the influence of compression ratios on combustion and performance characteristics has been thoroughly explored across the complete load circumstances at varied CR. Hence, it can be concluded that CR of 18.5:1 determined to be the most efficient with minimal or no knocking tendency and even with improved brake thermal efficiency as well as specific fuel consumption in comparison to other CR's in single fuel mode of operation.

References

1. Nayak SK, Mishra PC (2017) Analysis of a diesel engine fuelled with jojoba blend and coir pith producer gas. Int J Autom Mech Eng 14(4):4675–4689
2. Nayak SK, Mishra PC, Noor MM, Kadirgama K, Mamat R (2019) The performance of turbocharged diesel engine with injected calophylluminophyllum methyl ester blends and inducted babul wood gaseous fuels. Fuel 257:116060
3. Veza I, Afzal A, Mujtaba MA, Putra NR, Tamaldin N (2022) Review of artificial neural networks for gasoline, diesel and homogeneous charge compression ignition engine: review of ANN for gasoline, diesel and HCCI engine. Alex Eng J 61(11):8363–8391
4. Veza I, Karaoglan AD, Ileri E, Yatish KV, Idris M, et al (2003) Grasshopper optimization algorithm for diesel engine fuelled with ethanol-biodiesel-diesel blends. Case Stud Therm Eng 31:101817
5. Nayak SK, Hoang AT, Nižetić S, Nguyen XP, Le TH (2022) Effects of advanced injection timing and inducted gaseous fuel on performance, combustion and emission characteristics of a diesel engine operated in dual-fuel mode. Fuel 310:122232
6. Nayak SK, Nayak SK, Mishra PC, Tripathy S (2015) Influence of compression ratio on combustion characteristics of a VCR engine using calophyllum inophyllum biodiesel and diesel blends. J Mech Sci Technol 29(9):4047–4052
7. Hoang AT, Tran VD, Dong VH, Le AT (2022) An experimental analysis on physical properties and spray characteristics of an ultrasound-assisted emulsion of ultra-low-sulphur diesel and Jatropha-based biodiesel. J Marine Eng Technol 21(2):73–81
8. Nayak SK, Hoang AT, Nayak B, Mishra PC (2021) Influence of fish oil and waste cooking oil as post mixed binary biodiesel blends on performance improvement and emission reduction in diesel engine. Fuel 289:119948
9. Bui VG, Tu Bui TM, Hoang AT, Nguyen Thi TX, Vo AV, et al (2021) Hydrogen-enriched biogas premixed charge combustion and emissions in direct injection and indirect injection diesel dual fueled engines: a comparative study. J Energy Resour Technol Trans ASME: 143(12):120907

10. Nayak SK, Mishra PC (2016) Emission characteristics of jatropha oil blends using waste wood producer gas. Energy Sour Part A: Recovery, Utilization Environ Eff 38(14):2153–2160
11. Hoang AT, Ong HC, Fattah IMR, Sakthivel R, Ok YS (2021) Progress on the lignocellulosic biomass pyrolysis for biofuel production toward environmental sustainability. Fuel Proc Technol 223:106997
12. Tuan Hoang A, Nižetić S, Chyuan Ong H, Quang Chau M, Phuong Nguyen X (2021) A review on application of artificial neural network (ANN) for performance and emission characteristics of diesel engine fueled with biodiesel-based fuels. Sustain Energy Technol Assess 47:101416
13. Nayak SK, Mishra PC (2017) Application of neem biodiesel and dimethyl carbonate as alternative fuels. Energy Sour Part A: Recovery, Utilization Environ Eff 39(3):284–290
14. Tuan Hoang A, Viet Pham V (2021) 2-Methylfuran (MF) as a potential biofuel: a thorough review on the production pathway from biomass, combustion progress, and application in engines. Renew Sustain Energy Rev 148:111265
15. Vinayagam NK, Hoang AT, Solomon JM, EL-Seesy AI, Nguyen XP, et al (2021) Smart control strategy for effective hydrocarbon and carbon monoxide emission reduction on a conventional diesel engine using the pooled impact of pre-and post-combustion techniques. J Clean Prod 306:127310
16. Hoang AT (2021) Combustion behavior, performance and emission characteristics of diesel engine fuelled with biodiesel containing cerium oxide nanoparticles: a review. Fuel Process Technol 218:106840
17. Nayak SK, Mishra PC (2016) Application of Nagchampa biodiesel and rice husk gas as fuel. Energy Sour Part A: Recovery, Utilization Environ Eff 38(14):2024–2030
18. Nguyen XP, Hoang AT, Ölçer AI, Pham VV, Nayak SK (2021) Biomass-derived 2,5-dimethylfuran as a promising alternative fuel: an application review on the compression and spark ignition engine. Fuel Process Technol 214:106687
19. Nayak SK, Mishra PC (2016) Emission from utilization of producer gas and mixes of jatropha biodiesel. Energy Sour Part A: Recovery, Utilization Environ Eff 38(14):1993–2000
20. Nguyen DC, Hoang AT, Tran QV, Wattanavichien K, Pham VV, et al (2021) A review on the performance, combustion, and emission characteristics of spark-ignition engine fueled with 2,5-dimethylfuran compared to ethanol and gasoline. J Energy Resour Technol Trans ASME 143(4):1DMMY, 040801
21. Tuan Hoang A, Nižetić S, Viet Pham V, Ga Bui V, Vang Le V, et al (2021) Combustion and emission characteristics of spark and compression ignition engine fueled with 2,5-dimethylfuran (DMF): a comprehensive review. Fuel 288:119757

Energy Technology

Performance of Solar Collector Coupled with Three Fluid Heat Exchanger and Heat Storage System for Simultaneous Water and Space Heating

Rama Chandra Panda, Sudhansu S. Sahoo, Ashok K. Barik, Debjyoti Sahu, Taraprasad Mohapatra, and Auroshis Rout

Nomenclature

FPSC	Flat plate solar collector
STSS	Sensible thermal storage system
NTU	Normal transfer unit
$C_{p\,\min}, C_{p\,\max}$	Heat capacity rates
η_{optical}	Optical efficiency
F_{R}	Heat removal factor
U_{L}	Overall heat transfer coefficient
T_{a}	Ambient temperature
G	Mass flux
d	Equivalent spherical diameter
μ_{f}	Dynamics viscosity
A_{c}	Collector area

R. C. Panda
Biju Patnaik University of Technology, Rourkela, Odisha, India

S. S. Sahoo (✉) · A. K. Barik
Department of Mechanical Engineering, Odisha University of Technology and Research, Bhubaneswar, Odisha, India
e-mail: sudhansu@cet.edu.in; sahoo.sudhansu@gmail.com

D. Sahu
School of Mechanical Engineering, KIIT Deemed to be University, Bhubaneswar, India

T. Mohapatra
Department of Mechanical Engineering, CV Raman Global University, Bhubaneswar, India

A. Rout
Department of Mechanical Engineering, Odisha University of Technology and Research, Bhubaneswar, Odisha, India

© The Author(s), under exclusive license to Springer Nature Singapore Pte Ltd. 2023
S. Revankar et al. (eds.), *Recent Advances in Thermofluids and Manufacturing Engineering*, Lecture Notes in Mechanical Engineering,
https://doi.org/10.1007/978-981-19-4388-1_30

A_p	Aperture area
ω	Hour angle
I_T	Total radiation on horizontal surface
Q_u	Useful heat absorbed

1 Introduction

It is well-known fact that the solar power is free and used in various conventional and non-conventional processes for years [1]. Being a solar-rich country, India is considered one among the technically developed nations for solar PV and solar thermal-related applications for domestic as well as industrial applications. Usually, on solar thermal category, flat plate solar collector (FPSC) is used as energy harvesting device for water heating purposes during day time [2]. This device is being used in rooftops of houses or apartments. Apart from water heating, FPSC is used for space heating, drying, industrial process heating. Mostly, applications requiring less than 100 °C are best to make use of FPSC during daytime. This system is a combination of various sub-systems namely collectors, storage tanks, fluid distribution systems, fluid control and transport systems.

Many research work on heat transfer analysis of flat plate solar collectors and performance have been carried out by scientists and researchers [3]. Optimization of solar power harvesting vis-a-vis heat storage has been explored and reported widely [4]. Like many other researchers, Karim and Hawalder (2004) experientially investigated flat plate solar collectors along with soft computing approach [5]. It is obvious that effective solar energy is available during sunshine hours, however, with use of effective storage system, the solar energy can be used during night time as well [6, 7].

From the few related work on this field reported, it has been observed that few researchers were mainly focused on energy analysis on various models of FPSC or solar water heaters. However, still there are many literature gaps identified such system modifications, combination of storage system analysis with FPSC, etc. Keeping in mind of dilute source of energy and unavailability during night time and applications related to simultaneous water heating and space heating at few places, special kind of system needs to be implemented. Work-related to this kind of system is also rarely mentioned in literature. Hence, present study concentrates on heat transfer modelling and analysis of FPSC with three fluid heat exchangers (TFHE) and sensible thermal storage system (STSS) for region which demand hot water and hot air round the clock. Performance analysis of a thermal system like the flat plate solar collector, heat exchanger and sensible heat storage system is executed and analyzed here using the standard theoretical model. The solar and weather data are used the simulation purposes [8]. The next section describes on mathematical modelling of three sub-systems namely, FPSC, TFHE and STSS in detail.

2 Mathematical Modelling

2.1 Flat Plate Solar Collector (FPSC)

Thermal modelling of solar collectors is presented in this sub-section. FPSC has been considered here for thermodynamic modelling and its performance analysis. A typical FPSC is considered as shown in Fig. 1. Absorber plate, copper tubes, transparent or glazing cover sheets and an insulated box are main components of it. High thermal conductivity-based sheet metal is used as absorber metal brazed with tubes or ducts below it. The surface of absorber plate is either selective coated or black with the objective to maximize heat absorption and reducing emission.

The insulated box provides structural support to the whole system and minimizes the conductive heat losses in side and back sides of the FPSC. The transparent cover or glass cover sheets usually is made of low ferrite glass to maximize the transmissivity and allow sunlight to pass through to the absorber while simultaneously minimizing convective heat loss.

Usually, water like thermic fluid or nanofluid is circulated inside the tubes that are carrying the absorbed heat. The steady-state useful energy of the collector can be approximated using the conventional expression [9, 10].

$$Q_u = \dot{m} c_{pw} (T_{fo} - T_{fi}) = F_R A_p [S - U_L (T_{fi} - T_a)] \quad (1)$$

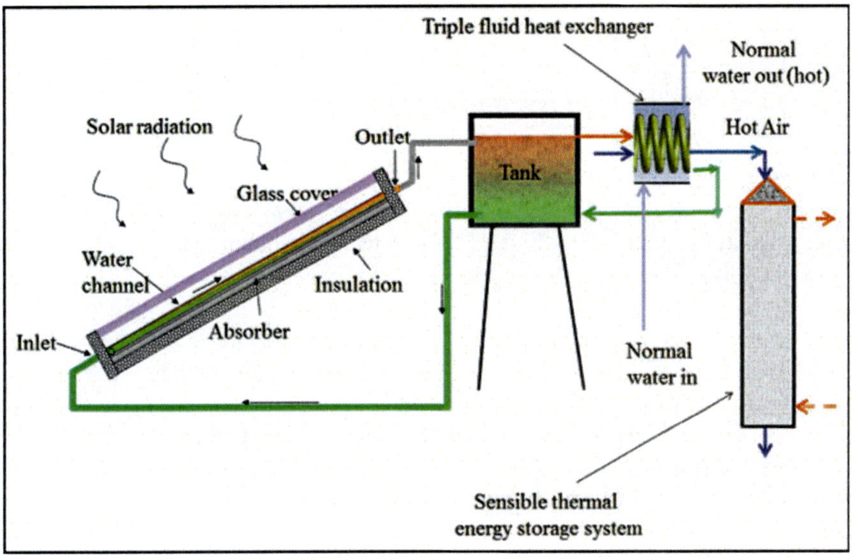

Fig. 1 Pictorial depiction of FPSC with TFHE and STES

Here, by the fluid m, T_{fi}, and T_{fo} are rate of mass flow, inlet and outlet temperature of fluid flowing through the tubes. Heat removal factor is denoted as F_{R} and overall heat transfer coefficient is mentioned as U_{L}. T_{a} can be the ambient temperature. Heat energy absorbed by the absorber plate can be calculated using the expression given in Eq. (2)

$$S = I_{\mathrm{T}} \eta_{\mathrm{opt}} \tag{2}$$

The total radiation on horizontal surface is I_{T} and the optical efficiency of the collector thermal system is η_{optical}.

Assume tilt factors are almost unity. I_{T} and H are related by the expression given in Eq. (3).

$$H = \frac{180}{\pi} \sum_{-\omega_{\mathrm{s}}}^{\omega_{\mathrm{s}}} I_{\mathrm{T}} \mathrm{d}\omega \tag{3}$$

The hour angle is ω here, can be 15° per hour. The instantaneous thermal efficiency of the collector here by the expression given in Eq. (4).

$$\eta_{\mathrm{i}} = \frac{q_{\mathrm{u}}}{A_{\mathrm{C}} I_{\mathrm{T}}} \tag{4}$$

A_{c} is the collector area of the FPSC. It is 10% more than aperture area, A_{p}. The dimensions are mentioned in Table 1.

2.2 *Three Fluid Heat Exchanger (TFHE)*

In the multi-fluid heat exchanger here, a helical tube is inserted between two straight concentric tubes for hot fluid flow from FPSC tank. Normal water flows through the outer tube/drum and air flows in the centrally located straight tube (Fig. 1). During movement of three fluids in three pipes, heat transfer occurs between them. Heat lost from the hot fluid in helical tube is transferred to water in the outer tube/drum and air in the straight tube. By giving up heat to water carried by outer tube and air, water comes back to FPSC for another cycle. Normal water after gaining heat can be used for domestic applications. Hot air from the TFHE is used for sensible thermal energy storage systems as well as for other applications. Hence the purpose of simultaneous heating of water and air is possible with addition of this TFHE with FPSC.

Energy balance for TFHE can be written as [10]

$$\begin{aligned} &\dot{m} c_{\mathrm{pw}} (T_{\mathrm{h1}} - T_{\mathrm{h2}}) \\ &\quad = \dot{m}_n c_{\mathrm{pn}} (T_{\mathrm{n2}} - T_{\mathrm{n1}}) + \dot{m}_{\mathrm{a}} c_{\mathrm{pa}} (T_{\mathrm{a2}} - T_{\mathrm{a1}}) \end{aligned} \tag{5}$$

Table 1 Operating and geometrical data of the FPSC with STSS in present work

Items	Value	Items	Value
FPSC		**TFHE**	
Absorber plate length	2 m	Outer pipe diameter	0.07 m
Absorber plate width	1.13 m	Outer pipe thickness	0.0025 m
Gap between plate and cover	0.025 m	Outer dia. of inner tube	0.0288 m
Plate thermal conductivity	350 W/mK	Thickness of inner tube	0.001 m
Plate thickness	0.00015 m	Outside dia. of helical tube	0.0065 m
Plate absortivity	0.94	Thickness of inner tube	0.001 m
Plate emissivity	0.14		
Tube outer diameter	0.0137 m	**Items**	**Value**
Tube inner diameter	0.0125 m	**STSS**	
Tube pitch below absorber plate	0.113 m	Cylindrical quartz tube height	1.5 m
glass cover's absorptivity	0.88	Cylindrical quartz tube diameter	0.7 m
Glass cover's Emissivity	0.88	Storage media	Alumina
glass cover thickness	0.04 m	Specific heat	0.88 kJ/kg-K
Adhesive thermal diffusivity	0	Density	3000 kg/m^3
Water flow rate	70 kg/s	Pebble size	3 mm
Water Inlet temperature	60 C	Void fraction	0.4
Ambient temperature	25 C	Initial temperature	20 C
Heat transfer coefficient due to wind effect	16.4 W/m^{2}K	Mass flow rate to the unit	0.5 kg/s

$$\left(\frac{\dot{m}c_{\mathrm{pw}}(T_{\mathrm{h1}}-T_{\mathrm{h2}})}{C_{\mathrm{p\,min}}(T_{\mathrm{h1}}-T_{\mathrm{n1}})}\right) = \frac{1-\exp\left(1-\mathrm{NTU}\left(1+C_{\mathrm{p\,min}}/C_{\mathrm{p\,max}}\right)\right)}{1+C_{\mathrm{p\,min}}/C_{\mathrm{p\,max}}} \tag{6}$$

Here T_{h1}, T_{h2}, T_{n1}, T_{n2}, T_{a1} and T_{a2} are temperatures at inlet and outlet of hot water stream, normal water and air streamline, respectively. $\dot{m}$, $\dot{m}_{\mathrm{n}}$, $\dot{m}_{\mathrm{a}}$ are mass flow rate of hot fluid, normal fluid and air, respectively. Corresponding specific heats are c_{pw}, c_{pn} and c_{pa}, respectively. $C_{\mathrm{p\,min}}$, $C_{\mathrm{p\,max}}$ are heat capacity rates. NTU is normal transfer unit for considered TFHE. The detailed analysis of TFHE is mentioned in Mohapatra et al. literature[11].

2.3 Sensible Thermal Storage System (STSS)

Thermal modelling of energy storage media is executed and the process, in brief, is presented here. Packed bed type sensible type of storage medium is considered in the present analysis (Fig. 1). It is least complicated compared with latent or chemical storage medium and it is inexpensive. There are drawbacks of course; sensible heat requires large quantities of materials and volumes. A quartz vessel homogeneously but randomly packed (ideally cylindrical) with alumina particles is used as STSS in the present analysis [6]. Alumina beads are mostly spherical. The dimensions are mentioned in Table 1. During charging time, heat supplied to the STSS is stored and during discharging time, the heat is carried by secondary fluid for any applications, mostly during no sun moments or night applications.

The charging and discharging period are in a transient condition. During the charging stage, energy is stored in the thermal energy storage system until the heat storage capacity of the STSS is fully used. Similarly in discharging period, energy recovery takes place. Alumina pebbles are considered as storage medium and hot air from TFHE is used as hot fluid.

Assume the bed material has infinite thermal conductivity in the radial direction and zero conductivity in the axial flow direction. Also, assume no varying heat transfer coefficient with time and space inside the bed. However, Thermal conductivity can be semi-infinite in the direction of flow. A separate energy balance for bed material and water can be approximated for an element and time as expressed in Eqs. 7 and 8 [10].

$$(1-\varepsilon)\rho_{\mathrm{s}}c_{\mathrm{ps}}\frac{\partial T_{\mathrm{s}}}{\partial t}=h_{\mathrm{v}}(T_{\mathrm{f}}-T_{\mathrm{s}}) \tag{7}$$

$$\varepsilon\rho_f c_{pf}\frac{\partial T_f}{\partial t}+\frac{4\dot{m}c_{pf}}{\pi D^2}\frac{\partial T_f}{\partial x}=h_v\left(T_s-T_f\right) \tag{8}$$

where ε is void fraction, h_{v} is the volumetric heat transfer coefficient, ρ_{s}, ρ_{f} are densities of the solid and fluid and c_{ps}. c_{pf} are respective specific heats. Diameter of the storage unit is denoted as D and $\dot{m}$ is rate of airflow in the storage unit. T_{s}, T_{f} are temperature of storage material (alumina particles in this case) and fluid (air in this case), respectively.

Assuming the storage system is well insulated and neglecting heat losses, two dimensionless parameters time τ and dimensionless distance X can be written as

$$\tau=\frac{h_{\mathrm{v}}t}{\rho_{\mathrm{s}}c_{\mathrm{ps}}(1-\varepsilon)} \tag{9}$$

$$X=\frac{\pi D^2 h_{\mathrm{v}}x}{4\dot{m}c_{\mathrm{pf}}} \tag{10}$$

With simplification, Eqs. (7) and (8) can be written as

$$\frac{\partial T_s}{\partial \tau} = T_f - T_s \tag{11}$$

$$\frac{\partial T_s}{\partial X} = T_s - T_f \tag{12}$$

With assumption of initial temperature of solid medium at T_i, dimensionless temperature distribution may be obtained

$$\frac{T_s - T_i}{T_{fi} - T_i} = 1 - \exp^{-(X+\tau)} \sum_{n=0}^{n=\infty} X^n M_n(X\tau) \tag{13}$$

$$\frac{T_f - T_i}{T_{fi} - T_i} = 1 - \exp^{-(X+\tau)} \sum_{n=1}^{n=\infty} X^n M_n(X\tau) \tag{14}$$

where,

$$M_n(X\tau) = \sum_{k=0}^{k=\infty} \frac{(X\tau)^k}{k!(k+n)!} \tag{15}$$

T_{fi} is initial fluid temperature while entering the storage unit. Volumetric heat transfer coefficient, h_v can be calculated with the correlation given in Eq. 16 [11].

$$\frac{h_v d^2}{k_f} = 1.45(\mathrm{Re_d})^{0.7} \tag{16}$$

where Reynolds number is as follows

$$\mathrm{Re_d} = Gd/\mu_f \tag{17}$$

where G is mass flux and d is equivalent spherical diameter. μ_f is the dynamic viscosity of the working fluid.

3 Results and Discussion

This section highlights key results of the simulation on the modelling equations mentioned in the previous section. The geometrical parameters and assumed operating parameters are mentioned in Table 1.

The solar and atmospheric data assumed for present analysis are based on weather data concerned to Indian climatic conditions. The absorber plate and tube data are considered based on the realistic conditions. Glass cover transmissivity and absorptivity-related optical efficiency value can be taken as 0.85. Based on the thermal modelling mentioned in earlier sections, temperature of the outlet fluid and efficiency are obtained and presented.

The variation of instantaneous efficiency of flat plate solar collectors with respect to the heat energy absorbed by the absorber plate is illustrated in Fig. 2. Here, the efficiency value is 31% initially and increases to 71%. In the meantime, the solar radiation increased from 500 to 900 W/m^2. Moreover, the atmospheric temperature surrounding the collector increases at a lower pace. The outlet fluid temperature for different levels of heat energy absorbed in absorber plate vis-a-vis varying atmospheric temperature levels is as shown in Fig. 3. The trend seems to be almost linear kind for the both cases. The fluid temperature at the outlet increases with variations in solar radiance and atmospheric temperature.

The temperature distribution obtained analytically of three fluids along the length of the TFHE for parallel flow regime is presented in Fig. 4. The mass flow rates in three pipes are varied in such a way so as to attain air temperature 50 °C. The hot air from TFPE is being used to charge the storage medium.

Figure 5 depicts the temperature variation in the storage medium. The bed and the fluid temperature variation have been established. The graph represents the temperature along the storage section after 10 min. It is seen that temperature decreases along the length of the flow for both the storage medium and fluid. Packed bed types mainly alumina pebbles were used for storage of thermal energy. Hot air at 50 °C was used for fluid temperature during charging time where initial temperature was maintained at 20 °C. Future work may be oriented on experimentation of such solar

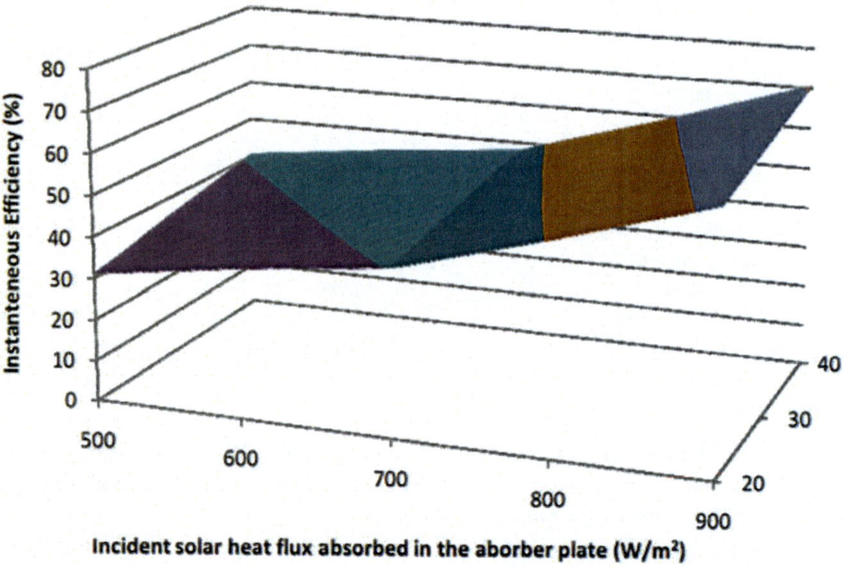

Fig. 2 Efficiency variation vs. Incident solar heat flux versus atmospheric temperature

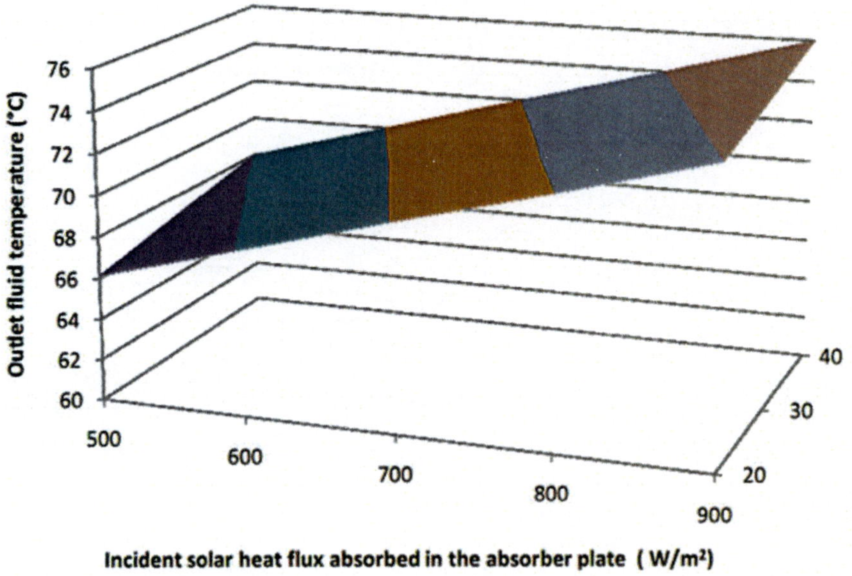

Fig. 3 Outlet fluid temperature variation versus incident solar heat flux versus atmospheric temperature

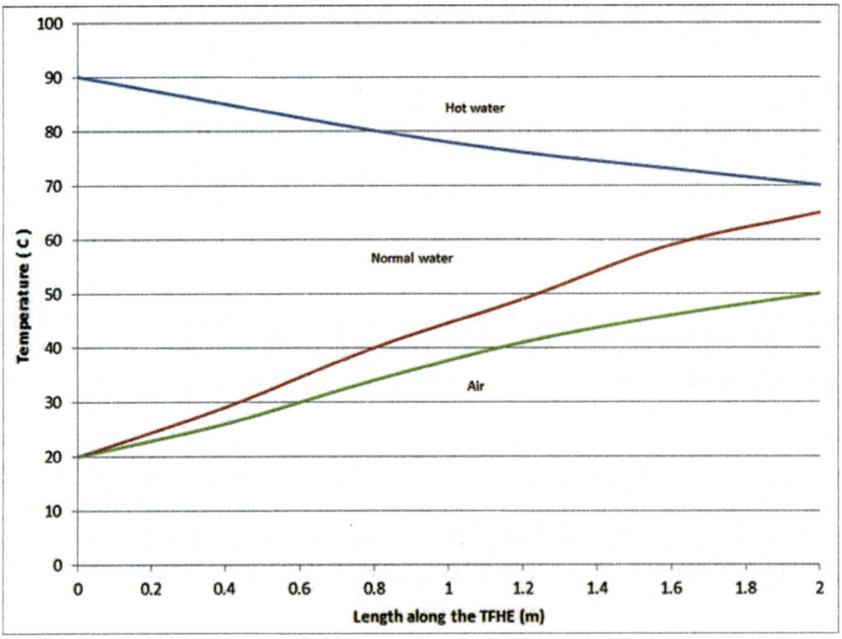

Fig. 4 Variation of temperature in multi-fluid heat exchanger

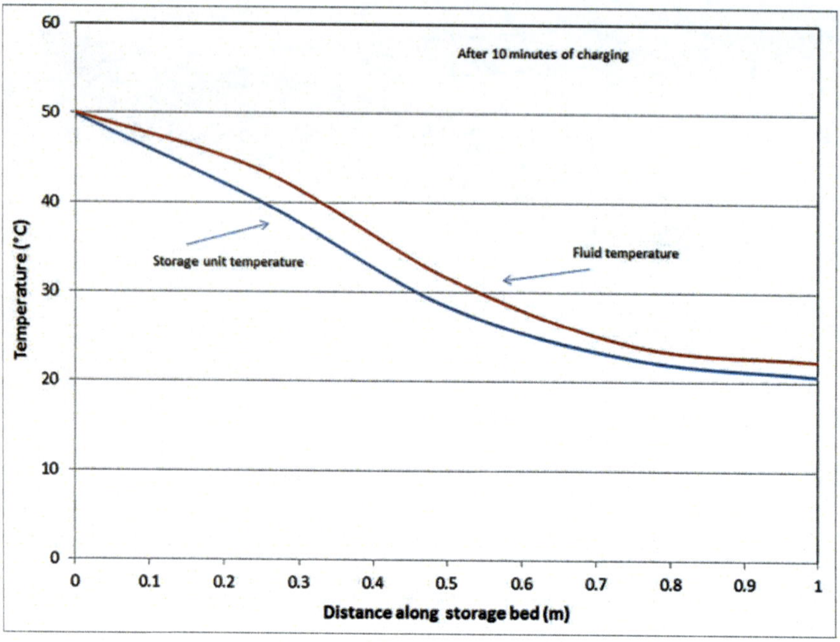

Fig. 5 Fluid temperature variation in the storage medium

thermal systems for simultaneous heating of water and air as well as storage for night time applications.

4 Conclusions

In search of efficient solar energy harvesting, combined effect can make the system most efficient. Analysis by heat transfer modelling of flat plate collector coupled with multi-fluid heat exchanger and heat storage system is interesting. Outlet fluid temperature for different solar energy absorbed by the plate along with the ambient temperature values were recorded accurately. FPSC was integrated with TFHE for attaining hot air and hot water. Hot air from TFPE was used for energy storage in STSS. Major observation in this work is that instantaneous efficiency can rise up to 71% at solar radiation of 900 W/m^2. Increased efficiency value can be attributed to atmospheric temperature rising at a slower pace.

References

1. Kalogirou SA (2004) Solar thermal collectors and applications. Prog Energy Combust Sci 30:231–295
2. Alghoul MA, Sulaiman MY, Azmi BZ, Wahab MAbd (2005) Review of materials for solar thermal collectors. Anti-Corros Methods Mater 52:199–206
3. Ayompe LM, Duffy A (2013) Analysis of the thermal performance of solar water heating system with flat plate collectors in a temperate climate. Appl Therm Eng 58:447–454
4. Farahat S, Sarhaddi F, Ajam H (2009) Exergetic optimization of flat plate solar collectors. Renew Energy 34:1169–1174
5. Karim MA, Hawalder MNA (2004) Development of solar air collectors for drying applications. Energy Convers Manage 45:329–344
6. Edwards J, Bindra H (2017) An experimental study on storing thermal energy in packed beds with saturated steam as heat transfer fluid. Sol Energy 157:456–461
7. Kalaiarasi G, Velraj R, Swami MV (2016) Experimental energy and exergy analysis of a flat plate solar air heater with a new design of integrated sensible heat storage. Energy 111:609–619
8. Solar Radiation Handbook (2008) MNRE and IMD, India
9. Mohanty S, Rout A, Patra PK, Sahoo SS (2017) Soft computing techniques for a solar collector using solar radiation data. Energy Proc 109
10. Sukhatme SP, Nayak JK (2008) Principle of thermal collection and storage, 3rd Edn. Tata McGraw-Hill Publishing Company Limited
11. Mohapatra T, Padhi BN, Sahoo SS (2017) Experimental investigation of convective heat transfer in an inserted coiled tube type three fluid heat exchanger. Appl Therm Eng 117:297–307

Effect of Water Flow Rate and Temperature on Humidification and Cooling of Air in Air Washer

Abhilas Swain, Radha Kanta Sarangi, Satya Prakash Kar, P. Chandrasekhar, and Siba Padarabinda Behera

1 Introduction

In the era of growing civilization and industrialization, the requirement of energy is growing. By looking at the energy consumption in domestic and commercial buildings, one can easily find that large amount of the major portion of the energy is consumed in refrigeration and air-conditioning. Most of the refrigeration and air-conditioning systems are based on the vapor compression systems which consume a lot more energy than the evaporative cooling system. Thus, the researchers are continuously working toward implementing the evaporative cooling-based techniques for air-conditioning purpose.

In low-humidity environmental conditions, evaporating water into the air produces a cooling effect. This phenomenon is implemented in various space conditioning applications. The humidifying and dehumidifying are generally carried out in an equipment called air washer in which the air flow allowed to pass through a packed bed of porous material wetted by spraying of water. The other advantage of the evaporative cooling system is lower initial cost. Therefore, nowadays, it is widely adopted in heating, ventilation, and air-conditioning (HVAC) industry [1, 2] and also for cooling of very small space which is called microclimate cooling [3]. Additionally, there are various advantages of evaporative-based cooling units such as lower energy consumption, no usage of CFCs, reduced environmental pollution and better control of indoor air quality. The evaporative cooling units may be divided into direct and indirect along with air-mediated or water-mediated evaporative systems.

The direct evaporative cooling system is a widely applied cooling system in many industries, such as electric power engineering (as cooling tower) [4–6] and in data center (Ndukaife and Nnanna 2019) [6–8].

A. Swain (✉) · R. K. Sarangi · S. P. Kar · P. Chandrasekhar · S. P. Behera
School of Mechanical Engineering, KIIT Deemed to Be University, Bhubaneswar 751024, India
e-mail: abhilas.swainfme@kiit.ac.in

© The Author(s), under exclusive license to Springer Nature Singapore Pte Ltd. 2023
S. Revankar et al. (eds.), *Recent Advances in Thermofluids and Manufacturing Engineering*, Lecture Notes in Mechanical Engineering,
https://doi.org/10.1007/978-981-19-4388-1_31

Different types of research already been conducted on the various aspects of the air washer. Santos et al. [9] carried out the analytical solution to the simultaneous mass and heat transfer process occurring in air washers and results are validated with the results from literature. A linear approximation was assumed for the variation between the enthalpy and temperature of air saturation curve. The model is validated with a case study discussed in ASHRAE handbook fundamentals [10]. In another study, Santos [11] conducted exergy analysis of the air washer system along with the effect of input temperature and humidity, washer length on the effectiveness and second law efficiency was determined. Table 1 shows some of the important studies on various aspects of the solid desiccant-based air-conditioning units.

Further, more research is going on the application of desiccant materials for achieving dehumidification and integration of renewable energy resources for the regeneration of the desiccant materials [3].

The present article is focused on the water mediated evaporating cooling system in which an air stream is cooled by evaporating sufficient amount of water into the air, and the same humidified air stream is directly supplied for the cooling purpose. Although the entire evaporative cooling system is widely studied, a focused study on the humidification is not explored in detail for better understanding.

Table 1 Outcomes from certain studies

Sl. No.	Article and year	Subject of study	Outcomes
1	Yang et al. (2019)	Review of different types of evaporative cooling systems	
3	Jing et al. (2020)	Multi-evaporator unit (both numerical and experimental)	The effect of the volume of air in the different stages is discussed
4	Chu et al. (2021)	Experimental investigation of a new regenerative dew point indirect evaporative cooler that uses a polymer plant fiber composite material and a pure counter flow heat exchange structure	Cooling performance is improved Wet bulb effectiveness is improved
5	Yan et al. (2020)	Sea water and fresh water are compared	Cooling performance is better for fresh water Empirical correlation is developed
6	Santos et al. (2011)	Analytical modeling of the heat and mass transfer phenomena in the desiccant material	Effect of various parameters is studied
7	Santos et al. (2013)	Energy and exergy analysis of desiccant-based air-conditioning systems	Optimal conditions for thermal comfort

2 Experimental Setup

Figure 1 shows the schematic of the air washer setup for the present investigation. Starting from the bottom, a reservoir is there at the bottom of the setup for storing the water needed for humidifying air. The tank is fitted with a heater associated with a temperature controller. By setting the temperature on the control panel, the heater is switched on or off perpetually to maintain that temperature. Importantly, a pump is present to allow the water to flow through the entire system with a flow control valve and a bypass valve. The flow rate is measured through a metallic rotameter. Then, the water is sprayed on to packed bed of porous material having cross channels for flow of air. During the flow of air through the cross channels, the air gets humidified and the specific humidity increases. The water collected at the bottom of the porous material is drained to the reservoir. After the air gets humidified, if required, the temperature may be increased by switching on a heater present in the delivery duct. The heater in the delivery duct is not used in the present investigation. A butterfly valve kind of arrangement is there at the exit to control the air flow rate. The blower is present in between the heater and the packed bed. The packed bed is made from the CELdek 7060 which is very much popular.

The water flow rate can be varied by changing the positions of the flow control and bypass isolation valves. The flow rate is measured by a metallic rotameter having range of 25–250 L per hour (LPH). The water temperature can be set by the digital controller on the panel. The air flow rate is controlled by the butterfly valve present in the delivery duct. The air temperatures (both DBT and WBT) and the humidity level are measured by a hygrometer (Fig. 2).

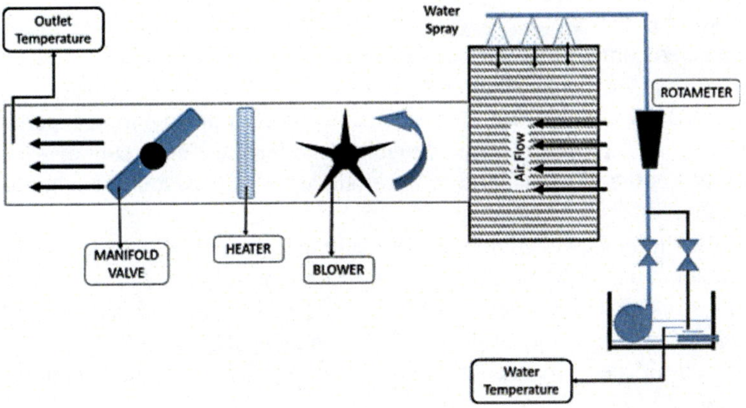

Fig. 1 Schematic of the experimental setup

Fig. 2 Photograph of the experimental Setup

3 Results and Discussions

In the present investigation, only humidification is focused. Therefore, the heater is switched off during these experiments. All the experiments are conducted on one day for keeping the inlet air conditions close to each other. Moreover, the conditions are also recorded as per the measurement through the hygrometer. The dry bulb temperature and the wet bulb temperature are recorded and plotted against the various parameters. Table 2 shows the range of parameters considered during experimentation. The parameters are measured variables, and thus, the uncertainties are due to the accuracy of the instruments. The uncertainties associated are also mentioned in Table 2 (Fig. 3).

Figure 3 shows the variation of the dry bulb temperature with respect to the flow rate of water at a fixed value of water temperature of 40 °C and for three values of the air flow rate. It can be observed that the DBT of the outlet air is increasing with the increase in the flow rate of the water. This may be due to the fact that at higher flow rate, the air gets more in contact with the water which is at a higher temperature

Table 2 Range of operating parameters

Parameters	Values	Uncertainty
Air flow rate	6.252 m/s, 7.35 m/s, 9.61 m/s	±0.5 m/s
Water temperature	40 °C and 60 °C	±1° C
Water flow rates	50, 75, 100 LPH	±10 LPH

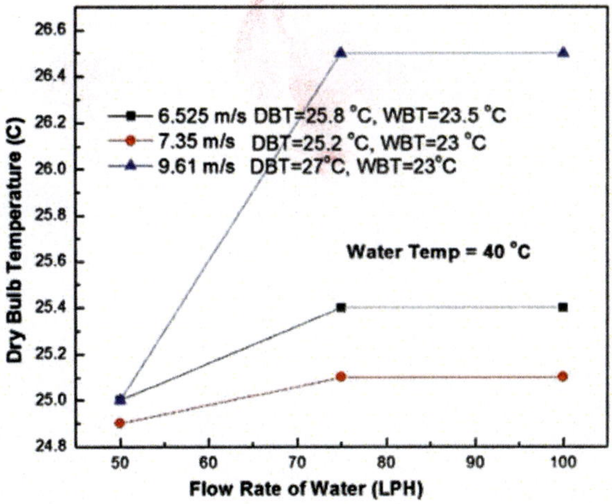

Fig. 3 Variation of the DBT with respect to flow rate of water for water temperature = 40 °C

of 40 °C with respect to the inlet DBT of around 25 °C. The inlet conditions for the air flow rate of 9.61 m/s are even higher and also at higher air flow rate, the heat transfer coefficient will be higher. Thus, the outlet air conditions have higher dry bulb temperature. Due to the higher water temperature than the supplied air, its increase in the temperature is achieved. However, the time available for the interaction between the air and the water sprayed is less as the flow rate increases which limit the change in temperature.

Similarly, Fig. 4 shows the variation of the wet bulb temperature of the outlet air with respect to the water flow rate for the same three cases considered in Fig. 4. Here also, it can be observed that the wet bulb temperature is also increasing with a water flow rate for the all the three air flow and inlet conditions. This, also reflects the increasing trend of the specific humidity value in the outlet air stream.

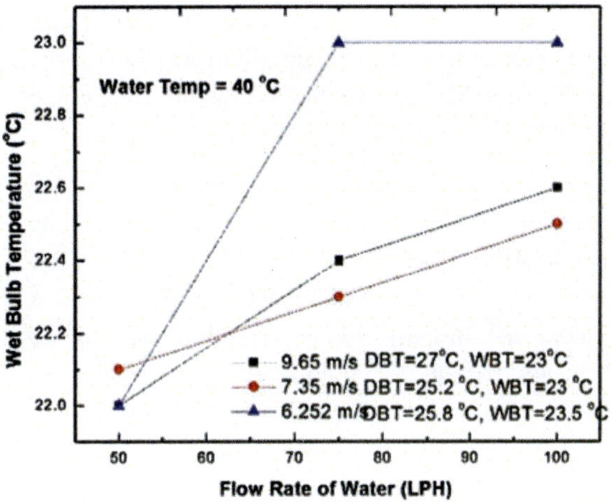

Fig. 4 Variation of the WBT with respect to flow rate of water for water temperature = 40 °C

Fig. 5 Variation of DBT with flow rate for water temperature of 60 °C

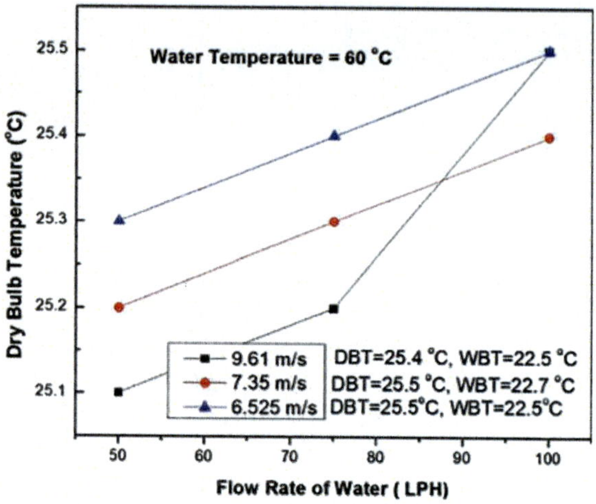

The evaporative cooling is the process in which the reduction in the temperature is achieved by taking away heat in the form of latent heat of evaporation. The water particle sprayed into the conditioned air stream is converted to water vapor. This process simultaneously increases the specific humidity of the air along with cooling effect. Here, in the present study, additional cooling is not considered. In accordance with the choice of cooling media, the evaporative cooling could be as air mediate and water mediated cooling.

Figure 5 shows the variation of the DBT with respect to water flow rate and for the above mentioned conditions of three air flow rates and different inlet conditions. Similar variations are also observed in water temperature of 60 °C. The DBT and WBT increase with increase in the flow rate. The variation of the WBT can also be observed in Fig. 6.

With the increase in the water temperature, the water vapor pressure of the air increases, and hence, the difference in the moisture level is lowered. Thus, the heat transfer and the mass transfer are lowered as the temperature of the water increases. Again, with the increase in the flow rate, the temperature of the outlet air increases because of more contact with the water.

4 Conclusion

The present investigation is carried out focusing on the effect of water temperature and flow rate on the cooling and humidification operation in the air washer setup under different operating conditions of spray water flow rate and its temperature. The inlet air conditions are measured through hygrometer and temperature and humidity sensor.

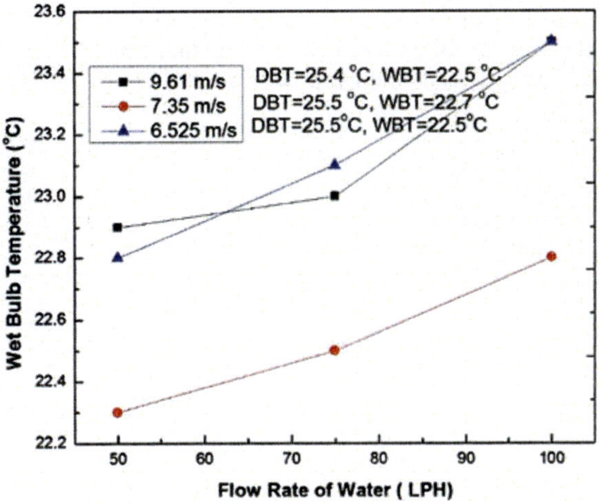

Fig. 6 Variation of WBT with flow rate for water temperature of 60 °C

It can be concluded from the study that the dry bulb temperature of the supplied air increases with the increase in the flow rate for the water temperature of 40, 50 and 60 °C. With the increase in the flow rate, more sensible heat transfer occurs and the dry bulb temperature increases. Similarly, the wet bulb temperature of the air supplied also increases. Further experiments are needed to be carried out in the air washer setup for dehumidification, heating and combination of these better understanding and industrial application.

References

1. ASHRAE (2007) ANSI/ASHRAE stand 62.1-2007, ventilation for acceptable indoor air quality. American Society of Heating, Refrigerating and Air-Conditioning Engineers, Inc., Atlanta
2. ASHRAE (2004) ANSI/ASHRAE standard 55-2004, thermal environmental conditions for human occupancy. American Society of Heating, Refrigerating and Air Conditioning Engineers, Inc., Altantaf
3. Yang Y, Cui G, Lan CQ (2019) Developments in evaporative cooling and enhanced evaporative cooling—A review. Renew Sustain Energy Rev 113:109230
4. Kumar K et al (20.7) Standalone photovoltaic (PV) module outdoor testing facility for UAE climate. Submitted to CSEM-UAE Innovation Centre
5. He S, Guan Z, Gurgenci H, Hooman K, Lu Y, Alkhedhair AM (2015) Experimental study of the application of two trickle media for inlet air pre-cooling of natural draft dry cooling towers. Energy Convers Manag 89:644–654
6. Dong P, Kaiser AS, Guan Z, Li X, Gurgenci H, Hooman K (2019) A novel method to predict the transient start-up time for natural draft dry cooling towers in dispatchable power plants. Int J Heat Mass Transf 145:118794
7. Yan M, He S, Gaoa M, Xu M, Miao J, Huang X, Hoomanc K (2021) Comparative study on the cooling performance of evaporative cooling systems using seawater and freshwater. Int J Refrig 121:23–32

8. Ndukaife TA, Nnanna AA (2018). Optimization of water consumption in hybrid evaporative cooling air conditioning systems for data center cooling applications.Heat Transfer Eng 40:559–573
9. Santos JC, Medeiros JM, dos Santos JC, Gurgel JM, Marcondes F (2011) Analytical solution for the simultaneous heat and mass transfer problem in air washers. Int J Refrig 34:353–361
10. ASHRAE Handbook Fundamentals, SI ed (1997) American Society of Heating, Refrigerating and Air Conditioning Engineers
11. Santos JC, Barros GDT, Gurgel JM, Marcondes F (2013) Energy and exergy analysis applied to the evaporative cooling process in air washers. Int J Refrig 36:1154–1161

Modelling and Performance Simulation of a Solar Powered e-rickshaw Retrofit in the North East Region of India

Richik Ghosh Thakur, Daya Shankar, Beauty Pandey, Santipada GonChaudhuri, Ajoyjit Ghosh, and Debdutta Choudhury

1 Introduction

The EV sector is snowballing in India. Public transportation (PT) and private transportation are also included in the EV industry (PV). However, the e-rickshaw is an example of a frequently overlooked means of urban transit. According to a recent analysis [5], e-rickshaws might be considered 'low hanging fruit' in terms of EV adoption due to their cheap life cycle costs. It happens due to a better understanding of the technology and its implementations. As a result, a greater understanding of this mode may lead to more effective technology deployment.

E-rickshaws provide a zero-emission modal choice for first/last-mile connectivity in cities. For commuters, this three-wheeled electric vehicle is expected to close the value chain of public transportation service gaps. These have a net power of less than 4000 W, a top speed of 25 kmph, and space for four passengers plus the driver. Commuters and operators alike are enthusiastic about e-rickshaws because they are demand-responsive, pleasant, inexpensive, flexible, and handy. However, as the number of e-rickshaws grows uncontrollably, authorities are finding it difficult to regulate them [16].

Environment-friendly e-rickshaws utilizing grid charging were introduced in Delhi in 2010. More than 1,00,000 e-rickshaws have been acquired in Delhi alone in less than five years, and they have become the principal source of income for more than 3,00,000 people in India [21].

R. Ghosh Thakur (✉) · S. GonChaudhuri · A. Ghosh
Department of Renewable Energy, NB Institute for Rural Technology, Kolkata, India
e-mail: mailer.richik@gmail.com

D. Shankar · B. Pandey
School of Technology, Woxsen University, Hyderabad, India

D. Choudhury
School of Business, Woxsen University, Hyderabad, India

© The Author(s), under exclusive license to Springer Nature Singapore Pte Ltd. 2023

S. Revankar et al. (eds.), *Recent Advances in Thermofluids and Manufacturing Engineering*, Lecture Notes in Mechanical Engineering,
https://doi.org/10.1007/978-981-19-4388-1_32

Since their introduction, Electric Auto Rickshaws have become the most regularly visible and used public vehicles in several parts of India. They are environmentally friendly and ideal for short-distance travel. As per e-rickshaw drivers, a complete charge takes roughly 4 to 5 h. It lasts for 7 to 8 h after being completely charged. A one-time charge will cost around Rs. 50/-. Charging is simple, does not cause any issues, and does not require any technical or professional qualifications or training. Whether a student, a homemaker, or an operator, anyone in the family can quickly charge an e-rickshaw. If the Electric Auto Rickshaw is run regularly by a trained driver for 6–7 months, the cost of the rickshaw is recovered soon. The benefits begin to arrive after seven months. This environmentally efficient car has a bright future as India aims to reduce fuel imports by 10% by 2022 [26].

The only disadvantage of e-rickshaws is the charging process based on electricity taken from the grid [24]. Grid electricity involves carbon emission into the environment as electricity is majorly produced by thermal power plants in India [8]. Compared with conventional fossil fuel powered vehicles, e-rickshaws are not far behind in emission rates. However, due to e-rickshaws being electrically powered, renewable energy is the most probable solution to mitigate the significant carbon footprint of e-rickshaws [9, 19, 23]. RES (Renewable Energy Sources) like Solar, Hydro, Wind, Biomass are very popular and feasible generation sources [12]. These sources can be utilized effectively to reduce the carbon footprint of the e-rickshaws. Out of these renewable sources, Solar PV comes out to be the most feasible solution for integration of RE (Renewable Energy) with an e-rickshaw [18].

In the current study, the operational feasibility of the standard e-rickshaw integrated with Solar PV has been simulated in the North-Eastern States of India. A payback period estimation due to the addition of Solar PV in an existing e-rickshaw has been simulated. The performance simulation has been done in PVsyst. A standard and widely available Solar PV panel have been considered for the simulation. After simulation of the performance of the retrofit, the results have been utilized to calculate the 'energy dynamics variables' of the work, including solar energy utilized, the runtime of the e-vehicle, and region-wise variation. From these results, the carbon footprint of the intervention is calculated to demonstrate its low carbon footprint as opposed to conventional e-rickshaws through the intervention of Solar PV [25].

2 Methodology

To simulate the performance of the e-rickshaw upon the addition of Solar PV, PVsyst, a popular software for simulation of PV power plants, pumping systems, and standalone systems, has been used. The software was used to import the meteorological data, and to import the parameters of the Solar PV module, controller, and battery. The software has a substantial database of PV components essential for PV systems modelling. Simulation of irradiation and shadow analysis was done Considering the

Perez-Ineichen model [22]. Figure 1, provided below, depicts the simulation interface done in PVsyst software. The 3D interface and sun path diagram are shown in the figure.

For practical system simulation, the area considered in this evaluation is 2 m^2, the typical roof area of a generic Electric Rikshaw where a Solar PV panel can be mounted. The optimization of PV capacity and control system selection was done based on this roof area. Based on this parameter, the 440 Wp generic half-cut cell monocrystalline solar module was chosen because of the market's most significant yield per unit area and availability. An albedo factor of 0.2 was taken as the PV module mounted on the FRP roof. The platform PVGIS API TMY was

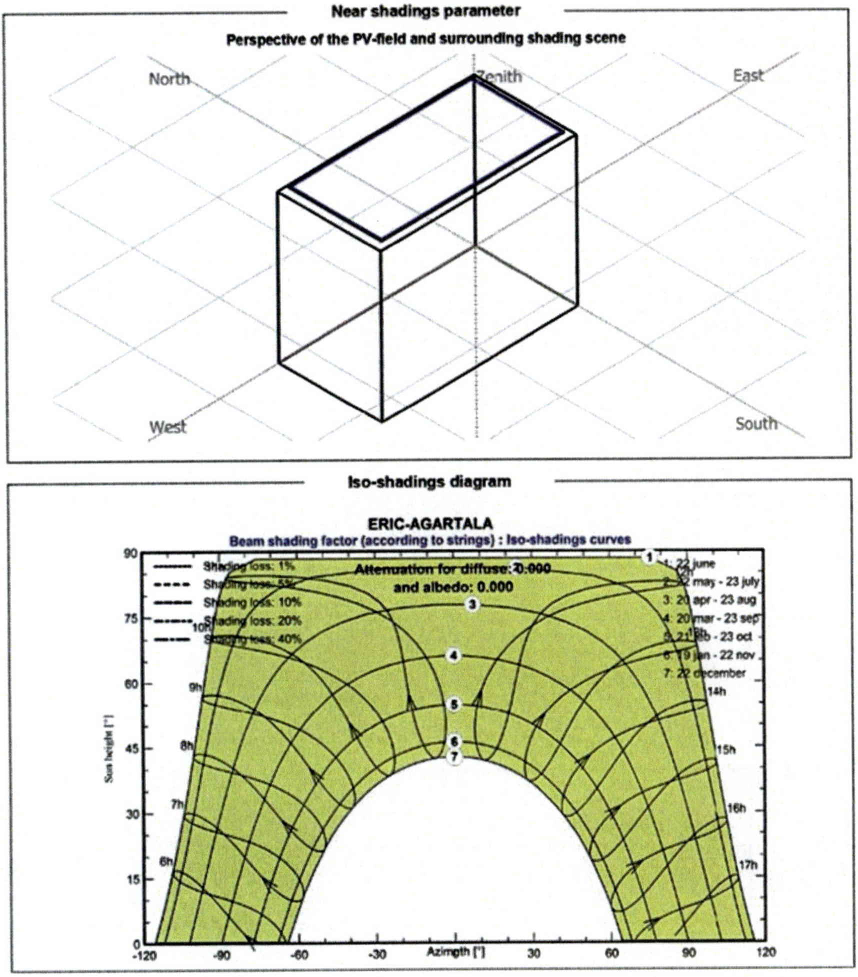

Fig. 1 Simulation interface

used [6]. This platform provides the latest captured values of meteorological data like Solar Global Horizontal Irradiance, Diffuse Horizontal irradiance, Temperature, wind velocity, and relative humidity. The data was utilized in the model to calculate SPV generation monthly and annually [27].

The simulation parameters had to be set up to obtain the PV generation results. After defining and locating each site in the North-Eastern region where the simulation will be done, the parameters such as module selection, battery selection were done. The SPV module has been selected as per the optimized model. The conventional battery system of 125 Ah, 48 V comprising 4 × 12V, 125 Ah lead-acid low maintenance batteries in series used in e-rickshaws has been selected [15]. The performance characteristics of the PV module obtained from PVsyst are provided in Fig. 2. The PV module, mounting structure, battery system, and controller have been kept identical for the entire simulation phase.

Apart from the PV module and the battery, an equally important component is the power controller. A generic MPPT controller with a built-in DC-DC controller has been chosen [17]. The MPPT controller follows the 'Perturb and Observe' MPPT algorithm to allow maximum utilization of PV power [20]. The PV module has 'Vmpp' to the tune of 41.6 V. For this purpose, a buck-boost DC-DC converter has been integrated to set the desired charging voltage automatically by the MPPT.

The MPPT controller has been used at the nominal output voltage set at 48 V with minimum MPP voltage at 20 V, maximum at 58 V, and a power threshold of 3.8 W. The nominal efficiency for the DC-DC converter lies between 95 and 97% around STC (Standard Testing Conditions). This value is also based on the type and

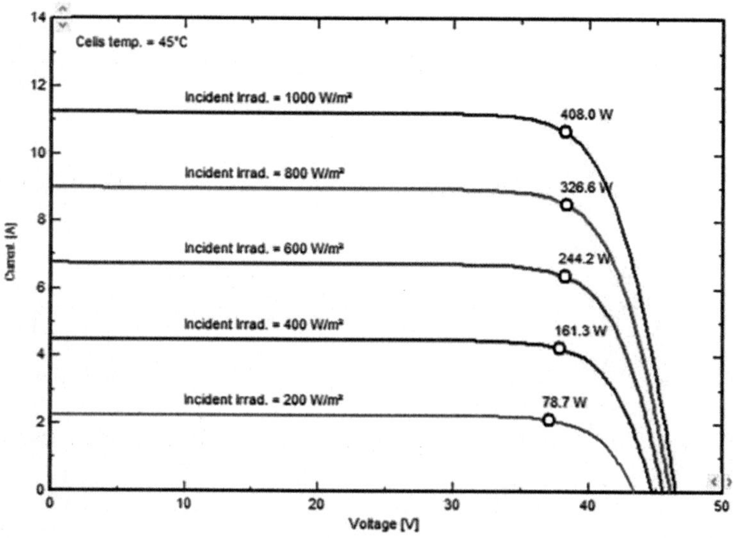

Fig. 2 I-V curve of the solar PV module at 45 °C PV cell temperature

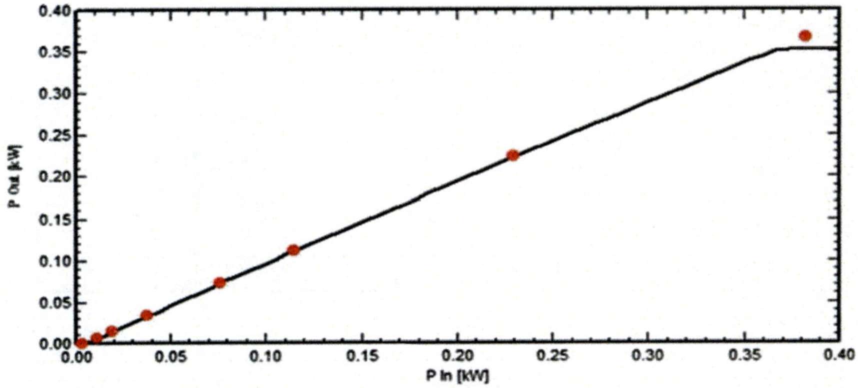

Fig. 3 Performance of the converter at a different input power

magnitude of generation, solar insolation, and other factors [1]. The graph in Fig. 3 depicts the converter's performance at different input power.

From Fig. 3, it is clear that the PV generation system can generate a maximum power of about 355 W with the help of the MPPT controller and DC-DC converter. The area available for installing the PV module is about 2 m^2 on the roof of the e-vehicle. A generic mounting structure has been included in the simulation. The mounting structure is made of galvanized iron to prevent corrosion, and aerodynamic loss is also kept to a minimum. Detailed shadow analysis has been omitted in this work because of its dependency upon the vehicle's route and area of operation. However, a constant shadow loss was considered assuming that the vehicle would not constantly be subjected to direct sunlight. A loss of 11% is regarded as the tilt angle loss; 30% Azimuth and shadow loss are also considered in the study. The power generation from the SPV module is plotted against the sun path diagram to compute the total cumulative power generation. After the generation values have been simulated, the load profiling for the vehicle has been done. An average e-rickshaw has a BLDC motor and lights as the baseload [14]. Other optional loads present in a few vehicle models are insignificant and neglected. A load of 1 kW peak is considered for the vehicle, and the hourly load distribution has also been provided.

As per Fig. 4, the motor is considered to operate at a rated capacity of 1 kW for 5 h each day. An average of 500 W per half-hour has been considered for simplifying the simulation.

A total of 5 h of operation of the vehicle has been considered [10] for the present work. The load profile is based on the vehicle's battery capacity considering 80% Depth of Discharge (DoD). Based on these factors, the PV generation, solar fraction, and PR have been computed in 'PV-Syst'.

The nine selected cities in North-Eastern India are based on population and 'TOTO' as an effective means of local short-distance transport. Assam, Guwahati, and Silchar are taken because they fall in separate climatic and insolation zones.

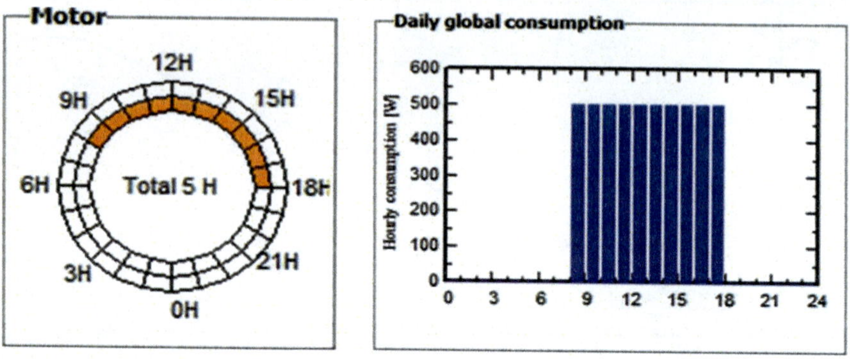

Fig. 4 Motor load and hourly electricity consumption of e-rickshaw

Other cities are categorized based on population and the availability of e-rickshaw as local transport.

1. Guwahati, Assam
2. Agartala, Tripura
3. Imphal, Manipur
4. Dimapur, Nagaland
5. Shillong, Meghalaya
6. Aizawl, Mizoram
7. Silchar, Assam
8. Gangtok, Sikkim
9. Itanagar, Arunachal Pradesh.

These 9 cities have been selected based on the operation of e-rickshaws in these regions. The availability of Solar PV modules and purchase rates of components have been considered in the work. An average rate of purchase of materials valid for the entire North-East Region has been considered for ease of calculation.

2.1 *Components of Calculation*

After simulation of the generation and load catering abilities of the Solar PV retrofit in the selected regions in NE India, some parameters need to be established and computed to access the techno-economic feasibility of the intervention. Thus, performance parameters like energy utilization and extension of operational duration of the e-rickshaw due to Solar Integration have been computed. The payback period has been calculated for the economic evaluation, and the net carbon reduction has been estimated for environmental assessment. The parameters established and defined for the current study has been described in the section that follows.

2.2 Energy Dynamics

1. PV energy utilized (P_u): PV energy utilized is the net energy obtained by the Solar PV setup after conversion, storage, and transmission losses which is available to the user for utilization in extending the duration of operation of the e-vehicle in a month. The simulation obtains this in the PVsyst interface.
2. Energy Required (E_{Req}): This is the total energy required by the motor and associated electronics in the e-vehicle during the entire operation time, taken as 5 h @ 1 kW in this study. It is obtained as follows:

$$E_{\text{Req}} = P_{\text{ar}} * T_d * D \quad (1)$$

where,
E_{Req} = energy required to operate the e-vehicle in a month
P_{ar} = Average Rated power of the e-vehicle while in operation
T_d = duration of operation per day
D = No. of days in a month.

3. Battery Energy (E_{Batt}): It is the maximum dispatchable energy that is stored in the battery considering 80% Depth of Discharge (DoD).
4. Duration of operation (h): The duration (total no. of hours) of which the e-vehicle operates while on battery in a month. It is calculated by the following equation:

$$h = \frac{E_{\text{Batt}}}{P_{\text{ar}}} \quad (2)$$

5. Duration of operation per day (T_{db}): is the no. of hours the e-vehicle operates per day from the energy stored in the battery. It is calculated by the equation:

$$T_{\text{db}} = \frac{h}{D} = \frac{E_{\text{Batt}}}{P_{\text{ar}} * D} \quad (3)$$

6. Extended duration due to Solar PD (T_s): This is the extension of duration (total no. of hours) of the operation of the e-vehicle only due to the incorporation of a Solar PV charging system per day of the month. It is calculated by the formula:

$$T_s = \frac{P_u}{D} \quad (4)$$

7. Total duration per day (T_{tot}): This is the total duration of the operation of the e-vehicle after incorporating Solar PV charging on a day of the month. It is calculated by:

$$T_{\text{tot}} = T_{\text{db}} + T_s = \frac{E_{\text{Batt}}}{P_{\text{ar}} * D} + \frac{P_u}{D} \quad (5)$$

8. Total Energy (E_{tot}): This is the total energy that can be utilized after incorporating Solar PV charging in the e-vehicles. It is determined by the expression:

$$E_{tot} = P_u + E_{Batt} \tag{6}$$

9. Additional gained energy (E_{ex}): It is the excess Energy or additional Energy when compared to the daily requirement of the e-vehicle. It is determined as:

$$E_{ex} = E_{Req} - E_{tot} = (P_{ar} * T_d * D) - (P_u + E_{Batt}) \tag{7}$$

2.3 *Economic Evaluation*

1. Duration Extension per year (D_e): It is the extension of the duration of operation of the e-vehicle for an entire year calculated by the summation of extended hours for each month. It is expressed as:

$$D_e = \sum_{M=1}^{12} P_u \tag{8}$$

2. Per hr average earnings (R_h): This is the earnings of the operator (driver) of the e-vehicle per hour.
3. Installation cost (R_{IC}): This is the cost of installing the solar setup in the e-vehicle as per the latest MNRE (Ministry of New and Renewable Energy) rates.
4. Earnings from Solar PY (R_{PV}): This is the earnings from the contribution of additional duration to the operation of the e-vehicle. It is calculated as:

$$R_{PV} = D_e * R_h \tag{9}$$

5. Maintenance Cost (R_m): This is the average yearly maintenance cost of the e-vehicle due to the retrofit, which includes a thorough cleaning of the solar panels, hydration of the battery, and maintenance of the Solar CCU and structure.
6. Payback in months (p_m): This is the simplified payback period calculation (in months) which is expressed as:

$$p_m = \frac{\frac{R_{IC}}{R_{PV} - R_m}}{12} = \frac{R_{IC}}{12 * (R_{PV} - R_m)} \tag{10}$$

2.4 Constants

Constants have been incorporated in the simplified cost–benefit calculation of Solar Integration of e-rickshaws. The no. of days in a month has been taken as per calendar.

Emission from Grid Charging of conventional e-rickshaws: Charging is primarily done through the grid for conventional electric rickshaws. A significant portion of grid power in India as of 2021 comes from thermal power plants which use coal as fuel [4]. Thus, for charging e-rickshaws, coal is burnt, and CO_2 is emitted and released into the atmosphere. It has been considered that the emission of CO_2 from the burning of coal in thermal powerplants is to the tune of 1.04 kg per kWh of Electricity Generated [13].

Now, for a conventional e-rickshaw, operational at any 9 of the selected cities, the average distance covered by the e-rickshaw per day is about 57 Kms [ref]. Considering the mechanical and electrical efficiencies of the system, the average power consumed by an e-rickshaw is about 6.5kWh per day. 6.5kWh per day accounts for the emission of about 6.7kgs of CO_2 by a single e-rickshaw in a single day without considering transmission line and charging losses. It is about 2 tons of CO_2 emissions from a single vehicle [2].

From studies conducted in analyzing entrepreneurial success am impact assessment of e-rickshaws [11], an average e-rickshaw driver earns about Rs.70/- per hour, and the average per day earnings of an e-rickshaw driver is about Rs.550/ to Rs.800/-.

2.5 Source of Meteorological Data

For simulation of Solar PV performance of e-rickshaws in PVsyst, meteorological data was required to simulate the results. Solar PV power output (PV Current) depends upon the intensity of Solar Radiation falling into the modules and on temperature, humidity, and relative wind velocity. These parameters were required for proper simulation of the PV Output power. TMY Generator, a Project of EU Science hub (Eu, no date), was utilized to obtain the meteorological data. The radiation database used is retrieved from PVGIS-SARAH, PVGIS-NSRB or PVGIS-ERA5. These databases collect data from the EUMETSAT geostationary satellites [3].

The simulation has been done for the selected eight North-Eastern (NE) states with nine regions. Assam, a geographically more extensive state, occupies two distinct climatic zones. Thus, two areas from the state of Assam have been chosen for the simulation. As per the objective of the study, Equations 1–10 has been specifically formulated for calculating and realizing the several performance indicators like Average Extension of duration of operation for the NE region, duration of extension in different states in the NE region, payback period due to the retrofit and annual reduction of CO_2 due to the retrofit. The individual results are discussed below.

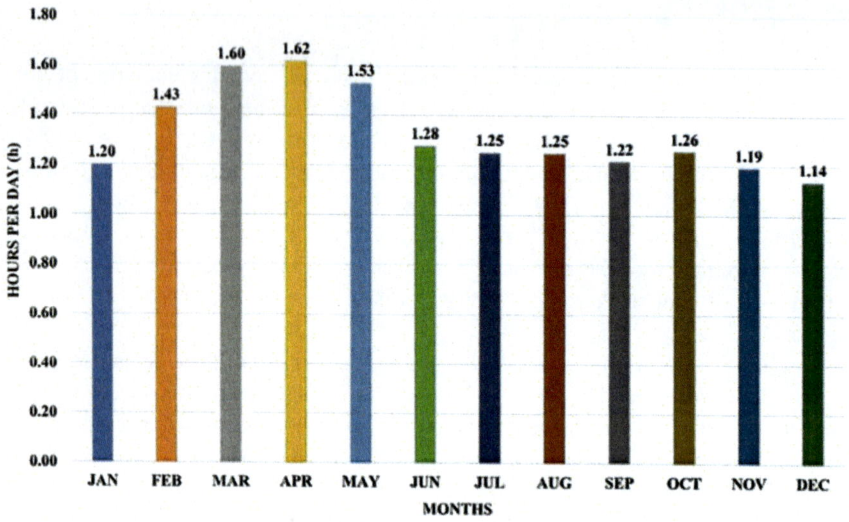

Fig. 5 Average extension of hourly duration in NE region

3 Results and Discussions

3.1 Average Extension of the Duration of Operation for the NE Region

The simulation results of the selected eight states in the NE region have been studied. As the individual performance of the solar PV retrofit for the e-rickshaws is different for the selected regions, a performance average of all these regions has been taken to account to find out the monthly variation of the total hourly extension per day due to the integration of the solar PV retrofit. The maximum extension of the duration of 1 h 37 min is observed in April. Similar values are observed in March due to the highest solar radiation during these two months, as shown in Fig. 5. The lowest value of extension is observed to be 1 h and 8 min in December. Overall, it can be inferred that due to the retrofit, a minimum of 1 h of extra operation time can be expected from the e-rickshaw.

3.2 Extension of Duration in Different States of NE Region

The simulation results of the nine selected regions indicate that all the regions are distinctly different in performance regarding the operation of the solar-powered e-rickshaw. A graph representing the annual extension hours of the nine selected regions

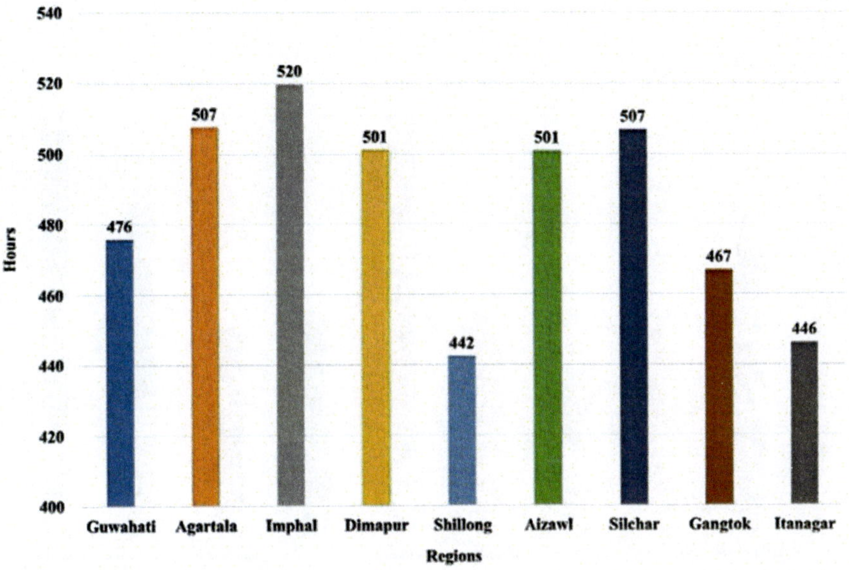

Fig. 6 Annual hours of extension in selected regions

has been plotted in Fig. 6. We observe the lowest performance in Shillong, having 442 h, and maximum extension in Imphal, about 520 h.

3.3 Payback Period of Installation (Retrofit)

Taking the cost of materials for installing the retrofit equipment for the solar-powered e-rickshaw, the payback period of the intervention has been calculated. The payback period calculation also includes the maintenance charges/costs for the e-rickshaw. For ease of analysis, inflation/depreciation rates of assets have not been considered. The no. of months in which the payback will be done is plotted concerning the selected regions in Fig. 7. It is observed that Shillong, which had the least annual extension hours, has the most significant payback period. It is expected as the revenue generated from the retrofit installation will be lesser than the other regions. Imphal shows the fastest payback in about eight months, and Shillong the longest in under ten months. Thus, the payback period should lie within 8 and 10 months to install this retrofit in the NE region.

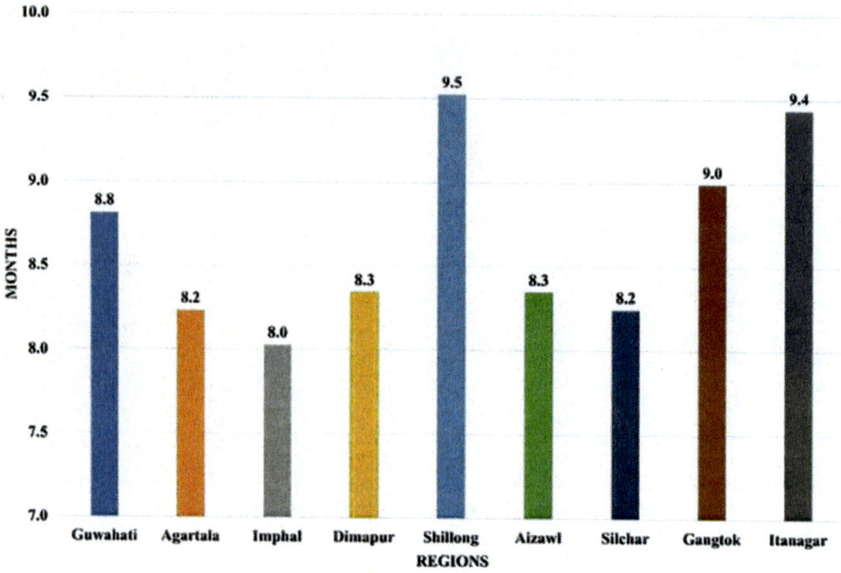

Fig. 7 Payback period of the retrofit amongst selected regions

3.4 *Reduction of CO_2 Per Year*

E-rickshaws are conventionally charged from the utility grid. The substantial amount of power in the electricity grid in the NE region is supplied from coal-fired thermal power generating stations and the rest from mainly Hydel power. Thus, conventional charging of the e-rickshaws involves the burning of coal and, as a result, leads to carbon emissions. Upon charging with Solar PV power, recurring carbon emissions can be stymied. In the present study, it is considered that the extra backup of power supplied by the Solar PV is emission-free clean electric power. If this power were to be provided from the grid, it would involve additional carbon emissions. Based upon the selected regions in NE India, the annual reduction of emissions due to the retrofit has been calculated and represented in Fig. 8. It is observed that a solar-powered e-rickshaw will lead to a decrease of about 0.48 tonnes of CO_2 in Shillong annually, which is the maximum reduction observed out of all the selected regions. In contrast, a minor reduction of 0.41 tonnes is observed in Shillong.

The overview of the results is provided in Table 1. This overview is vital for researchers, engineers, and e-rickshaw manufacturers to consider integrating the retrofit at the production level and studying the market potential.

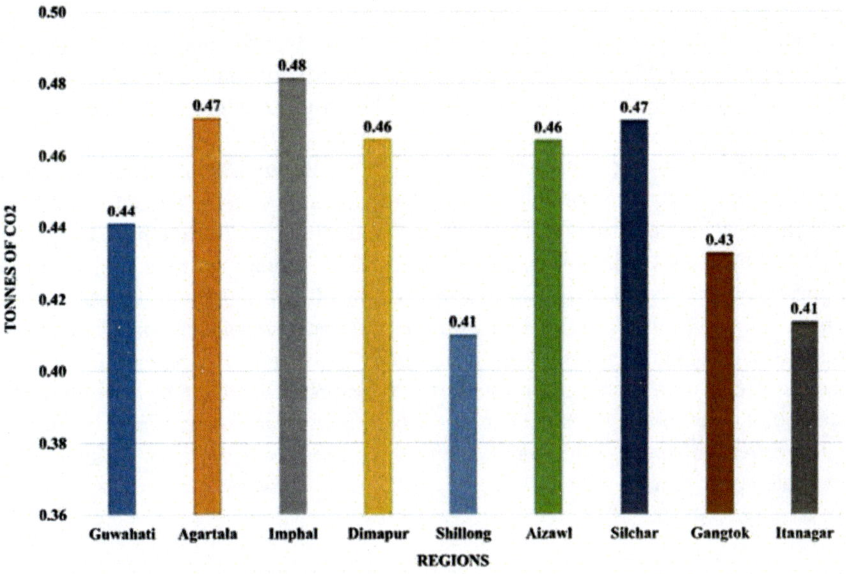

Fig. 8 Region-wise annual carbon reduction

Table 1 Results overview

Sl. No	State	Duration of extension (h)	Payback period (months)	Annual reduction of CO_2 (tonnes)
1	Guwahati	476	8.8	0.44
2	Agartala	507	8.2	0.47
3	Imphal	520	8.0	0.48
4	Dimapur	501	8.3	0.46
5	Shillong	442	9.5	0.41
6	Aizawl	501	8.3	0.46
7	Silchar	507	8.2	0.47
8	Gangtok	467	9.0	0.43
9	Itanagar	446	9.4	0.41

4 Conclusion

In the present work involving the study and simulation of the performance prediction of Solar PV Integrated e-rickshaws in selected areas of the NE region, the team has arrived at meaningful conclusions. An average of 1 h operation time is expected in the NE region due to the retrofit installation in an e-rickshaw. The use of standard simulation techniques and variables that effectively calculate the performance of the intervention lead to results that could be validated by models incorporating the same technology currently operational. The analysis shows that the highest performance is predicted in April and the lowest in December in the North-East region. Considering the retrofit performance in the selected regions, Imphal offers the highest annual hours of extension and Shillong the lowest.

In contrast, the average yearly extension duration due to the retrofit was about 485 h. As the payback period and the CO_2 reduction due to the retrofit installation are related to the performance of the Solar PV modules, an average payback period of 8 months and 19 days is estimated in the NE region. Shillong will have the highest payback period of 9 months and 15 days, while Imphal will have the lowest within eight months. Thus, this intervention's performance and payback vary among the selected regions. However, an e-rickshaw operator in the NE region will be benefitted from this intervention. Benefits are also observed from an environmental standpoint. It has been estimated that, in the NE region, implementation of the retrofit will ensure an average reduction of about 0.45 tonnes of CO_2 from a single solar integrated e-rickshaw annually.

References

1. Abdel-Salam M, El-Mohandes MT, Goda M (2018) An improved perturb-and-observe based MPPT method for PV systems under varying irradiation levels. Sol Energy 171(June):547–561. https://doi.org/10.1016/j.solener.2018.06.080
2. Bhawan S, Puram RK (2011) CO_2 baseline database for the indian power sector user guide government of india ministry of power central electricity authority. Report by ministry of power government of India
3. Carrer D et al (2021) Surface albedo retrieval from 40-years of earth observations through the eumetsat/lsa saf and eu/c3s programmes: the versatile algorithm of pyalus. Rem Sens 13(3):1–32. https://doi.org/10.3390/rs13030372
4. Dasgupta D, Sarangi GK (2021) Meeting India's electricity demand in 2030. Energy Clim Change 2(February):100038. https://doi.org/10.1016/j.egycc.2021.100038
5. Deloitte (2019) Recharging India's electric vehicle ambition by electrifying public transport. In: Future of mobility
6. Engineering IT et al (2020) Design and prototyping of an automatic quoting tool for domestic solar systems
7. Eu (no date) TMY generator
8. Guttikunda SK, Jawahar P (2014) Atmospheric emissions and pollution from the coal-fired thermal power plants in India. Atmos Environ 92:449–460. https://doi.org/10.1016/j.atmosenv.2014.04.057
9. Khan EA, Quaddus M (2020) E-rickshaws on urban streets: sustainability issues and policies. Int J Sociol Soc Policy 41(7–8):930–948. https://doi.org/10.1108/IJSSP-07-2020-0315
10. Majumdar D, Jash T (2015) Merits and challenges of e-rickshaw as an alternative form of public road transport system: a case study in the state of West Bengal in India. Energy Procedia. Elsevier B.V. https://doi.org/10.1016/j.egypro.2015.11.492
11. Malik Y et al (2018) Impact assessment of e-rickshaws while analyzing entrepreneurial success of rickshaw pullers. Ind J Econ Bus 17(3):287–294
12. Mathur J, Bansal NK (2001) Analysis of selected renewable energy options for India. Energy Sources 23(10):877–888. https://doi.org/10.1080/009083101317071324
13. Mittal ML (2010) Estimates of emissions from coal fired thermal power plants in India. 39:1–22

14. Murali N et al (2020) Sizing and performance analysis of an electric motor in an e-rickshaw. In: 2020 IEEE international conference on power systems technology, POWERCON 2020, pp 0–5. https://doi.org/10.1109/POWERCON48463.2020.9230564
15. Pandey R, Singh B (2020) Canonical switching cell (CSC) converter-based power factor-corrected battery charger for e-rickshaw. IEEE Trans Ind Appl 56(5):5046–5055. https://doi.org/10.1109/TIA.2020.2996539
16. Roy A (2016) E-rickshaw service in Barddhaman town: importance, problems and future prospects. Int J Sci Res Publications 6(9):702–706. Available at: www.ijsrp.org
17. Sahu D, Nayak PSR (2020) Design and analysis of solar e-rickshaw charging system. In: Proceedings of 2020 IEEE-HYDCON international conference on engineering in the 4th industrial revolution, HYDCON 2020. https://doi.org/10.1109/HYDCON48903.2020.9242723
18. Sameeullah M, Chandel S (2016) Design and analysis of solar electric rickshaw: a green transport model. In: 2016 International conference on energy efficient technologies for sustainability, ICEETS 2016, pp 206–211. https://doi.org/10.1109/ICEETS.2016.7582927
19. Scheffknecht G (2008) Fossil fuel-based power generation and greenhouse gas emissions: state of the art and perspective. Int J Energy Clean Environ 9(1–3)
20. Shahid H et al (2018) Implementation of the novel temperature controller and incremental conductance MPPT algorithm for indoor photovoltaic system. Sol Energy 163(February):235–242. https://doi.org/10.1016/j.solener.2018.02.018
21. Singh S (2014) A study of the battery operated e-rickshaws in the state of Delhi. Centre for Civil Society. www.ccs.in
22. Solar F (2013) © Copyright 2013, First Solar, Inc.
23. Soni MS et al (2017) Thermo-economic analysis of a solar-coal hybrid power plant. In: Proceedings of the 24th national and 2nd international ISHMT-ASTFE heat and mass transfer conference (IHMTC-2017)
24. Tirpude S (2020) An investigation into awareness and usage of lead acid batteries in E-rickshaws: a field survey in New Delhi. Indian J Sci Technol 13(33):3467–3483. https://doi.org/10.17485/ijst/v13i33.1065
25. Verma PK et al (2017) Operating experience of a 20kW rooftop solar power plant. In: Proceedings of the 24th national and 2nd international ISHMT-ASTFE heat and mass transfer conference (IHMTC-2017)
26. Vermani S, Rajput A, Sharma P (2020) E-rickshaw: shared micro mobility—the green revolution on Indian roads. Int J Innov Appl Stud 29(4):9324
27. Ziane A et al (2021) Photovoltaic output power performance assessment and forecasting: impact of meteorological variables. Sol Energy 220(March):745–757. https://doi.org/10.1016/j.solener.2021.04.004

Comparative Study of Performance and Emission of Biodiesel Produced from Water Hyacinth and Salvinia Molesta: A Critical Review

Akshay Jain, Bhaskor Jyoti Bora, Rakesh Kumar, and Abdulrajak Buradi

Nomenclatures and Abbreviations

B0	0% Biodiesel in fuel + 100% Diesel
B10	10% Biodiesel in fuel + 90% Diesel
B20	20% Biodiesel in fuel + 80% Diesel
B30	30% Biodiesel in fuel + 70% Diesel
B40	40% Biodiesel in fuel + 60% Diesel
B100	100% Biodiesel in fuel + 0% Diesel
CI	Compression ignition
CO	Carbon monoxide
CO_2	Carbon dioxide
HRR	Heat release rate
PM	Particulate matter
O_2	Oxygen

A. Jain · B. J. Bora (✉)
Energy Institute Bengaluru (Centre of RGIPT, Jais, Amethi), Bengaluru, India
e-mail: bjbora@rgipt.ac.in

R. Kumar
RGIPT, Jais, Amethi, India

A. Buradi
Nitte Meenakshai Institute of Technology, Bengaluru, India

© The Author(s), under exclusive license to Springer Nature Singapore Pte Ltd. 2023

S. Revankar et al. (eds.), *Recent Advances in Thermofluids and Manufacturing Engineering*, Lecture Notes in Mechanical Engineering,
https://doi.org/10.1007/978-981-19-4388-1_33

1 Introduction

Need of energy has been continued to be a concern for both developing and developed countries for their economic and cultural growth. Fossil fuels such as mineral oils, coal and gas satisfy most of the energy need till date [1]. In the search of alternative energy sources, solar energy, wind energy, energy obtained from biomass directly or in the form of biofuels are playing an important role for supplying increased demand of energy [2, 3]. Rudolf diesel invented the very first internal combustion engine which worked on peanut oil [4]. Vegetable oils possess high viscosity and unfavourable cold flow properties which limits their direct usage in modern diesel engines. Some of the alternative biodiesel feedstocks are waste plastic oil, waste pyrolysis oil and waste cooking oil. Some of the conventional feedstocks of biodiesel are palm, sunflower, olive oil, etc. which are considered as first-generation feedstocks [5, 6]. First-generation feedstocks have some limitations as their oils are edible which creates a threat to food security and could lead to inflation of food prices [7]. Hence a need of feedstock arose which should be non-edible and easily available [8–11]. Second generation of biodiesel feedstock consists of no-edible sources such as jatropha oil, karanja oil, pongamia oil, tamarind oil, etc. [12]. Algae and seaweeds are considered as third-generation feedstocks for biodiesel production [13].

A perennial aquatic weed Water Hyacinth is a Brazilian native aquatic plant that was introduced in West Bengal in India as an ornamental plant in early twentieth century but now it is an invasive species and it got named as 'Terror of Bengal' in the subcontinent of Asia. Water Hyacinth can grow at very fast rate and it can double itself in less than two weeks. It can tolerate temperature variation from minimum of 12 °C to maximum 35 °C. It can survive in the pH range between 5 and 7.5. Biomass obtained from dry Water Hyacinth can serve a heating value of 22 MJ/m^3 [14, 15]. Water Hyacinth (Eichhornia Crassipes) belongs to monocotyledonous freshwater aquatic plant which is related to Pontederiaceae family related to Liliaceae lily family [16]. The floating mats of water weeds cause considerable biodiversity impacts by decreasing light penetration, displacing native vegetation and by preventing oxygen exchange in the water which results in unsuitable habitat for fish and other aquatic fauna [17]. Mats also prevent birds and other fauna to access the water [18]. Biofuels such as biodiesel and bioethanol can be produced at an economic level [19]. Third-generation biofuels include biodiesel production from microalgae and aquatic weeds such as Salvinia Molesta and Water Hyacinth [20, 21]. Mubarak et al. [22]tested Salvinia Molesta biodiesel blends with diesel and reported that B20 (20% biodiesel and 80% Diesel) blends can be the alternative fuel for diesel engine. Alagu et al. [15] and Venu et al. [14] experimented Water Hyacinth biodiesel blends with diesel and concluded that the characteristics of 20% Water Hyacinth biodiesel + 80% diesel fuel (B20) blend were in par with diesel for CO, HC, smoke emission and thermal efficiency. Shanab et al. [23] mentioned 6.79–10.45% lipid content in Water Hyacinth which produced biodiesel 3.22–6.36% after transesterification of lipids. Rahman and Aziz [24] used combination of Water Hyacinth biodiesel with hydroxyl. Biomass

Table 1 Biomass productivity of different aquatic weeds feedstocks

Researchers	Feedstocks	Productivity (tonne/ha/year)
Aquatic weeds		
Mishima et al. [26]	Water Hyacinth (Eichhornia Cressipes)	60–100
Mishima et al. [26]	Pistiastatiotes	60–100
Xu et al. [27]	Duckweed	33
Abbasi et al. [28]	Salvinia Molesta	53
Other feedstocks		
Lime et al.	Eucalyptus	15–25
Jayus et al. [29]	Sugarcane	60–74
McAloon et. al. [30]	Corn Stover	40
Williams et al. [31]	Corn	6.61

productivity plays an important role in selection of feedstock for continuous biofuel production. Table 1 shows productivity of biomass of different feedstock [25].

Transesterification of lipids extracted from the biomass produces biodiesel and glycerol. It contains methyl or ethyl ester of fatty acids. Table 2 shows the properties of biodiesel obtained from Salvinia Molesta and Water Hyacinth. Properties of biodiesel obtained from microalgae, cyanobacteria, waste cooking oil and waste plastic oil are described in Table 3.

Table 2 Properties of diesel, ASTM standards, Water Hyacinth biodiesel and Salvinia Molesta biodiesel

Source	Diesel	ASTM 6751-02	Salvinia Molesta B100	Water Hyacinth B100
Researchers	Mubarak et al. [22]			Alagu et al. [15]
Density (kg/m^3)	821	870–890	870	887
Kinematic viscosity, 40 °C (mm^2/s)	2.716	1.9–6	5.013	3.96
Flash Point (°C)	79	>130	194	212
Pour point (°C)	−7	−35 to −15	−5	7
Fire point (°C)	86	>130	199	–
Calorific value (MJ/kg)	42.200	–	36.935	36.9
Acid number (mgKOH/g)	–	<0.8	0.561	0.42
Cetane number	–	–	–	52.5

Table 3 Properties of biodiesel

Researchers	Satputley et al. [32]			Nautiyal et al. [33]	Meng et al. [34]	Mani et al. [35]
Feedstocks/standards	ASTM Biodie-sel std	Diesel	Micro-algae	Cyano-bacteria (Spirulina)	Waste cooking Oil	Waste Plastic Oil
Density (kg/lit)	862–900	830	864	860	890	835.5
Kinematic viscosity, 40 °C (cSt)	3.5–5.5	1.9–4.1	4.41	–	4.23	2.52
Flash point (°C)	Min 100	75	115	–	171	42
Calorific value (MJ/kg)	37.5	43.5	37.5	41.36	32.9	44,340
Cetane number	>47	>49	–	–	54.5	51

2 Materials and Methods

Alagu et al. [15] explained the production procedure of biodiesel from Water Hyacinth in laboratory and the process is simple [15]. Mubarak et al. [22] took fresh Salvinia Molesta which was washed then dried in sun. Pulverization was done to make powder of dried biomass of Salvinia Molesta. Total lipid content was estimated as 16% of dry weight by Bligh and Dyer's method. Detailed procedure is also available in their paper [22]. Engine specifications for Water Hyacinth biodiesel and Salvinia Molesta biodiesel testing are mentioned in Table 4.

Table 4 Engine specification

Type of biodiesel	Water Hyacinth biodiesel Alagu et al. [15]	Salvinia Molesta biodiesel Mubarak et al. [22]
Type of engine	Air-cooled, single-cylinder DI diesel engine	Water-cooled, single-cylinder DI diesel engine
Bore × Stroke	87.5 mm × 110 mm	80 mm × 110 mm
Compression ratio	17.5: 1	17.5: 1
Injection timing	23°bTDC (static)	23°bTDC
Injection pressure	–	200 bar
Rated power	4.4 kW	3.73 kW
Rated speed	1500 rpm	1500 rpm
Number of nozzles	3	–
Holes in nozzle	–	3

3 Engine Performance Studies

3.1 *Brake Thermal Efficiency (BTE)*

Fuel conversion efficiency or BTE represents the potential of conversion of chemical energy into useful mechanical work [15]. The BTE for Water Hyacinth biodiesel blends B10, B20, B30, B40, B100 was reported as 29.6%, 28.3%, 27.9%, 26.85% and 24.8%, respectively in comparison to 32.6% in diesel mode as found by Alagu et al. [15]. The study indicated that BTE values of B10 are closer to diesel throughout the different loading because of inbuilt oxygen content in the blend. However, the BTE decreases when concentration of biodiesel increases in the blends. This was due to poor energy content and higher viscosity of biodiesel as claimed by authors.

Mubarak et al. [22] observed that all blends of Salvinia Molesta biodiesel had lower BTE when they were compared with pure diesel due to their higher viscosity and lower calorific value which resulted in less atomization and mixing of fuel inside the cylinder. BTE of Salvinia Molesta biodiesel blend B10, B20 and B30 was reported as 30.54%, 29.51% and 28.86%, respectively and diesel showed the maximum BTE of 30.72% at the maximum load of 3.73 kW.

3.2 *Exhaust Gas Temperature (EGT)*

Water Hyacinth biodiesel blends showed higher EGT when compared with diesel due to their high oxygen content and high cetane number as reported by Alagu et al. [15]. The study indicated that the high cetane number, short ID and high O_2 content resulted in complete combustion which leads to higher exhaust temperature. The highest EGT that was reported on combustion of B40 at full load was 530 °C. Mubarak et al. [22] observed the EGT at the maximum load of 3.73 kW of 267 °C for B10, 265 °C with B20, 263 °C with B30 and 272 °C in case of diesel. Lower calorific value of Salvinia Molesta biodiesel blends was responsible for lowering the EGT which resulted in reduction in total energy generation, peak cylinder pressure and exhaust temperature (Figs. 1 and 2).

4 Combustion and Emission

4.1 *Heat Release Rate (HRR)*

In case of Water Hyacinth biodiesel blends, Alagu et al. [15] reported earlier start of combustion for blends of Water Hyacinth biodiesel due to lowered ID which lowered HRR significantly.

Fig. 1 Change in EGT for change in engine load for engine running with Water Hyacinth biodiesel blends [15]

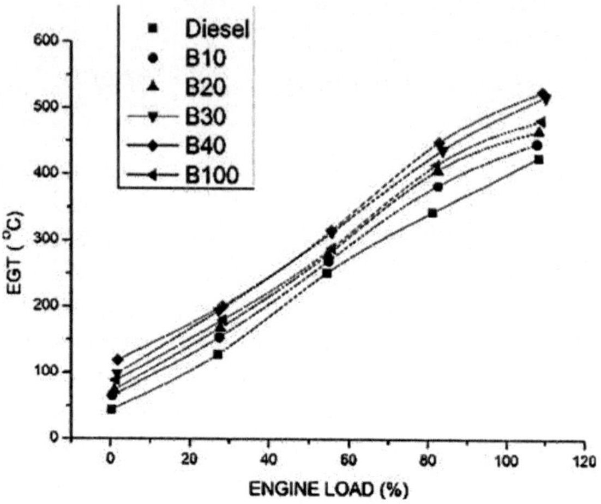

Fig. 2 Change in EGT for change in brake power when engine running on Salvinia Molesta biodiesel blends [22]

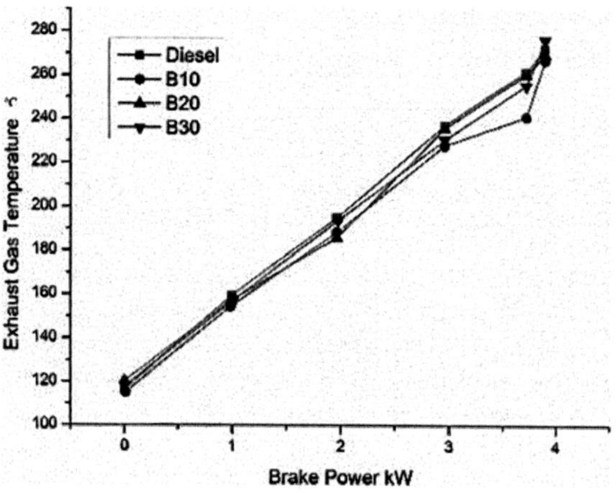

Mubarak et al. [22] reported negative value of HRR curve in earlier stage for all the fuels used due to vaporization of fuel takes place while ID period. Salvinia Molesta biodiesel blends shown reduced premixed stage when compared with pure diesel. This is due to the fact of shorter ID. The maximum values of HRR for B10, B20 and B30 were reported as 69.87 J/°CA, 67.39 J/°CA and 65.92 J/°CA, respectively, whereas HRR for diesel was 72.30 J/°CA which concludes that decrement in HRR for Salvinia Molesta biodiesel blends.

4.2 Ignition Delay (ID)

Alagu et al. [15] found that Water Hyacinth biodiesel B100 have lower ID (5.8 °CA) when compared to pure diesel (9.2 °CA). ID of Water Hyacinth biodiesel blends B10, B20, B30 and b40 had ID as 8.5 °CA, 7.6 °CA, 6.9 °CA and 6.6 °CA, respectively at full load. Mubarak et al. [22] reported reduced ID for Salvinia Molesta biodiesel blended fuels when compared with diesel.

4.3 Peak Cylinder Pressure (PCP) and Crank Angle Diagram

The investigation on Water Hyacinth biodiesel blends by Alagu et al. [15] indicated that biodiesel blends resulted in lower peak cylinder pressure in comparison with pure diesel. The reason for this may be the low ID of biodiesel resulted in lower amount of accumulation of fuel quantity. This caused lowered cylinder pressure. Mubarak et al. [22] reported that the average pressure was higher with diesel in comparison with Salvinia Molesta biodiesel blends. Salvinia Molesta biodiesel blends resulted reduced peak in-cylinder pressure. This could be due to low volatility caused by the higher density which leads to improper mixing of air and fuel. Average peak in-cylinder pressure for different biodiesel blends are 63.15 bar for B30, 63.68 bar for B20, 64 bar for B10 whereas 64.91 bar is for diesel.

4.4 Oxide of Nitrogen (NO_x) Emission

Alagu et al. [15] observed that due to availability of oxygen, the combustion rate improved in case of Water Hyacinth biodiesel. This resulted in an increase of NO_x emissions. However, there was reduction in NO_x emission in case of B100 due to lower calorific value which leads to lower rate of combustion and in-cylinder temperature which resulted in low NO_x emissions.

Mubarak et al. [22] reported NO emission for Salvinia Molesta biodiesel blends B10, B20 and B30 as 820 ppm, 752 ppm and 794 ppm, respectively and for diesel, it was 863 ppm at the full load of 3.73 kW. The presence of saturated fatty acids in Salvinia Molesta biodiesel blends resulted in decreased amount of NO emissions (Figs. 3 and 4).

4.5 Unburned Hydrocarbon Emission (UBHC)

The study on Water Hyacinth biodiesel blends conducted by Alagu et al. [15] reported that HC emissions for B10, B100, B40 and B20 were 63 ppm, 59 ppm, 58 ppm and

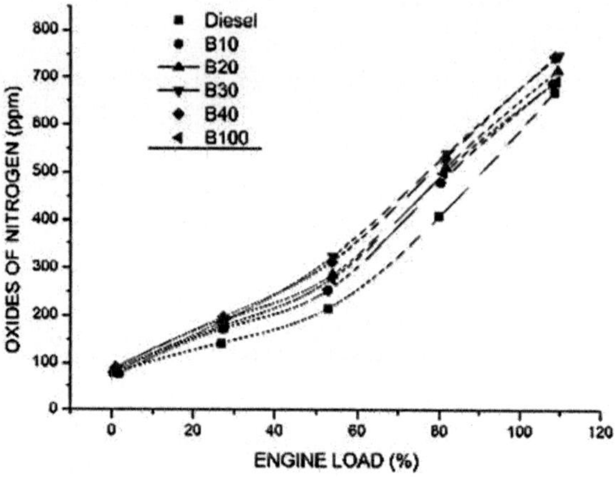

Fig. 3 Changes in oxides of nitrogen versus engine load when engine running on Water Hyacinth biodiesel blends [15]

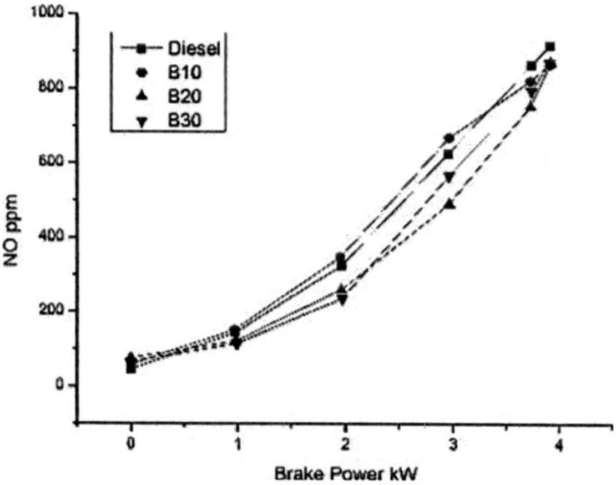

Fig. 4 Changes in NO emissions versus brake power when engine running on Salvinia Molesta biodiesel blends [22]

56 ppm, respectively and in case of diesel it was 67 ppm at full engine load. The low HC emissions for Water Hyacinth biodiesel blends can be attributed to more oxygen availability in blends, thereby, improving both the mixing rate and vaporization rate.

Mubarak et al. [22] stated UBHC emissions for B10, B20 and B30 as 21 ppm, 19 ppm and 22 ppm, respectively and for diesel it was 24 ppm at full load. Availability of oxygen in Salvinia Molesta biodiesel blends enhances the oxidation of hydrocarbons which reduces the UBHC emission in B10 and B20. In case of B30 higher UBHC was observed due to improper mixing and atomization. With the increase in engine load, UBHC was increased for all four fuels which may be due to increased amount of fuel admission inside the cylinder causing rich fuel zone formation at high load which resulted in localized deficiency of oxygen.

4.6 Carbon Monoxide (CO) Emission

Experimental study reported by Alagu et al. [15] stated that the emission of carbon monoxide in diesel was higher as compared to Water Hyacinth biodiesel blends due to availability of oxygen improves the combustion. Lowest emission of CO was observed for B40 when compared with diesel at full load. Increase in load causes increase in emission of CO due to high amount of fuel involved in combustion process which leads to generation of rich fuel zones due to less oxygen. Emission of CO for B100 was lower by 3.7%, 6.9% and 8.2% for B20, B30 and B40, respectively at full load.

Mubarak et al. [22] reported CO emissions for B10, B20 and B30 as 0.046% vol., 0.043% vol. and 0.047% vol., respectively and for diesel it was 0.05% vol. at the maximum load of 3.73 kW. All Salvinia Molesta biodiesel blended fuels resulted in reduced CO emissions when compared with pure diesel for all load conditions due to the presence of oxygen in the biodiesel which enhances complete combustion.

4.7 Smoke, Soot, Particulate Matter (PM) Emission

In the experimental study by Alagu et al. [15] on Water Hyacinth biodiesel blends, it was found that lower emissions of smoke were observed in biodiesel mode when compared with diesel. Maximum smoke was observed in case of B100 when compared with other blends due to large size droplets, poor air–fuel mixing rate, lower rate of evaporation and animosity of oxidation in case of high viscous fuel. Mubarak et al. [22] observed the percentage smoke opacity of 25.30 for B10, 23.32 for B20, 27.23 for B30 and 26.20 was observed with diesel at full load. In case of B10 and B20, reduced opacity of smoke is observed which may be due to utilization of available oxygen in the diffusion phase.

Therefore the combination of Water Hyacinth and Salvinia Molesta with other aquatic weeds [36] and algae as feedstock for biodiesel production may increase the overall yield from the production process and it might be economically more feasible. Properties of such biodiesel may be improved to make it comparable with diesel for its utilization in unmodified diesel engines [25]. Biodiesel obtained from biomasses generally have higher content of oxygen which becomes a reason for complete combustion [37].

5 Conclusion

Rapid growth of aquatic weeds creates problem for environment. Use of this aquatic weed biomass in making biodiesel gives a promising solution for increasing demand of diesel. Biodiesel can be used in the form of blend with pure diesel for enhancement in the engine performance and for reduction in harmful exhaust emissions. Improved results are observed with 20% biodiesel blend with 80% diesel. Removing waste from environment and converting it into useful fuel may be considered as cogeneration process in many ways as it solves following purposes simultaneously:

- By absorbing greenhouse gas CO_2 in photosynthesis these plants reduce greenhouse gas pollution.
- By providing promising solutions for biofuel generation throughout the year which contributes to generation of fuel and energy security in our country.
- Use of biofuels that are generated from aquatic biomass has shown reduction in emission.

References

1. Konwar LJ, Boro J, Deka D (2014) Review on latest developments in biodiesel production using carbon-based catalysts. Renew Sustain Energy Rev 29:546–564. https://doi.org/10.1016/j.rser.2013.09.003
2. Yusuf NNAN, Kamarudin SK, Yaakub Z (2011) Overview on the current trends in biodiesel production. Energy Convers Manage 52:2741–2751. https://doi.org/10.1016/j.enconman.2010.12.004
3. Demirbas A (2009) Progress and recent trends in biodiesel fuels. Energy Convers Manage 50:14–34. https://doi.org/10.1016/j.enconman.2008.09.001
4. Issariyakul T, Dalai AK, Desai P (2011) Evaluating esters derived from mustard oil (Sinapis alba) as potential diesel additives. JAOCS, J Am Oil Chem Soc 88:391–402. https://doi.org/10.1007/s11746-010-1679-6
5. Saladini F, Patrizi N, Pulselli FM, Marchettini N, Bastianoni S (2016) Guidelines for emergy evaluation of first, second and third generation biofuels. Renew Sustain Energy Rev 66:221–227. https://doi.org/10.1016/j.rser.2016.07.073
6. Bhuiya MMK, Rasul MG, Khan MMK, Ashwath N, Azad AK, Hazrat MA (2014) Second generation biodiesel: potential alternative to-edible oil-derived biodiesel. Energy Procedia 61:1969–1972. https://doi.org/10.1016/j.egypro.2014.12.054
7. Atadashi IM, Aroua MK, Aziz AA (2010) High quality biodiesel and its diesel engine application: a review. Renew Sustain Energy Rev 14:1999–2008. https://doi.org/10.1016/j.rser.2010.03.020
8. Umdu ES, Tuncer M, Seker E (2009) Transesterification of Nannochloropsis oculata microalga's lipid to biodiesel on Al_2O_3 supported CaO and MgO catalysts. Biores Technol 100:2828–2831. https://doi.org/10.1016/j.biortech.2008.12.027
9. Huang GH, Chen F, Wei D, Zhang XW, Chen G (2010) Biodiesel production by microalgal biotechnology. Appl Energy 87:38–46. https://doi.org/10.1016/j.apenergy.2009.06.016
10. Raju V, Kishore PS, Nanthagopal K, Ashok B (2018) An experimental study on the effect of nanoparticles with novel tamarind seed methyl ester for diesel engine applications. Energy Convers Manage 164:655–666. https://doi.org/10.1016/j.enconman.2018.03.032
11. Atabani AE, Mahlia TMI, Anjum Badruddin I, Masjuki HH, Chong WT, Lee KT (2013) Investigation of physical and chemical properties of potential edible and non-edible feedstocks for biodiesel production, a comparative analysis. Renew Sustain Energy Rev 21:749–755. https://doi.org/10.1016/j.rser.2013.01.027
12. Zhong W, Xuan T, He Z, Wang Q, Li D, Zhang X et al (2016) Experimental study of combustion and emission characteristics of diesel engine with diesel/second-generation biodiesel blending fuels. Energy Convers Manage 121:241–250. https://doi.org/10.1016/j.enconman.2016.05.033

13. Srinophakun P, Thanapimmetha A, Rattanaphanyapan K, Sahaya T, Saisriyoot M (2017) Feedstock production for third generation biofuels through cultivation of Arthrobacter AK19 under stress conditions. J Clean Prod 142:1259–1266. https://doi.org/10.1016/j.jclepro.2016.08.068
14. Venu H, Venkataraman D, Purushothaman P, Vallapudi DR (2019) Eichhornia crassipes biodiesel as a renewable green fuel for diesel engine applications: performance , combustion , and emission characteristics. Environ Sci Pollut Res 26:18084–18097. https://doi.org/10.1007/s11356-019-04939-z
15. Alagu K, Venu H, Jayaraman J, Raju VD, Subramani L, Appavu P et al (2019) Novel Water Hyacinth biodiesel as a potential alternative fuel for existing unmodified diesel engine: performance , combustion and emission characteristics. Energy 179:295–305. https://doi.org/10.1016/j.energy.2019.04.207
16. Ganguly A, Chatterjee PK, Dey A (2012) Studies on ethanol production from Water Hyacinth—A review. Renew Sustain Energy Rev 16:966–972. https://doi.org/10.1016/j.rser.2011.09.018
17. Bote MA, Naik VR, Jagadeeshgouda KB (2020) Materials science for energy technologies review on Water Hyacinth weed as a potential bio fuel crop to meet collective energy needs. Mater Sci Energy Technol 3:397–406. https://doi.org/10.1016/j.mset.2020.02.003
18. Awasthi M, Kaur J, Rana S (2013) Bioethanol production through Water Hyacinth, Eichhornia crassipes via optimization of the pretreatment conditions. Int J Emerg Technol Adv Eng 3:42–46
19. Gielen D, Boshell F, Saygin D, Bazilian MD, Wagner N, Gorini R (2019) The role of renewable energy in the global energy transformation. Energy Strateg Rev 24:38–50. https://doi.org/10.1016/j.esr.2019.01.006
20. Mubarak M, Shaija A (2016) Ultrasonication: An effective pre-treatment method for extracting lipid from Salvinia Molesta for biodiesel production. Resour Technol 2:126–132. https://doi.org/10.1016/j.reffit.2016.07.005
21. Mubarak M, Shaija A, Suchithra TV (2016) Ultrasonication: An effective pre-treatment method for extracting lipid from Salvinia molesta for biodiesel production. Resour Efficient Technol 2(3): 126–132. https://doi.org/10.1016/j.reffit.2016.07.005
22. Mubarak M, Shaija A, Suchithra TV (2020) Experimental evaluation of Salvinia Molesta oil biodiesel/diesel blends fuel on combustion, performance and emission analysis of diesel engine. Fuel 287: 119526. https://doi.org/10.1016/j.fuel.2020.119526
23. Shanab SMM, Hanafy EA, Shalaby EA (2016) Water Hyacinth as non-edible source for biofuel production. Waste Biomass Valor. https://doi.org/10.1007/s12649-016-9816-6
24. Rahman MA, Aziz MA (2021) Biodiesel from Water Hyacinth biomass and its influence on CI engine performance , emission , combustion and heat loss characteristics with the induction of hydroxy. Energy 24:120151. https://doi.org/10.1016/j.energy.2021.120151
25. Alam NS, Singh B, Guldhe A (2021) Aquatic weed as a biorefinery for biofuels and value-added products: challenges and recent advancements. Cleaner Eng Technol. https://doi.org/10.1016/j.clet.2021.100235
26. Mishima D, Kuniki M, Sei K, Soda S, Ile M, Fujita M (2008) Ethanol production from candidate energy crops: Water hyacinth (Eichhornia crassipes) and water lettuce (Pistiaatratiotes L.). Bioresour Technol 99:2495–2500. https://doi.org/10.1016/j.biortech.2005.09.029
27. Xu J, Cui W, Cheng JJ, Stomp AM (2011) Production of high-starch duckweed and its conversion to bioethanol. Biosyst Eng 110:67–72. https://doi.org/10.1016/j.biosystemseng.2011.06.007
28. Abbasi SA, Naseema A, Nipaney PC (1995) Productivity (net primary production) of salvinia (Salvinia molesta, Mitchell) in nature environments. Int J Resour Conserv, pp 215–220
29. Jayus L, Nurhayati N, Mayzhuroh A, Arindhani S, Caroenchai C (2016) Studies on bioethanol production of commercial baker's and alcohol yeast under aerated culture using sugarcane molasses as the media. Agric Agric Sci procedia 9:493–499. https://doi.org/10.1016/j.aaspro.2016.02.168
30. McAloon A, Taylorr F, Yee W, Ibsen K, Wooley R (2000) Determining the cost of producing ethanol from corn starch and lignocellulosic feedstock. National Renewable Energy Lab. Golden CO(US). https://doi.org/10.2172/766198

31. Williams J, Brammer J, Llewelyn R, Bergtold J (2016) An economic analysis of harvesting biomass from sorghums and corn. J AASFMRA, pp 89–101. https://doi.org/10.22004/ag.econ.236656
32. Satputaley SS, Zodpe DB, Deshpande NV (2018) ScienceDirect performance, combustion and exhaust emissions analysis of a diesel engine fuelled with algae oil and algae biodiesel. Mater Today Process 5:23022–23032. https://doi.org/10.1016/j.matpr.2018.11.031
33. Nautiyal P, Subramanian KA, Dastidar MG, Kumar A (2020) Experimental assessment of performance , combustion and emissions of a compression ignition engine fuelled with Spirulina platensis biodiesel. Energy 193. https://doi.org/10.1016/j.energy.2019.116861
34. Meng X, Chen G, Wang Y (2008) Biodiesel production from waste cooking oil via alkali catalyst and its engine test. Fuel Process Technol 9:5–11. https://doi.org/10.1016/j.fuproc.2008.02.006
35. Mani M, Subash C, Nagarajan G (2009) Performance, emission and combustion characteristics of a DI diesel engine using waste plastic oil. Appl Therm Eng 29:2738–2744. https://doi.org/10.1016/j.applthermaleng.2009.01.007
36. Arefin MA, Rashid F, Islam A (2021) A review of biofuel production from floating aquatic plants : an emerging source of bio-renewable energy. Biofuels, Bioprod Biorefin 15:574–591. https://doi.org/10.1002/bbb.2180
37. Thiruvenkatachari S, Saravanan CG, Geo VE, Vikneswaran M, Udayakumar R (2022) Experimental investigations on the production and testing of azolla methyl esters from Azolla microphylla in a compression ignition engine. Fuel 119448. https://doi.org/10.1016/j.fuel.2020.119448

Design and Validation of Fault Ride Through Controller Design for Grid Connected Large Photovoltaic System

Ramu Srikakulapu, Vivekanandan Subburaj, G. Charan Kumar, M. Dileep Krishna, Abhishek Dasore, and Gyan Sagar Sinha

1 Introduction

The sunlight, wind, rain, tides and geothermal heat are some of the natural from which an energy can be extracted such energy is known as renewable energy. The growth and development of clean and renewable energy sources has been provided with a renewed impetus by the global energy crunch. Organizations across the globe had adopted clean development mechanisms. Pollution associated with combustion is another major factor working against fossil fuels in addition to the decrease of fossil fuels in the world. Unlike their conventional counter parts renewable energy sources are much cleaner and produce energy with no harmful effects. The photovoltaic (PV) controller is set up in such a way that it can stay linked to the system under abnormal

R. Srikakulapu
Department of Electrical & Electronics Engineering, Vignan's Institute of Information Technology, Visakhapatnam, Andhra Pradesh, India

V. Subburaj
School of Electrical & Electronics Engineering, REVA University, Bangalore, India

G. C. Kumar
Department of Mathematics, Koneru Lakshmaiah Education Foundation, Guntur, Andhra Pradesh 522502, India

M. D. Krishna
Department of Electrical & Electronics Engineering, Vidya Jyothi Institute of Technology, Hyderabad, India

A. Dasore (✉)
Department of Mechanical Engineering, RGM College of Engineering and Technology, Nandyal, Andhra Pradesh 518501, India
e-mail: dasoreabhishek@gmail.com

G. S. Sinha
School of Mechanical Engineering, KIIT Deemed University, Bhubaneswar, Odisha, India

© The Author(s), under exclusive license to Springer Nature Singapore Pte Ltd. 2023

S. Revankar et al. (eds.), *Recent Advances in Thermofluids and Manufacturing Engineering*, Lecture Notes in Mechanical Engineering,
https://doi.org/10.1007/978-981-19-4388-1_34

operating conditions and contribute to voltage support during and after fault operation. The grid integrated large scale PV power plant system is facing major issues as low-voltage conditions due to fault conditions. In that context, researchers are worked on various control strategies to improve the fault ride through (FRT) capability of the system. A reduced-order nonlinear model is constructed to further the converters' dynamic synchronization characteristic. Examining the impact of grid impedance and researching the spatial vector tracking relation [1] reveals the resynchronization principle of the converters during low-voltage ride through (LVRT).

Formulated a new swing equation to examine the transient stability of renewable energy conversion systems during LVRT [2]. A unique sliding mode control (SMC)-based feedback linearization is described [3] to boost PV system response speed under LVRT situations. The continuous mixed p-norm-based adaptive filtering technique is used to update the gains of the proportional integral (PI) controller in grid connected PV systems [4]. A novel control mechanism is employed for the three wire three-phase two-stage PV converter to improve power quality in abnormal conditions. The minimization of double grid frequency oscillations in the DC-link voltage and active power under unbalanced faults is one of the primary contributions of the control strategy [5]. For operating in both normal and faulty grid situations, a simple control technique with an adaptive cost function is proposed [6]. In a typical controller, it eliminates the necessity for a multi–nested–loop. An LVRT control strategy based on positive/negative sequence droop management is proposed for grid-interactive microgrids to ride through voltage sags with not only inductive/resistive, but also complicated line impedance [7]. The LVRT control plan divides a PV system's front and back stages into two separate controls, opts for model predictive control for inverter control, and advances MPPT technology [8].

PV System modelling is explained in Sect. 2 and it contains the maximum power point tracker (MPPT) technique with flowchart and mathematical modelling of PV with circuit diagram is explained. Conventional PI control and sliding mode control are detailed with block diagram. After that, PI-based controller and Hybrid controller design are illustrated with proper design. Section 3 contains the simulation and results of PV system with various fault cases for FRT analysis is detailed and finally concluded the work in last section.

2 System Modelling

The grid interconnected PV system is modelled and it consists of PV array, side voltage source inverter (VSI), VSI controller, MPPT controller, LC filter and grid. This includes the PV array, MPPT used to extract maximum available power from the PV array and it is connected to the boost converter with capacitor. The boost converter is interconnected to VSI through the DC-link and whole system is interconnected to grid. In addition, it is controlled by a three-phase VSI, RL filter which connected to the AC grid. The RL filter is connected AC side to reduce the harmonics.

2.1 PV Modelling

Figure 1. depict the PV module's equivalent circuit and made up of a number of PV cells. In this work, incremental conductance MPPT algorithm is applied to extract the maximum power from PV array and its flow chart is given in Fig. 2.

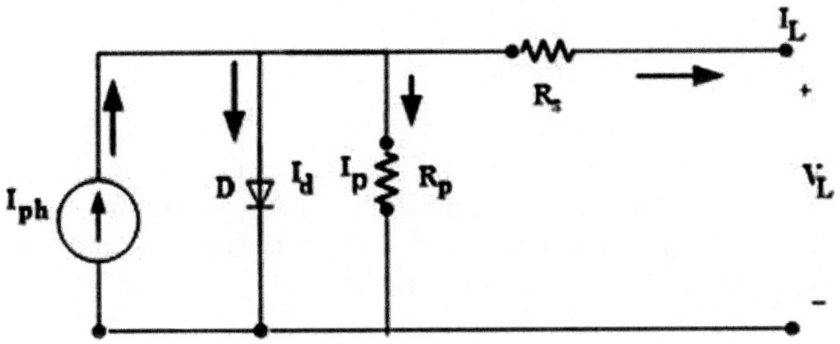

Fig. 1 Equivalent circuit of solar cell

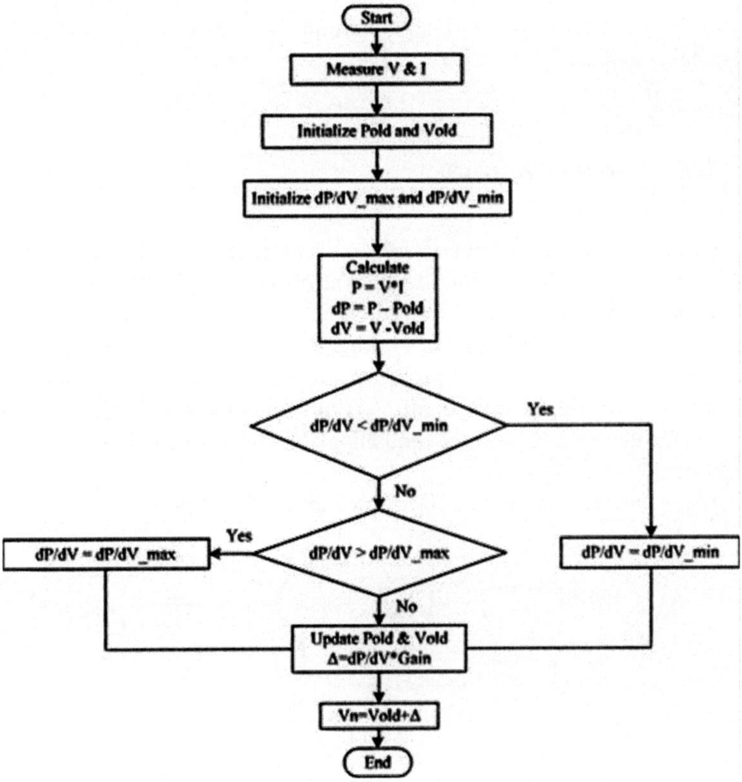

Fig. 2 Flow chart of incremental conductance MPPT algorithm

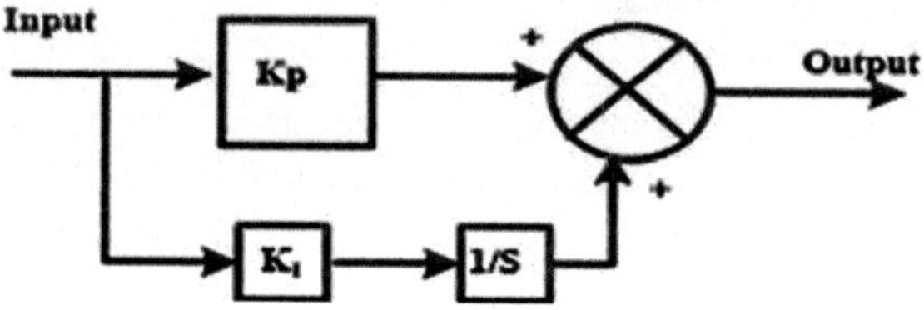

Fig. 3 The block diagram of PI controller

2.2 Conventional PI Controller

The problem of proportional control is offset is more and PID controller both offset and study state error more and PI controller is slow response and stability of the system is slow. The basic block diagram of PI controller is shown in Fig. 3. The basic equation of PI controller is given in Eq. (1).

$$K_p + K_{I/s} \tag{1}$$

where K_p denotes the proportional gain and K_I denotes integral gain. Although PI control is faster than merely integral action, it may not be as fast as an only proportional controller.

2.3 Sliding Mode Controller

Sliding mode control is a type of nonlinear control that is stable in the face of uncertainty and disturbance. The sliding mode control approach is widely considered as one of the most effective methods for designing reliable controllers for complicated high-order nonlinear dynamic plants working in uncertain environments. The SMC takes a methodical approach to the problem of maintaining consistency and stability. The insensitivity of sliding mode controllers to fluctuation and disturbance parameters eliminates the need for precise modelling. The basic block diagram of SMC is shown in Fig. 4.

The mathematical equations of SMC are explained as below,

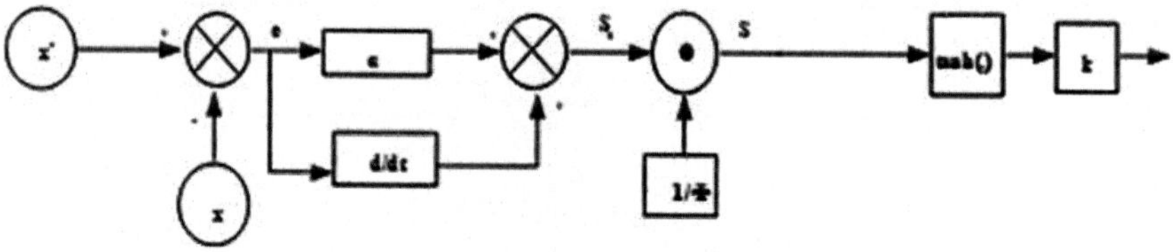

Fig. 4 The basic block diagram of sliding mode controller

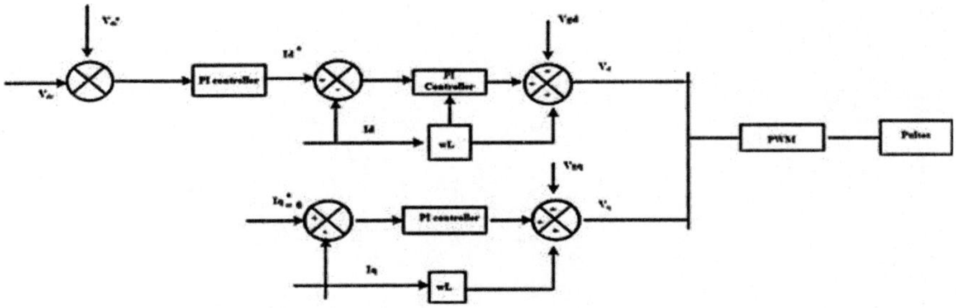

Fig. 5 PI-based VSI controller

$$S_s = \alpha e + \dot{e} \tag{2}$$

$$e = x^* - x \tag{3}$$

$$\tau_c = K \tanh(S_s/\emptyset) \tag{4}$$

Equations (2)–(4) are sliding surface S_s, e is error and τ_c is discontinue control action and α is a sliding function.

2.4 PI-Based VSI Controller

Proportional integral-based voltage source inverters are shown in the below block diagram. In this diagram there are two loops. It is made up of two loops: an inner current loop and an outer voltage loop. The grid current and the DC-link voltage V_{dc} are controlled by standard PI controllers. During normal operation, the voltage loop output serves as the active current reference I_d*, whilst the reactive current reference I_q* is set to zero and the grid inverter will operate to maintain the stable DC-link voltage. Finally, V_{abc} is linked to the Pulse-width modulation (PWM) signal generator to generate VSI switching pulses. Figure 5 shows the control diagram of a PI-based VSI.

2.5 Hybrid Controller

The hybrid controller is a mix of traditional PI and SMC techniques. Figure 6 depicts a VSI based on a hybrid controller. Inner and outer loops make up the hybrid VSI controller. The grid current is regulated by the inner loop, which is constructed using a PI controller. DC-link voltage is controlled by the outer loop, which is designed by SMC controller.

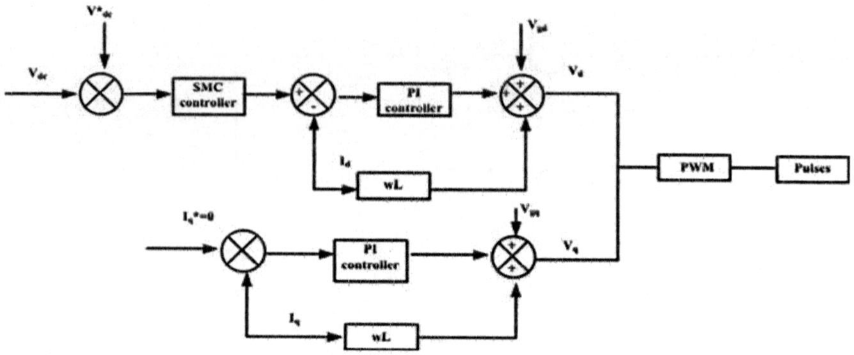

Fig. 6 Hybrid control based VSI

3 Simulation and Results

Figure 7 shows a single line diagram of a grid connected PV system. The system contains the PV source and grid. Both are interconnected by boost converter and VSI with DC-links. Table 1 lists the specifications for the specified system.

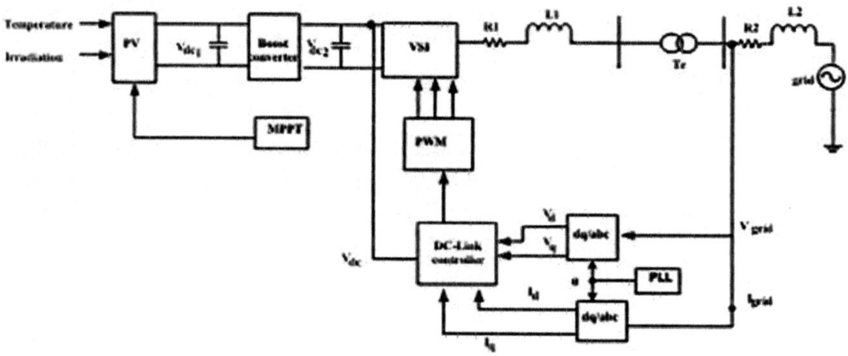

Fig. 7 Single line diagram of a PV integrated to the grid

Table 1 Parameters of block diagram

Specification	Rating
PV array	100 kW
C_1 and C_2	100 μF and 12,000 μF
Boost converter	$L = 2.5$ mH and $R = 10\ \Omega$
Grid	120 kV

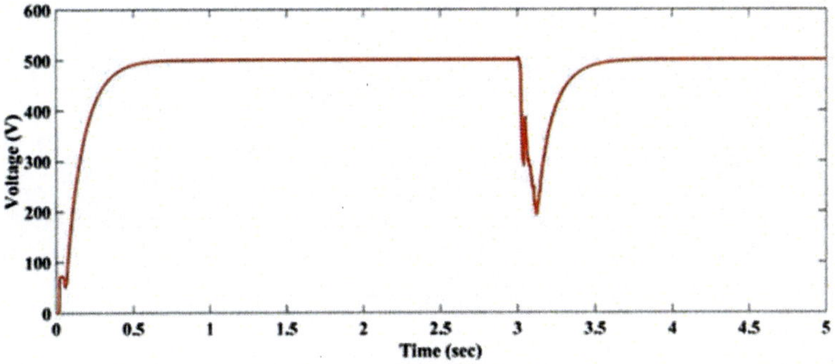

Fig. 8 DC-link voltage in case of LLLG fault in PI based VSI controller

3.1 PI Based VSI Controller for Grid Connected PV System

In this study, FRT capability is investigated for grid connected PV with PI controller-based VSI. Various fault conditions are applied at the grid terminal for the system. Those are symmetrical and asymmetrical faults. The simulation results of the specified system observed as in terms of DC-link voltage, active power and current waveforms.

3.2 Case 1: Single Fault Analysis

FRT capability is examined for grid connected PV system by applying the symmetrical and unsymmetrical faults at near to grid. The applied faults are line-line-line to ground (LLLG) and line to ground (LG). The fault analysis is observed through the DC-link voltage responses. If applied LLLG fault is applied at grid side with 0.1 variation the duration at which system is stabilized at 4.4 s. Figure 8 shows the DC-link voltage waveform. The fault analysis is observed through the grid current in explain Fig. 9.

3.3 Case 2: Multiple Fault Analysis

In this case, FRT capability is examined for Grid connected PV system by applying the symmetrical and unsymmetrical faults at near to grid. The multiple faults are given at instant of 3 s and 6 s with 0.1 s time period. The LG and LLLG faults are applying in system. The analysis is observed through the DC-link voltage and current responses given in Figs. 10 and 11.

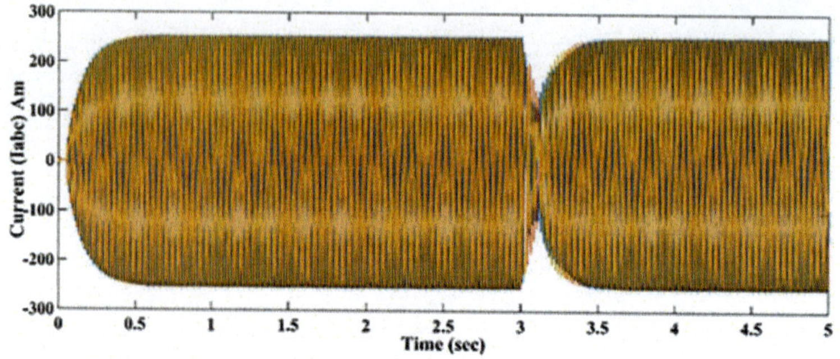

Fig. 9 Grid current in case of LLLG fault in PI based VSI controller

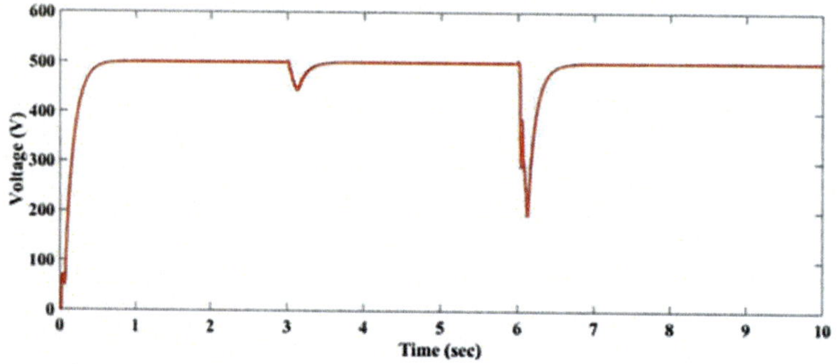

Fig. 10 DC-link voltage in case of multiple faults (LG and LLLG) in PI based VSI controller

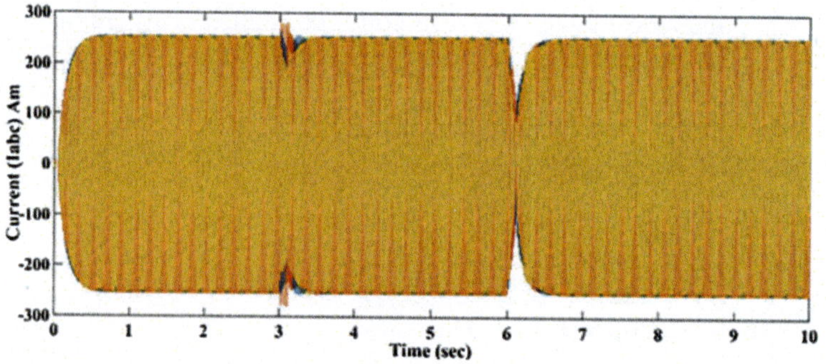

Fig. 11 Current in case of multiple faults (LG and LLLG)

3.4 Hybrid VSI Controller Grid Integrated PV System

FRT capability is investigated for grid connected PV with hybrid controller-based VSI. Various fault conditions are applied at the grid terminal for the system. Those are symmetrical and asymmetrical faults. The simulation results are.

3.5 Case 1: Single Fault Analysis

In this case, FRT capability is examined for grid connected PV system by applying the symmetrical and unsymmetrical faults at near to grid. The fault is applied at instant of 3 s with 0.1 s time period. The applied faults are LLLG and LG. The FRT analysis of PI and hybrid controller for grid connected system faults are given in Table 2 (Figs. 12 and 13).

Table 2 System stable time for PI and hybrid controller-based VSI with faults

Fault	Stable time (s)	
	PI controller	Hybrid controller
LG	3.7	4.4
LLLG	3.7	4.1

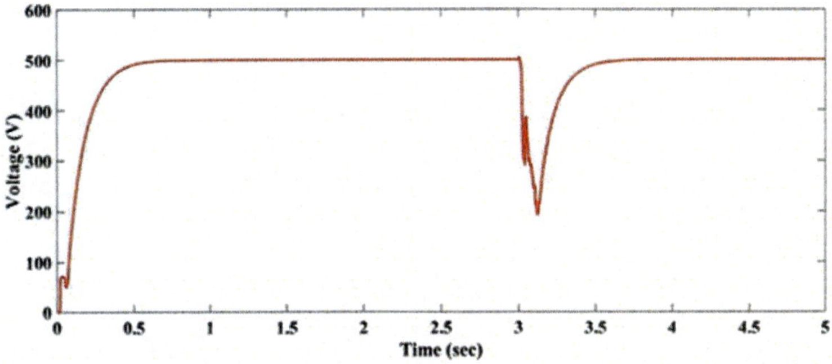

Fig. 12 DC-link voltage in case of LLLG fault

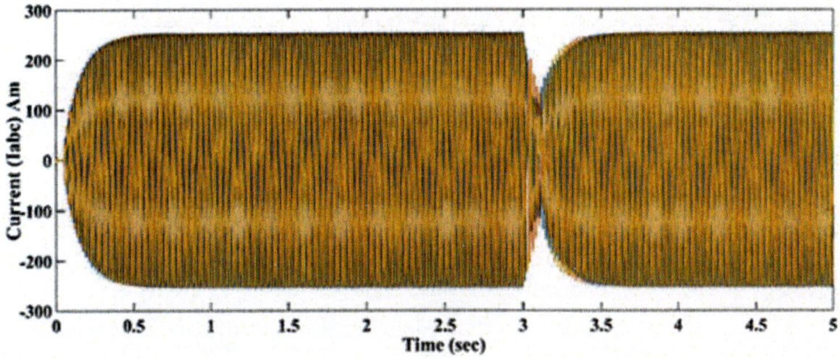

Fig. 13 Grid current in case of LLLG fault

3.6 *Case 2: Multiple Fault Analysis*

In this case, FRT capability is examined for Grid connected PV system by applying the symmetrical and unsymmetrical faults at near to grid. The multiple faults are applied at instant of 3 s and 6 s with 0.1 s time period. The LG and LLLG faults are applying in system. The analysis is observed through the DC-link voltage and current responses given in Figs. 14 and 15.

The outcomes of multiple faults case of PI and hybrid controller-based system are listed below in Table 3.

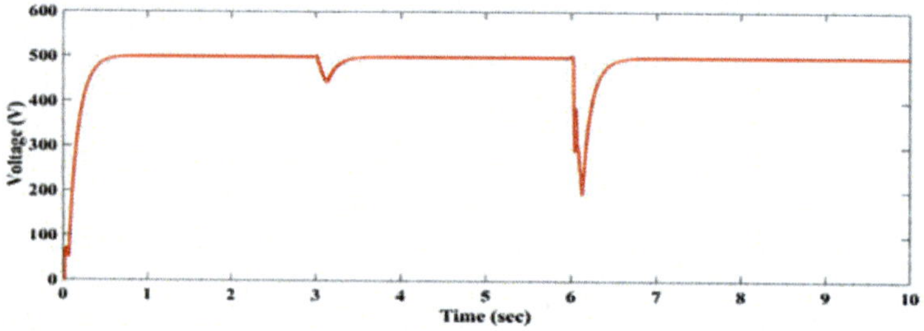

Fig. 14 DC-link voltage in case of multiple faults (LG and LLLG)

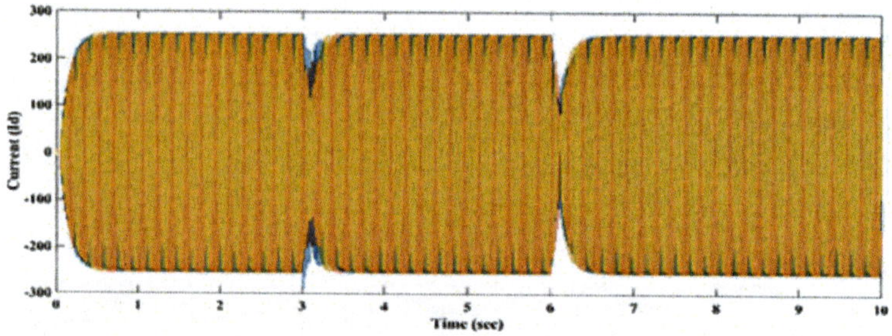

Fig. 15 Grid current in case of multiple faults (LG and LLLG)

Table 3 System stable time for PI and hybrid controller-based VSI with multiple faults

Multiple fault types	Stable time (s) PI controller Hybrid controller			
	Fault 1	Fault 2	Fault 1	Fault 2
LG and LLLG	3.8	7.4	3.7	6.9
LLG and LLLG	3.8	7.0	3.8	6.9
LG and LLG	3.8	7.3	3.6	6.8

4 Conclusions

The grid connected PV system is studied thoroughly in this work. The complete modelling is presented in steps for the switching model of a single PV system.

1. The importance of the inverter control during FRT is explained, considering the future connection requirements.
2. The complete model is designed in the MATLAB environment. The inverter control is designed for the DC-link voltage oscillations elimination.
3. The effectiveness is validated by running the LG and LLLG faults on the grid. The model of the complete PV system is designed for increasing the simulation speed and for the integration study of PV with the grid.
4. The FRT capability of PI and Hybrid controller-based systems are examined by the two different ways: individual fault analysis and multiple faults analysis.
5. In the both cases, LG, LLG and LLLG faults are induced to the system and the corresponding DC-link voltage and Grid power are observed. The above cases are examined for both conventional PI controller and hybrid controller.
6. The results promote that hybrid controller-based system has better fault clearing time and system stabilization time in comparison to conventional PI controller.

References

1. He X, Geng H, Ma S (2019) Transient stability analysis of grid-tied converters considering PLL's nonlinearity. CPSS Trans Power Electron Appl 4(1):40–49
2. He X, Geng H, Li R, Pal BC (2020) Transient stability analysis and enhancement of renewable energy conversion system during LVRT. IEEE Trans Sustain Energy 11(3):1612–1623
3. Zhang Y, Wang J, Li H, Zheng TQ, Lai J-S, Li J, Wang J, Chen Q (2020) Dynamic performance improving sliding-mode control-based feedback linearization for PV system under LVRT condition. IEEE Trans Power Electron 35(11):11745–11757
4. Ushasree P, Sudhakar A (2013) Fault Ride-through capability enhancement of PV system with voltage control strategy. Open J Appl Sci 3(2B):30–34
5. Afshari E, Moradi GR, Rahimi R, Farhangi B, Yang Y, Blaabjerg F, Farhangi S (2017) Control strategy for three-phase grid connected PV inverters enabling current limitation under unbalanced faults. IEEE Trans Ind Electron 64(11)
6. Sajadian S, Ahmadi R (2018) ZSI for PV systems with LVRT capability. IET Renew Power Gen 12(11):1286–1294
7. Zhao X, Guerrero JM (2017) Low-voltage ride-through operation of power converters in grid-interactive microgrids by using negative-sequence droop control. IEEE Trans Power Electron 32(4):3128–3142
8. Yuan J, Chen Y, Chen N, Zhang Y. Lin Y (2017) Research on low-voltage ride through control based on model predictive control. J Eng 2017(13):2114–2118

Design Development of a Heat Storage System at Small Scale for Solar Thermal Collectors

Sandeep Kumar, Pawan Kumar Kuldeep, and Sunita Mahavar

1 Introduction

The clean and long-lasting sources of the energies are the present need. Solar thermal energy is a promising source to meet-up the thermal energy demand of the world for low to high-temperature requirements. The solar thermal energy storage (STES) systems have special relevance in the solar thermal systems (STS). There are various types of STSs that directly collect the heat and serve the purpose of any heating applications in the range of 60–1000 °C. The variable, intermittent and unpredictable nature of solar radiation limit the operation of these systems due to the heat collection mismatch between the load and the supply. These storage systems provide heat backup for the system application in the absence of daylight or during insufficient solar radiation. Thermal energy storage (TES) systems are being widely used in solar thermal collectors for water heaters and as building heating systems [1, 2]. Among three kinds of storage systems (i) sensible heat storage (store energy while heating and release energy when cooling of a liquid or solid material), (ii) latent heat storage (store or release energy during phase change of material) and (iii) thermo-chemical storage (TCS) (uses chemical reactions to store and release thermal energy), the sensible method is the most useful. Due to the low density, the sensible heat storage requires a larger volume than that of PCM and TCS systems, respectively [2]. The properties of various storage materials are summarized in Table 1 [3–14].

Although various mediums have been tested and reported, water is the most used medium for storing sensible energy to use at low or medium temperature solar thermal systems. Storage at more than the boiling temperature of water needs pressure vessels which makes the system costly. For this, other liquids, viz. octane, isopentanol and

S. Kumar · P. Kumar Kuldeep · S. Mahavar (✉)
Solar Energy Research Laboratory (SERL), Department of Physics, University of Rajasthan, Jaipur 302004, India
e-mail: smjpr1986@gmail.com

© The Author(s), under exclusive license to Springer Nature Singapore Pte Ltd. 2023
S. Revankar et al. (eds.), *Recent Advances in Thermofluids and Manufacturing Engineering*, Lecture Notes in Mechanical Engineering,
https://doi.org/10.1007/978-981-19-4388-1_35

Table 1 Storage materials and their properties

Medium	Type	Melting point (°C)	Heat of fusion (kJ/kg)	Thermal conductivity (W/mK)	Specific heat capacity (kJ/kg K)	References
Mineral oil	Sensible	310 (boiling point)	–	0.12	1.966	[3]
Concrete	Sensible	–	–	1.5	0.85	[3]
Rocks	Sensible	–	–	0.48	0.96	[4]
Sand	Sensible	–	–		0.8	[5]
Parrafin wax	PCM	64	173	0.167 (liq.) 0.346 (solid)	–	[6]
Palmitic acid (C-16)	PCM	64	185	0.162	–	[7]
Stearic acid (C-18)	PCM	69	209	–	2.07	[7]
KNO_3/KCl	PCM	320	74	0.5	1.21	[8]
$MgCl_2/KCl$	PCM	435	351	0.81	0.8(solid) 0.96(liq.)	[9]
Al-Si	PCM	576	560	160	1.038	[10]
LiF/CaF_2	PCM	767	790	–	–	[11]
Acetamide	PCM	82	263	–	1.94	[12]

butanol with lower density and specific heat than water are used, but, those are flammable, so requires additional safety of the system [13, 15]. The present work is emphasized for low-temperature, so water is used as storage medium. A significant number of research work has been published on the design of heat storage systems for solar thermal collectors. Parsad and Muthukumar [16] have numerically investigated a sensible heat storage unit for high-temperature application using a 3D mathematical model. The storage is configured with cast steel, iron and concrete. The predicted values are in good match with the reported values. Nkhonjera et al. [17] discussed in detail various storage units for cooking in different research work and presented a good review on STSs. Celador et al. [18] described a methodology to select the different design parameters for thermal storage systems to decrease the system cost.

It is understandable that a small-scale solar thermal system, a heat storage system with reasonable cost will increase the utility and adoption of this technology among potential users. The present work is aimed to design an effective small capacity storage tank at low cost. Two storage tanks are designed for this purpose, and a thermal performance study is conducted with the help of a lab scale experimental set-up.

2 Development of Heat Storage System

A storage system extracting heat from a heat source is depicted in Fig. 1. This heat source is a solar collector when the storage is used to provide thermal backup or diurnal storage of solar energy. In the present study, this heat source is an electric heater with a controlled energy input to a metallic heat exchanger immersed in a heat transfer fluid. This arrangement is analogues to the solar thermal energy system for testing of heat storage system under indoor laboratory conditions.

The energy balance equation for this set-up is written as:

$$q_{\mathrm{us}} = q_{\mathrm{af}} - q_{\mathrm{ls}} \tag{1}$$

$$q_{\mathrm{us}} = (m_w C_w + m_s C_s)\frac{\Delta T_s}{\mathrm{dt}} \tag{2}$$

$$q_{\mathrm{ls}} = U_{\mathrm{ls}} A_s \left(\overline{T_s} - \overline{T}_a\right) \tag{3}$$

$$q_{\mathrm{af}} = q_{\mathrm{ah}} - q_{\mathrm{lh}} = m_f C_f \frac{\Delta T_f}{\mathrm{dt}} \tag{4}$$

$$q_{\mathrm{ah}} = q_e \eta_e \tag{5}$$

$$q_{\mathrm{lh}} = (m_h C_h)\frac{\Delta T_f}{\mathrm{d}t}. \tag{6}$$

where q_{us} is the rate of energy utilization by the storage system. q_{af} and q_{ah} are the rates of available heat to the heat transfer fluid and the heat source, resptively. q_{ls} and q_{lh} are the rates of energy loss from heat storage and heat source, respectively. q_e is the rate of electric energy supplied to electric heater and η_e is the conversion (electric

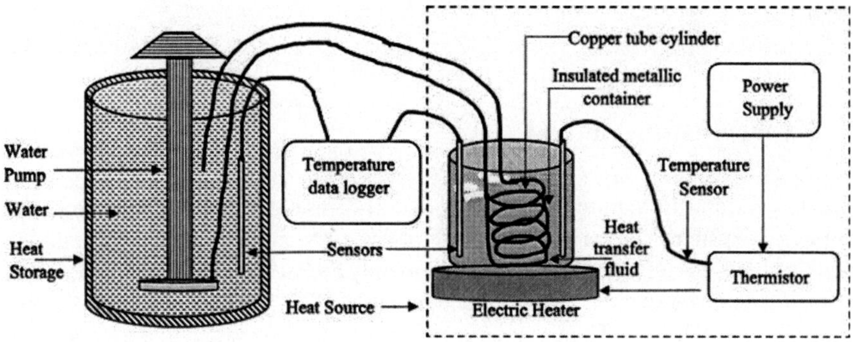

Fig. 1 Charging set-up (heat extraction) of heat storage systems from a heat source

energy to thermal energy) efficiency. m_w, m_s, m_f and m_h are masses of water (sensible storage medium), storage tank, heat transfer fluid and heat exchanger, respectively. C_w, C_s, C_f and C_h are the specific heats of water (sensible storage medium), storage tank material, heat transfer fluid and heat exchanger material, respectively. ΔT_s and ΔT_f are the rise of temperatures of storage system (i.e. water) and heat transfer fluid in dt time interval. U_{ls} and A_s are overall heat loss coefficient and total area of the storage system, respectively. $\overline{T_s}$ and $\overline{T}_a$ are average storage and ambient temperatures during the time period Δt. In the above equation, temperature stratification is neglected due to low capacity of storage system.

Using Eqs. (1)–(6)

$$q_{\text{us}} = q_e \eta_e - m_f C_f \frac{\Delta T_f}{\mathrm{d}t} - U_{\text{ls}} A_s \left(\overline{T_s} - \overline{T}_a\right) \tag{7}$$

The above equation is for charging cycle of the system. For discharging cycle energy balance equation is:

$$q_{\text{ls}} = m_s C_s \frac{\Delta T_s'}{\mathrm{d}t} \tag{8}$$

where $\Delta T_s'$ is decrease in storage system temperature in the dt time period. Clearly, charging and discharging of a the heat storage system are govern by the various heat losses occurred in a storage system. For low charging and high discharging time, these losses should be minimized while keeping the cost of the system reasonably fair. These heat losses mainly depend on the (i) storage structure elements, (ii) operation temperature range, (iii) temperature stratification and (iv) thermal losses [19, 20].

Here, focus is on minimize the heat losses by keeping (i) low operation temperature range (50–90 °C), (ii) low capacity of storage system to avoid temperature stratification and (iii) use of locally available material for base and layered insulation material to reduce fabrication cost. Design parameters of the system are given in Table 2. All the systems are cylindrical in shape having a maximum water capacity of 8 L.

3 Experimental Method

The charging and discharging curves of storage systems are compared with a commercial storage system. The component details of these systems are given in Table 2. The laboratory set-up consists (Figs. 1 and 2) of mainly four sections (i) a heat source that is an electric heater with a controlled energy input, (ii) a metallic heat exchanger to receive direct heat from the heat source, (iii) a heat storage system that receives heat from the heat exchanger and (iv) the temperature data logger that records temperature through the sensors which are kept at various positions in the system. An electric

Table 2 Design parameters of the heat storage systems

System	Material	Diameter (m)		Useful Height (m)	Insulation and casing material	Insulation thickness (m)
		Inner	Outer			
Commercial HS	Steel	0.215	0.285	0.275	Polyurethane, Plastic casing	0.035
Designed HS-1	Aluminium	0.24	0.31	0.31	Glass wool, cardboard casing	0.035
Designed HS-2	Aluminium	0.24	0.30	0.32	Cardboard, and theramcole, cardboard casing	0.03

heater (power consumption 1.5 kWh) is used to give thermal energy to HTF. The heat exchanger is an insulated metallic cylinder container (capacity 2 L) made of copper and it contains the heat transfer fluid. A copper tube cylinder is immersed in it for heat exchange from HTF. The metallic container is an insulated container with an inner diameter of 0.115 m, outer diameter of 0.18 m and height of 0.175 m.

In the present study, Hytherm (1 L) is used as HTF. The copper tube cylinder (CTC) is designed using a copper tube of 1 m length. The tube is rounded keeping the diameter 0.10 m to make spiral rings of height 0.08 m. This copper tube cylinder is kept in the HTF container. This heat exchanger having HTF and CTC is placed upon

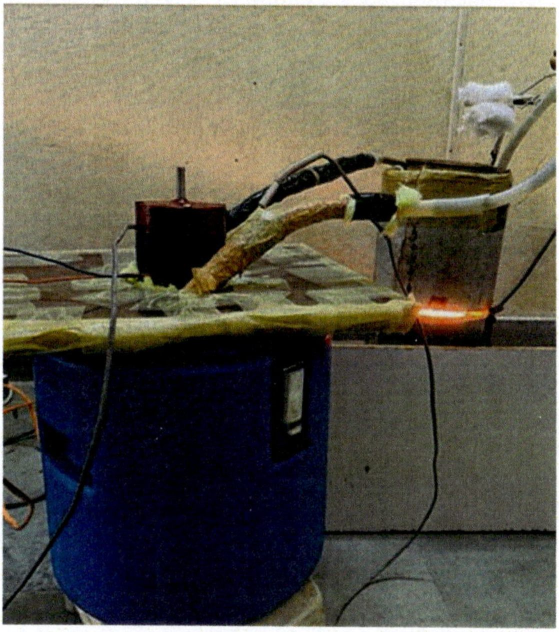

Fig. 2 Lab scale testing set-up with commercial heat storage system

the electric heater to receive the heat directly from the electric heater. By this heat, the temperature of HTF increases continuously. To obtain the constant temperature of HTF, an auto cut of power supply is set using a thermistor (±5 °C). In the present study, this temperature is kept at 150 °C in all the experiments. The water is chosen as a heat storage material. The water in the heat storage is circulated through the copper tube cylinder. Both the ends of this cylinder is connected with silicon tubes, and these tubes are immersed in the heat storage system. A water pump is dipped in the storage tank to circulate the water in the copper tube cylinder and exchange the heat from HTF. The temperatures of HTF and HS are measured by using k-type thermocouples dipped in the respective containers. The sensors' inputs are recorded through a data logger (85XX + Masibas data logger, LC 0.1 °C). For the charging experiments, the temperatures are recorded in 01 min interval while for the discharging set-up this time interval is 15 min.

4 Results and Discussion

The representative thermal profiles are shown in Figs. 3, 4, 5, 6 and 7. Figure 3 and 4 depict the charging of heat storage systems (HS, HS-1, HS,2) up to 90 °C while keeping all experimental conditions identical, i.e. HTF material and its temperature (initial and saturation), power source (heat input), initial room temperature and flow rate. The load capacity is 6 L and 8 L in the charging cycle. The discharging cycle under environmental conditions is also recorded for the different load capacities of the heat storage systems. These are depicted in Figs. 5 and 6. The temperature of heat transfer fluid (hytherm) is plotted in Fig. 7 for the different sets of experiments.

A comparative curve (Fig. 3) for charging of heat storage tank up to 90 °C indicates that the lab designed heat storage systems (HS-1 and HS-2) took around 6–10 min extra to reach 90 °C in comparison with the commercial HS for 6 L water, while for 8 L (Fig. 4), the time taken by systems HS and HS-1 is almost same. For all the systems, the heating of the 8 L water load took a larger time (5 to 8 min) than the 6 L load. The discharging plots shown in Fig. 5 infer that after 15 h, the temperatures of HS, HS-1 and HS-2 are 51.1 °C, 42.9 °C and 39.6 °C, respectively, for 6 L water load. The temperature difference of about 10–12 °C is found in this. In the other experiment for higher load capacity (8L), the temperature falls after 15 h are found 53.1, 47.6 and 44.5 °C for systems HS, HS-1 and HS-2, respectively. Overall, in the 15 h discharging time period, a little difference of about 6–12 °C is noted in the comparison with the commercial system. The rate of increase of the temperature for HS-1 is better than HS-2 due to good insulation material, and this rate is also equal to the commercial system (HS). The rate of decrease in temperature for both the developed system is slightly higher than HS. This is due to the marginally larger diameter of HS-1 and HS-2 about 0.03 m than the commercial one (HS). Using Eq. (8), Table 2, Figs. 5 and 6, the overall loss coefficients for the designed systems are computed and are given in Table 3. The curves between ΔT_s and $\left(\overline{T_s} - \overline{T}_a\right)$ are plotted to obtain the curve slopes for each system under 6 and 8 L water conditions.

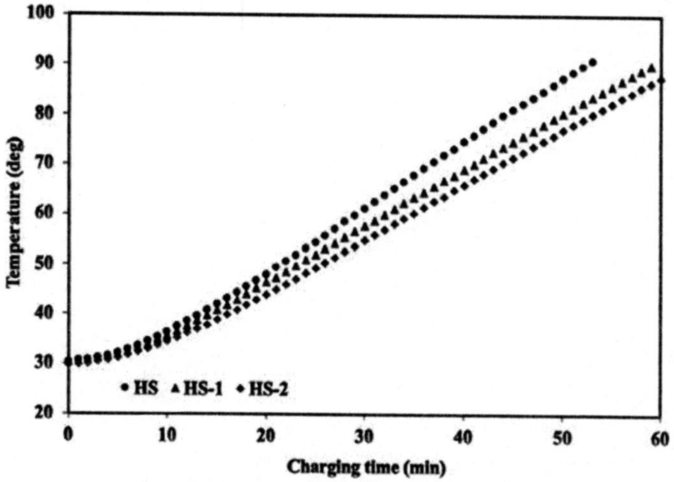

Fig. 3 Charging time period vs temperature curve for heat storage systems (HS, HS-1 and HS-2) for 6 L water load capacity

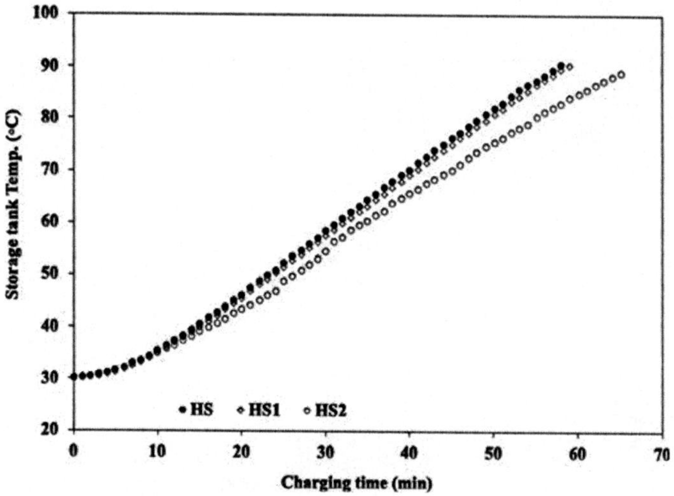

Fig. 4 Charging time period vs temperature curve for heat storage systems (HS, HS-1 and HS-2) for 8 L water load capacity

The time interval to compute ΔT_s is taken 1 h. The specific heat of water is taken to be 4186 J/kg °C.

Table 3 infers that the minimum heat loss coefficient is found for HS-1 system, while highest value is obtained for HS-2. This is a clear indication of better performance of the system HS-1 in comparison with the HS and HS-2. The costs of developed systems are about 60–65% less than the commercial ones due to use of different

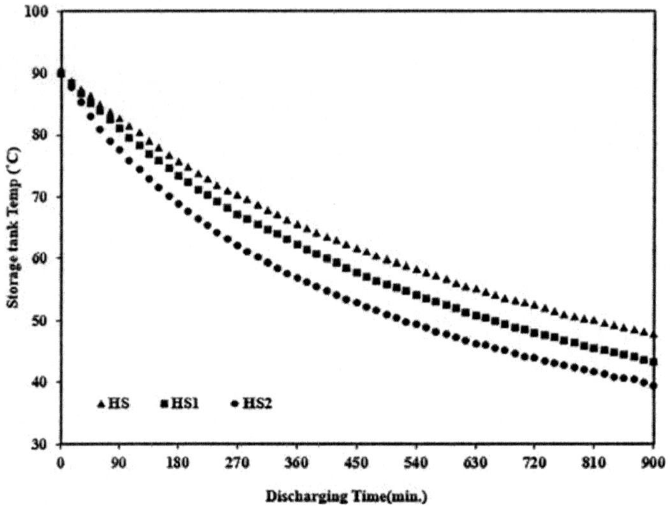

Fig. 5 Discharging time period vs temperature curve for heat storage systems (HS, HS-1 and HS-2) for 6 L water load capacity

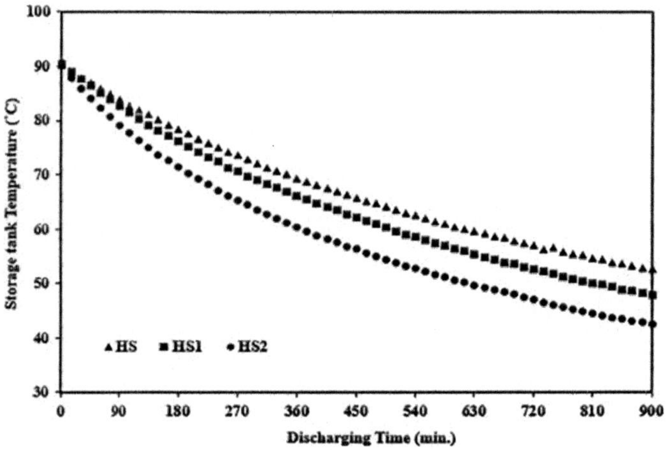

Fig. 6 Discharging time period vs temperature curve for heat storage systems (HS, HS-1 and HS-2) for 8 L water load capacity

layers of cost-effective insulation materials. These initial results obtained at the lab scale infer that the lab designed systems have significant potential.

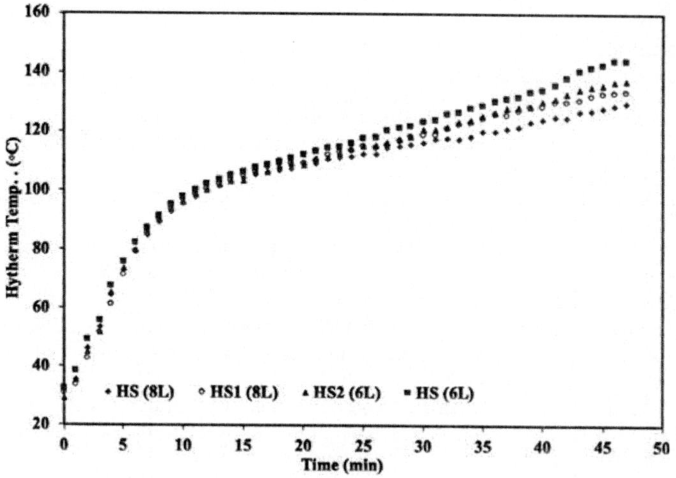

Fig. 7 Temperature profile of Hytherm (HTF) in different set of experiments

Table 3 Heat loss coefficients of heat storage systems

System	A_s (total area m^2)	Curve slope		Regression coefficient of curve		U_{ls} (overall heat loss coefficient (W/m^2 °C)	
		6 L	8L	6 L	8L	6 L	8L
Commercial HS	0.25	0.1126	0.1027	0.96	0.99	3.14	3.82
Designed HS-1	0.32	0.1255	0.1124	0.99	0.98	2.73	3.26
Designed HS-2	0.33	0.1768	0.1589	0.97	0.97	3.73	4.47

5 Conclusion

The heat storage systems are the important segment of any solar thermal collector system. Two low-cost solar thermal storages are fabricated to test low-temperature applications, and their experimental studies are performed and compared to the commercial one. The different charging and discharging tests are performed with the systems with water as heat storage material. The load capacities of all the systems are the same. The thermal performance of the fabricated systems are found to be competitive to commercial ones. The costs of the fabricated systems are reasonable.

Acknowledgements This research is carried out at solar energy research laboratory (SERL), Department of Physics, University of Rajasthan, Jaipur as a part of two project works and these projects (DST/TMD/SERI/S91 and EEQ/2018/000811) are supported by the Department of Science and Technology (DST) and Science and Engineering Research Board (SERB), New Delhi.

References

1. Guana C, Lub H, Zhang L, Yu Z (2020) Regulation of the output temperature in a novel water heating system using solid graphite as sensible heat thermal energy storage medium: effects of water tank. Energy Rep 6(7):160–171
2. Sarbu I, Sebarchievici C (2016) Solar heating and cooling systems: fundamentals, experiments and applications. Elsevier, Oxford, UK
3. Gil A, Medrano M, Martorell I, La´zaro A, Dolado P, Zalba B, Cabeza LF (2010) State of the art on high temperature thermal energy storage for power generation. Part 1 concepts, materials and modellization. Renew Sustain Energy 14:31–55
4. Hänchen M, Brückner S, Steinfeld A (2011) High-temperature thermal storage using a packed bed of rocks–heat transfer analysis and experimental validation. Appl Therm Eng 31(10):1798–1806
5. Yadav V, Yadav A (2013) Experimental investigation of novel design of solar cooker with dual thermal storage unit based on parabolic dish-type collector. Int J Energy Clean Environ 14(4):295–310
6. Farid MM, Khudhair AM, Razack SAK, Al-Hallaj S (2004) A review on phase change energy storage: materials and applications. Energy Convers Manag 45:1597–1615
7. Hasnain SM (1998) Review on sustainable thermal energy storage technologies, part1: heat storage materials and techniques. Energy Convers Manag 39:1127–1138
8. Michels H, Pitz-Paal R (2007) Cascaded latent heat storage for parabolic trough solar power plants. Sol Energy 81:829–837
9. Kenisarin MM (2010) High-temperature phase change materials for thermal energy storage. Renew Sustain Energy Rev 14(3):955–970
10. Sun JQ, Zhang RY, Liu ZP, Lu GH (2007) Thermal reliability test of Al–34%Mg–6%Zn alloy as latent heat storage material and corrosion of metal with respect to thermal cycling. Energy Convers Manage 48:619–624
11. Xing Y-M, Xu X, Yuan X-G, Cui H-T, Zhang Y-H (2004) Numerical simulation of high-temperature phase change heat storage system. Heat Transf Asian Res 33(1):32–41
12. Agrawal H, Yadav V, Kumar Y, Yadav A (2014) Comparison of experimental data for sensible and latent heat storage materials for late-evening cooking based on a dish-type solar cooker. Int J Energy Clean Environ 15(1):47–72
13. Basecq V, Michaux G, Inard C, Blondeau P (2013) Short-term storage systems of thermal energy for buildings: a review. Adv Build Energy Res 7(1):66–119
14. Vedrtnam A, Upadhyay MP, Kalauni K (2019) Experimental and theoretical studies of the heat transfer characteristic of the lab-scale sensible heat storage system. Int J Energy Clean Environ 20(3):1–26
15. International Energy Agency (IEA) (2016) Energy conservation through energy storage (ECES) programme. Brochure, International Energy Agency, Paris, France
16. Prasad L, Muthukumar P (2013) Design and optimization of lab-scale sensible heat storage prototype for solar thermal power plant application. Sol Energy 97:217–229
17. Nkhonjera L, Ochende TB, John G, King'ondu CK (2017) A review of thermal energy storage designs, heat storage materials and cooking performance of solar cookers with heat storage. Renew Sustain Energy Rev 75:157–167
18. Campos-Celador A, Diarce G, Larrinaga P, Romero AAG (2020) A simple method for the design of thermal energy storage systems. Energy Storage 2:140
19. Duffie JA, Beckman WA (2006) Solar engineering of thermal process. Wiley & Sons, Inc., USA
20. Sukhatme SP (2007) Solar energy: principles of thermal collection and storage. Tata McGraw-Hill Book Co., Delhi

Manufacturing Science

Structural, Morphological and Mechanical Property Analysis of TiAlN Thin Film Coating Deposited by CVD Technique

Spandan Guha and Soham Das

1 Introduction

The applications of hard coatings for industrial purposes (e.g. cutting of hard materials, machining of hardened tools, etc.) came in existence in the early 1970s [1]. Even the application of TiAlN can be extended up to dry and minimum quantity lubrication (MQL) machining [2]. The development and growth of titanium aluminium nitride (TiAlN) was taken place to fulfil the need of high-speed machining and machining of hard materials as this alloy possesses higher resistance to oxidation and higher thermal stability than TiN alloy. TiAlN alloy has a wide acceptability in various industrial applications in the form of hard coatings over cutting tools, various moulds [3], electronic devices and optical coatings [4]. TiAlN coatings come under the category of nanostructure coatings which specifically allow higher coating hardness with low residual stress and lower Young's modulus [5]. TiAlN nanocrystals were found to exhibit face-centred cubic (FCC) crystal structure [6]. The synthesis of TiAlN nanocomposite coatings can be synthesized by various techniques [6–9].

The properties of nanocomposites greatly depend upon the crystal size and these crystals are subjected to intrinsic residual strain due to variation in crystal size confinement. The presence of intrinsic strain (a crucial elastic property) in the crystals depends upon various deposition parameters and can be controlled by choosing an optimal set of synthesis parameters [10]. X-ray diffraction (XRD) is a characterization technique that is employed to estimate the crystallite size and crystallinity of the nanocrystals by analyzing the diffraction peaks observed at different diffraction

S. Guha (✉)
School of Mechanical Engineering, KIIT University, Bhubaneswar, Odisha 751024, India
e-mail: spandan.guhafme@kiit.ac.in

S. Das
Department of Mechanical Engineering, Sikkim Manipal Institute of Technology, Sikkim Manipal University, Majhitar, Rangpo 737136, India

© The Author(s), under exclusive license to Springer Nature Singapore Pte Ltd. 2023
S. Revankar et al. (eds.), *Recent Advances in Thermofluids and Manufacturing Engineering*, Lecture Notes in Mechanical Engineering,
https://doi.org/10.1007/978-981-19-4388-1_36

angles [11]. It has been reported that the presence of intrinsic strain which primarily takes place due to crystal size confinement resulted in XRD peak broadening [12]. This peak broadening is divided in two categories as size broadening and strain broadening [12]. The lattice strain is mainly caused due to various lattice defects such as contact stress, stacking faults, grain boundary defects, dislocation density, etc. [13–16].

Therefore the detailed investigation of x-ray diffraction peak broadening gives a very clear understanding of stress-induced in the crystals, energy density, intrinsic strain and many more [17]. For the estimation of crystallite size and other elastic properties research groups have reported various mathematical models such as Williamson Hall model (W–H), Balzar Method, etc. [18–22]. In W–H model the full width at half maxima (FWHM) of the XRD peak is taken into consideration for calculating the various elastic properties along with average crystallite size. W–H model basically includes uniform deformation model (UDM), uniform stress deformation model (USDM) and uniform deformation energy density model (UDEDM). Similarly, another model known as Size- Strain Plot (SSP) and Halder-Wagner (H–W) model is also employed to evaluate the various elastic properties of nanocrystals.

In this work, a comparison has been carried out between the particle size of CVD deposited TiAlN coating obtained from atomic force microscopy (AFM) characterization with the crystallite size obtained from Scherrer's equation and its various models.

2 Experimental Details

TiAlN single-layer thin film coating has been deposited over Si (100) substrate using thermal CVD technique. Before placing the substrate inside the CVD chamber the substrate was thoroughly cleaned using Radio Corporation of America (RCA) cleaning technique whose detailed explanation can be found in other literature [23]. After thorough cleaning of substrate it was placed inside the CVD tube furnace by placing it over a ceramic boat. Titanium dioxide (TiO_2) and Aluminium (Al) powder were mixed thoroughly in a ceramic boat in 1:1 ratio and placed inside the CVD chamber 6 cm behind the ceramic tube coating the substrate. The base pressure and working pressure of the CVD chamber were maintained at 3×10^{-6} Torr and 300 mTorr, respectively. Nitrogen (N_2) was used as precursor gas and was passed through the CVD chamber at 12 sccm flow rate. The furnace heating and cooling rate were maintained at 5 and 3 °C/min. The deposition was carried out for 90 min and the deposition temperature was fixed at 1100 °C. The chemical reaction involved is mentioned below:

$$TiO_2 + Al \xrightarrow{N_2} TiAlN + O_2 \tag{1}$$

2.1 Characterization of Samples

In this work SEM, EDS, AFM, XRD and nanoindentation processes are used for sample characterization. The details about the instrument is same as used in [24–27].

3 Results and Discussion

3.1 Compositional Study of TiAlN Coating Using EDS

The elemental composition in (atomic % and weight %) of TiAlN coating deposited using CVD technique is shown in Fig. 1 and Table 1. Results revealed successful deposition of TiAlN coating over the substrate. However, during EDS measurement the elemental composition of parent substrate (Si) was not considered. Here traces of oxygen (O_2) were also found in the coating. The presence of small amount of O_2 in the coating could be due to the presence of residual oxygen left in the deposition chamber during synthesis. The at. % of Ti was found to be maximum followed by Al and N.

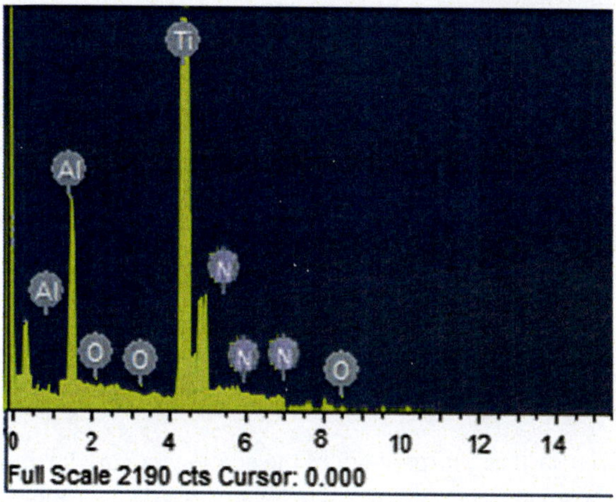

Fig. 1 EDS analysis of TiAlN coating prepared by CVD

Table 1 EDS results of TiAlN coating

Sample code	TiAlN	
	Wt%	At%
Ti	35.56	69.45
Al	57.88	21.54
N	6.56	9.01

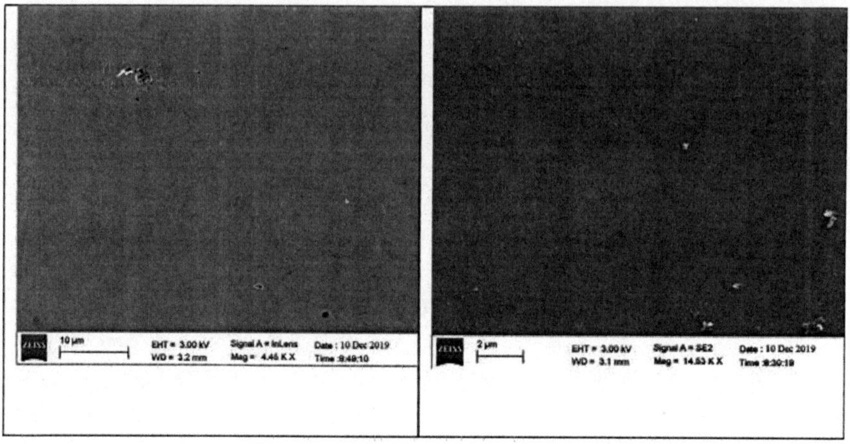

Fig. 2 SEM morphology of TiAlN coating captured at 2 and 10 μm microscopic levels

3.2 Surface Morphological Analysis Using SEM

The surface morphology of TiAlN coating deposited by CVD technique has been investigated using SEM technique. The SEM images (Fig. 2) of TiAlN coating was captured at two different microscopic scales of 2 and 10 μm, respectively. The SEM images revealed smooth coating over the substrate. However, it has also been observed that the morphological image of TiAlN coating captured at 2 μm showed the presence of small, agglomerated particles distributed over the surface. These agglomerations could be a mixture of Ti and Al powder which could have been deposited over the sample due to partial melting and evaporation at deposition temperature. The TiAlN coating thickness was observed as 6.45 ± 0.54 μm.

3.3 Morphological Analysis Using AFM

The surface morphology of the coating captured using the AFM technique has been shown in Fig. 3. The result indicated that the morphology of coating recorded at 5 μm consists of pyramidal-shaped grains distributed throughout the coating in which the maximum and average particle size was observed as 36.83 nm and 5 nm, respectively. The result also reveals the surface roughness of TiAlN coating as 12.38 nm.

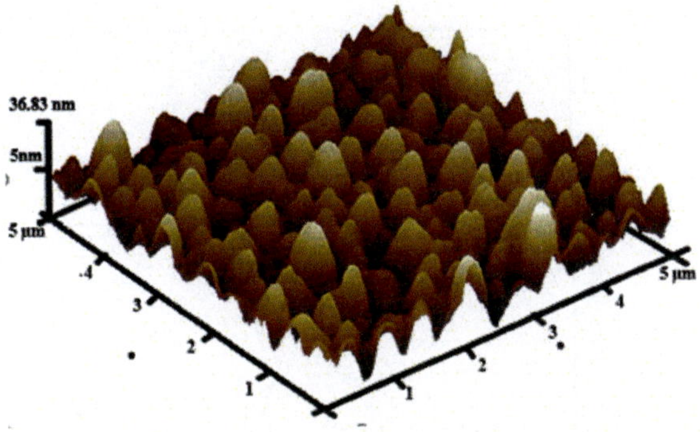

Fig. 3 AFM image of TiAlN coating

3.4 *Structural Analysis of TiAlN Coating Using XRD Technique*

The structural analysis such as investigation of crystallite size, evaluation of full width at half maxima (FWHM), interplanar spacing of TiAlN coating has been carried out using the XRD technique (Fig. 4). The diffraction pattern of TiAlN coating was recorded for 2θ ranging from 5° to 90°. The crystallite size of TiAlN coating was prepared at 1100 °C was calculated using the Scherrer's equation given by Eq. 2 Where, D_p—Crystallite size (nm), λ—wavelength ~ 0.154 nm, β—FWHM and is calculated in radian by considering the value of instrumental peak broadening (β_{is}) and physical broadening (β_{ip}) whose governing equation is Eq. 3.

$$D_p = \frac{0.9\lambda}{\beta Cos\theta}. \tag{2}$$

$$(\beta)^2 = (\beta_{\text{ip}})^2 - (\beta_{\text{is}})^2. \tag{3}$$

The XRD results revealed the diffraction peaks of TiAlN at 36.12°, 42.5°, 62.22° and 74.27° corresponding to (111), (200), (220) and (311) diffraction phases. The FWHM of the diffraction peak was evaluated as 2.72°, 1.05°, 1.90° and 1.56°, respectively. The average crystallite size of TiAlN coating considering all the diffraction peaks was evaluated as 4.7 nm. The average interplanar spacing and lattice constant of TiAlN coating were observed as 1.84 nm and 4.25 nm, respectively. The TiAlN coating was found to possess FCC B1 NaCl type crystal structure. The diffraction pattern obtained in the present case is in good agreement with the past published literature [23].

The crystallite size obtained from using Scherrer's equation has been compared with various theoretical models such as the Williamson Hall (W–H) model which mainly comprises of uniform deformation model (UDM), uniform stress deformation model (USDM), uniform deformation energy density model (UDEDM). These models are used to estimate the elastic properties of the deposited coating such as

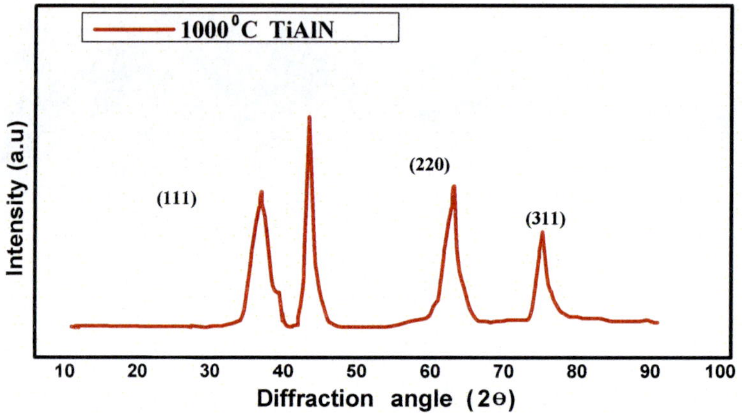

Fig. 4 XRD plot of TiAlN coating

induced strain, induced stress, energy density and many more. There exists another model known as a size-strain plot which is proved to be a very useful model in calculating the various elastic properties of the coating extracted from low angle XRD pattern [28–30]. The Halder-Wagner method is also a suitable theoretical model to estimate the elastic properties of the thin film coating through XRD peak broadening using the Voigt function [31].

3.4.1 Estimation of Elastic Properties Using Williamson's Hall Model

The application of Scherrer's equation is restricted up to the estimation of crystallite size. It does not explain anything related to micro-strain induced in the nanocrystal due to grain boundary, stacking faults, point defects, etc. [32–34]. According to W–H model, the peak broadening of the XRD pattern takes place due to varying nano crystallite sizes and presence of micro-strain and based on this the overall peak broadening can be written as Eq. 4.

$$\beta_{\mathrm{hkl}_{(\mathrm{overall})}} = \beta_{\mathrm{micro\ strain}} + \beta_{\mathrm{crystallite\ size}}. \tag{4}$$

here the average crystallite size, induced micro-strain have been evaluated using UDM, USDM and UEDEM model which has been presented one by one in the further sections.

3.4.2 Uniform Deformation Model

In this model there presents intrinsic uniform strain in the crystals in all over the crystallographic directions mainly because of crystal imperfections. Based on this

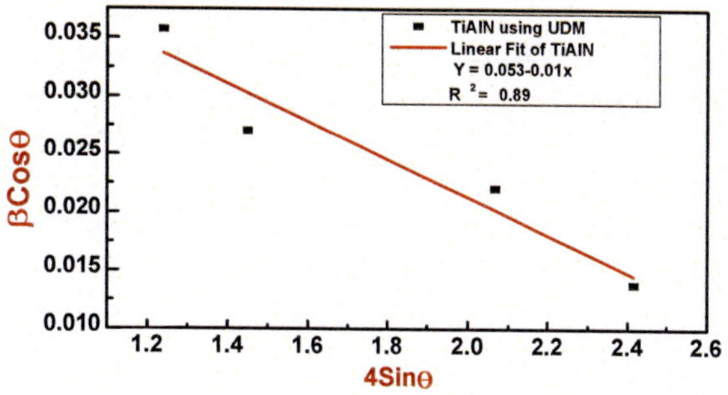

Fig. 5 UDM plot of CVD deposited TiAlN coating

UDM works with the theoretical assumption of the isotropic nature of strain [35]. Therefore, the peak broadening of the diffraction pattern of TiAlN caused due to induced strain can be deduced as Eq. 5. Where the overall peak broadening in the crystallographic directions can be given by Eq. 6.

Here,$\beta_{(\text{hkl})_{(\text{overall})}}$ FWHM of TiAlN diffraction peak obtained from XRD experimental data and is governed by Eq. 7 on shifting of numerator and denominator, the equation can be arranged as Eq. 8.

$$\beta_{(\text{hkl})_{\text{strain}}} = 4\varepsilon \tan\theta. \tag{5}$$

$$\beta_{(\text{hkl})_{(\text{overall})}} = \beta_{(\text{hkl})_{\text{strain}}} + \beta_{(\text{hkl})_{\text{size}}} \tag{6}$$

$$\beta_{(\text{hkl})_{(\text{overall})}} = \frac{k\lambda}{D} \times \frac{1}{\cos\theta} + 4\varepsilon \tan\theta. \tag{7}$$

$$\beta_{(\text{hkl})_{(\text{overall})}} \cos\theta = \frac{k\lambda}{D} + 4\varepsilon \sin\theta \tag{8}$$

Comparing Eq. 8 with a straight-line equation a plot between 4sin θ (as the x-axis) and $\beta_{(\text{hkl})_{(\text{overall})}} \cos\theta$ (as—y-axis) has been prepared and shown in Fig. 5. For this fitted line, the slope represents the induced strain and the intercept of the theoretical average crystallite size of the nanocrystals. Here, the value of R^2 was observed as 0.89 and the presence in the coating at the nanocrystal level could be due to lattice expansion or contraction which is primarily caused due to continuous alteration of atomic arrangements [36]. The negative slope of the fitted line indicates lattice contraction [12].

3.4.3 Uniform Stress Deformation Model

The basic limitation of UDM is its assumption regarding the homogenous and isotropic nature of the coating. However, in actuality, the arrangement of crystals in a coating is anisotropic due to which the stress distribution among the crystals cannot be neglected and needs to be considered [11]. Due to the presence of stress and strain in the coating, according to Hook's law stress is proportional to strain and is given by Eq. 9.

$$\sigma_{\text{coating}} = \mathrm{Y}_{\text{hkl (coating)}} * \varepsilon_{\text{coating}}. \tag{9}$$

where $Y_{\text{hkl (coating)}}$ is Young's modulus.

Based on USDM model it is assumed that XRD peak broadening is a result of stress-induced and anisotropic nature of Young's modulus [12]. Therefore, USD model is represented as Eq. 10.

$$\beta_{(hkl)_{\text{(overall)}}} \cos\theta = \frac{k\lambda}{D} + 4 * \sigma * \frac{\text{Sin}\theta}{Y_{\text{hkl (coating)}}} \tag{10}$$

$Y_{\text{hkl (coating)}}$ is calculated as Eq. 11.

$$\frac{1}{Y_{hkl}} = S_{11} - \left[(S_{11} - S_{12}) - \frac{1}{2} * S_{44}\right] * \left[\frac{h^2k^2 + l^2k^2 + h^2l^2}{h^2 + k^2 + l^2}\right] \tag{11}$$

S_{11}, S_{12}, S_{44}, indicate elastic compliance of TiAlN and can be estimated from elastic stiffness constant $C_{11'}$, $C_{12'}$, $C_{44'}$, [11]. Equations 12–14.

$$S_{11} = \frac{C_{11'} + C_{12'}}{(C'_{11} - C_{12'}) * (C_{11'} + 2C_{12'})}. \tag{12}$$

$$S_{12} = \frac{-C'_{12}}{(C'_{11} - C_{12'}) * (C_{11'} + 2C_{12'})}. \tag{13}$$

$$S_{44,} = \frac{1}{C_{44',}}. \tag{14}$$

Here, the value of $C_{11'}$, $C_{12'}$, $C_{44'}$, was referred from published literature as 489.81 GPa, 151.90 GPa and 204.81 GPa, respectively [37]. The corresponding values of elastic compliance calculated from Eqs. (12), (13) and (14) were 0.00239×10^{-10}, 0.000566×10^{-10} and 0.00488×10^{-10}, respectively. At diffraction phase of (111), (200), (220) and (311) the corresponding Y_{hkl} was were observed as 24.39, 39.03, 24.66 and 28.27 MPa, respectively. So, the average Y_{hkl} was calculated as 29.08 MPa. Therefore the slope and intercept of the fitted line for the plot between $4^*\sin\theta/Y_{hkl}$ (as the x-axis) and $\beta_{(hkl)_{\text{(overall)}}} \cos\theta$ (as—y-axis) was obtained as –0.1 and 0.033, respectively and the R^2 value of the fitted line was observed as 0.86. Here again, the plot (Fig. 6) indicated lattice contraction among the crystals.

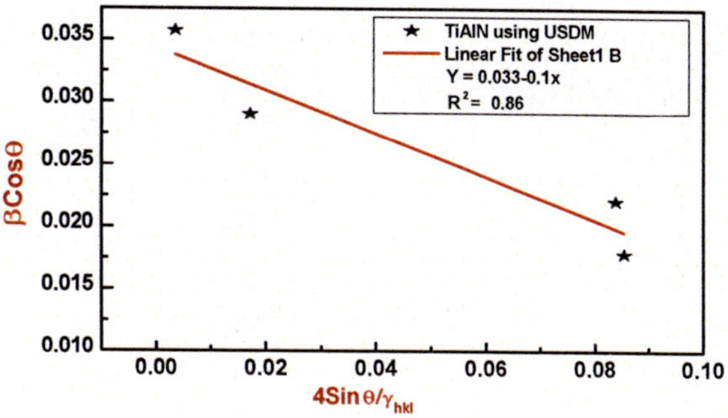

Fig. 6 USDM plot of CVD deposited TiAlN coating

Uniform Deformation Energy Density Model

In the UDED model, the estimation of crystallite size is carried out based upon the theory of uniform anisotropic lattice strain in all directions of crystallographic orientation which is mainly caused due to density of deformation energy [38].

Therefore, based on the above assumption the energy density (U') and strain can be interrelated as Eq. 15 As it is well known that $\sigma = Y_{\text{hkl (coating)}} * (\varepsilon_{\text{coating}})$, therefore the intrinsic strain as a function of energy density can be expressed as Eq. 16 here, $Y_{\text{hkl (coating)}}$—anisotropic modulus of elasticity. Therefore, by placing Eq. 15 in Eq. 7 we get Eq. 17.

$$U' = \left(\varepsilon_{\text{coating}}\right)^2 * \frac{Y_{hkl(\text{coating})}}{2}. \tag{15}$$

$$\varepsilon_{\text{coating}} = \sigma * \sqrt{\frac{2 \times U'}{Y_{hkl(\text{coating})}}}. \tag{16}$$

$$\beta_{(hkl)_{(\text{overall})}} \cos\theta = \frac{k\lambda}{D} + 4\sigma * \sin\theta \sqrt{\frac{2 \times U'}{Y_{hkl(\text{coating})}}}. \tag{17}$$

The slope of the fitted line in the graph (Fig. 7) plotted between $4\sin\theta * \sqrt{\frac{2*u}{Y_{hkl}}}$ (x-axis) and $\beta_{hkl\ (\text{total})} \cos\theta$ (y-axis) was calculated as −0.03 indicating lattice contraction [12] and the R^2 value of the linear fit was obtained as 0.91.

Size Strain Plot (SSP)

The basic difference between the W–H model and the SSP model is that in the case of the W–H method peak broadening is a function of a combination of size and strain-induced broadening. However, in the SSP model, the peak broadening of the diffraction peaks is considered to be a combination of Lorentzian and Gaussian functions. In the SSP model size broadened XRD peak is a Lorentz function and

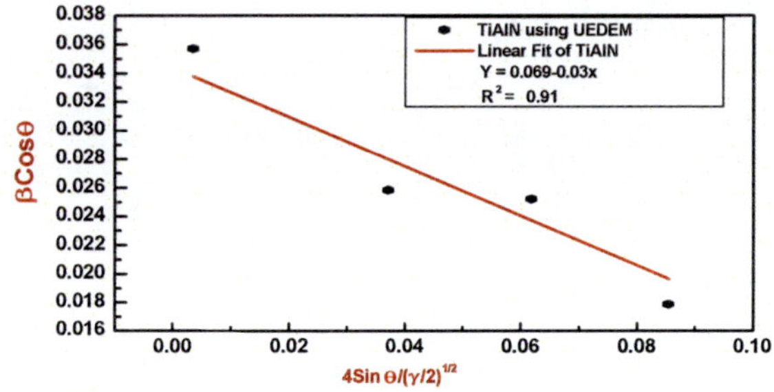

Fig. 7 UDEDM plot of CVD deposited TiAlN coating

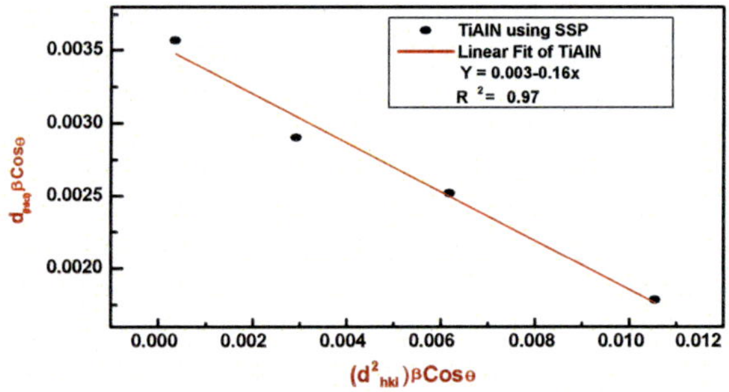

Fig. 8 SSP plot of CVD deposited TiAlN coating

strain broadened XRD profile is a function of Gaussian function [34]. Therefore, the governing equation for the SSP model can be expressed as the SSP model provides higher flexibility in estimating the various elastic properties of a coating as it emphasizes more for lower angle reflections which always provides higher accuracy and precision as compared to higher angle reflection. Therefore the governing equation in the SSP model is given as [38].

$$\beta_{(\mathrm{hkl})_{(\mathrm{overall})}} = \beta_{\mathrm{Gaussian}} + \beta_{\mathrm{Lorenztian}}. \tag{18}$$

$$(d_{\mathrm{hkl}}.\beta_{(hkl)_{(\mathrm{overall})}}\mathrm{Cos}\theta)^2 = \frac{k\lambda}{D}(d_{\mathrm{hkl}}^2.\beta_{(hkl)_{(\mathrm{overall})}}\mathrm{Cos}\theta) + \frac{\varepsilon^2}{4} \tag{19}$$

Here d_{hkl} represents lattice distance between (hkl) planes. The slope of the fitted line in the graph (Fig. 8) plotted between ($d_{\mathrm{hkl}}{}^2$. β_{hkl} . Cosθ) (x-axis) and (d_{hkl} . β_{hkl} . Cosθ)^2 (y-axis) was calculated as −0.16 indicating lattice contraction [12] and the R^2 value of the linear fit was obtained as 0.97. For all the nomenclatures of Eqs. 4–19 (see Ref. [26]) (Table 2).

Table 2 Average crystallite size and elastic properties of CVD TiAlN thin film estimated by using different model

Coating	Scherrer equation	W–H method									SSP		AFM avg. particle size (nm)
		UDM	USDM				UDEDM						
	D (nm)	*D* (nm)	ε	*D* (nm)	ε	Σ (MPa)	*D* (nm)	ε	σ (MPa)	*u* (KJ/m^3)	*D* (nm)	ε	
TiAlN	4.7	5.5	0.0041	4.81	0.0064	51.41	6.47	0.0081	63.12	19.52	5.12		5.5

3.5 Analysis of Mechanical Properties by Nano Indentation

The nanomechanical properties (Hardness and Young's Modulus) of TiAlN coating have been evaluated using the nanoindentation technique. The nanoindentation graph was plotted between load and displacement where the indenter was allowed for a maximum load of 55 mN. The curve AB is known as loading where the indenter is made to penetrate in the coating. The portion CD indicates the unloading. As the maximum load is here 55 mN and after reaching this maximum load indenter was held for a few seconds to prevent continuous sinking of the indenter on the surface because of time-dependent deformation [39].

The flat portion in between B and C represents the thermal drift as it is a hard coating material [39]. The unloading part is represented here by C to D, after a certain time the applied load started to decrease at a uniform rate. The whole plot as shown in Fig. 9 represents the elastic deformation of the materials. As per the Oliver-Pharr model, this loading–unloading plot follows the simple power-law expression [40, 41].

$$P = \alpha(h - h_f)^{m,n}. \tag{20}$$

where P is the applied load, α is a loading–unloading fitting parameter, h & h_f are indenter displacement and indenter fitting parameters, respectively, m and n represent displacement exponent in P–h relation for loading and unloading, respectively. By using the following equations [22, 23, 28–31] the different nanomechanical properties of TiAlN have been obtained as given in Table 3.

$$H = \frac{P_{\max}}{A}. \tag{21}$$

$$S = \frac{dp}{dh} = 2\beta\sqrt{\frac{A}{\pi}}E_r. \tag{22}$$

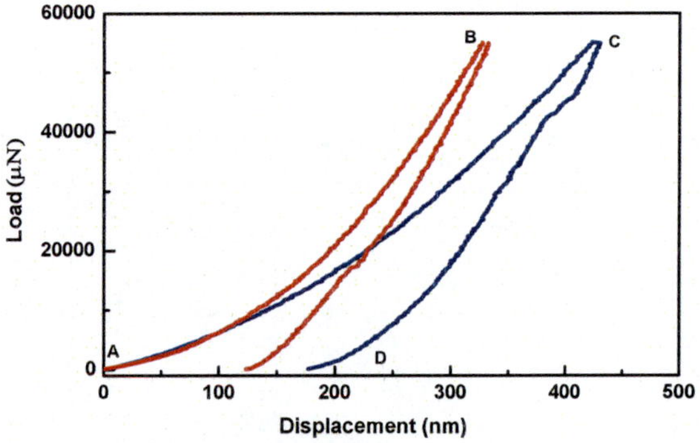

Fig. 9 Loading–unloading plot of CVD deposited TiAlN coating

Table 3 Nanomechanical properties of TiAlN

Hardness (H) (GPa)	Young's modulus (E_f) (GPa)	Reduced modulus (E_r) (GPa)	Plastic deformation index (H^3/E^{*2})	Plasticity index (H/E_r)	Yield strength (σ_y) (GPa)
32.97	277.98	221.34	0.278	0.1489	867.36

$$\frac{1}{E_r} = \frac{1 - \upsilon_f^2}{E_f} + \frac{1 - \upsilon_{id}^2}{E_{id}}. \tag{23}$$

$$\mathrm{E_r} = \frac{S\sqrt{\pi}}{2\sqrt{A_C}}. \tag{24}$$

$$E^* = \frac{E_f}{1 - \upsilon^2}. \tag{25}$$

$$\sigma_\mathrm{y} = 0.78\ \mathrm{a}^2\left(\frac{H^3}{E^{*2}}\right). \tag{26}$$

4 Conclusion

The TiAlN thin film coating has been deposited at 1100 °C in the N_2 gas atmosphere by a thermal CVD process. It's observed from EDS analysis that the composition of Al is highest within the TiAlN coating. The comparative analysis of structural properties with morphological properties has been extensively done by using a different model. The obtained nanomechanical properties suggested that the developed TiAlN coating has good mechanical characteristics.

References

1. Damond E (2000) Revêtements PVD. Durs Pour la mécanique. Développement d'un réacteur PVD Industriel et de Deux revêtements Multicouches associés. Mém. DPE En Génie Procédés
2. Das RK, Sahoo AK, Kumar R, Roy S, Mishra PC, Mohanty T (2019) MQL assisted cleaner machining using PVD TiAlN coated carbide insert: comparative assessment. Indian J Eng Mater Sci 26:311–325
3. Jin N, Yang Y, Luo X, Xia Z (2013) Development of CVD Ti-containing films. Prog Mater Sci 58:1490–1533
4. Shum P, Li K, Zhou Z, Shen Y (2004) •Structural and mechanical properties of titanium–aluminium–nitride films deposited by reactive close-field unbalanced magnetron sputtering. Surf Coat Technol 185:245–253

5. Veprek S, Veprek-Heijman MGJ, Karvankova P, Prochazka J (2005) Different approaches to super hard coatings and nanocomposites. Thin Solid Films 476:1–29
6. Barshilia HC, Surya Prakash M, Jain A, Rajam KS (2005) Structure, hardness and thermal stability of TiAlN and nano layered TiAlN/CrN multilayer films. Vacuum 77:169–179
7. Barshilia HC, Rajam KS (2004) Raman spectroscopy studies on the thermal stability of TiN, CrN, TiAlN coatings and nano layered TiN/CrN, TiAlN/CrN multilayer coatings. J Mater Res 19:3196–3205
8. Bhaduri D, Kuar AS, Sarkar S, Biswas SK, Mitra S (2009) Electro discharge machining of titanium nitride-aluminium oxide composite for optimum process criterial yield. Mater Manuf Processes 24:1312–1320
9. Posti E, Nieminen I (1989) Coating thickness effects on the life of titanium nitride PVD coated tools. Mater Manuf Processes 4:239–252
10. Schneider R, Weigert F, Lesnyak V, Leubner S, Lorenz T, Behnke T, Dubavik A, Joswig J-O, Resch-Genger U, Gaponikc N, Eychmüllerc A (2016) pH and concentration dependence of the optical properties of thiol-capped CdTe nanocrystals in water and D2O. Phys Chem Chem Phys 18:19083–19093
11. Sarkar S, Das R (2018) Determination of structural elements of synthesized silver nano-hexagon from X-ray diffraction analysis. Indian J Pure Appl Phys 56:765–772
12. Sarkar S, Das R (2018) Shape effect on the elastic properties of Ag nanocrystals. Micro Nano Lett 13:312–315
13. Thool GS, Singh AK, Singh RS, Gupta A, Susan MABH (2014) Facile synthesis of flat crystal ZnO thin films by solution growth method: a micro-structural investigation. J Saudi Chem Soc 18:712–721
14. Mahmood NB, Al-Shakarchi EK (2011) Three techniques used to produce BaTiO3 fine powder. J Mod Phys 2:1420–1428
15. Cullity BD, Stock SR (2001) Elements of X-ray diffraction. Addison-Wesley Publishing Company, Mass
16. Guinier A (1994) X-ray diffraction in crystals imperfect crystals and amorphous bodies. Dover, New York
17. Chen HJ, Wu K-L, Yan B-H (2013) Dry electrical discharge coating process on aluminum by using titanium powder compact electrode. Mater Manuf Processes 28:1286–1293
18. Santos CIL, Carvalho MS, Raphael E, Dantas C, Ferrari JL, Schiavon MA (2016) Synthesis, optical characterization, and size distribution determination by curve resolution methods of water-soluble CdSe quantum dots. Mater Res 9:1407–1417
19. Mohamed MB, Tonti D, Al-Salman A, Chemseddine A, Chergui M (2005) Synthesis of high quality zinc blende CdSe nanocrystals. J Phys Chem B 109:10533–10537
20. Delhez R, de Keijser TH, Mittemeijer EJ (1982) Determination of crystallite size and lattice distortions through X-ray diffraction line profile analysis. Fresenius Z Anal Chem 312:1–16
21. Das R, Nath SS, Bhattacharjee R (2010) Preparation of linoleic acid capped gold nanoparticles and their spectra. Physica E 43:224–227
22. Dey PC, Das R (2018) Effect of silver doping on the elastic properties of CdS nanoparticles. Indian J Phys 92(9):1099–1108
23. Das S, Guha S, Ghadai R, Kumar D, Swain BP (2017) Structural and mechanical properties of CVD deposited titanium aluminium nitride (TiAlN) thin films. Appl Phys A 123:412
24. Das S, Guha S, Ghadai R, Sharma A (2021) Influence of nitrogen gas over microstructural, vibrational and mechanical properties of CVD Titanium nitride (TiN) thin film coating. Ceram Int 47:16809–16819
25. Das S, Guha S, Ghadai R, Swain BP (2021) A comparative analysis over different properties of TiN, TiAlN and TiAlSiN thin film coatings grown in nitrogen gas atmosphere. Mater Chem Phys 258:123866
26. Das S, Guha S, Ghadai R, Sharma A, Chatterjee S (2022) Morphological, Mechanical property analysis and comparative study over structural properties of CVD TiN film grown under different substrate temperature in nitrogen gas atmosphere. SILICON 14:183–199

27. Das S, Kumar D, Borah R, Dutta A, Guha S (2022) Impact of elevated temperature over different properties of CVD SiCN coating developed in Nitrogen gas atmosphere. Silicon
28. Tagliente MA, Massaro M (2008) Strain-driven (002) preferred orientation of ZnO nanoparticles in ion-implanted silica. Nucl Instrum Methods Phys Res B 266:1055–1061
29. Jacob R, Isac J (2015) X-ray diffraction line profile analysis of Ba0.6Sr0.4FexTi(1-x) O3-δ, (x=0.4). Int J Chem Stud 2:12–21
30. Mahalingam T, Dhanasekaran V, Chandramohan R, Rhee JK (2012) Microstructural properties of electrochemically synthesized ZnSe thin films. J Mater Sci 47:1950–1957
31. Zak AK, Abrishami E, Majid WHA, Yousefi R, Hosseini SM (2011) Effects of annealing temperature on some structural and optical properties of ZnO nanoparticles prepared by a modified sol–gel combustion method. Ceram Int 37:393–398
32. Das R, Sarkar S (2015) Determination of intrinsic strain in poly(vinylpyrrolidone)-capped silver nano-hexapod using X-ray diffraction technique. Curr Sci 109:775–778
33. Warren BE, Averbach BL (1952) The Separation of Stacking Fault Broadening in Cold-Worked Metals. J Appl Phys 23:497
34. Balzar D, Ledbetter H (1993) Voigt-function modeling in Fourier analysis of size-and strain-broadened X-ray diffraction peaks. J Appl Crystallogr 26:97–103
35. Hall WH (1949) X-ray line broadening in metals. Proc Phys Soc Sect A 62:741–743
36. Nath D, Singh F, Das R (2020) X-ray diffraction analysis by Williamson-Hall, Halder-Wagner and size-strain plot methods of CdSe nanoparticles-a comparative study. Mater Chem Phys 239:122021
37. Tasnádi F, Abrikosov AI, Rogström L, Almer J, Johansson MP, Odén M (2010) Significant elastic anisotropy in Ti1− xAlxN alloys. Appl Phys Lett 97:231902
38. Mote VD, Purushotham Y, Dole BN (2012) Williamson-Hall analysis in estimation of lattice strain in nanometer-sized ZnO particles. J Theor Appl Phys 6:6–14
39. Guha S, Das S, Bandyopadhyay A, Das S, Swain BP (2018) Investigation of mechanical properties of CVD grown titanium silicon nitride thin films under reduced atmosphere. Appl Phys A 124:35
40. Das S, Guha S, Das PP, Ghadai RK (2020) Analysis of morphological, microstructural, electrochemical and nano mechanical characteristics of TiCN coatings prepared under N2 gas flow rate by chemical vapour deposition (CVD) process at higher temperature. Ceram Int 46:10292–10298
41. Guha S, Das S, Bandyopadhyay A, Das S, Swain BP (2018) Investigation of structural network and mechanical properties of Titanium silicon nitride (TiSiN) thin films. J Alloy Comp 731:347–353

An Investigation on Turning of AISI 4340 Steel Using Innovative Eco-Friendly Cutting Fluids

Krishnendu Mondal, Subhasish Chatterjee, Santanu Das, and Bijoy Mandal

1 Introduction

Historically, hardened steel parts have been finish machined using grinding. However, hard turning has become a focus of attention of researchers to shape a workpiece with above 45 HRC hardness. A variety of ceramic cutting tool inserts was tried to produce comparable surface finish at a remarkably higher MRR than grinding [1]. Sahoo and Sahoo experimented on the machinability aspects while doing hard turning to have high surface finish. They employed coated as well as uncoated carbide tools and achieved machining at high cutting velocity [2].

An investigation on surface roughness and power requirement in turning of hard steel rod was made utilizing coated carbide inserts. A mixed technique, applying orthogonal array and ANOVA was used to find out the contribution and effects of cutting conditions on power consumption and roughness of turned surface. Cutting velocity imparted large effect on roughness and the need of power [3]. Uhlmann et al. experimented over cemented carbide tool by putting the coating of TiA1N using some Sputtering techniques. The result shows that HiPIMS coating was far better than DCMS coating considering wear resistance and coating adhesion [4]. Another study between uncoated and TiN-coated carbide inserts was performed [5]. Titanium nitride-coated carbide inserts gave 2.6 times better life than uncoated inserts. Many researchers have been used carbide inserts with Al_2O_3, TiCN, and TiN coating [6–13].

K. Mondal · S. Chatterjee
Department of Mechanical Engineering, Bankura Unnayani Institute of Engineering, Bankura, West Bengal, India

K. Mondal · S. Chatterjee · S. Das (✉) · B. Mandal
Department of Mechanical Engineering, Kalyani Government Engineering College, Kalyani, Nadia, West Bengal, India
e-mail: santanu.das@kgec.edu.in

© The Author(s), under exclusive license to Springer Nature Singapore Pte Ltd. 2023

S. Revankar et al. (eds.), *Recent Advances in Thermofluids and Manufacturing Engineering*, Lecture Notes in Mechanical Engineering,
https://doi.org/10.1007/978-981-19-4388-1_37

Conventionally, flood cooling systems had been expensive compared to other cooling processes. It may be varied up to 17% of total cost of machining [14, 15]. Also, in the machining zone, the cutting fluid cannot reach due to obstruction created by chips. Minimum quantity lubrication (MQL) techniques were used successfully at this situation [16].

Gajrani et al. [17] made a comparison between mineral oil-based cutting fluid and bio-cutting fluid when they conducted machining of hardened steel using coated carbide inserts. Flood cooling and minimum quantity cutting were used in this experiment. Pressurized air mixed with minute cutting fluid particles was also used for both the cutting fluids. The result shows the significant effect in minimization of force, coefficient of friction, surface quality for MQCF. Present-day to avoid hazardous effect of conventional cutting fluid, dry or near to dry (like MCQF) machining are in demand. Vegetable oil-based cutting fluids are used to minimize the hazardous effect and it possesses good lubricating ability compared to other fluids [18]. In another work, vegetable-based oil and mineral oil were chosen as a cutting fluids to compare the machining in MQL, MQCL, wet and dry environments. A286 super alloy and cemented uncoated carbide inserts were taken for this experiment. Vegetable-based oil in MQL has given an 11.4% reduction in Ra value compared to mineral-based oil. Better surface finish and tool life (up to 200%) were found in the MQCL method throughout the experiment [19].

Raw coconut oil and water were mixed with some emulsifiers to make green cutting fluid (GCF) to perform machining of AISI H-13 steel. In terms of good thermal conductivity 1:16 emulsion was given optimum results. Compared to dry machining, MQCF has given better performance concerning feed force, cutting force, and surface roughness [20]. Mia et al. [21] investigated tool life and surface finish of AISI 1060 using coated carbide insert under various environmental conditions. MQL system could be able to tackle heat transfer problems and reported improved machinability with clean production. Pervaiz et al. investigated the benefit of MQL utilizing biodegradable vegetable-based oils for machining titanium alloys, while Sen et al. [23] reviewed various research works and concluded that vegetable oils in MQL gave better surface, needed lower forces, and had improved tool life than conventional flood cooling system [22, 23].

AISI 4340 has high strength and toughness with medium carbon content. This steel is employed in military aircraft, automotive systems, oil and gas industries, forged hydraulic systems, etc. Since it is quite hard to machine, numerous researchers took keen interest on finding out suitable machining conditions to machine AISI 4340 steel smoothly [24].

In this work, turning test is done on AISI 4340 steel having Rockwell hardness of 42 in C scale under varying cutting conditions with a coated carbide insert. As far as wet turning is concerned, an attempt is made to develop new environment-friendly cutting fluids toward green manufacturing. This is the novelty of the work undertaken. Under various cutting conditions, machinability with this work-tool combination is explored.

2 Experimental Procedure

For hard turning experiments, an 11 kW NH22-HMT Lathe (Fig. 1a) is used. The AISI 4340 steel is considered a workpiece with 42 HRC hardness. Table 1 shows the setup and experimental conditions. Response Surface Methodology (RSM) is used to design experiments following Central Composite Design (CCD). Three levels of cutting velocity (V_c) and feed (S_o) are resorted to at a constant depth of cut of 2 mm. Coded values of V_c and S_o used to construct RSM are given in Table 2. Turning is performed on 150 mm diameter rods with 300 mm long rods (Fig. 1b) with the use of new cutting edges. SNMG 120416 PR carbide inserts (Fig. 1c) with Al_2O_3, TiCN, and TiN coating (make: Sandvik Asia Ltd, India) is employed. Tool signature has been: inclination and normal rake angle of −6°, principal and auxiliary cutting-edge angles of 75° and 15°, respectively, and nose radius of 1.6 mm. Two environmental conditions are tried. Refined rice bran oil, water (2:13 ratio), and 85 g of Ariel Matic detergent powder are mixed to formulated one of the cutting fluids which is used in this experiment. Another cutting fluid is formulated by mixing Clinic plus shampoo and water at the ratio of 1:20. Small quantity lubrication (150 cc/min) is used for wet conditions using above said cutting fluid. In this work, cutting force components, chip formation, chip reduction coefficient, formation of BUE is noted under various cutting conditions to determine tool performance. Also, wear test is done at some conditions. It is obvious that machinability of a tool–work combination can be judged by taking into account the force value, chip formation, surface roughness, and wear characteristics. Point-end-micrometer is utilized for measuring chip thickness (a_2). Response variables examined are chip thickness, type of chip, chip reduction coefficient (ξ), and build-up of edge formation. From this, chip reduction coefficient, CRC [$a_2/a_1 = a_2/(S_o \sin\Phi)$] [25] is calculated, where a_1 is uncut chip thickness and it is given by $S_o \sin\Phi$, where S_o is feed and Φ is principal cutting-edge angle. Cutting force, surface roughness is measured. The presence of a built-up edge is also observed after each turning test visually. In the second set of experiments (2a & 2b) as mentioned in Table 3, tool wear experiments with a carbide tool insert at 2 mm depth of cut (t), 0.08 mm/rev feed (S_o), and a cutting velocity of over 200 m/min (V_c) were carried out under the mentioned wet environment. Average flank wear (V_b) and nose wear (V_n) were measured on a Mitutoyo, Japan-made tool makers' microscope. A digital camera attached to the tool maker's microscope and connected to a PC takes a picture of the insert with a magnification of ×20 .

3 Results and Discussion

3.1 Experiment Set 1a

Experiments are performed under wet environment (2:13 oil & water +85 g Ariel detergent powder in 4 L) using SNMG 120416 PR coated carbide Inserts on AISI

Table 1 Experimental details

Exp. set	Cutting velocity, V_c (m/min)	Feed, S_o (mm/rev)	Depth of cut, t (mm)	Environment
1a	144, 186, 244	0.08, 0.12, 0.16	2	Refined rice bran oil & water at the ratio of 2:13 and Ariel Matic detergent powder (85 g in total 4 lit solution)
1b	136, 175, 230	0.08, 0.12, 0.16	2	An alkaline solution of Clinic plus shampoo and water at the ratio of 1:20

Table 2 Coded value of V_c and S_o

V_c	V_c (m/min)		S_o	S_o (mm/rev)
Coded value	Exp. set 1a	Exp. set 1b	Coded value	Exp. set 1a, b
+1	244	230	+1	0.16
0	186	175	0	0.12
−1	144	136	−1	0.08

Table 3 Experimental condition for tool wear test in wet environment

Exp. set	Cutting velocity, V_c (m/min)	Feed, S_o (mm/rev)	Depth of cut, t (mm)	Environment
2a	227	0.08	2	Refined rice bran oil & water at the ratio of 2:13 and Ariel detergent powder (85 g in total 4 L solution)
2b	213	0.08	2	Alkaline solution of clinic plus shampoo and water at the ratio of 1:20

Fig. 1 **a** Experimental setup, **b** close view of SQL system, and **c** insert used

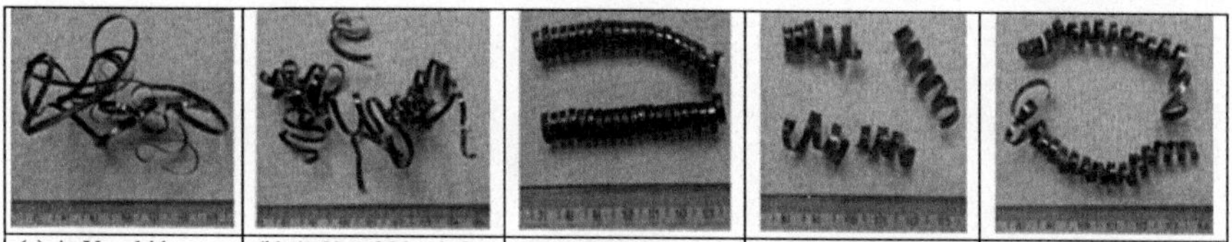

(a) At Vc= 144 m/min, So= 0.08 mm/rev	(b) At Vc= 244 m/min, So= 0.08 mm/rev	(c) At Vc= 140 m/min, So= 0.16 mm/rev	(d) At Vc= 237 m/min, So= 0.16 mm/rev	(e) At Vc= 186 m/min, So= 0.12 mm/rev

Fig. 2 Photographs of collected chips at experiment set 1a

4340 steel of 150 mm in diameter and 300 mm length workpiece. The insert nose radius is 1.6 mm. Small quantity cutting fluids 150 ml/min are supplied in tool-tip.

The value of chip thickness for every machining condition is measured using a point-end-micrometer. Chip reduction coefficient has been calculated and the main cutting force, horizontal cutting force, surface roughness is measured. Chip reduction coefficient (CRC) varies from 1.29 to 1.42. Observed chip formation is given in Fig. 2a–e. At a low feed rate (0.08 mm/rev), long continuous ribbon-type chip is found. At cutting velocity (V_c) 186 m/min, 237 m/min and 0.12 mm/rev, 0.16 mm/rev feed rate (S_o), coil types chips are seen. Similar type of observation has also been made by Kumar et al. [26]. Chip breaking effect is not observed at such in this experiment set. No build-up edge (BUE) is observed throughout the experiment.

At 0.08 mm/rev feed rate, the cutting force is found low compared to other conditions due to ribbon type of chip formation. But for the other two feeds (0.12 mm/rev and 0.16 mm/rev) coil type chips are found which consume more force due to the curling of the chip. Surface roughness (R_a) value is varied within 0.535 μm to 0.627 μm and R_z value is varied within 2.545 μm to 2.921 μm. At 237 m/min cutting velocity and 0.16 mm/rev feed, highest 499.54 N main cutting force and highest 79.70 N horizontal force (P_{xy}) is observed.

3.1.1 Response Surface Methodology (RSM) for Experimental Set 1a

Multiple regression equations given below are obtained using Response Surface Methodology (RSM) in "MINITAB 17" software. Using the experimental data of experiment set 1aand they are represented in Eqs. (3.1), (3.2), (3.3), (3.4) which are the relationship of main cutting forces (P_z), horizontal cutting forces (P_{xy}), and surface roughness (R_a & R_z) with cutting velocity (V_c) and feed rate (S_o), respectively. These equations are generated by eliminating (so*so) terms that do not have a significant effect on responses.

Regression Equation for P_Z

$$P_Z = 379.62 - 0.71\ V_c + 117.79\ S_o - 2.98\ V_c \times V_c + 5.83\ V_c \times S_o \quad (3.1)$$

The value of the square of R is 99.92%

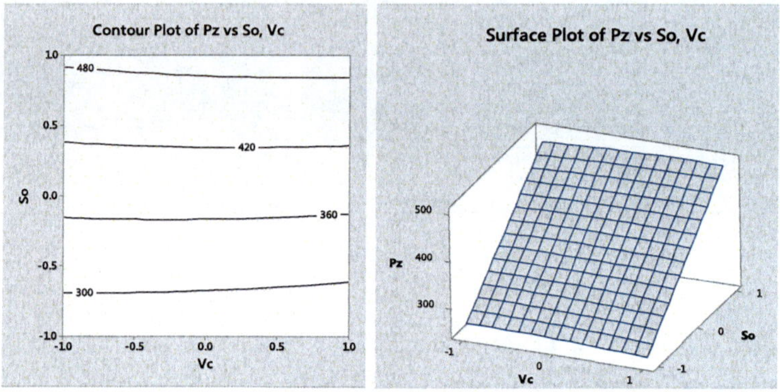

Fig. 3 Contour and surface plot of P_z at different V_c and S_o for experimental set 1a

Regression Equation for P_{XY}

$$P_{XY} = 64.74 - 0.71\ V_c + 12.83\ S_o - 2.98\ V_c \times V_c + 5.83\ V_c \times S_o \tag{3.2}$$

The value of the square of R is 94.61%

Regression Equation for R_a

$$R_a = 0.59562 + 0.00825\ V_c + 0.03797\ S_o - 0.01429\ V_c \times V_c - 0.00024\ V_c \times S_o \tag{3.3}$$

The value of the square of R is 93.43%

Regression Equation for R_Z

$$R_Z = 2.8560 - 0.0955\ V_c + 0.0922\ S_o - 0.2098\ V_c \times V_c - 0.0866\ V_c \times S_o \tag{3.4}$$

The value of the square of R is 94.75%

The contour plot and surface plot of main cutting force (P_Z) horizontal cutting force (P_{XY}), surface roughness (R_a& R_Z) is shown as a function of cutting velocity (V_c) and feed (S_o) from Figs. 3, 4, 5 and 6.

3.1.2 Effect of Cutting Forces

Figure 3 shows the contour plot and surface plot of the main cutting force (P_z) with the variation of cutting velocity (V_c) and feed (S_o). The main cutting force (P_z) is low at 237 m/min cutting velocity (Vc) and 0.08 mm/rev feed due to low uncut chip thickness. But P_z is increasing with an increase in feed (S_o). Increasing uncut shear area with feed rate may be the cause of high cutting force.

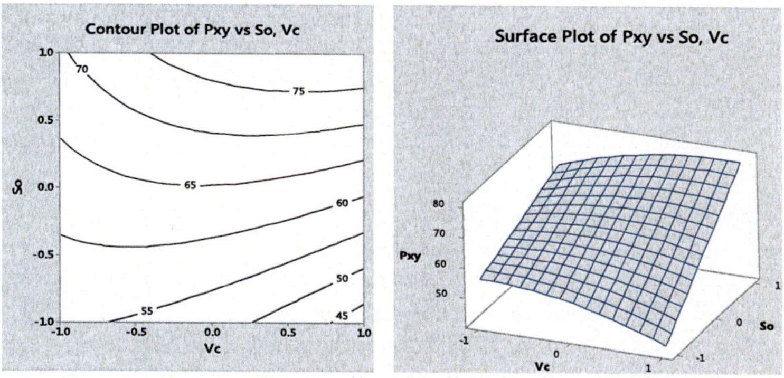

Fig. 4 Contour and surface plot of P_{xy} at different V_c and S_o for experimental set 1a

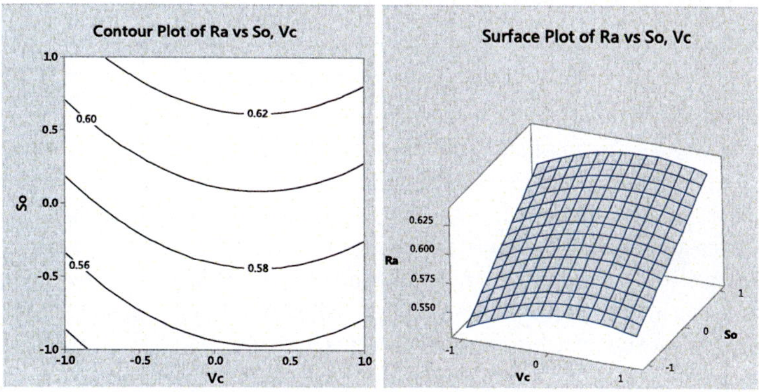

Fig. 5 Contour and surface plot of R_a at different V_c and S_o for Experimental set 1a

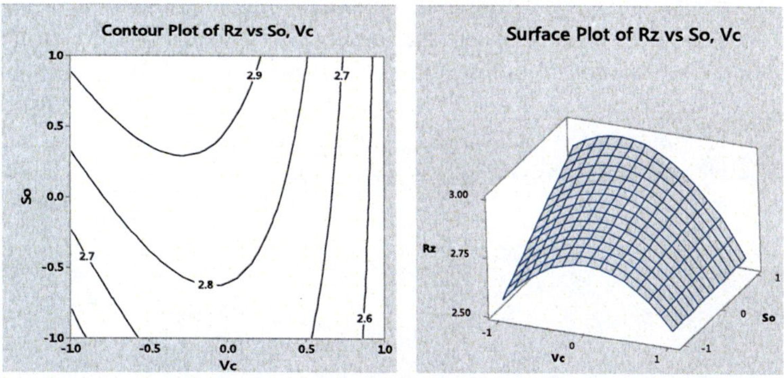

Fig. 6 Contour and surface plot of R_z at different V_c and S_o for Experimental set 1a

Similarly, Fig. 4 shows contour plot and surface plot of the horizontal cutting force (P_{xy}). In that case, horizontal cutting force (P_{xy}) is maximum at high cutting velocity (V_c) and high feed (S_o).

3.1.3 Surface Roughness of Set 1a

Figures 5 and 6 show the contour plot and surface plot of the avg. surface roughness value (R_a & R_z) with the variation of cutting velocity (V_c) and feed (S_o). It is noticed that the avg. surface roughness (R_a) value increases with the increase in feed (S_o) value. But R_z shows the maximum value at low cutting velocity and high feed rate.

3.2 Experiment Set 1b

This experiment is done under a wet machining environment; alkaline solution (1:20-shampoo & water) is supplied in the form of SQL (150 cc/min) at the tool-tip with the help of a miniature pump on AISI 4340 workpiece. Depth of cut has been kept constant as 2 mm throughout the experiment. Cutting velocity (V_c) and feed rate (S_o) have been taken as process parameters. Each of the process parameters has been taken of three levels (low-moderate-high) and Response Surface Methodology (RSM) was taken into consideration for the results obtained during the experiment.

The results are noted. Chip reduction coefficient (CRC) varies from 1.190 to 1.423 in this experimental work [26]. No built-up edge (BUE) has been observed throughout the experiment. The cutting velocity is selected much higher in this experiment. This may be the cause of the disappearance of BUE. Figure 7a–e shows the chip formation in the experiment. It shows that a long continuous ribbon-type chip is formed at a low feed rate (S_o) of 0.08 mm/rev. Long continuous closed coil type chips are formed at 0.16 mm/rev feed and 136 m/min cutting velocity (V_c) [2]. Concentric open coil, 5–9 turns type chip is formed at 0.16 mm/rev feed and 230 m/min cutting velocity (V_c). At 0.12 mm/rev feed rate, a long continuous close coiled type chip is found. Here effectiveness of the chip breaker is not found.

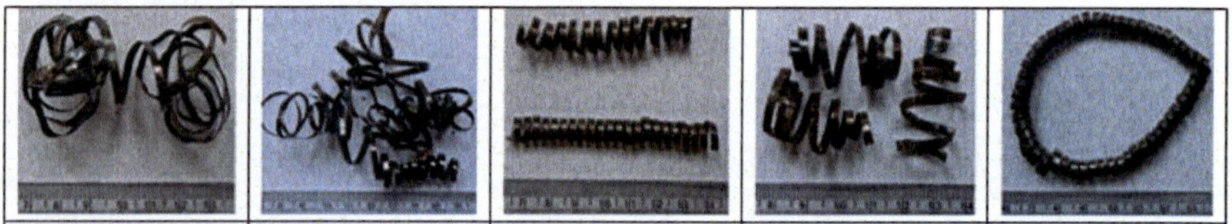

(a) At Vc= 136 m/min, So= 0.08 mm/rev	(b) At Vc= 230 m/min, So= 0.08 mm/rev	(c) At Vc= 136 m/min, So= 0.16 mm/rev	(d) At Vc= 230 m/min, So= 0.16 mm/rev	(e) At Vc= 175 m/min, So= 0.12 mm/rev

Fig. 7 Photographs of the collected chips at experimental set 1b

In this investigation, observation of cutting force and surface roughness is tabulated. Main cutting force (P_z) are varying from 265.39 N to 483.89 N and horizontal cutting force (P_{xy}) varies from 55.47 N to 65.55 N. Lowest cutting forces are found at 136 m/min cutting velocity and 0.08 mm/rev feed rate. Surface roughness (R_a) is varying from 0.544 μm to 0.661 μm and R_z are varied from 2.259 μm to 3.095 μm. Lower surface roughness (R_a) is found at 136 m/min cutting velocity and 0.08 mm/rev feed rate.

3.2.1 Response Surface Methodology (RSM) for Experimental Set 1b

Multiple regression Equations no. 3.5, 3.6, 3.7, and 3.8 are represented the relationship of main cutting forces (P_z), horizontal cutting force (P_{xy}), and surface roughness (R_a & R_z) which are found out using Response Surface Methodology (RSM).

Regression Equation for P_Z

$$P_Z = 376.899 + 2.015\ V_c + 109.629\ S_o - 0.24\ V_c \times V_c + 0.381\ V_c \times S_o \quad (3.5)$$

The value of square of R is 99.98%

Regression Equation for P_{XY}

$$P_{XY} = 62.019 + 2.015\ V_c + 4.669\ S_o - 0.24\ V_c \times V_c + 0.381\ V_c \times S_o \quad (3.6)$$

The value of the square of R is 91.57%

Regression Equation for R_a

$$R_a = 0.59355 + 0.04198\ V_c - 0.01132\ S_o - 0.00208\ V_c \times V_c - 0.01651\ V_c \times S_o \quad (3.7)$$

The value of the square of R is 99.08%

Regression Equation for R_z

$$R_Z = 2.88377 + 0.3757\ V_c - 0.0422\ S_o - 0.2351\ V_c \times V_c - 0.0285\ V_c \times S_o \quad (3.8)$$

The value of the square of R is 99.74%.

The contour plot and surface plot of main cutting force (P_z), horizontal cutting force (Pxy), surface roughness (R_a & R_z) is shown as a function of cutting velocity (V_c) and feed (S_o) in Fig. 8 through Fig. 11.

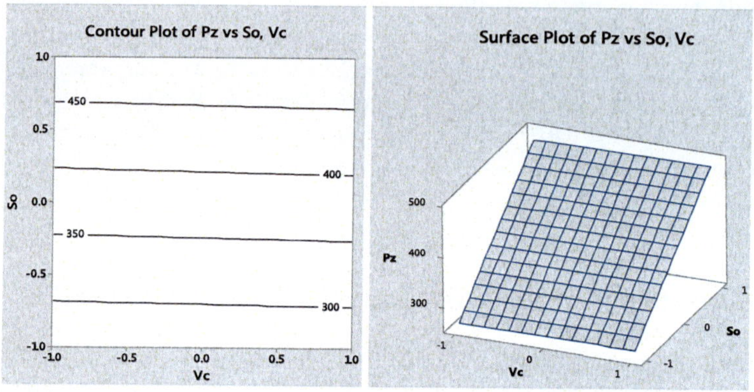

Fig. 8 Contour and surface plot of P_z at different V_c and S_o for experimental set 1b

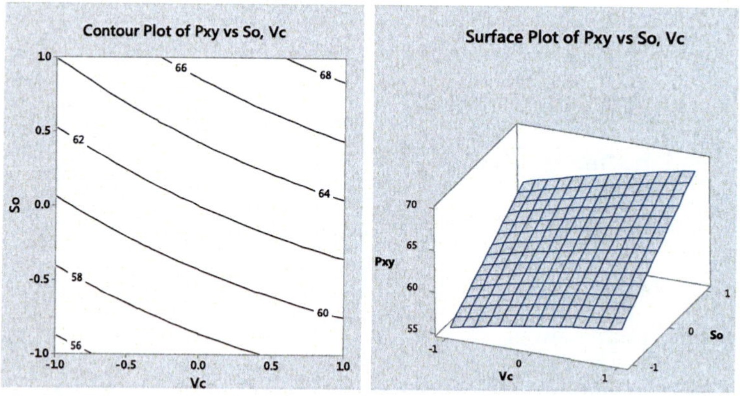

Fig. 9 Contour and surface plot of P_{xy} at different V_c and S_o for experimental set 1b

3.2.2 Discussion on Cutting Forces

The contour plot and surface plot of P_z and P_{xy} are plotted in Figs. 8 and 9, respectively with the variation of cutting velocity (V_c) and feed (S_o). It is noticed that due to increases in the shear area the main cutting force increases with increasing of feed rate. However, with the increase of cutting velocity (V_c), main cutting force (P_z) is slightly decreased. Similar trends are found for horizontal cutting force (P_{xy}) also.

3.2.3 Discussion on Surface Roughness

Figures 10 and 11 show the contour plot and surface plot of the avg. surface roughness value (R_a & R_z) with the variation of cutting velocity (V_c) and feed (S_o). It is noticed that the avg. surface roughness (R_a) value increases with the increase offeed (S_o)

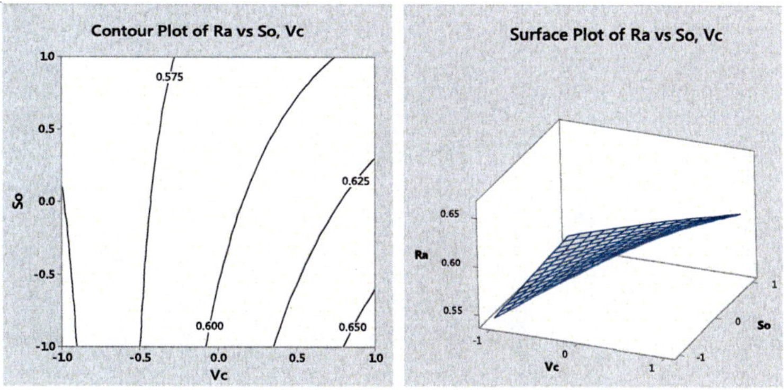

Fig. 10 Contour and surface plot of R_a at different V_c and S_o for Experimental set 1b

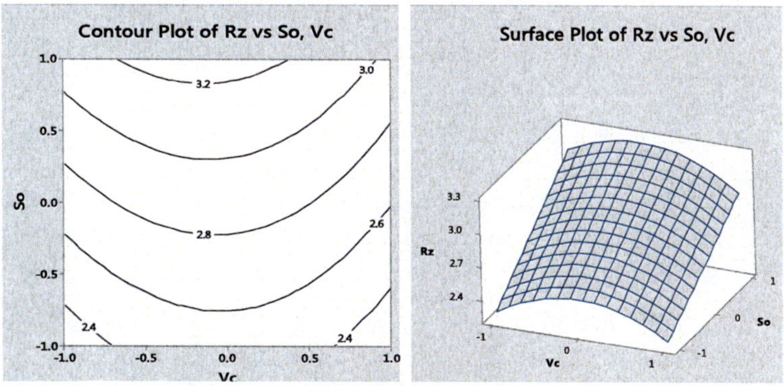

Fig. 11 Contour and surface plot of R_z at different V_c and S_o for Experimental set 1b

value. At 136 m/min cutting velocity (V_c) and 0.16 mm/rev feed (S_o), R_a value is maximum then decreases up to a higher cutting velocity (V_c) of 230 m/min.

3.3 Comparison Between Experimental Set 1a, 1b

Main cutting force (P_z) is compared in between experimental setups 1a and 1b in Fig. 12 of SNMG 120416 PR coated carbide inserts. It is observed that the value of the main cutting force (P_z) is mostly similar for all conditions. Figure 13 shows the comparison of horizontal forces (P_{xy}) for the above two experiment set. In that case, rice bran oil-based environment is taking more horizontal cutting force (P_{xy}) for all conditions except conditions No.1 & 2 where machining is carried out at low feed. Figures 14 and 15 show the surface roughness comparisons. In that case, R_a & R_z are mostly similar for all conditions.

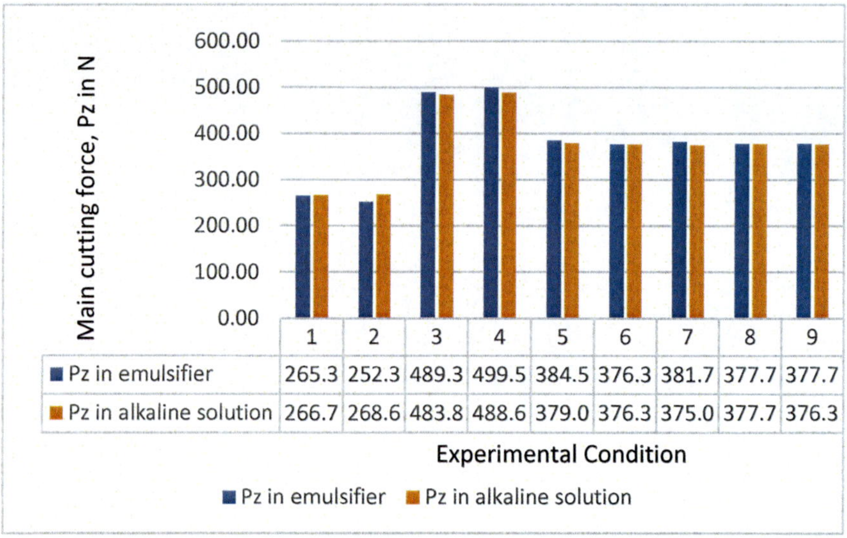

Fig. 12 Comparison of main cutting force (P_z) in between different environmental conditions

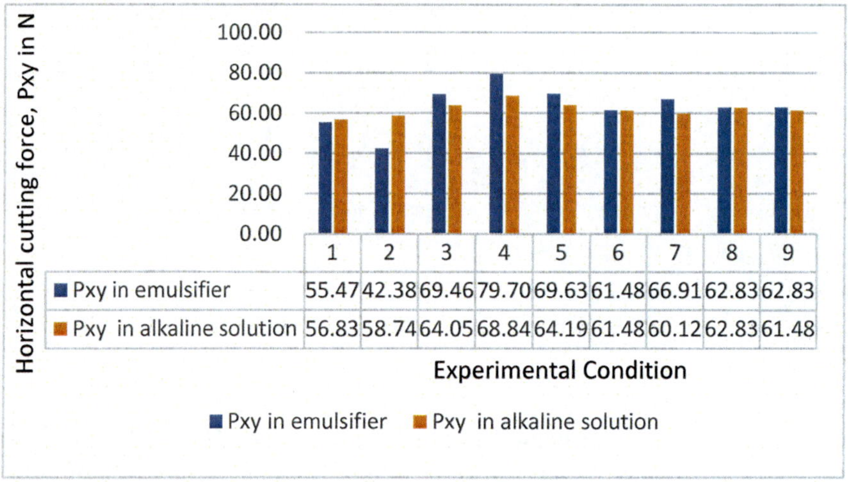

Fig. 13 Comparison of horizontal cutting force (P_{xy}) under different environment conditions

3.4 Experiment Set 2a

Experiment set 2a is carried out for the wear test of the coated carbide inserts under two different wet environmental conditions. The cutting environment chosen is SQL (150 cc/min).

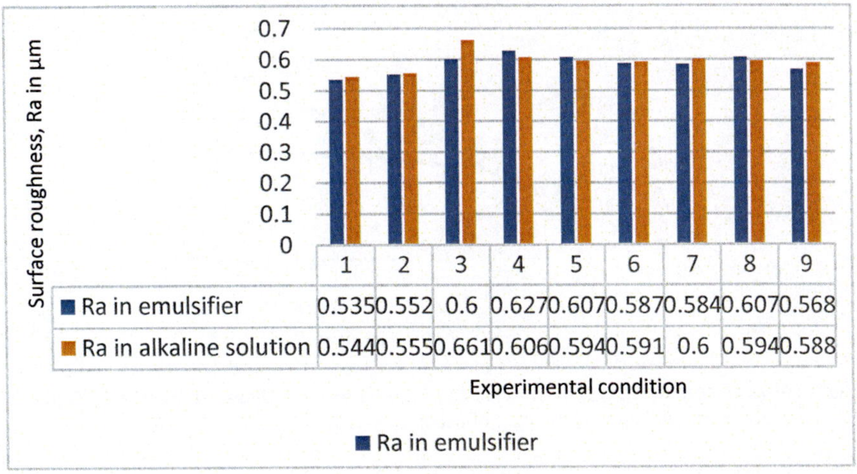

Fig. 14 Comparison of surface roughness (R_a) under different environmental conditions

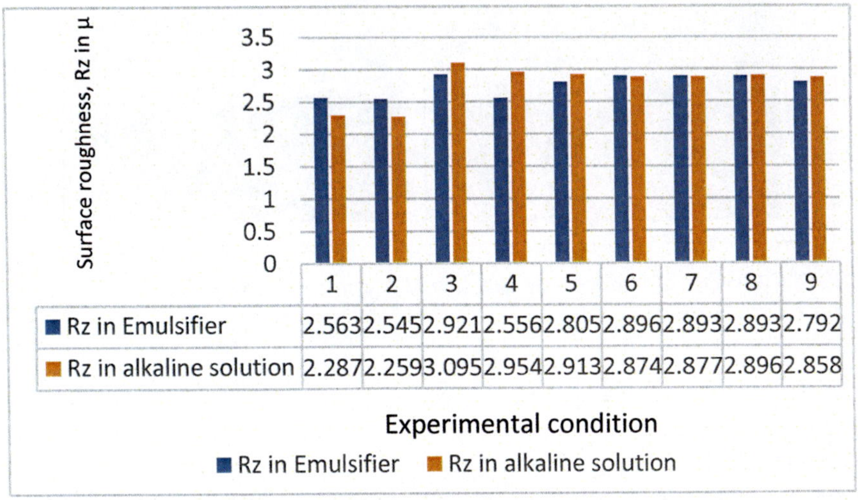

Fig. 15 Comparison of surface roughness (R_z) under different environment conditions

3.4.1 Wear Test of Experiment Set 2a

Wear test is carried out using 2:13 oil & water +85 g Ariel Matic detergent powder in 4 L in exp. set 2a. During tool wear measurement in hard turning, one of eight edges of the tool insert has been used. At different time intervals, tool wear measurement has been performed in an Advanced Stereo Microscope. Under the microscope, flank wear and nose wear are measured.

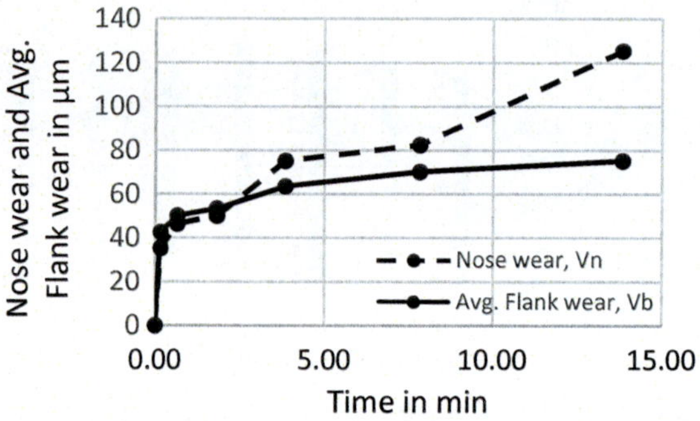

Fig. 16 Plot of variation of avg. flank wear and nose wear with machining time at $V_c = 227$ m/min, $S_o = 0.08$ mm/rev and 2 mm depth of cut of experiment set 2a

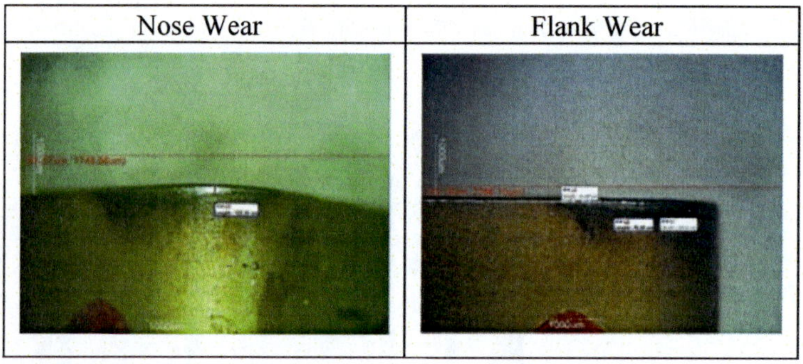

Fig. 17 Photograph of nose and flank wear after 13.83 min of machining

Progress of average flank wear (V_b) and nose wear (V_n) with machining time are shown in Fig. 16. It is observed that the progress of flank wear and nose wear are almost the same up to 1.83 min of machining. After that, the increment of nose wear is higher than flank wear. Up to 13.83 min of machining, nose wear (V_n) gives the value 125.33 µm and flank wear (V_b) gives the value 75.2 µm. Photograph of nose wear (V_n) and flank wear (V_b) of SNMG 120416PR tool insert after 13.83 min of machining is given in Fig. 17.

3.4.2 Wear Test for Experimental Set 2b

Wear test is carried out using alkaline solution (1:20- Shampoo & water) in exp. set 2b. The rest of the parameters are the same as exp. set 2a. Progress of average flank wear (V_b) and nose wear (V_n) with machining time are shown in Fig. 18. It is observed that the progress of flank wear and nose wear are almost the same up to 14 min of machining. After that, the increment of nose wear is higher than flank

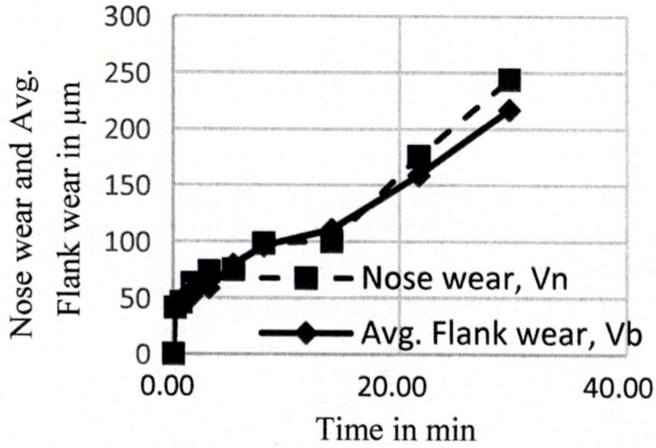

Fig. 18 Plot of variation of avg. flank wear and nose wear with machining time at $V_c = 213$ m/min, $S_o = 0.08$ mm/rev and 2 mm depth of cut for experiment set 2b

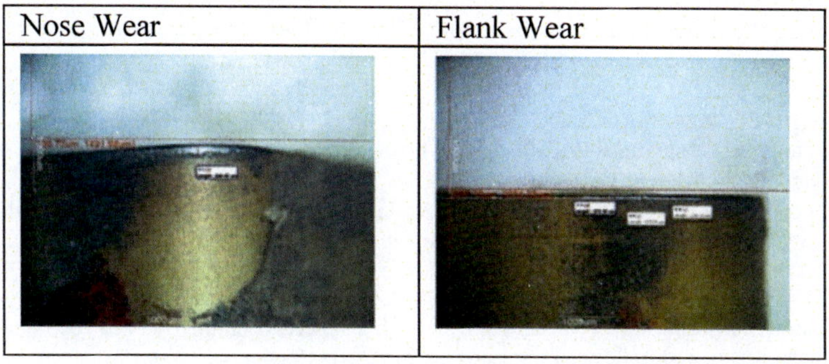

Fig. 19 Photograph of Flank wear and nose wear of SNMG 120416-PR tool insert after 29.67 min of machining

wear. Up to 29.67 min of machining, nose wear (V_n) gives the value 244.4 µm and flank wear (V_b) gives the value 217.28 µm. Photograph of nose wear (V_n) and flank wear (V_b) of SNMG 120416PR tool insert after 29.67 min of machining is given in Fig. 19. Comparing exp. set 2a and 2b, it is observed that up to 14 min of machining rice bran oil base cutting fluid performs better than an alkaline solution.

4 Conclusion

From the experimental findings using coated carbide tool insert, the following conclusions may be drawn:

- Chip breaking action is not seen throughout the experiment while using a groove-type insert.

- Low feed (0.08 mm/rev) gives continuous ribbon-type chip.
- Main cutting force (P_z) does not change significantly in both environmental conditions.
- Surface roughness is observed mostly similar in both wet conditions.
- Progress of tool wear with machining time can be reduced significantly by the use of rice bran oil-based cutting fluid in SQL mode as a lubricator compares to alkaline-based cutting fluid.
- From regression analysis, value of R^2 is found to be about 1 for forces and avg. surface roughness value, and hence, its validity.

The future scope of work in this area may be innovating different eco-friendly cutting fluids and their suitable concentration for machining difficult-to-machine materials at appropriate cutting conditions effectively so that it can be recommended to apply to industries.

References

1. Hosseini SB, Beno T, Klement U, Kaminski J, Ryttberg K (2014) Cutting temperatures during hard turning-measurements and effects on white layer formation in AISI 52100. J Mater Process Technol 214:1293–1300
2. Sahoo AK, Sahoo B (2012) Experimental investigations on machinability aspects in finish hard turning of AISI 4340 steel using uncoated and multilayer coated carbide inserts. Measurement 45:2153–2165
3. Bhattacharya A, Das S, Majumder P, Batish A (2009) Estimating the effect of cutting parameters on surface finish and power consumption during high-speed machining of AISI 1045 steel using Taguchi design and ANOVA. Prod Eng Res Dev 3:31–40
4. Uhlmann E, Stawiszynski B, Leyens C, Heinze S, Sammler F (2016) Hard turning of hot work and cold work steels with HiPIMS and DCMS TiA1N coated carbide inserts. Procedia CIRP 46:591–594
5. Vasilko K, Murcinkova Z, Murcinko J (2021) Evaluation of performance of uncoated cemented carbide cutting tools at longitudinal turning at cutting velocity 3–500 m/min and influence of coating. Mat Today: Proc 44.https://doi.org/10.1016/j.matpr.2020.12.641
6. Mondal K, Adak AK, Chatterjee S, Mandal B, Das S (2019) An investigation on turning of hardened steel using coated carbide insert. Processes national conference on trends and advances in mechanical engineering. Kalyani, West Bengal, India, p 76
7. Mondal K, Adak AK, Das S, Mandal B (2020) Hard turning of AISI 4340 steel using two types of carbide inserts. Processes international conference on advance in mechanics engineering. Kolkata, India, p 43
8. Mondal K, Das S (18–20 Dec 2015) An investigation on machinability during hard turning with narrow groove tool insert in different environmental conditions. Processes international conference on mechanical engineering. Dhaka, Bangladesh, p 566
9. Mondal K, Das S (2018) An investigation on machinability during turning hardened steel in dry condition. J Inst Eng India: Ser C 99(6):637–644
10. Mondal K, Das S (2016) Investigation on machinability in finish turning of AISI 52100 steel using different tool inserts in dry machining. Manuf Technol Today 15:9–20
11. Mondal K, Das S (2017) Wet turning of AISI 52100 steel—an experimental investigation. J Assoc Eng, India 87(3–4):46–59

12. Mondal K, Rahaman J, Das S, Mandal B (15–16 Feb 2019) Machinability study of AISI 1045 steel using different coated carbide insert. Processing national conference on trends and advances in mechanical engineering. Kalyani, West Bengal, India, p 75
13. Mondal K, Das S, Mandal B, Sarkar D (2016) An investigation on turning hardened steel using different tool inserts. Mater Manuf Proc 31:1770–1781
14. Hamdan A, Sarhan AAD, Hamdi M (2012) An optimization method of the machining parameters in high-speed machining of stainless steel using coated carbide tool for best surface finish. Int J Adv Manuf Technol 58(1–4):81–91
15. Amiril SAS, Rahim AE, Syahrullail S (2017) A review on ionic liquids as sustainable lubricants in manufacturing and engineering: recent research, performance and applications. J Cleaner Prod 168:1571–1589
16. Goindi GS, Sarkar P (2017) Dry machining: a step towards sustainable machining challenges and future directions. J Cleaner Prod 165:1557–1571
17. Gajrani KK, Ram D, Sankar MR (2017) Biodegradation and hard machining performance comparison of ecofriendly cutting fluid and mineral oil using flood cooling and minimum quantity cutting fluid techniques. J Cleaner Prod 165:1420–1435
18. John J, Bhattacharya M, Raynor PC (2004) Emulsions containing vegetable oils for cutting fluid application. Collids Surf A: Physicochemical Eng Aspects 237(1–3):141–150
19. Musavi SH, Davoodi B, Niknam SA (2018) Environmental-friendly turning of A286 superalloy. J Manuf Proc 32:734–743
20. Gajrani KK, Suvin PS, Kailas SV, Ravi Shankar M (2019) Hard machining performance of indigenously developed green cutting fluid using flood cooling and minimum quantity cutting fluid. J Cleaner Prod 206:108–123
21. Mia M, Gupta MK, Singh G, Krolczyk G, Pimenov DY (2018) An approach to cleaner production for machining hardened steel using different cooling-lubrication conditions. J Cleaner Prod 187:1069–1081
22. Pervaiz S, Anwar S, Qureshi I, Ahmed N (2019) Recent advances in the machining of Titanium alloys using minimum quantity lubrication (MQL) based techniques. Int J Precis Eng Manuf-Green Technol 6:133–145
23. Sen B, Mia M, Krolczyk GM, Mandal UK, Mondal SP (2021) Eco-friendly cutting fluids in minimum quantity lubrication assisted machining: a review on the perception of sustainable manufacturing. Int J Prec Eng Manuf- Green Technol 8:249–280
24. Bag R, Panda A, Sahoo AK, Kumar R (2020) A Comprehensive review on AISI 4340 hardened steel: emphasis on industry implemented machining settings, implications, and statistical analysis. Int J Integr Eng 12(8):61–82
25. Roy S, Kumar R, Sahoo AK, Panda A (2020) Cutting tool failure and surface finish analysis in pulsating MQL-assisted hard turning. J Fail Anal Prev 20:1274–1291
26. Kumar R, Shoo AK, Mishra PC, Das RK (2018) Comparative investigation towards machinability improvement in hard turning using coated and uncoated carbide inserts: part I experimental investigation. J Adv Manuf 6:52–70

Hot Corrosion Behaviour of Incoloy 800H Superalloy with Various Molten Salts Environment

M. Kamatchi Hariharan, A. Anderson, and K. Ravikumar

1 Introduction

In recent years, energy requirements across the globe are rising in a very strong fashion, which leads the researchers to continuously work on the energy storage and effective utilization of the energy. Solar and thermal-based energy storage technologies are evolving day by day with different techniques where the exploitation can be done at the maximum. The molten salts technology is one of the key factors where the heat energy obtained from the concentrated solar tower is stored for a period of time, and it can be possibly recovered at any time without much losses [1–3].

Among the wide spread of different salts, the chloride-based salts such as sodium and potassium chloride, in the nitrate-based salts such as sodium and potassium nitrate, in the sulphate-based salts such as sodium sulphate, in the fluoride-based salts such as lithium fluoride and in the carbonate-based salts such as calcium carbonate are termed as molten salts because of its high melting point and boiling point [4–7]. In this present investigation the salts, sodium and potassium chloride, calcium carbonate and sodium sulphate salts are used. The melting point and boiling point of the molten salts were approximately 850 °C and 1400 °C, respectively.

Alloys materials having high tolerances for oxidising environments, high performance in aggressive temperature environments are better known as superalloys. These alloys exhibit exceptional mechanical strength and creep resistance at high temperatures, good surface stability and corrosion and oxidation resistance [8, 9].

M. K. Hariharan (✉) · A. Anderson
School of Mechanical Engineering, Sathyabama Institute of Science and Technology, Chennai, India
e-mail: hariharanj7@gmail.com

K. Ravikumar
Department of Mechanical Engineering, SRM Institute of Science and Technology, Vadapalani Campus, Chennai, India

© The Author(s), under exclusive license to Springer Nature Singapore Pte Ltd. 2023
S. Revankar et al. (eds.), *Recent Advances in Thermofluids and Manufacturing Engineering*, Lecture Notes in Mechanical Engineering,
https://doi.org/10.1007/978-981-19-4388-1_38

The corrosion-resistant superalloys are widely used in extreme environments where incredible heat and corrosion resistance is supreme to the integrity of the end product. Nickel-based, cobalt-based and ferrous-based superalloys have found their distinctive applications in the industry sector [10–13].

Nickel-based superalloys are highly enriched with alloying elements including chromium (Cr), aluminium (Al), titanium (Ti), molybdenum (Mo), tungsten (W), niobium (Nb), tantalum (Ta) and cobalt (Co). Due to these elements' availability, nickel-based superalloys such as Inconel, Hastelloy and Incoloy can be preferred for high temperature energy storage applications. Incoloy 800H is one among the above-mentioned, and this superalloy is chosen in the form of sheets for the present study. All the superalloy specimens are heated to 1000 °C with the selected salts for a different period of time in a furnace. Then the mechanical testing and material characterization were performed to know the impacts of the salt corrosion on the specimens. Since the material is enriched with chemical compounds which could resist the corrosion in a healthier mode, the novel approach of heating the Incoloy 800H specimens with molten salts was completed.

2 Materials and Methods

Inconel 625, Incoloy 800H and Hastelloy x are nickel-based superalloys which are highly suitable for storage materials to be used with different molten salts. In this present study, Incoloy 800H specimens were preferred. Incoloy 800H is an alloy enriched with nickel, iron and chromium. The presence of high amounts of nickel and chromium makes the alloy capable of remaining stable and maintaining its austenitic structure even after long exposure to high temperatures. Due to these superior properties, Incoloy 800H found its application in high-temperature environmental applications, such as solar power plants, nuclear power plants, etc. The material was procured from the supplier m/s Raghavendra metals, Chennai. The chemical composition testing was done to check the genuineness of the material used for this investigation. The material composition was found with the OES FOUNDRY PRO ANALYSER setup.

From Table 1, the chemical composition such as 35.85% as nickel, 21.18% as chromium and 38.4% as ferrous ensures that, the material taken up for this investigation is Incoloy 800H superalloy. Specimens for tensile test, hardness, porosity test, SEM and XRD analysis were prepared as per the ASTM standards. Figure 1 depicts the specimens prepared by using laser beam machining technique.

A eutectic mixture of sodium and potassium chloride (NaCl & KCl, Frank Alkanes), with 800 °C as melting point and 1400 °C as boiling point (approx.) along with sodium sulphate (Na_2SO_4, Frank Alkanes) with 850 °C as melting point 1430 °C as boiling point and calcium carbonate (CaCO3, Frank Alkanes) with 825 °C as melting point 1300 °C as boiling point molten salts were taken for this investigation. The specimens were cleaned with kerosene and acetone solution to remove the impurities which could affect the material properties. All the alloy specimens were

Table 1 Chemical composition of incoloy 800H superalloy

Name of the element	Fe	Ni	Cr	Mn	C	Cu	Si	S	Al	Ti	P
Chemical composition	38.4	35.85	21.18	1.01	0.29	0.75	0.89	0.31	0.67	0.58	0.07

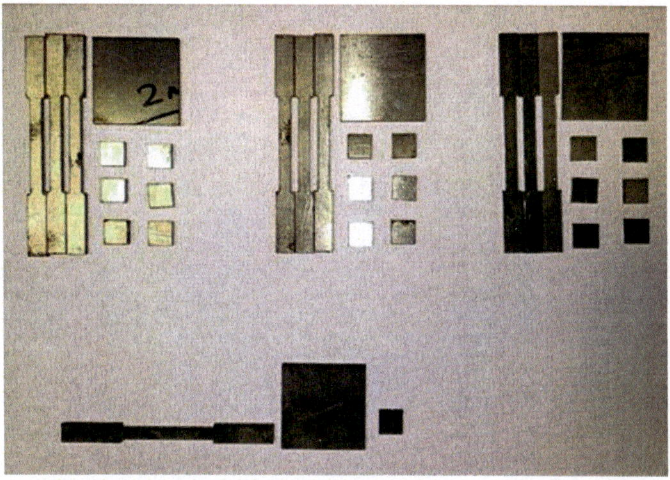

Fig. 1 Incoloy 800H specimens for testing and characterization

taken with 100 g of molten salts in a silicon crucible. For the sodium and potassium chloride mixture, the molten salts were taken as 7:3 ratio in weight fraction. Silicon crucibles were preferred because of their high temperature withstanding capacity. The specimens were heated to an aggressive temperature of 1000 °C in a muffle furnace for 8 h, 12 h and 16 h.

The heated specimens were cleaned to remove the salt crystals which had formed over the specimens due to aggressive heating. The polished specimens were used for the mechanical testing and characterization [14–16].

3 Results and Discussions

Tensile test was carried out in a universal testing machine, thereby yield strength, ultimate tensile strength and % of elongation in 25 mm GL were found for the different hour heated specimens. Hardness values of the specimens have been found using a Vickers hardness tester [17–20].

Before testing the heated specimens, the un-heated specimens were tested and resulted in 318 MPa as yield strength, 631 MPa as ultimate tensile strength and

Table 2 Tensile test results of specimens heated with sodium sulphate molten salt

Incoloy 800H specimens	Yield strength (MPa)	Ultimate tensile strength (MPa)	% of elongation in 25 mm GL
8 h heated	188	429	21.00
12 h heated	180	369	20.00
16 h heated	156	338	16.50

39.00 as % of elongation in 25 mm gauge length. The hardness value was found to be 144.8 HV.

For specimens heated with sodium sulphate molten salts, the tensile test results are as follows (Table 2).

The Incoloy 800H specimens heated with sodium sulphate molten salts result in a significant decrease in the tensile properties of the specimens. With reference to Fig. 2, the yield strength of the heated specimens was reduced by 40.88%, 43.39% and 50.94% for 8 h, 12 h and 16 h heated specimens, respectively. The ultimate tensile strength was reduced by 32.01%, 41.52% and 46.43%, and percentage of elongation was reduced by 46.15%, 48.71% and 57.69% for 8 h, 12 h and 16 h heated specimens, respectively (Table 3).

For specimens heated with sodium and potassium chloride-based molten salts, the tensile test results are as follows.

The superalloy specimens heated with sodium and potassium chloride molten salts result in a notable reduction in the tensile properties of the specimens. With reference to the Fig. 3, the yield strength of the heated specimens was reduced by

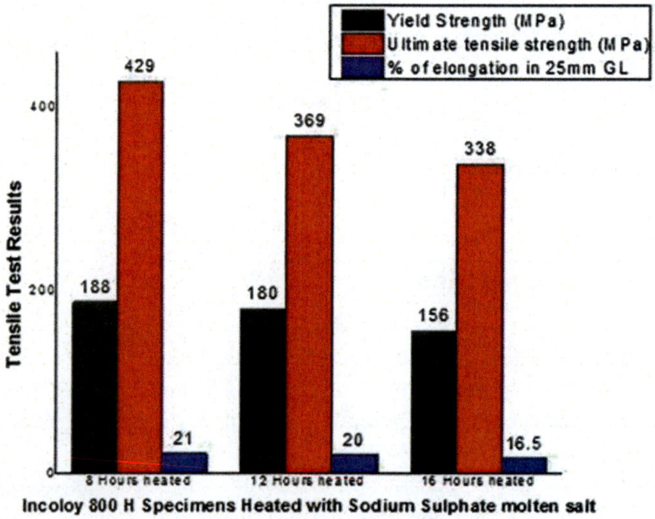

Fig. 2 Tensile test results of specimens heated with sodium sulphate molten salt

Table 3 Tensile test results of specimens heated with chloride-based molten salt

Incoloy 800 H specimens	Yield strength (MPa)	Ultimate tensile strength (MPa)	% of elongation in 25 mm GL
8 h heated	280	620	36.50
12 h heated	274	592	31.00
16 h heated	267	608	38.50

11.94%, 13.83% and 16.03% for 8 h, 12 h and 16 h heated specimens, respectively. The ultimate tensile strength was reduced by 1.7%, 6.1% and 3.6%, and percentage of elongation was reduced by 6.4%, 20.5% and 1.2% for 8 h, 12 h and 16 h heated specimens, respectively.

For specimens heated with calcium carbonate molten salts, the tensile test results are as follows (Table 4).

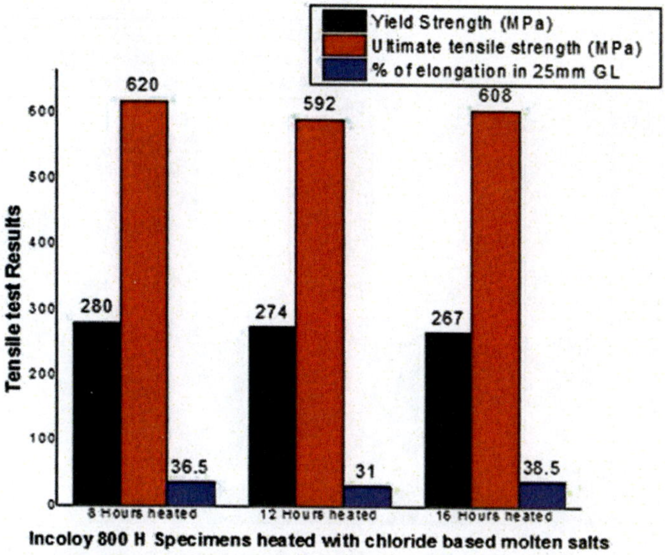

Fig. 3 Tensile test results of specimens heated with chloride-based molten salt

Table 4 Tensile test results of specimens with calcium carbonate molten salt

Incoloy 800H specimens	Yield strength (MPa)	Ultimate tensile strength (MPa)	% of elongation in 25 mm GL
8 h heated	287	601	36.00
12 h heated	276	594	31.50
16 h heated	274	575	31.50

The specimens heated with calcium carbonate molten salts result in a drop in the tensile properties of the specimens in the same fashion as present in the former investigations. With reference to the Fig. 4, the yield strength of the heated specimens was reduced by 9.7%, 13.2% and 35.5% for 8 h, 12 h and 16 h heated specimens, respectively. The ultimate tensile strength was reduced by 4.7%, 5.8% and 8.8%, and the percentage of elongation was reduced by 7.6%, 19.2% and 19.2% for 8 h, 12 h and 16 h heated specimens, respectively.

The hardness values for the different hourly heated specimens are tabulated as follows (Table 5).

With reference to the Fig. 5, the hardness values of the 16 h heated specimens decrease to the maximum of 24.82% with sodium sulphate, 15.11% with sodium and potassium chloride and 18.65% with calcium carbonate molten salt. From the results of the tensile test and hardness test, it is evident that sodium sulphate molten salt is highly reactive with superalloy specimens followed by calcium carbonate molten salt, and the least reaction is present with the chloride-based molten salts [21, 22].

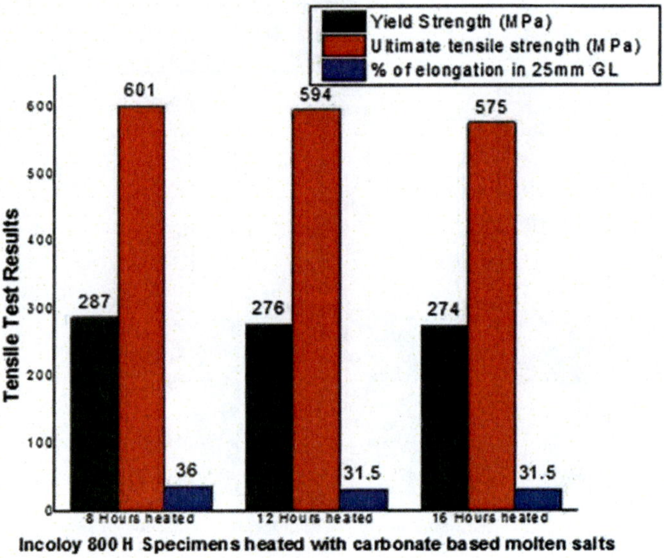

Fig. 4 Tensile test results of specimens heated with calcium carbonate molten salt

Table 5 Hardness test results of specimens with various molten salts

Incoloy 800H specimens- hardness values in HV	8 h heated	12 h heated	16 h heated
Sodium sulphate	121.2	112.3	108.4
Sodium and potassium chloride	132.3	127.2	122.4
Calcium carbonate molten salt	130.9	124.6	117.3

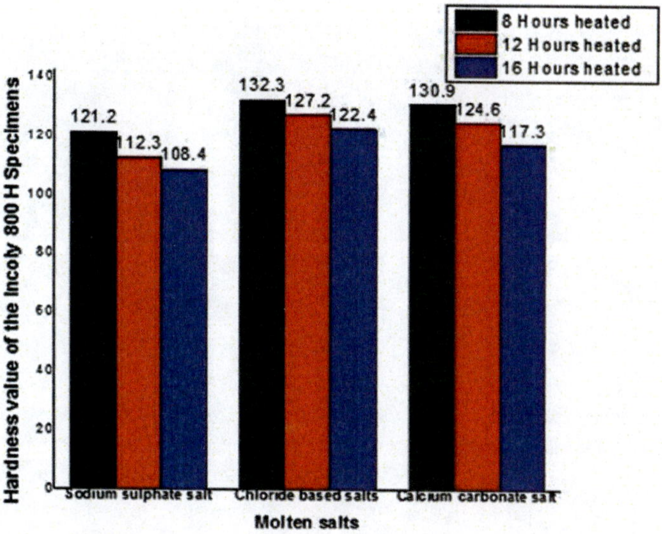

Fig. 5 Hardness test results of specimens with various molten salts

3.1 Morphology

To study in detail about the corrosion process on the specimen surface microstructural analysis such as porosity test, SEM images and XRD analysis were done (Fig. 6).

From the scanning electron microscope images, the salt crystal formation is clearly evident for all the molten salts used for the study. Among them, sodium sulphate salt leads to more corrosion on the specimen surfaces followed by carbonate-based salts and least corrosion with chloride-based salts, respectively (Fig. 7).

The porosity test results and images reveal that the pores and void spaces were minimally formed on all the heated specimens for which the Incoloy 800H material can be preferred for thermal energy storage applications (Fig. 8).

From the XRD analysis, the salt deposits on the specimen surface are well evident by chlorine, carbon and sulphur deposition on the heated specimen surfaces. The sodium sulphate salt leads to a high amount of corrosion, and the sulphur element is highly formed on the heated specimens compared with the other specimens.

4 Conclusion

In view of using the Incoloy 800H superalloy as the molten salts storage material in thermal energy storage systems, Incoloy 800H superalloy specimens were prepared as per the ASTM standards for mechanical testing and material characterization. Sodium sulphate, sodium and potassium chloride eutectic mixture and

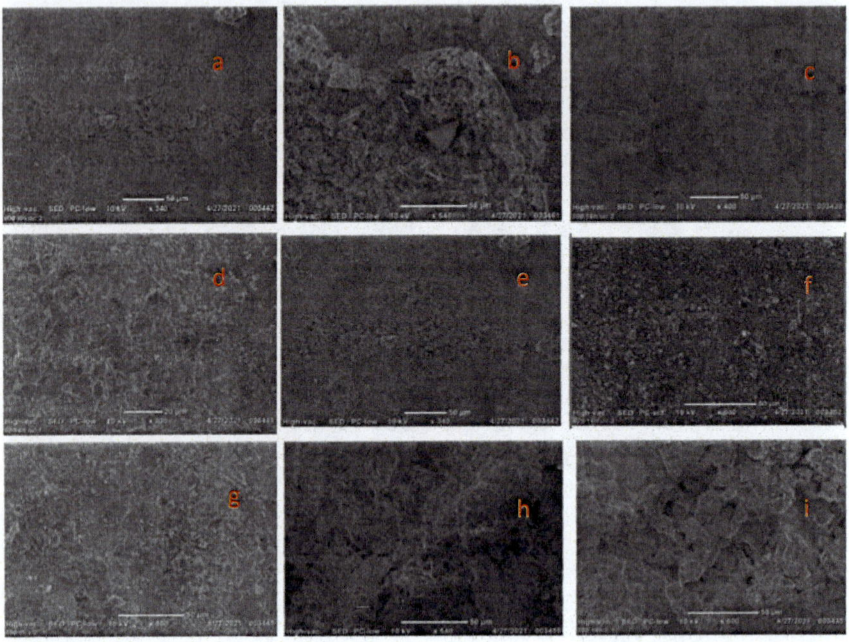

Fig. 6 **a**, **b**, **c**—SEM images of specimens heated with sodium sulphate for 8 h, 12 h and 16 h. **d**, **e**, **f**—SEM images of specimens heated with sodium and potassium chloride for 8 h, 12 h and 16 h. **g**, **h**, **i**—SEM images of specimens heated with calcium carbonate for 8 h, 12 h and 16 h

calcium carbonate with high melting point and boiling point molten salts were taken for the investigation. The specimens were heated upto 1000 °C for 8 h, 12 h and 16 h, respectively, with molten salts. Based on the tensile test and characterization study, the following conclusions were made.

- The sodium sulphate molten salt leads to high amounts of oxidation corrosion where calcium carbonate and chloride-based molten salts lead to minimal effects, respectively.
- High amount of sulphur deposits is well evident on the specimen surfaces since its high amount of corrosion behaviour.
- From the results obtained and the minor effects of hot corrosion, Incoloy 800H superalloy can be used with sodium chloride and potassium chloride mixture for thermal energy storage applications.

In addition to that the storage material can be coated with thermal barrier coating such as yttria stabilized zirconia and other rare earth elements to enhance the mechanical properties which will further reduce the corrosion effects on the specimens. For

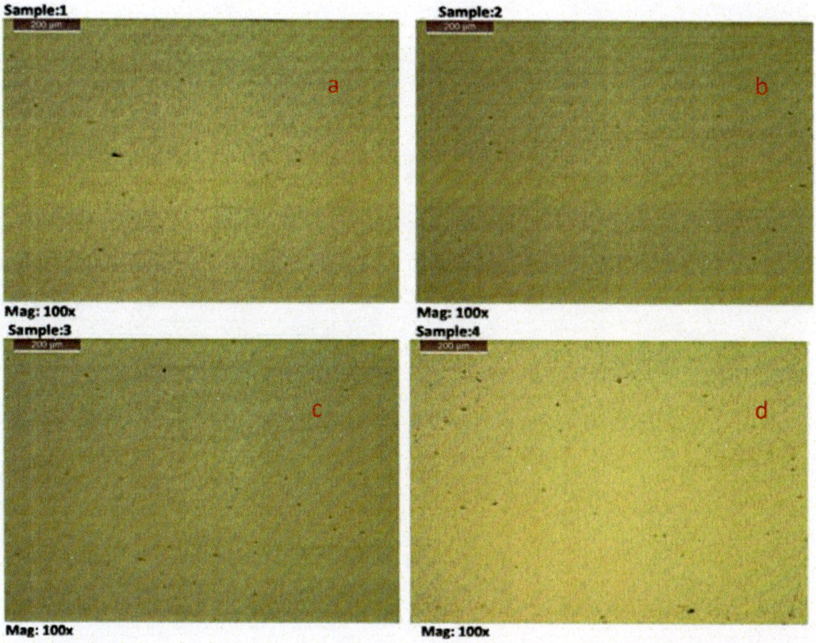

Fig. 7 **a**—Porosity images of incoloy 800H un-heated specimen, **b**—16 h heated specimen with sodium sulphate, **c**—16 h heated specimen with sodium and potassium chloride and **d**—16 h heated specimen with calcium carbonate

future work, varieties of superalloys with different thermal barrier coatings can be preferred for further investigations.

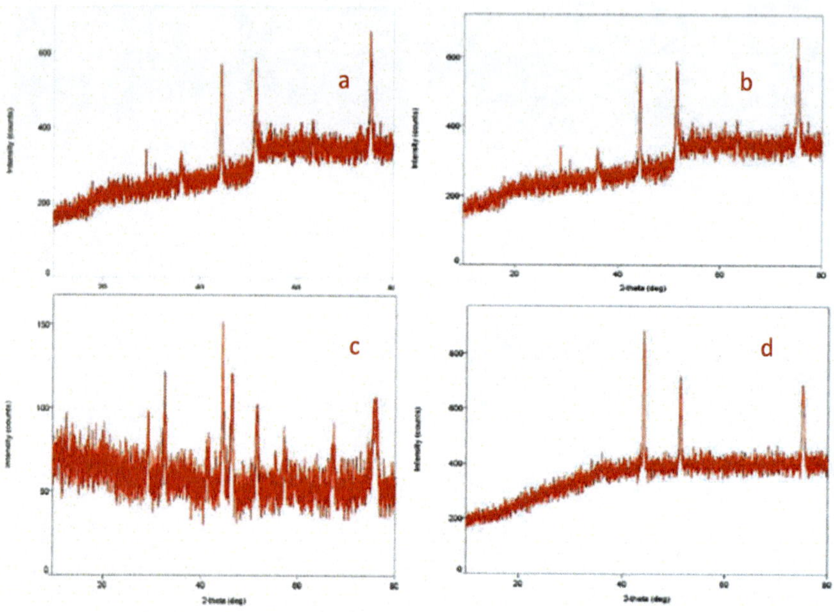

Fig. 8 **a**—XRD results of incoloy 800H un-heated specimen, **b**—16 h heated specimen with sodium sulphate, **c**—16 h heated specimen with sodium and potassium chloride and **d**—16 h heated specimen with calcium carbonate

References

1. Persdotter A et al (2021) Minor element effect on high temperature corrosion of a low-alloyed steel: insight into alkali- and chlorine induced corrosion by means of atom probe tomography. Corros Sci 192: 109779. https://doi.org/10.1016/j.corsci.2021.109779. ISSN 0010-938X.
2. Villada C et al (2021) Engineering molten $MgCl_2$–KCl–NaCl salt for high-temperature thermal energy storage: review on salt properties and corrosion control strategies. Solar Energy Mater Solar Cells 232: 111344. https://doi.org/10.1016/j.solmat.2021.111344. ISSN 0927-0248
3. Koszelow D et al (2021) High temperature corrosion evaluation and lifetime prediction of porous Fe22Cr stainless steel in air in temperature range 700–900 °C. Corros Sci 189:109589. https://doi.org/10.1016/j.corsci.2021.109589. ISSN 0010-938X
4. Wu D et al (2021) Influence of water vapor on the chlorine-induced high-temperature corrosion behavior of nickel aluminide coatings. Corros Sci 190:109689. https://doi.org/10.1016/j.corsci.2021.109689. ISSN 0010-938X
5. Li F et al (2021) "tudy on cyclic thermal corrosion behavior of APS-7YSZ thermal barrier coating at room- and high temperature. Ceram Int 47(20):29490–29498. https://doi.org/10.1016/j.ceramint.2021.07.117. ISSN 0272-8842
6. Chen F et al (2021) Cavitation erosion behaviour of incoloy alloy 865 in NaCl solution using ultrasonic vibration. J Alloys Compd 831:154783. https://doi.org/10.1016/j.jallcom.2020.154783. ISSN 0925-8388
7. Prashar G et al (2020) Hot corrosion behavior of super alloys. Mater Today: Proc, Part 2 26: 1131–1135. https://doi.org/10.1016/j.matpr.2020.02.226. ISSN 2214-7853

8. Dai J et al (2021) High temperature oxidation and hot corrosion behaviors of Ti2AlNb alloy at 923 K and 1023 K. Corros Sci 184: 109336. https://doi.org/10.1016/j.corsci.2021.109336. ISSN 0010-938X
9. Kim J et al (2021) Hot corrosion behavior of Yb_2O_3–Gd_2O_3–Y_2O_3 co-stabilized zirconia-based thermal barrier coatings covered with a Lewis neutral layer. Surf Coat Technol 428:127911. https://doi.org/10.1016/j.surfcoat.2021.127911. ISSN 0257-8972
10. Joshi KK et al (2018) An experimental investigations in turning of incoloy 800 in dry, MQL and flood cooling conditions. Procedia Manuf 20:350–357. https://doi.org/10.1016/j.promfg.2018.02.051. ISSN 2351-9789
11. Kumar S et al (2020) Hot corrosion behaviour of CNT reinforced zirconium yttrium coatings in molten salt environment. J Bio- Tribo Corros 6:81. https://doi.org/10.1007/s40735-020-00378-3
12. Manuela Nimmervoll et al (2021) Surface sulphide formation on high-temperature corrosion resistant alloys in a H_2S-HCl-CO_2 mixed atmosphere. Corros Sci 181:109241. https://doi.org/10.1016/j.corsci.2021.109241. ISSN 0010-938X
13. Natesh M, Sumanth BV et al (2021) Effect of heat treatment on the properties of solution annealed and hot rolled incoloy 20. Mater Today: Proc Part 17 46:7265–7270. https://doi.org/10.1016/j.matpr.2020.12.855. ISSN 2214-7853
14. Sayyar N et al (2020) EBSD observations of microstructural features and mechanical assessment of INCOLOY 825 alloy/AISI 321 stainless steel dissimilar welds. J Manuf Processes 60:86–95. https://doi.org/10.1016/j.jmapro.2020.10.042. ISSN 1526-6125
15. Abu-warda N et al (2021) High temperature corrosionbehavior of Ni and Co base HVOF coatings exposed to NaCl-KCl salt mixture. Surf Coat Technol 418:127277. https://doi.org/10.1016/j.surfcoat.2021.127277. ISSN 0257-8972
16. Ishibashi R et al (2021) Corrosion-resistant metallic coating on silicon carbide for use in high-temperature water. J Nucl Mater 557:153214. https://doi.org/10.1016/j.jnucmat.2021.153214. ISSN 0022-3115
17. Wang W et al (2022) Specimen size effect on evaluation of strength properties of 3Cr1MoV and incoloy 800H using small punch test. Mater Sci Eng: A 832:142400. https://doi.org/10.1016/j.msea.2021.142400. ISSN 0921-5093
18. Sun X et al (2021) Study on high temperature corrosion mechanism of water wall tubes of 350 MW supercritical unit. Eng Fail Anal 121:105131. https://doi.org/10.1016/j.engfailanal.2020.105131. ISSN 1350-6307
19. Wang Y et al (2021) Effect of ZnS/PbS deposits on high temperature corrosion of waterwall tubes in reducing atmosphere. Fuel Process Technol 216:106793. https://doi.org/10.1016/j.fuproc.2021.106793. ISSN 0378-3820
20. Cai Y et al (2021) High-temperature oxidation behavior and corrosion behavior of high strength Mg-xGd alloys with high Gd content. Corros Sci 193:109872. https://doi.org/10.1016/j.corsci.2021.109872. ISSN 0010-938X
21. Zou Z et al (2021) Thermal properties, oxidation corrosion behavior, and the in situ ceramization mechanism of SiB6@BPR/HF composites under high-temperature corrosion. Corros Sci 193:109913. https://doi.org/10.1016/j.corsci.2021.109913. ISSN 0010-938X
22. Liang Z et al (2021) High-temperature corrosion of an Fe–Ni-based alloy HR6W under various conditions at 750 °C and 810 °C: effect of the temperature, water vapor, simulated ash and SO_2. Mater Chem Phys 256:123670. https://doi.org/10.1016/j.matchemphys.2020.123670. ISSN 0254-0584

Exploration of Material Removal Mechanism in Micro-Nano-Electrical Discharge Machining

Abira Mukherjee, Nagahanumaiah, and Santanu Das

1 Introduction

Micro-Nano Systems Technology (MNST) is used to design, integrate and manufacture miniature devices in the ranges from few nanometres to tens of microns. In recent times, the ability to create minute components, instruments and systems prompted a varied range of scientific research leading to the evolution of novel miniaturized devices and systems for wider applications. These micro devices integrate mechanical, electronic and optical elements with computational, chemical, biological protocols processed over a range of advanced materials coatings) over multi-materials. The processing techniques being used to create these nano surfaces include material ablation, deposition synthesis and modification (functionalization), forming or hybrid techniques. However, in the present status, cost effective manufacturing of nanometric features in bulk (batch scale), within the limit of industrial engineering attributes such as reproducibility, control of the size (range of features), shape, homogeneity and robustness of the manufactured structures imposes several challenges for the researchers. In the light of this pressing need for creating micro-nanometric scale features, a top-down fabrication (material removal) process using low energy short pulsed electric discharge widely known as micro-EDM (μ-EDM) offers significant potential in processing difficult-to-machine conductive materials.

A. Mukherjee (✉)
Department of Mechanical Engineering, NMIMS, Mukesh Patel School of Technology Management and Engineering, Mumbai 400056, India
e-mail: abira.mech@gmail.com

Nagahanumaiah
Central Manufacturing Technology Institute, Tumkur Road, Bengaluru 560022, India

S. Das
Department of Mechanical Engineering, Kalyani Government Engineering College, Kalyani, West Bengal 741235, India

© The Author(s), under exclusive license to Springer Nature Singapore Pte Ltd. 2023

S. Revankar et al. (eds.), *Recent Advances in Thermofluids and Manufacturing Engineering*, Lecture Notes in Mechanical Engineering,
https://doi.org/10.1007/978-981-19-4388-1_39

These enhanced capabilities, prompted several research groups to adopt the electrical discharge machining concept for nano-scale machining. Few researchers have introduced a dielectric liquid of very small volumes between a Platinum-Iridium erosion tool and work piece surface, and demonstrated steady machining of features as small as 10 nm [1]. Whilst this work has put forth the future potential for using EDM concept in nano-scale machining, no stand-alone machines are developed yet. This is because the technology with respect to material removal mechanism and the gap phenomena is not fully understood for low energy short pulsed electric discharge machining conditions. Therefore, in recent years, molecular dynamics (MD) simulation that can track the movement of atoms with time is being used in such thermo-mechanically induced material processing such as ultra-short pulsed (pico and femto second) laser ablation, micro machining. Yang et al. [2, 3] have studied the formation of bulge and craters during micro-EDM. This study concludes the formation of bulges caused due to shear flow of molten material caused by very high pressure in the superheated material, and accumulation of ejected material. The material removal efficiency is found to be ranging between 0.02 and 0.05, since most of the metal remains re-solidified. These researchers also performed MD simulation of residual stress produced during conventional EDM. It is found that most of residual stress existed at the melting and solid region besides the molten area. The residual stress on the electrode surface is tensile, whilst inside it is compressive. MD simulation has also been used in recent times to understand the material removal in conventional EDM using deionized water and gas. Whilst compared to discharge in gases, discharge in deionized water generates expanding bubble and flowing liquid tending to blocking effects during material removal [4]. Yang and Yue [5] in their recent article discussed that the previous research model being very small, the effect of free electrons on conduction of heat could not be considered. A two-temperature model has been created to overcome this problem and MD simulation is used to understand the discharge process and its effect on lattice vibration. Roy et al. [6] in their article has proposed a new method in which a ratio of certain material removal at higher spark energy and at lower energy can be calculated.

Understanding the thermo-mechanical effects upon nano-EDM by experimental investigations is extremely challenging and difficult owing to operational difficulties and instrumentation limitations. This paper therefore focuses on the thermo-mechanical effects at a molecular level after nano-scale removal of material by ultra-short pulsed low energy electric discharge between copper electrodes using molecular dynamics (MD)-based simulations. Spectral temperature profile, crater formation and re-solidification process for 2 ps pulsed electric discharge over anode copper substrate at different power densities have been explored.

2 Materials and Methods

2.1 *Molecular Dynamics Simulation Hypothesis*

The computational method that uses a set of classical equations of motion to study the progression of a system is molecular dynamics (MD) simulation [2]. The classical equation of motion can be written as:

$$m_i \frac{d^2 r_i}{dt^2} = \sum_{i \neq j} F_{ij} \tag{2.1}$$

where m_i and r_i are the mass and position of particle i, respectively. F_{ij} is the force acting on the particle i due to influence of particle j. Initial positions and masses of particles are defined according to lattice structure and type of material, respectively. A potential function is used to define the inter particle interaction, and forces between each atom.

Once the initial conditions such as positions, velocities of each atom and potential functions are defined, the equation of motion is solved numerically. The MD simulation generates results in the form of trajectories (positions and velocities) of each particle as a function of time. In this work the spatial and time evolutions of structural and thermodynamic properties of copper substrate are studied with this output. The schematic representation of computational scheme up is shown in Fig. 1. In actual EDM, the work piece (pure copper) is submerged in dielectric liquid and is mostly covered with bubbles, here for modelling simplicity the working gap is assumed to be vacuum. The sample is having a total number of 790591 atoms. The non-periodic boundary condition is given in x and z direction and free boundary condition in y direction. The open-source code Large-scale Atomic/Molecular Massively Parallel Simulator (LAMMPS) [7] is used for molecular dynamic simulation.

Initially the work piece of the specified size is created as a face centred cubic (FCC) cell structure and the initial velocities are randomly allocated to each atom based on Gaussian distribution. The embedded-atom method (EAM) [8] is used to represent the potential of different interacting atoms. Compared to other pair potentials, the

Fig. 1 Schematic of computational scheme

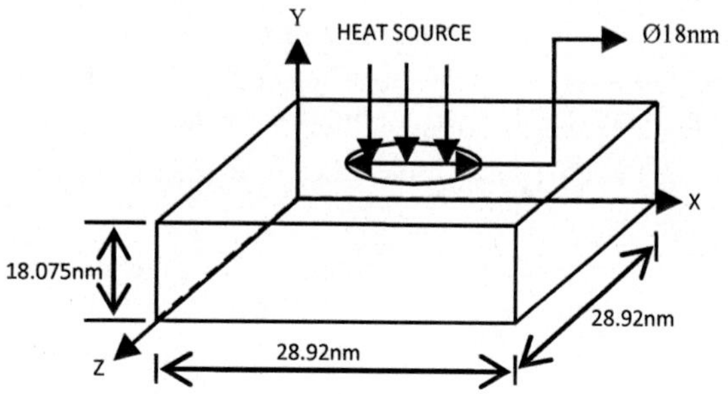

EAM potential has a term, which is related to many-body interaction. According to this potential, total energy E for atom i is given by,

$$E_i = F_\alpha\left(\sum_{j\neq i}\rho_\beta\left(r_{ij}\right)\right) + \frac{1}{2}\sum_{j\neq i}\varphi_{\alpha\beta}\left(r_{ij}\right) \tag{2.2}$$

where F is the embedding energy which is a function of the atomic electron density ρ, φ is a pair potential interaction, α and β are the element types of atoms i and j. In metals, heat transfer or heat distribution of free electrons takes place by both thermal movement and the lattice vibration. But since there is no scope of considering electron movement in classical molecular dynamics, we assume that the kinetic energy of the atoms is sufficient to include all the electron effects. So, in this study the material discharge forming craters due to lattice vibration through heat transfer has been investigated. The outcome of the simulation illustrates elucidation of the material removal mechanism for the cognizance and comprehension of the micro-EDM.

2.2 *Representation of Process Conditions in MD*

It is very well appreciated that true representation of process conditions becomes a greater challenge in simulation study. The assumptions made on process conditions to make the computational process easy and to obtain a solution under the given constraints often derive very approximate results. For example, several heat input methods such as circular disc, point source, have been assumed in various research to forecast the heat distribution in electrodes in conventional EDM, moreover such studies have not been reported in case of micro-nano-scale-EDM. Therefore, in order to imitate this process condition and the input of thermal energy for MD simulation, systematic experiments were conducted at discharge energy of (25–150 μJ) with pulse-on-time up to 3microseconds (experiments were not practicable at much shorter pulse durations). Figure 2a, b shows the experimental setup built by the authors, where the discharge between 300 μm tungsten carbide cathode electrode over copper anode work piece were recorded using high speed camera at 5000 frames per second.

Based on the experimental observations on discharge pattern and other complexity associated with the material removal by ultrafast low energy discharge, certain assumptions have been taken into consideration to elucidate the proposed model mathematically. The assumptions are made as follows:

- The model is made for a spark occurring during a pulse-on-time at a given discharge energy.
- The work piece is considered to be homogeneous, isotropic and free from internal residual stresses before machining.
- The free surface of the anode is receiving heat from a circular source which is assumed to emulate the heat flux from the discharge column.

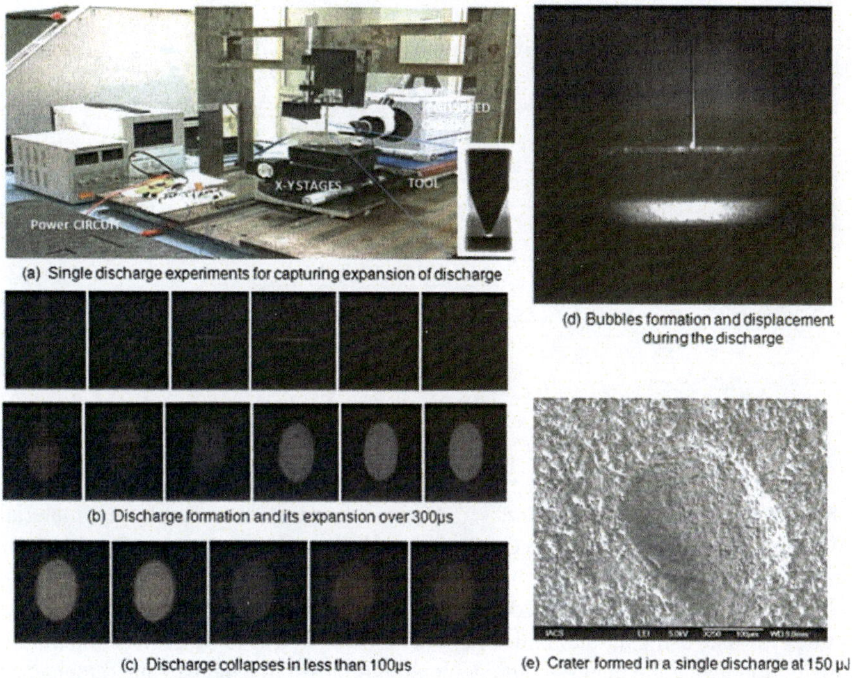

Fig. 2 Experimental investigation

- It is assumed that the radius of the heat source is equal to that of the discharge column. The temperature in the discharge column is assumed to be constant at 7000 K based on spectroscopic measurement results obtained in an earlier work [9, 10].
- Although, linear expansion of discharge column during pulse duration was observed experimentally, the time scale used in MD simulation is in the order of few picoseconds. Therefore, the discharge column was assumed to be static at the centre of the electrode.
- As the tiny electrodes separated by a few microns are immersed inside the dielectric, bubbles formed between the electrodes at this narrow gap push the liquid dielectric leaving vacuum at the heat input source.

To emulate the input of thermal energy, the velocity for individual atom on the surface of the anode is considered to follow the Maxwell–Boltzmann distribution at 7000 K. The discharge column was considered stationary means that temporal evolution does not take place during the discharge, whereas spatial distribution takes place and the environment at the gap between electrodes is vacuum. In this model, a circular Gaussian heat source whose radius expands linearly with time was placed on the surface of the work piece at the centre to imitate the heat flux from the arc column. The Gaussian heat flux $q(R)$ expression can be given as in Eq. (4) [2].

Table 1 Simulation parameters

Sl. No.	Simulation parameters	Values
1	Discharge duration	2 ps
2	Power density	0.6 GW/cm^2
3	Discharge column temperature	7000 K
4	Discharge column radius	9.0375 nm
5	Atomic mass (m_i)	63.54 g/mol
6	Lattice constant (a)	0.3615 nm
7	Time step	1 fs
8	Number of atoms	790591

$$q(R) = \frac{4.45EIU}{\pi r^2}\exp\left[-4.5(R/r)^2\right] \quad (2.3)$$

where q = Heat flux [W/mm^2], E = Fraction of energy absorbed by the material [W], I = Pulse current [A], U = Gap voltage [V], R = Radial distance from the axis of the spark [μm], r = Plasma channel radius [μm].

The MD study for the selected copper computational model shown in Fig. 1 has been performed using the simulation parameters [2] listed in Table 1. The temperature distribution inside the material lattice, material removal mechanism and dimensions of the crater formed at given process conditions are discussed in the next section.

3 Results and Discussion

3.1 *Temperature Distribution Inside Material Lattice*

The temperature gradient across the material lattice changes as the thermal energy input into work surface increases. In Fig. 3, it is observed that the material removal starts at about 3 ps just after the discharge and eroded material starts to be ejected out and much of which re-solidified at the edge of the crater in the form of bulge.

The temperature distribution along the discharge axis has four distinguishable regions as shown in Fig. 4. The region A is the top surface at a temperature greater than 6000 K, region B is the melting zone where temperature varies from 3000 to 4200 K and molten metal pool starts evaporating, the region C, where temperature is slightly below the melting zone without any phase change and unaffected region D has slightly elevated temperature. It has been found that compared to radial direction the increase of temperature at discharge axis is extended to greater depth.

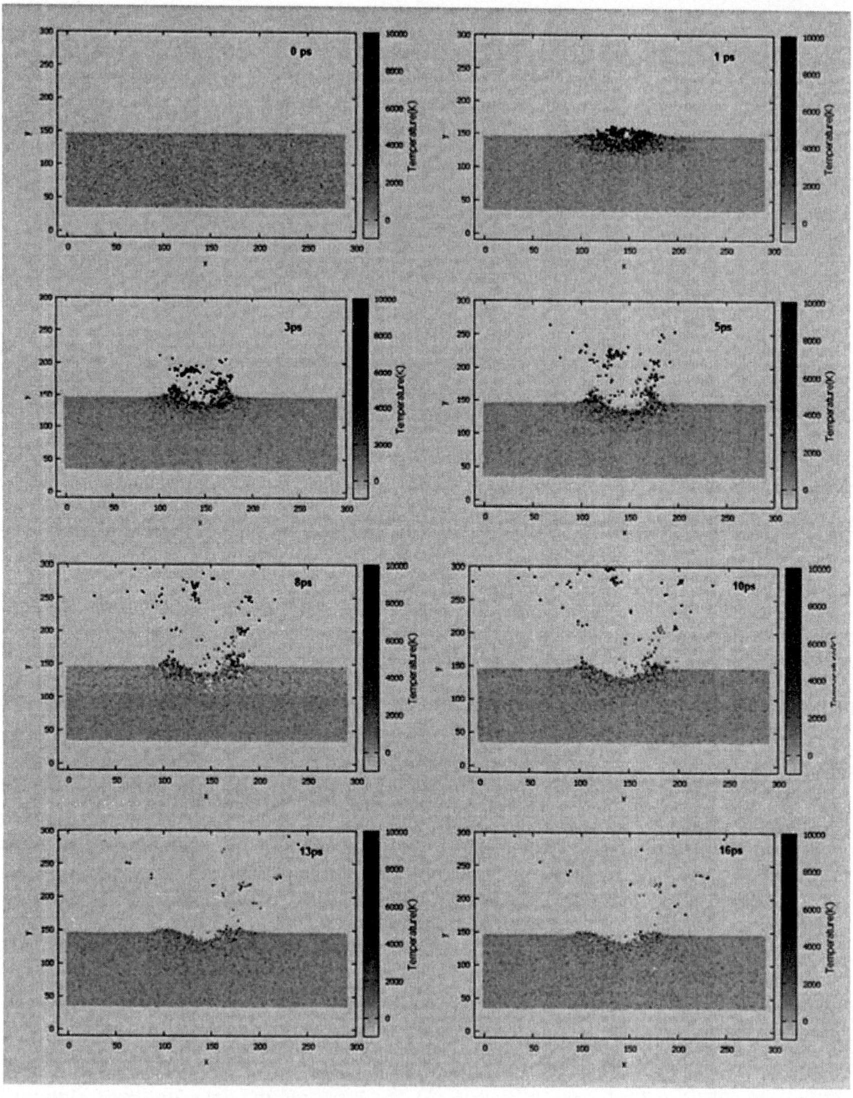

Fig. 3 Temperature distribution in the target at different time steps

3.2 Thermal Stress Induced in Material Lattice

The rapid rise in temperature maybe attributed to the local heating by a discharge column induced thermal stress in the material lattice [3]. In this simulation study per-atom thermal stress has been calculated by virial stress and its distribution in the target at different time steps have been shown in Fig. 5. The results show that maximum thermal stress up to 5×10^6 N/m^2 is induced during the material removal

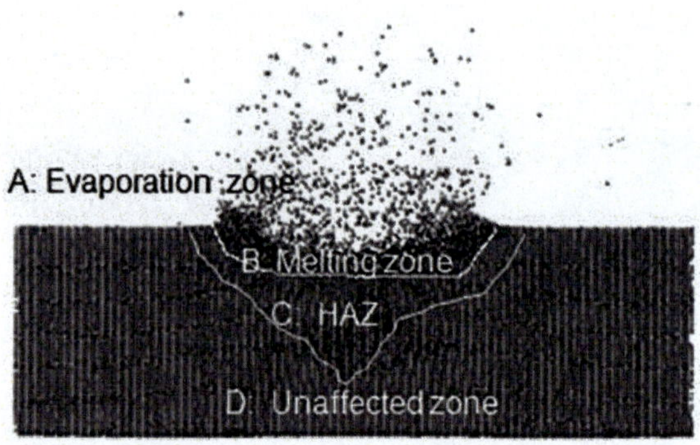

Fig. 4 Thermally induced damage in material lattice at 5 ps

at different material lattice beneath the crater formed. The significant increase in axial stress and absence of shear stress after 3 ps along the discharge axis confirms higher material removal rate. It is also noted that between 3 and 10 ps with increase in thermal stress maximum crater is formed and gradually the stress reduces thereafter.

3.3 Material Removal Mechanism

The outcome of the simulation on removal of material over time is shown in Fig. 6. The temperature gradient across the electrode surface increases due to the energy generated during discharge. This local heating increases the surface temperature rapidly and creates thermal waves propagating downwards. With further heating, the crack propagates downwards through the material lattice. Further rise in temperature results in thermal expansion of material lattice before it starts melting. As shown in Fig. 6 at 1 ps, initiation of detachment of layer of atoms from the parent material is caused by propagation of thermal waves. When temperature on the surface rises above the melting point, the detached material undergoes phase change, atoms in the melt get accelerated further by an input of heat energy leading to vaporization and ejection of particle at 3 ps and onwards. At $t = 5$ ps, the pressure in the molten and vaporized material increases to a greater extent resulting in rapid increase of bubbles as shown in Fig. 6. These eruptions separate the material melted from the surface of the anode mostly in nano-clusters.

It can be observed that depth of crater is more along the discharge axis. This is because of the fact that temperature rise along the axis of discharge column is more rapid than in radial direction [11, 12]. Additionally, the melt detached from the parent material lattice, flows towards the centre in the absence of external pressure. This sudden rise in heat input rapidly vaporizes the melt leading to explosion. This high pressure exerted over the fresh surface layer of material originates cracks below the melt pool resulting in further removal of material. This is one of the main reasons

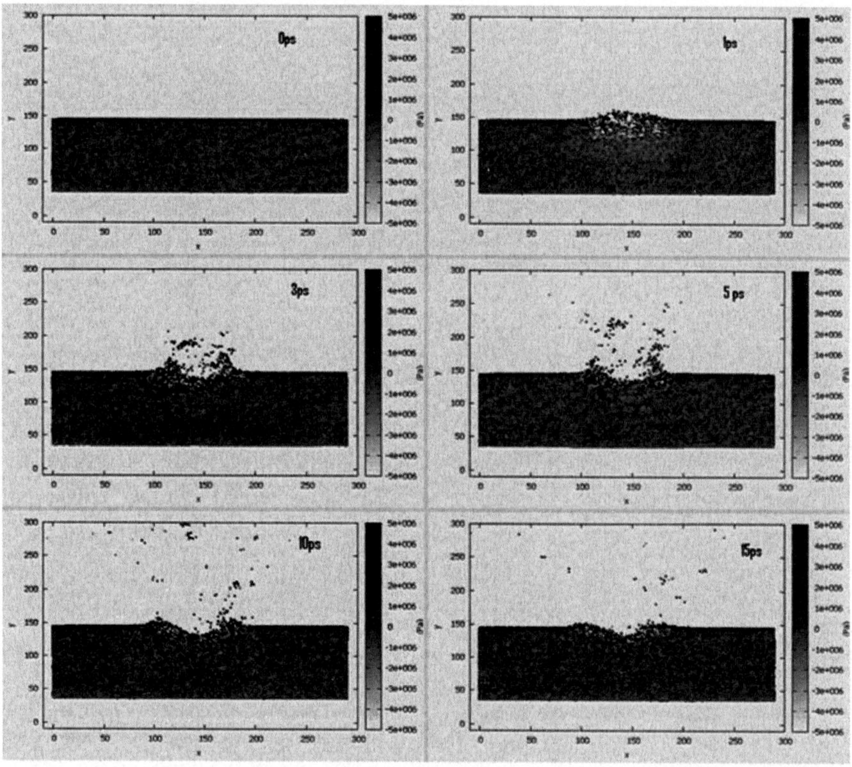

Fig. 5 Stress distribution along X-direction in the target and geometry of the target at different time steps

for the bowl shape of craters formed by a single discharge. The material ejected out of the discharge zone in the form of a jet of melt spray initially in the form of large clusters (as observed at 5 ps) and reducing down to the atoms (as observed at 8 ps) falls over the cold surface. With the combined effect of surface tension and rapid cooling of jet of tiny particle forms balls deposited around the crater edge. In the absence of external forces such as hydrodynamic force caused by breakdown of liquid dielectric, the centrifugal force and electrostatic force pushes the melt liquid towards the edge. This displaced layer of melt and deposition of ejected material at the edge of the crater which is relatively cold, results in re-solidification of the material forming a bulge shaped recast layer around the edge of the crater created by this single discharge. Based on this simulation study, it has been confirmed that material can be removed by applying ultra-short pulsed electric discharge even in the absence of dielectric liquid. The rapid process chain that includes thermal wave propagation, delamination of layer of atoms, melting, explosion and re-solidification of part of removed material takes place rapidly, which has been explained in this work by monitoring the process in different time steps in Fig. 6.

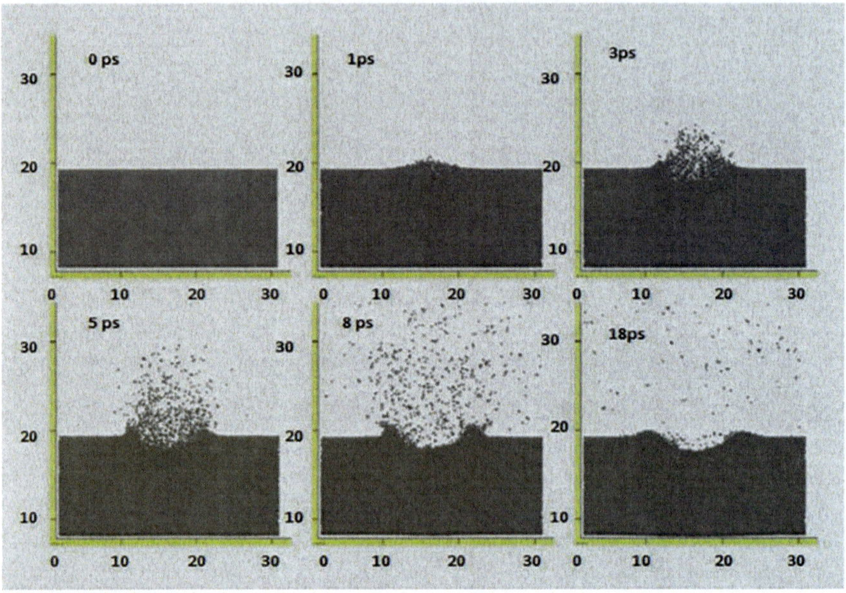

Fig. 6 Material removal process of anode

This simulation has been run up to 40 ps. It has been found that much of the material removal process is completed at around 20–25 ps, and the machined crater surface cools slowly leaving the crater. Further, dimensions of the crater have been measured using different parameters. With the simulation results performed for different conditions, it is found that crater dimensions are functions of power density. For example, at a low power density of 0.3 GW/cm^2 ($t = 5$ ps), debris starts forming large clusters as shown in Fig. 7a, whereas at 2.0 GW/cm^2($t = 5$ ps) power densities, the debris is ejected out much farther away from the crater in the form of smaller clusters, leaving large craters as shown in Fig. 7b.

Fig. 7 Material removal at **a** 0.3 GW/cm^2 and **b** 2.0 GW/cm^2 power densities at $t = 5$ ps

4 Conclusion

In this paper, molecular level study has been performed to understand heat distribution in material lattice. The outcome of the simulation is stated below:

- The rapid heat transfer along the discharge axis melts the material at centre of discharge axis, the high pressure generated beneath this melt pool leads to the explosion of metal.
- A part of the melt is ejected out from the crater and remaining melt is displaced towards the crater edge and re-solidifies to form a recast layer at the crater edge in the form of a bulge.
- The actual diameter of the discharge crater formed by a single discharge is smaller than the diameter of the discharge column and increase in power density increases the melt pool area.
- Whilst this study demonstrates the removal of material by ultra-short pulsed electric discharge with low energy, there is a need to incorporate few features into the real process to minimize the recast layer formation that enhances material removal rate and controls the crater geometry.

References

1. Guo C, Thomson D (1992) Material transfer between metallic tips and surface in the STM. Ultramicroscopy 1452–1458
2. Yang X, Guo J, Chen X, Kunieda M (2011) Molecular dynamics simulation of the material removal mechanism in micro-EDM. Precis Eng 35:51–57
3. Yang X, Han X, Zhou F, Kunieda M (2013) Molecular dynamics simulation of residual stress generated in EDM. Procedia CIRP 6:432–437
4. Yue X, Yang X (2016) Study on the distribution of removal material of EDM in deionized water and gas with molecular dynamics simulation. Procedia CIRP 42:691–696
5. Yang X, Yue X (2020) Molecular dynamics simulation of material removal process and mechanism of EDM using a two-temperature model. Appl Surf Sci 528
6. Roy T, Sharma A, Datta D, Balasubramaniam R (2019) Molecular dynamics simulation of single discharge and dimensionless correlation with actual material removal in micro electrical discharge machining. Mol Simul 45(13):985–995
7. Plimpton S, Thompson A, Crozier P (2013) Large-scale atomic/molecular massively parallel simulator. Sandia National Laboratory, USA
8. Foiles S, Baskes M, Daw M (1989) Embedded-atom-method functions for the FCC metals Cu, Ag, Au, Ni, Pd, Pt, and their alloys. Phys Rev B (Condens. Matter) 33:7983–799
9. Nagahanumaiah J, Ramkumar N, Glumac SK, Richard E (2009) Characterization of plasma in micro-EDM discharge using optical spectroscopy. J Manuf Proc 11:82–87
10. Shimada S, Ikawa N, Tanaka H, Ohmori G, Uchikoshi J, Yoshinaga H (1993) Feasibility study on ultimate accuracy in microcutting using molecular dynamics simulation. Ann CIRP 1:91–94
11. Singh AK, Singhal D, Kumar R (2020) Machining of aluminum 7075 alloy using EDM process: an ANN validation. Mater Today Proc 26(2):2839–2844
12. Kumar R, Roy S, Gunjan P, Divya AS, Sarkar D, Kumar R (2018) Analysis of MRR and surface roughness in machining Ti-6Al-4V-ELI titanium alloy using EDM process. Procedia Manuf 20:358–364

Fabrication of Mechanical Barrier Using Waste Polymer and Wood Saw Dust and Turnings

Sonu K. Samuel, M. Govindaraju, and R. Vaira Vignesh

1 Introduction

In a few decades, humans witnessed a leaping advancement in the development of technologies for the production of polymeric materials. Currently, the global production of polymeric material stands at 270 million tons per year. More than 55% of polymeric wastes are discarded and only 20% of polymeric wastes are recycled every year. Central Pollution Control Board (CPCB, Government of India) estimated that ~ 4059 tons of polymeric waste are generated in major cities of India per day, where 25,940 tons of polymer waste are generated per day in India only. The degradation of polymers through biological activities in terrestrial environments takes more than five centuries and in the marine environment takes two decades. Hence, polymers pose a high degree of pollution and hence the consequences on the environment [1, 2]. In most wood mills, sawdust is considered commercial waste. The commercial value of the sawdust waste could be increased by the possible creation of new products. It is estimated that around 4 million to 6 million tons of wood waste is generated per year, which is mostly from the furniture manufacturing, construction, and packing sectors [3, 4]. Responsible consumption and production would ensure the sustainable use of natural resources.

Also, the practice of recycling wastes would sustainable environment for us and our future generation [5]. Turning waste into composite material presents a major technical, economic, and ecological challenge and importance both internationally and nationally [6–8]. Fabrication of composite has been an interesting research area, as evidenced from the previous research work. Basalp et al. [9] investigated the influence of wood flour composition and processing temperature on the properties

S. K. Samuel · M. Govindaraju · R. V. Vignesh (✉)
Department of Mechanical Engineering, Amrita School of Engineering, Amrita Vishwa Vidyapeetham, Coimbatore, India
e-mail: r_vairavignesh@cb.amrita.edu

© The Author(s), under exclusive license to Springer Nature Singapore Pte Ltd. 2023

S. Revankar et al. (eds.), *Recent Advances in Thermofluids and Manufacturing Engineering*, Lecture Notes in Mechanical Engineering,
https://doi.org/10.1007/978-981-19-4388-1_40

of wood-plastic composite. The decrease in the property of water absorption in the specimen indicates that there was an increase in interfacial adhesion between recycled polymer and wood flour. The study concluded that processing at 185 °C improves the tensile strength of the wood-plastic composite.

Hernandez et al. [10] studied the mechanical behavior of wood polymer composites that were made from polystyrene and white oak wood flour. The particle size, the ratio of the wood, and the wood content influence the mechanical and water absorption property in the specimen. Nami et al. [11] studied water absorption, fungal resistance in natural fiber (wood, bamboo), and polypropylene-based wood-plastic composite. The wood particle size (30 mesh to 100 mesh) and particle content affected the water absorption and fungal resistance of the composite specimen. Baptista et al. [12] studied the microstructure of thermoplastic composite made of sawdust, wool, and high-density polyethylene or polypropylene. A less-intensive bonding was observed between wood, wool, and high-density polyethylene or polypropylene, because of the absence of a coupling agent. Jumadi et al. [13] studied the microhardness of the composite, which was prepared with recycled material (polypropylene and polyethylene) and natural fiber (kenaf fiber). The polished specimen with 40 wt.% kenaf fiber had better visibility and a maximum microhardness of 9.2 HV.

Taghi et al. [14] investigated the fabrication of wood-plastic composite from completely recycled material (wood flour and granules of polypropylene). The results indicate that the addition of wood flour beyond 35 wt.% deteriorated the physical and mechanical properties. However, a low-water absorption percentage (0.5%) was observed in the wood-plastic composite with 60 wt.% polypropylene. Youssef et al. [15] investigated the structural and mechanical characteristics of the wood-plastic composite (high-density polyethylene waste and camphor wood waste). Maximum bending strength and Young's modulus were observed in the wood-plastic composite that had 40% plastic content. Also, the composite exhibited better impact bending strength (30.75 MPa) than other specimens. Fabrice et al. [16] investigated mechanical and thermal properties of bio-based material to develop an insulating material using miscanthus. At 50 wt.% miscanthus and 50 wt.% binder, better compressive strength (3.5 MPa) was observed.

Beatriz et al. [17] studied the mechanical behavior and vibroacoustic behavior of rubber granules with rice husk. The composite with 75% rice husk content exhibited a maximum compressive strength of 5.951 MPa. An increase in density of stiffened the composite. Behzad et al. [18] studied the influence of immidazolium-based green solvents on the water absorption of wood-plastic composite. The wood flour was treated chemically with imidazolium-based ionic liquids and that in turn was combined with plastic (high-density polyethylene). It was observed that imidazolium-based ionic liquids increased the quality of adhesion between the polymer matrix and the cellulosic materials. Nourbakshi et al. [19] studied the properties of the wood-plastic composite made from high-density polyethylene, wood fiber (55, 70, 80 wt.%), and coupling agent (polyethylene-grafted maleic anhydride) (0, 2, 4 wt.%). The results indicated that addition of more wood fiber and coupling agent improved the mechanical and physical properties of the composite.

Omoniyi et al. [20] studied the influence of saw dust (*Gmelina Arborea*) and plastic (waste water bottles) content on the mechanical behavior of wood-plastic composite. An increase in wood fiber decreased the tensile strength of the composite. The mechanical characteristics of composites were significantly affected by an increase in the sawdust-plastic ratio. Faith et al. [21] investigated the mechanical properties of wood-plastic composite that was fabricated pine wood fiber and thermoplastic. It was observed that the high-density and polypropylene-based composite had a maximum tensile strength of 37.02 MPa. Polypropylene-based composite exhibited a maximum flexural strength of 35.82 MPa. Harish et al. [22] studied the influence of polymer waste (high-density polyethylene, low-density polyethylene) mixed with different natural filler materials (concrete, clay, natural residue, rice husk), and coupling agents (maleic anhydride polypropylene, natural oil, silane, and alkaline treatment) in the properties of wood-plastic composite. The wood-plastic composite with 70% rice husk and 30% high-density polyethylene had a high-tensile strength of 36 MPa. Qiang et al. [23] investigated the characteristics of wood-plastic composite made from polyethylene with maple wood fiber utilizing twin-screw extruder method. The inclusion of maleic anhydride-grafted polyolefin (0.9 wt.%) to the composite increased the amount of adhesion between the wood fiber and the polyolefin matrix. The fractography confirmed the fracture of the cellulose fiber and lesser voids that in turn indicated improved mechanical performance. Prakash et al. [24] studied the composite material (coconut fiber, sansevieria cylindrica fiber, artificial fibers) that would be suitable for tsunami-resistant design. The mechanical strength of composite was evaluated and which was reinforced with coconut fiber, sansevieria cylindrica fiber, glass fiber with resin (isophthalic polyester resin) was evaluated. The composite material made from coconut fiber exhibited an impact strength of 0.022 J/mm^2 and moisture percentage comparable to that of glass fiber.

In this study, wood polymer composite was fabricated through powder metallurgy processing technique. The microstructure, tensile strength, compressive strength, water absorption, and thickness swelling were studied for various wood polymer composites.

2 Materials and Method

2.1 Materials

The high-density polyethylene (HDPE) bottles were collected from the domestic waste. The bottles were shredded using a mechanical shredder. The shredded HDPE bottles were termed waste polymer. The wood husk was collected from a local sawmill. The average particle size of wood husk was ~1500 μm. The wood husk was dried in shade at room temperature for 72 h to eliminate the moisture content.

Table 1 Composition of wood and polymer used for making composite

Specimen	Wood husk (wt%)	Waste polymer (wt%)	Number of specimens	Dimension (mm)
A	40	60	3	100 × 100 × 25
B	45	55	3	100 × 100 × 25
C	50	50	3	100 × 100 × 25

2.2 Experimental Layout

The composite specimen was prepared by varying the composition of wood husk and waste polymer. The composition of the fabricated composite specimen is shown in Table 1.

2.3 Composite Preparation

The wood husk and waste polymer were dried under direct sunlight for 24 h. The dried contents were weighed in the specified ratio, as given in Table 1, using a high-precision weighing balance with a readability of 0.0001 g. The weighed contents were homogeneously blended using a mechanical ball milling unit at 250 rpm for 6 h. The blended contents were molten in a graphite crucible using a resistance-heating furnace at a temperature range of 160 °C to 180 °C. The agglomeration and segregation were avoided by an automatic stirring setup. The melt was poured into a rectangular mold of dimension 100 × 100 × 25 mm^3. Before pouring, mineral oil was applied to the walls of the rectangular mold to lessen the adhesive effects. In the course of cooling, the specimens were compacted using a hydraulic compaction machine. The specimens were compacted such that the volume reduction of the composite at the end of compaction was 10%. Subsequently, the specimen was cured in the atmosphere to obtain the fabricated specimen.

2.4 Microhardness

The composite specimens were polished using silicon carbide paper to reduce surface uniformity. The Vicker's microhardness of the composite specimen was measured as per the standard ASTM E 384 [25]. The specimen was indented with a diamond in an axial load of 10 g for a dwell time of 15 s. An average of five measurements was reported.

2.5 Water Absorption Test

The water absorption of the specimens was determined as per the standard ASTM D 5790 [26]. The mass of the specimens was measured using a precision weighing balance with the readability of 0.0001 g. The specimens were immersed in water at room temperature. The specimens were removed after 2 h and 24 h of immersion. The specimens were wiped to remove the water on the surface and weighed immediately. The mass gain of the specimen was measured, and water absorption (%) was calculated using the Eq. (1) [26].

$$\text{Water Absorption } (\%) = \frac{Mc - Mo}{Mo} \times 100 \tag{1}$$

where Mc is the mass of the specimen after immersion (g) and Mo is the mass of the specimen before immersion (g).

2.6 Thickness Swell Test

The swelling of the specimens after the immersion test was measured. The thickness of the specimen before and after the immersion test was measured using a digital vernier caliper with a least count of 0.01 mm. The swelling thickness was calculated using Eq. (2) [26].

$$\text{Thickness Swell } (\%) = \frac{Tc - To}{To} \times 100 \tag{2}$$

where Tc is the thickness of the specimen after immersion and To is the thickness of the specimen before immersion.

2.7 Tensile Test

The tensile test specimens were prepared as per standard ASTM D638 [27]. The specimens were tested for tensile strength and elongation at room temperature using a tensile testing machine (Make: Tinius Olsen) at a crosshead speed of 1 mm/min.

2.8 Compressive Test

The compression strength of the specimens was determined as per the standard ASTM D 3410 [28]. The compressive test of the specimens was carried under room temperature using a universal testing machine at the crosshead speed of 5.6 mm/min.

2.9 Fracture Morphology Analysis

The fracture morphology of the tensile-fractured specimen was observed under a field emission scanning electron microscope (Make: Zeiss Sigma). The fractography were captured at EHT of 5 kV and various magnifications.

3 Results and Discussion

3.1 Fabricated Specimen

The wood husk and waste polymer of specified proportionate composition were ball milled to a homogeneous mixture. The mixture was melted in a graphite crucible induction furnace and was poured into a rectangular mold. Subsequently, the semi-solid melt was compacted using a hydraulic press. The properties of the composite specimens are described below.

3.2 Microhardness Test

Figure 1 shows the Vicker's microhardness of the fabricated composite specimen. The polished specimens had a visible indentation mark, which enabled easy and effective determination of microhardness. The microhardness of specimen A (9.1 HV) was higher than specimen B (6.4 HV). In turn, the microhardness of specimen C (4.3 HV) was lesser than that of specimen A and specimen B. A linear decreasing trend was observed in hardness with a decrease in waste polymer content of the composite. The results explicited the significant influence of waste polymer content on the microhardness of the composite specimen.

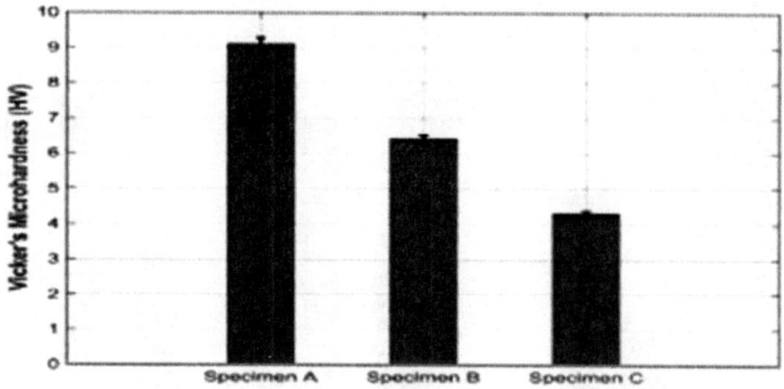

Fig. 1 Microhardness of the composite

3.3 Water Absorption Test

Figure 2 shows the water absorption of the specimens based on various content of wood husk content after 2 h and 24 h of immersion in water. Specimen A had 10% of water absorption after 24 h of immersion test. Specimen B had 13.9% of water absorption after 24 h of immersion, which is 3.9% higher than specimen A. Specimen C with an equal proportion of wood husk and waste polymer had 16.3% of water absorption after 24 h of immersion. The water absorption of specimen C was 6.3% higher than specimen A. The water absorption of the specimens increased with an increase in immersion time. Also, it was observed that the water absorption of the specimens increased with an increase in wood husk content. The reason could be attributed to the natural hydrophilic phenomenon of wood husk with water. Wood in turn is a hydrophilic porous composite that consists of cellulose, lignin, and hemicellulose polymers. These polymers are rich in functional groups such as hydroxyls, which readily interact with water molecules by hydrogen bonding. Hence, the specimens with more wood husk content have the potentiality to absorb water. However, the reason for low-water absorption in the specimens with high-waste polymer could be attributed to the hydrophobic character of HDPE. Also, poor joint interface bonding between waste polymer and wood husk enabled the formation of voids. The voids acted as a medium for stored the absorbed water. Therefore, the water absorption increases with an increase in immersion time from 2 to 24 h, as well as with the proportionate increase in wood content increase in water absorption.

3.4 Thickness Swell Test

Figure 3 shows the thickness swelling of the specimens based on various content of wood husk content after 2 h and 24 h of immersion in water. Specimen A had 4.9% of thickness swelling after 24 h of immersion test. Specimen B had 5.2% of thickness swelling after 24 h of immersion, which is 0.3% higher than specimen A. Specimen

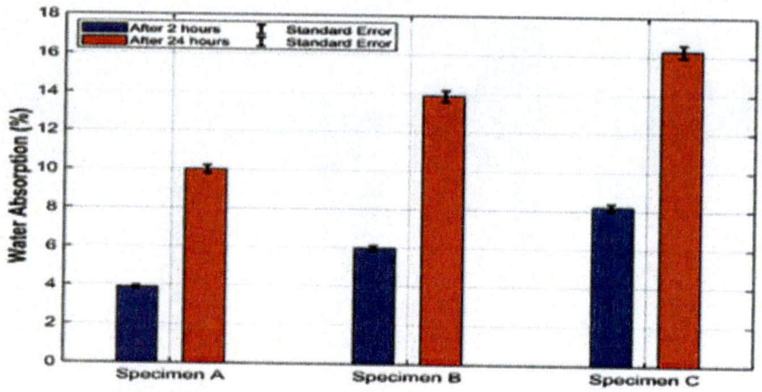

Fig. 2 Water absorption of the specimen

C with an equal proportion of wood husk and waste polymer had 5.8% of thickness swelling after 24 h of immersion. The thickness swelling of specimen C was 0.9% higher than specimen A. The thickness swelling of the specimens had a similar tendency as that of water absorption. The results indicated that the specimens with more fractions of wood husk had higher thickness swelling. Also, the results illustrated that the specimens with more fractions of waste polymer had lower thickness swelling. The primary reason for thickness swelling lies with the hydrophilic nature of wood husk and the hydrophobic nature of waste polymer (HDPE). The wood has a critical surface energy between 40 mJ/m^2 and 60 mJ/mm^2 which is higher than waste polymer. The substantial difference in surface energy between wood and waste polymer might make waste polymer to be water repellent or hydrophobic (16).

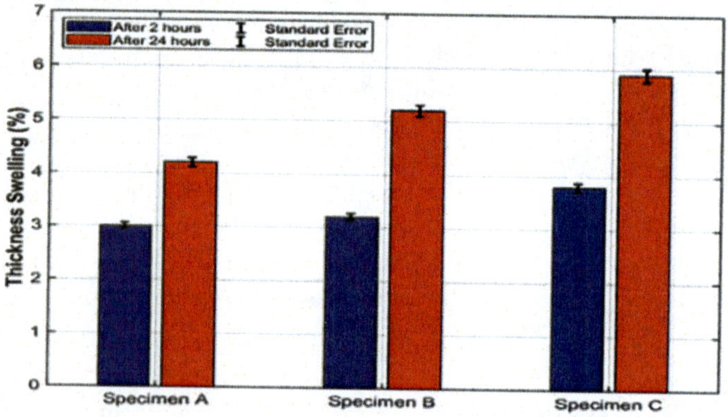

Fig. 3 Thickness swelling of the specimen

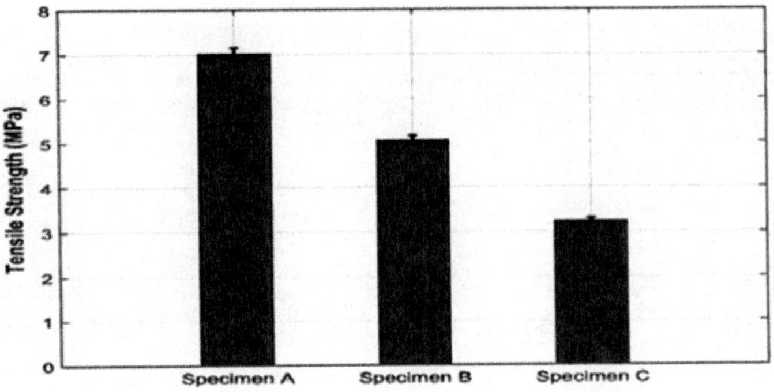

Fig. 4 Tensile strength of WPC

3.5 Tensile Strength

The tensile test results are as shown in Fig. 4. Specimen A had a peak load of 350.52 N and tensile strength of 7.0104 MPa. Specimen B had a peak load of 252.5 N and tensile strength of 5.05 MPa, which was lower than specimen A. The tensile strength of specimen C was 3.233 MPa, with a peak load of 161.68 N, which was lesser than specimen A. The tensile results indicate that the specimen with low-wood husk content had higher tensile strength than the specimen with high-wood husk content.

3.6 Fractography Analysis

The fractograph of the specimen A is shown in Fig. 5. The transition region from matrix to fiber (cusp) and cracks was observed in Fig. 5. Figure 5 shows the pores in matrix. The pore formation could be attributed to chemical interaction of the waste polymer with the external environment in the course of processing. Figure 6 shows the fractography of the specimen B. Fiber (wood husk) pulled out region and deformation zone are extensively observed in specimen B. This was attributed to the poor adherence of incompatible wood husk and the waste polymer matrix. More pores and cracks were observed in the fractograph of the specimen C, as shown in Fig. 7. Also, the fractographs confirmed the presence of individual fibers in the matrix. Some wood husks could be seen partially embedded in the matrix, demonstrating poor adherence. The bonding is not complete since some debonding may be seen on the interface of the wood husk and the waste polymer matrix. The strength decreases because of the inadequate interfacial interaction between waste polymer and wood husk. That in turn hindered the effective stress transfer between the materials, despite the fact that natural fibers have a greater modulus than polymeric materials. Also, the binding capability of the waste polymer used appeared to be decreasing over time. Besides, the homogeneity of lignocellulosic materials disseminated in the polymeric

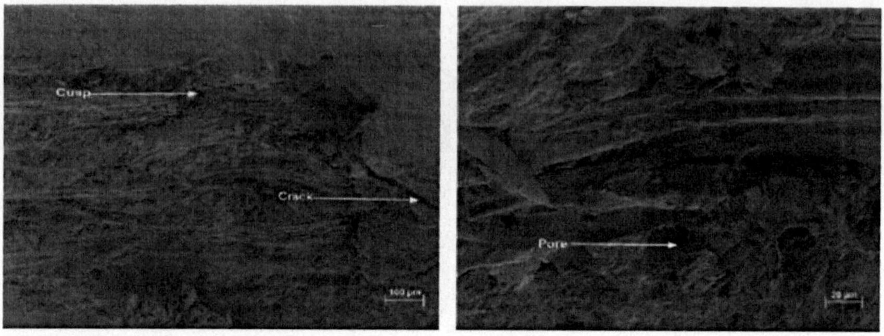

Fig. 5 Fractograph of 40% WF and 60% HDPE (specimen A)

matrix has a significant impact on the mechanical performance of wood polymer composite.

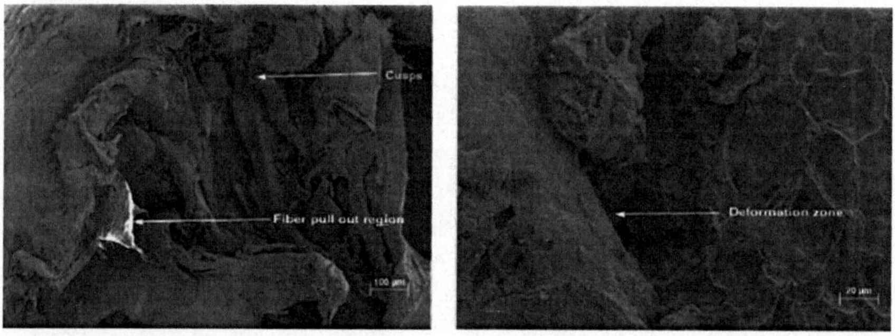

Fig. 6 Fractograph of 45% WF and 55% HDPE (specimen B)

Fig. 7 Fractograph of 50% WF and 50% HDPE (specimen C)

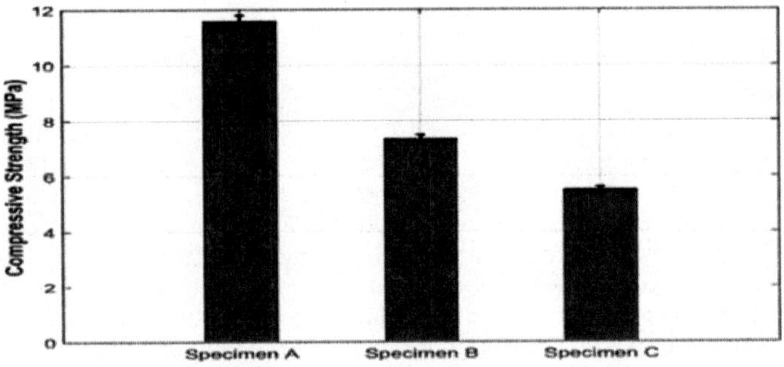

Fig. 8 Compressive strength of WPC

3.7 Compressive Strength

Specimen A had a compressive strength of 11.57 MPa, whereas specimen B with 45 wt.% wood husk exhibited a compressive strength of 7.33 MPa. With a further increase in the wood husk to 50 wt.% in specimen C, compressive strength of 5.50 MPa was achieved. The decreasing trend of the compressive strength was similar to that of the decreasing trend of tensile strength of the composite, as shown in Fig. 8. With an increase in wood husk content, the compressive strength decreased. The decrease in compressive strength could also be attributed to the ineffective interfacial interaction between waste polymer and wood husk. That in turn hindered the efficient stress transfer between the materials. In macroscopic point of view, an increase of wood husk percentage might have increased the void fraction, causing a decrease in compressive strength of the specimen.

4 Conclusion

The technology for the fabrication of a composite material from wood husk and polymer was demonstrated. The prepared composite exhibited good physio-mechanical properties. The results demonstrated the following:

- The wood husk and waste polymer synergistically influence the microstructure and physio-mechanical properties of the wood polymer composite.
- The specimen A with of 50% wood husk and 50% waste polymer exhibited a maximum tensile strength of 7.01 MPa, compressive strength of 11.57 MPa, microhardness of 9.1 HV, and least water absorption.

References

1. Rajesh K, Govindaraju M, Vignesh RV (2020) Development of fly ash-based friction material for wind turbines by liquid phase sintering technology. Proc Inst Mech Eng, Part J: J Eng Tribol. 235. https://doi.org/10.1177/1350650120963998
2. Rafey A, Siddiqui FZ (2021) A review of plastic waste management in India—challenges and opportunities. Int J Environ Anal Chem. https://doi.org/10.1080/03067319.2021.1917560
3. http://www.calrecycle.ca.gov/Organics/
4. Keshav M, Hemchandran C, Dharsan B, Pradhin K, Vignesh RV, Govindaraju M (2020) Manufacturing of continuous fiber reinforced sintered brake pad and friction material. Mater Today: Proc. https://doi.org/10.1016/j.matpr.2020.09.686
5. https://en.m.wikipedia.org/wiki/Sustainable_Development_Goal_12
6. Ghongade G, Kalyan KP, Vignesh RV, Govindaraju M (2021) Design, fabrication, and analysis of cost-effective steel honeycomb structures. Mater Today: Proc (Published Online). https://doi.org/10.1016/j.matpr.2020.09.694. Elsevier
7. Jagadeep R, Vignesh RV, Sumanth P, Sarathi V, Govindaraju M (2021) Fabrication of fly-ash based tiles using liquid phase sintering technology. Mater Today: Proc (Published Online). https://doi.org/10.1016/j.matpr.2020.12.348. Elsevier
8. Pamula V, Kalyan KP, Vignesh RV, Govindaraju M (2021) Development and characterization of aluminum matrix composite reinforced with continuous stainless-steel fibers. Mater Today: Proc (Published Online). https://doi.org/10.1016/j.matpr.2020.12.201. Elsevier
9. Basalp D, Tihminlioglu F, Sofuoglu S, Inal F, Sofuoglu A (2020) Utilization of municipal plastic and wood waste in industrial manufacturing of wood plastic composites. Waste Biomass Valorization 11. https://doi.org/10.1007/s12649-020-00986-7
10. Flores-Hernández M, González I, Lomeli Ramirez MG, Fuentes-Talavera F, Silva-Guzmán J, Cerpa-Gallegos M, Garcia S (2014) Physical and mechanical properties of wood plastic composites polystyrene-white oak wood flour. J Compos Mater 48:209
11. Kartal S, Aysal S, Terzi E, Yilgor N, Yoshimura T, Tsunoda K (2013) Wood and bamboo-PP composites: fungal and termite resistance, water absorption, and FT-IR analyses. Bioresources. 8. https://doi.org/10.15376/biores.8.1.1222-1244
12. Batista C, Martins G, Santos C, Mateus A, Antunes F (2019) Microstructure of thermoplastic composites reinforced with wool and wood. Appl Mech Mater 890:98–112. https://doi.org/10.4028/www.scientific.net/AMM.890.98
13. SAKURA symposium on mechanical science and engineering (Nov 2017) © Centre for advanced research on energy, the study on microhardness of natural fibre composite made by recycled material, Muhammad Taufiq Jumadi1, Muhd Ridzuan Mansor, Zaleha Mustafa, Takayuki Tokoroyama, Noritsugu Umehara, Faculty of Mechanical Engineering, Universiti Teknikal Malaysia Melaka, Hang Tuah Jaya, pp 17–19
14. (2011) Key engineering materials, vols 471–472. Trans Tech Publications, Switzerland. https://doi.org/10.4028/www.scientific.net/KEM.471-472.62
15. Youssef P, Zahran K, Nassar K, Darwish M, El-Haggar S (2019) Manufacturing of wood-plastic composite boards and their mechanical and structural characteristics. J Mater Civ Eng 31. https://doi.org/10.1061/(ASCE)MT.1943-5533.0002881
16. Ntimugura F, Vinai R, Harper A, Walker P (2020) Mechanical, thermal, hygroscopic and acoustic properties of bio-aggregates—lime and alkali—activated insulating composite materials: a review of current status and prospects for miscanthus as an innovative resource in the South West of England. Sustain Mater Technol 26:e00211. https://doi.org/10.1016/j.susmat.2020.e00211
17. Marques B, Antonio J, Almeida J, Tadeu A, Brito J, Dias S, Pedro F, Sena J (2020) Vibro-acoustic behaviour of polymer-based composite materials produced with rice husk and recycled rubber granules. Constr Build Mater 264:120221. https://doi.org/10.1016/j.conbuildmat.2020.120221

18. Kord B, Ghalehno M, Movahedi F (2020) Effect of immidazolium-based green solvents on the moisture absorption and thickness swelling behavior of wood flour/polyethylene composites. J Thermoplast Compos Mater 089270572096217. https://doi.org/10.1177/0892705720962170
19. Nourbakhsh A, Ashori A (2009) Preparation and properties of wood plastic composites made of recycled high-density polyethylene. J Compos Mater 43:877–883. https://doi.org/10.1177/0021998309103089
20. Omoniyi T (2018) Development of processing equipment for Souvenir production from wood and plastics wastes
21. Mengeloglu F, Kurt R, Gardner D, O'Neill S (2007) Mechanical properties of extruded high density polyethylene and polypropylene wood flour decking boards. Iran Polym J 16:477–487
22. Harish M, Jayanarayanan K, Mini KM (2021) Recent trends in utilization of plastics waste composites as construction materials. Constr Build Mater 271:121520
23. Yuan Q, Wu D, Gotama J, Bateman S (2008) Wood fiber reinforced polyethylene and polypropylene composites with high modulus and impact strength. J Thermoplast Compos Mater 21:195–208. https://doi.org/10.1177/0892705708089472
24. Meyyappan PL, Chandrasekaran S, Sekar T, Marimuthu P (2015) Suitability of composite material in tsunami resistant design. Aquatic Procedia 4:397–403. https://doi.org/10.1016/j.aqpro.2015.02.053
25. ASTM E384-17 (2017) Standard test method for microindentation hardness of materials, ASTM international, West Conshohocken, PA. www.astm.org
26. ASTM D5790-18 (2018) Standard test method for measurement of purgeable organic compounds in water by capillary column gas chromatography/mass spectrometry, ASTM international, West Conshohocken, PA. www.astm.org
27. ASTM D638-14 (2014) Standard test method for tensile properties of plastics, ASTM international, West Conshohocken, PA. www.astm.org
28. ASTM D3410/D3410M-16e1 (2016) Standard test method for compressive properties of polymer matrix composite materials with unsupported gage section by shear loading, ASTM international, West Conshohocken, PA. www.astm.org

Spinelization from Nitrate Precursors of Stoichiometric Ratio by Sol–gel Process Using Urea as Fuel

Soumya Mukherjee

1 Introduction

Stoichometric spinel magnesium aluminate possesses high melting point (2135 °C), high chemical inertness for both acidic and basic conditions, low thermal expansion behavior, resistance toward chemical attack of molten metal and slag, thus making it fit for high temperature applications. In addition absence of low temperature eutectics in $MgO–Al_2O_3$ along with high hardness value of spinel (Mohs Scale 7.5–8) make it prominent material for structural application at elevated temperatures. Spinel is noted to have optical properties and found to be fit for transparent ceramics [1, 2]. Spinel magnesium aluminate is represented by general formula AB_2O_4 where 'A' represents divalent metal ions, 'B' represent trivalent metal ions. Stoichometrically spinel contains 28.2 wt% MgO and 71.8 wt% Al_2O_3 as indicated from $MgO-Al_2O_3$ phase diagram [3]. Spinel is found to have versatile application as refractory materials in steel ladle, cement rotary kilns, vacuum induction furnaces, continuous casting, degasser snorkels and lances, glass industries, furnace linings for cement, aluminum and other metallurgical industries along with application as humidity sensor, photo-catalyst and electronic devices [4]. Spinel formation is accompanied by 7% volume expansion when oxide precursors are used for spinel formation making sintering technology for spinel a challenge [5]. Magnesium aluminate is an important refractory material due to its high corrosion resistance and thermal shock against basic slag [6]. For application as a refractory, corrosion problems by slag, liquid metal abrasion, thermal spalling, oxidation of carbon layer, strength deterioration at elevated temperature need to be checked along with purity and grade of raw material used for preparation. Special focus is needed on thermal shock resistance, creep reduction

S. Mukherjee (✉)
Kazi Nazrul University, Asansol 713340, India
e-mail: smmukherjee4a@gmail.com

© The Author(s), under exclusive license to Springer Nature Singapore Pte Ltd. 2023

S. Revankar et al. (eds.), *Recent Advances in Thermofluids and Manufacturing Engineering*, Lecture Notes in Mechanical Engineering,
https://doi.org/10.1007/978-981-19-4388-1_41

at higher temperature and better sintering process for application in high temperature ceramics [7]. Spinel formation is found to be dependent on process parameters, chemical precursors, particle size, purity of precursors utilized for synthesis [8]. Some popular techniques for spinel processing are sol–gel [9, 10], precipitation [11], co-precipitation [12], hydrothermal, combustion synthesis, freeze drying, decomposition of an organometallic complex in super critical, plasma spray decomposition of powders [13–18], microwave assisted sintering [19], mechanochemical synthesis [20] and others [21].

In the present article synthesis of spinel magnesium aluminate is carried by precursors of magnesium aluminum nitrates in 1:2 molar ratio with urea as both fuel and reducing agent at temperature far lower than solid state sintering process. The process is simple and does not involve much complex systems, chemical precursors with formation at lower temperature and optimum soaking period.

2 Experimental

The present spinel compound was synthesized using stoichiometric ratio of Magnesium nitrate: Al-Nitrate (AR Grade) in 1:2 ratio along with Urea (AR grade) as fuel and reducing agent. Initially nitrate mixtures were prepared by adding DI water. After obtaining the resultant solution urea was added in 1.75 molar ratios with respect to nitrate precursors followed by magnetic stirring of media. Final solution obtained was then undergone drying in an oven at 80 °C for 3–4 h to obtain the gel like mass. Dried gel mass was put into DSC-TGA analyzer (Model Pyris Diamond, Make Perkin Elmer, Japan) with purging of inert gas Nitrogen at the flow rate of 15 ml/min to study crystallization for proper phase development. After analyzing the thermal range for required phase formation annealing was carried at about 700 °C for 5 h soaking duration. Phase analyses was carried by XRD (Rigaku, ultima III) having wavelength of Cu Kα, 1.54Å 40 kV in the scan range of 10–80° with scan rate of 5°/min. FTIR analysis (IR, Prestige-21 Shimadzu) was carried to determine the bonds and M–O coordination present in the sample after pressing to form pellet with KBr. SEM analysis (Jeol, JAX 840A) was carried to observe the morphological features of synthesized samples.

3 Results and Discussion

The thermal analysis curve exhibits two prominent exothermic peaks, one minor exothermic peak along with one endothermic minor trough for the entire range of thermal analyses. Prominent exothermic peak is noted at about 100 °C while 2nd exothermic peak is observed at about 250 °C. Both exothermic peaks may be attributed to urea decomposition as noted by Schaber et al. [22] along with nitrate salt decomposition. Weight losses are also noted corresponding to urea decomposition and volatilization along with complex intermediate reaction with nitrate salts used as precursors. Within this range upto 250 °C, urea decomposition, weight loss is quite prominent while biuret mass production is also noted by Schaber et al.[22]. Endothermic peak at 200 °C suggests removal of physically absorbed and loosely bonded water and also suggests removal of structural water from nitrates (comparatively more strong). Then around 380 °C there is a small endothermic peak due to formation of γ-alumina. The above analyses are in correspondence with experimental findings of Fernandes et al. [23], Mukherjee et al. [24], Ghosh et al. [21]. The researcher observed minor endothermic peak at around 350 °C due to γ-alumina formation. After that from 500 °C to 700 °C there is a weak broader exothermic peak which refers to start of crystallization process. Transformation of gel to crystal involves weight loss and exothermic reactions. Presence of multicomponent and their interaction will yield such weight loss followed by stability after initiation of crystallization (Figs. 1 and 2).

XRD study as confirmed from JCPDS analyses indicate only spinel phase as card no #01-077-1193, #01-077-0437 and #01-075-1800. The XRD spectra is noted from 10 to 80° having scan rate of 5°/minute with corresponding planes of (111), (220), (311), (400), (511), (440) and (533) which is possible due to thermodynamic stability of phase growth orientation along indicated planes with minimum free energy change. No peaks correspond to presence of precursors or intermediate oxides as noted from analysis with respect to JCPDS of possible intermediates. Further deconvolutions of main peaks are not observed hence indicating thermodynamic feasible phase

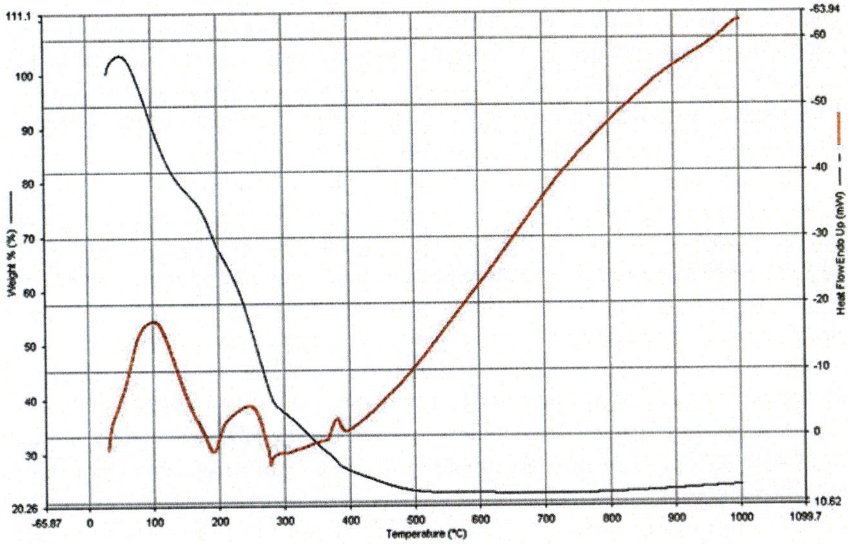

Fig. 1 DSC-TGA graph of stoichiometric ratio of magnesium nitrate: Al-nitrate precursors by Sol–gel route using urea 1.75 molar ratio for spinelization

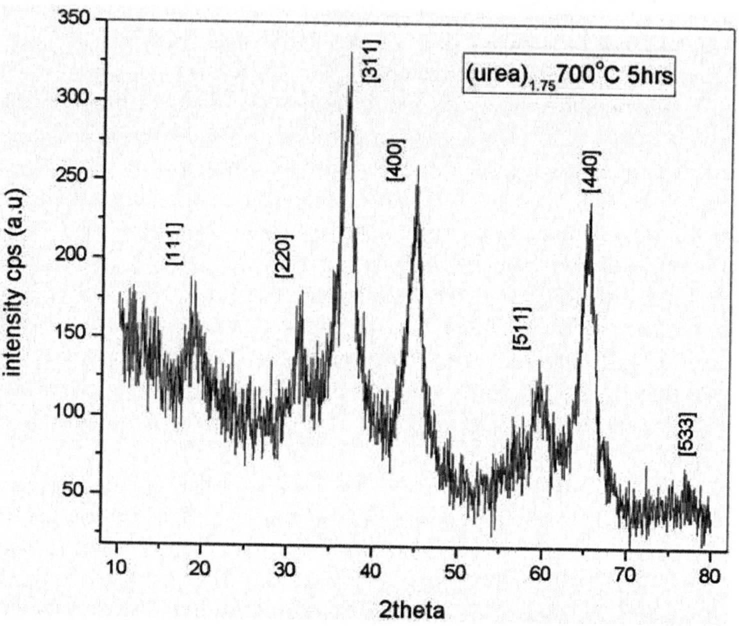

Fig. 2 XRD of stoichometric ratio of spinel magnesium aluminate from magnesium nitrate: Al-nitrate precursors by Sol–gel route using urea 1.75 molar ratios as fuel after annealing at 700 °C for 5 h

formation at required annealing conditions. The present experimental findings are noted with in correspondence with findings by the author in previous work [2]. Crystallite size determination using Scherrers formula $t = 0.9\ \lambda/\beta\cos\theta$ where 't' is crystallite size, $K =$ shape factor, 0.9 and $\beta =$ Full width Half Mean. Using Scherrer's formula crystallite size is estimated to be around 70 nm (Fig. 3).

FTIR analysis exhibits presence of inorganic bonds required for spinelization within 1000 cm^{-1}. Most important bonds present in the synthesized samples are Al–O stretching at 539 cm^{-1}, Mg–O–Al vibrations at 677 cm^{-1}, Vibration of NO_3^- group at about 1403 cm^{-1} while that of O–H bonding vibration at about 3435 cm^{-1}. M–O stretching and vibration corresponding to Al–O and Mg–O–Al gives a strong indication of successful bonding information on spinelization of compound. The above finding tallies with experimental results of authors previous work on synthesis and characterization of spinel prepared by using different fuel and non-stoichiometric precursors [2, 24]. Experimental findings from FTIR and XRD confirmed the formation of required spinel compound without any complex process, precursors in the above mentioned experimental condition (Fig. 4).

SEM morphology indicates dense agglomerate with negligible porosity. Morphology reveals Agglomerates are polygonal shaped having step like serrations in some portion. Individual particulates are spherical and polygonal shape with size approximately close to 0.2 μm. Agglomerates are devoid of sharp corners and have

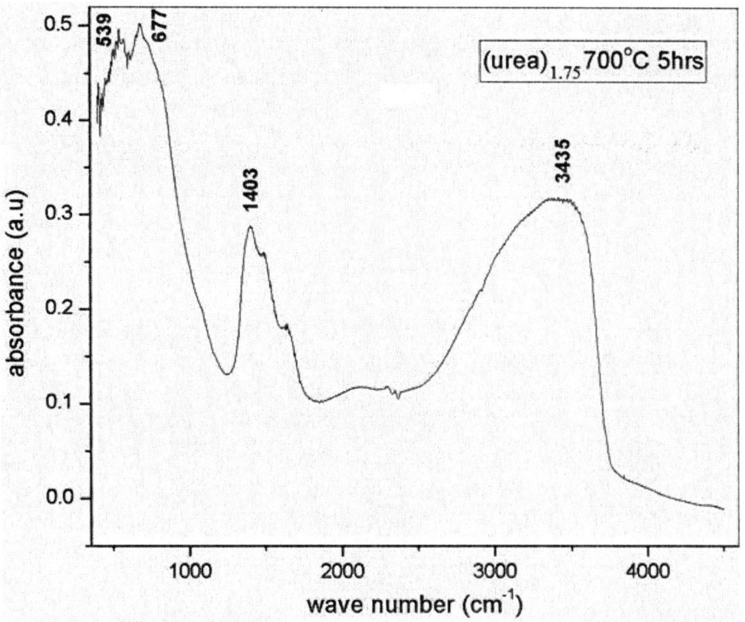

Fig. 3 FTIR of stoichiometric ratio of magnesium nitrate: Al-nitrate precursors by Sol–gel route using urea 1.75 molar ratio after annealing at 700 °C for 5 h

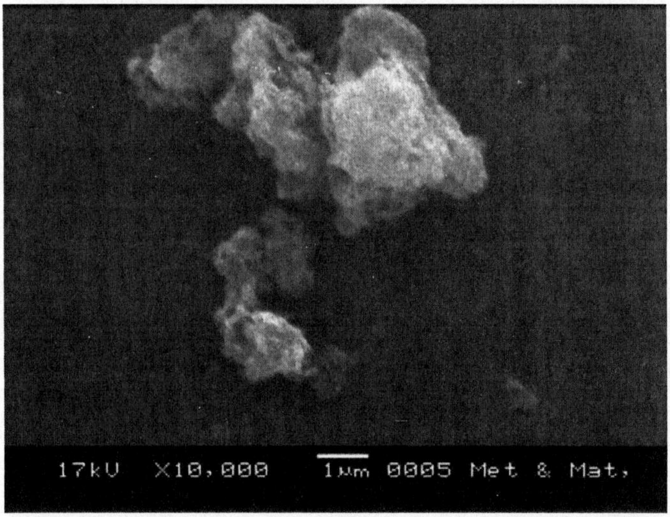

Fig. 4 SEM morphology of stoichiometric ratio of magnesium nitrate: Al-nitrate precursors by Sol–gel route using urea 1.75 molar ratio as fuel after annealing at 700 °C for 5 h

blunt edges. Top view of agglomerate exhibits conchoidal fracture like impression along the z axis or height of agglomerate. Dense micro structural feature indicates proper sintering of synthesized spinel magnesium aluminate at lower temperature by the proposed chemical route.

4 Conclusions

Thermal analyses results in phase transition to spinelization at about 500 °C while annealing carried at 700 °C for 5 h soaking period yielded spinel phase as per XRD study. XRD studies estimate crystallite size to be about 70 nm and phase study as per JCPDS card no #01-077-1193, #01-077-0437 and #01-075-1800. No findings of intermediate phase was noted while FTIR yields required M–O coordination within 1000 cm^{-1} with important bonding information obtained were Al–O stretching at 539 cm^{-1}, Mg–O–Al vibrations at 677 cm^{-1} while presence of nitrate and O–H stretching was also noted at about 1403 cm^{-1} and at about 3435 cm^{-1}. Morphology study was carried by SEM which results in dense agglomerates with negligible porosity. Individual particulates were noted to be about 0.2 μ having spherical and polygonal shape.

References

1. Radishevskaya NI, Nazarova AY, Lvov OV, Kasatsky NG, Kitler VD (2019) Synthesis of magnesium aluminate spinel in the MgO–Al2O3–Al system using the SHS method. In: 7th international conference on current issues of continuum mechanics and celestial mechanics 26–28 November 2018. Tomsk State University, Tomsk, Russia IOP Publisher Tomsk, Russia, pp 1–7
2. Mukherjee S (2020) Evolution of spinel magnesium aluminate by combustion route using glycine as fuel and its characterization. Int J Emerg Trends Sci Technol 6(1):24–27
3. Ṕacurariu C, Lažau I, Ecsedi Z, Lažau R, Barvinschi P, Ḿarginean G (2013) New synthesis methods of $MgAl_2O_4$ spinel. J Eur Ceram Soc 27:707–710
4. Khalil NM, Hassan MB, Ali FS, Shakdofa MME (2013) Preparation and characterization of Nano spinel powder via co-precipitation and sol gel techniques. Main Group Chem 12:331–347
5. De Aza AH, Pena P, De Aza S (1999) Ternary system Al_2O_3–CaO–MgO: I, primary phase field of crystallization of spinel in the subsystem $MgAl_2O_4$–$CaAl_4O_7$–CaO–MgO. J Am Ceram Soc 82:2193
6. Moritz K, Aneziris GC (2016) Enhancing the thermal shock resistance of alumina-rich magnesium aluminate spinel refractories by aluminium titanate phase. Ceram Int 42(12):14155–14160
7. Ghosh SR, Mukherjee S, Banerjee S (2018) Solution combustion synthesis of alumina spinel and its characterization. Interceram 67(1–2):34–41
8. Saberi A, Fard-Golestani F, Sarpoolaky H, Willert-Porada M, Gerdes T, Simon R (2008) Chemical synthesis of nanocrystalline magnesium aluminate spinel via nitrate-citrate combustion route. J Alloys Compd 462:142–146
9. Debsikdar JC (1985) Preparation of transparent non crystalline stoichometric magnesium aluminate gel-monolith by the sol–gel process. J Mater Sci 20:4454

10. Naskar MK, Chatterjee M (2005) Magnesium aluminate ($MgAl_2O_4$) spinel powders from water based sols. J Am Ceram Soc 88(1):38–44
11. Li JG, Ikegami T, Lee JH, Mori T (2000) Fabrication of translucent magnesium aluminium spinel ceramics. J Am Ceram Soc 83(11):2866–2868
12. Guo J, Lou H, Zhao H, Wang X, Zheng X (2004) Novel Synthesis of high surface area $MgAl_2O_4$ spinel as catalyst support. Mater Lett 58(12–13):1920–1923
13. Barj M, Bocquet JF, Chhor K, Pommier C (1992) Submicronic $MgAl_2O_4$ powder synthesis in supercritical ethanol. J Mater Sci 27:2187–2192
14. Bickmore CR, Waldner KF, Treadwell DR, Laine RM (1996) Ultrafine spinel powders by flame pyrolysis of a magnesium aluminium double hydroxide. J Am Ceram Soc 79(5):1419–1423
15. Bratton RJ (1969) Coprecipitates yielding $MgAl_2O_4$ powders. Am Ceram Soc Bull 83:759–762
16. Pommier C, Chhor K, Bocquet JF, Barj M (1990) Reactions in supercritical fluids, a new route for oxide ceramic powder elaboration, synthesis of the spinel $MgAl_2O_4$. Mater Res Bull 25(2):213–221
17. Ning Y, Li C (1992) Structural inhomogenity and crystallization behavior of aerosol reacted MgAl2O4 powders. Mater Lett 15(1–2):84–88
18. Varnier O, Hovnanian N, Larbot A, Bergez P, Cot L, Charpin J (1994) Sol-Gel synthesis of magnesium aluminium spinel from a heterometallic alkoxide. Mater Res Bull 29(5):479–488
19. Torkian L, Amini MM, Bahrami Z (2011) Synthesis of nanocrystalline $MgAl_2O_4$ powder by microwave assisted combustion. J Inorg Mater 26(5):550–554
20. Domanski D, Urretavizcaya G, Castro JF, Gennari CF (2004) Mechanochemical synthesis of magnesium aluminate spinel powder at room temperature. J Am Ceram Soc 87(11):2020–2024
21. Ghosh SR, Mukherjee S, Banerjee S (2020) Development of spinel magnesium aluminate by solution combustion route using thiourea and urea as fuel. J Eng Res 17(2):135–141
22. Schaber PM, Colson J, Higgins S, Thielen D, Anspach B, Brauer J (2004) Thermal decomposition (pyrolysis) of urea in an open reaction vessel. Thermochim Acta 424(1–2):131–142
23. Macêdo MLF, Bertran CA, Osawa CC (2007) Kinetics of the γ-α alumina phase transformation by quantitative X-ray diffraction. J Mater Sci 42:2830–2836
24. Mukherjee S, Ghosh SR (2021) Synthesis and characterization of magnesium aluminate spinel by combustion route. Indian Sci Cruiser 35(1):12–15

Effect of Three-Body Abrasion Wear Behavior on B_4C/BN-reinforced Al6061 Alloy Fabricated Through Powder Metallurgy Method

Sudipta Chand, P. Chandrasekhar, Joyjeet Ghose, R. K. Sarangi, S. P. Kar, and A. Swain

1 Introduction

Boron carbide (B_4C) is a promising ceramic material in the nuclear industries due to its superior properties like high neutron absorption potential (3800 bar) of the ^{10}B isotope, and high chemical stability with wide applications in nuclear shielding materials [1–7]. It also exhibits superior mechanical properties such as high toughness (greater than 30 GPa), high melting point (2450 °C), high elastic modulus (445 GPa), strong thermal stability, and (2.51 g cm^{-3}) low density. Such characteristics make it ideal for bulletproof vests, refractory applications, and ballistic protection [8, 9]. B_4C ceramic particles have poor toughness and plasticity and are usually added to a non-metal/metal matrix such as Al, stainless steel, polyethylene, and glass fiber to fabricate composites. Aluminum matrix emerged as a suitable material for B_4C particles in neutron shielding application due to its low density (2.7 g cm^{-3}), lightweight, good mechanical properties, and excellent corrosion resistance. The interfacial bonding between B_4C and the aluminum matrix is stronger than the bonding of other reinforcing materials such as Al_2O_3, SiC, and Si_3N_4. However, the presence of B_4C in the aluminum matrix results in poor forge ability and machinability. Thus, it was a

S. Chand
Department of Mechanical Engineering, Chaitanya Deemed to Be University, Warangal, Telangana 506001, India

P. Chandrasekhar (✉) · R. K. Sarangi · S. P. Kar · A. Swain
School of Mechanical Engineering, KIIT Deemed to Be University, Bhubaneswar, Odisha 751024, India
e-mail: csekharfme@kiit.ac.in

J. Ghose
Department of Production Engineering, Birla Institute of Technology, Mesra, Jharkhand 835215, India

© The Author(s), under exclusive license to Springer Nature Singapore Pte Ltd. 2023

S. Revankar et al. (eds.), *Recent Advances in Thermofluids and Manufacturing Engineering*, Lecture Notes in Mechanical Engineering,
https://doi.org/10.1007/978-981-19-4388-1_42

challenge to find ways to enhance the durability, ductility, and machinability of the composites.

The novelty of this work includes improving the ductility and wear-resistant properties of the B_4C ceramic-reinforced metal matrix composites with the addition of soft solid lubricating reinforcements like boron nitride (BN) to enhance the secondary fabrication techniques like rolling and extrusion in nuclear shielding applications. In the present paper, the aluminum matrix is reinforced by x wt.% B_4C ($x = 2$, 4, 6, 8), BN (2 wt.%) and was fabricated using powder metallurgy. A high amount of B_4C (8 wt.%) decreases the sinterability and compressibility of the composites, decreasing the mechanical strength [10]. A low amount of boron nitride solid lubricant was added to maintain the ductility to retain its mechanical and wear-resistant properties of the fabricated hybrid composites. The influence of the control parameters such as B_4C wt.%, sliding velocity, sliding distance, and the applied load on the specific wear rate of Al6061/B_4C/BN hybrid composites was investigated using Taguchi L_{16} DOE.

2 Experimental Methodology

Various elemental powders were used to develop aluminum 6061 alloy. BN and B_4C powder with a mesh size of 325 and 140 mesh, respectively, was reinforced in the aluminum matrix. Loba Chemie Pvt. Ltd (Mumbai) has supplied all the elemental powders and reinforcements. The various volume fractions of B_4C/BN-reinforced powders were processed in a high-energy ball mill (Retsch PM 400) in which the Al6061 alloy matrix uniformly mixed with the B_4C and BN reinforcements. The BPR (ball-to-powder ratio) was maintained at 10:1; milling speeds were set to 300 rpm, and the 250-ml WC jars were loaded with 35.52 g Al6061/B_4C/BN powders with WC balls of diameter 10 mm with a total mass of 355.26 g. Toluene ($C_6H_5CH_3$) was added to prevent any chemical reaction. Mechanical alloying was performed for 10 h, and after each hour of milling, 30 min of cooling was done to prevent overheating. The powders obtained were dried after the milling process and stored in the containers to prevent further oxidation. Closed cavity powder compaction die-sets of hardened alloy steel, with a bore diameter of 40 mm, were used to compact the powders. The die-sets fabricated includes three separate parts, the lower counter and the flat upper punch, with a central container. The powders were compacted into 40 mm diameter preforms with zinc stearate as a lubricant. For better compressibility, compaction was done at 800 MPa [10] on a 100-ton capacity universal testing machine (UTM). Preforms were further sintered at 600 °C for 3h [10] to obtain better sinterability at a heating rate of 10 °C per min after compaction.

2.1 Abrasive Wear Testing

The present paper elucidates the three-body abrasive wear analysis which is based on ASTM-G65 technical standard. The abrasion tester (TR-50, DUCOM Instrument, Bangalore), as shown in Fig. 1, was used to conduct the three-body abrasion using dry sand as an erodent, chlorobutyl rubber wheel as the abrading material, and the test samples as abrading surface creating a three-body abrasive wear environment. The constituents of the abrasion test include a nozzle for sand flow, sample holder, waste particle collection bag, and arrangement for load rings. The sample dimension was taken as 40 × 25 × 3.5 mm and was weighed accurately using an electronic weighing balance. The abrasive sand particle with a size ranging from 215 to 300 μm (AFS 50/70) silica was loaded to the abrasive hopper. The sand flow rate was regulated at 363 g/min between the specimen and the abrasive wheel of 228 mm dia. The RPM of the wheel can be varied between 0 and 200 with a step of 1 rpm. The sample holder is connected to a lever arm which applies a normal load ranging from 10 to 130 N on the test sample against the abrading wheel. The specimen was abraded with the required control parameters and was cleaned and weighed using an electronic balance. The specific wear rate was calculated using Eqs. (1) and (2). The conversion parameters are presented in Eqs. (3), (4), and (5). Table 1 presents the machine setting parameters. The specific wear rate was studied as a function of control parameters and was calculated using the formula:

Volume Loss:

$$\Delta V = \Delta m/\rho \tag{1}$$

Specific Wear Rate:

$$K_s = \frac{\Delta V}{D \times L} \tag{2}$$

where Δm is the mass loss, ρ is the density, K_s is the specific wear rate in mm^3/Nm.

Conversion of machine parameters to test parameters:

Sliding Velocity:

$$V = \frac{2\pi}{60} \times r \times N \tag{3}$$

Sliding Distance:

$$D = 2\pi r N \tag{4}$$

Load:

$$L = W \times 9.806 \tag{5}$$

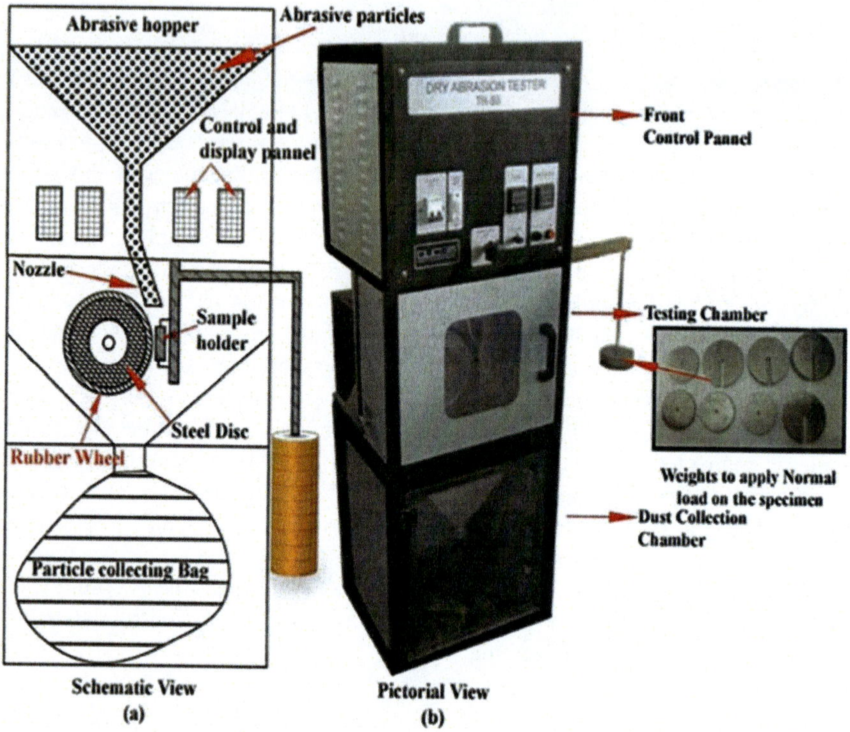

Fig. 1 **a** Schematic and **b** Pictorial view of three-body abrasion tester

Table 1 Parameters for three-body abrasive wear test rig

Fixed parameters	
Sample size	40 × 25 × 3.5 mm
Measurement scale	Least count: 0.01 mg
Abrasive wheel	Material—chlorobutyl, (with A-60 shore hardness) width of 12.7 mm, and diameter of 228 mm
Test condition	Ambient
Erodent flow rate	363/min

where 'V' is the sliding velocity (m/s), 'N' is the RPM of the rubber wheel, 'r' is the radius of the rubber wheel, 'D' is the sliding distance in meters, 'L' is the load in newton, and 'W' is the load weights in kg.

Table 2 Levels and factors for Taguchi L_{16} DOE

Control variables	Symbols	Units	Values			
B_4C wt.%	A	%	2	4	6	8
Load	B	N	10	20	30	40
Velocity	C	$m\ s^{-1}$	1	1.5	2	2.5
Sliding distance	D	m	500	1000	1500	2000

2.2 Design of Experiment

Various statistical methods are used in engineering and related research to increase the efficiency of the product and process. The three-body abrasive wear test evaluates the performance of several control variables collectively. In this context, the Taguchi L_{16} design of experiments analyzed the impact of more than one variable and its interactions on the performance of the method. B_4C wt.%, load, sliding velocity, and sliding distance were selected among several significant parameters after an exhaustive literature review as the most influential parameters on dry sliding abrasive wear [11]. Design factors and levels used for dry sliding abrasive wear based on the orthogonal array of L_{16} are represented in Table 2.

Signal-to-noise ratio (*S*/*N*) has been calculated using the equation below:

$$\frac{S}{N} = 10log\left[\frac{1}{n}\sum_{i=1}^{n} y_i^2\right] \tag{6}$$

The response table and output of the Taguchi L_{16} analysis were calculated using MINITAB software and are given in Tables 3 and 4, respectively.

3 Results and Discussion

3.1 Effect of Applied Load

From Fig. 2, it is inferred that increase in the applied load (coded as B) leads to decrease in the specific wear rate. This is mainly due to the presence of uniformly distributed boron carbide particles within the aluminum matrix, which enables the top layer of the composites to withstand the applied loads reducing the contact area between the abrading wheel and the composite, improving the wear resistance. Increase in the contact pressure and the applied load, the specific wear rate due to adhesive wear also decreases, where the worn-out particles are welded back to the composite surface rather than being completely removed. The formation of tribolayer controls the specific wear rate as the applied load increases. Increased load on the hybrid composites leads to an increase in temperature even at low sliding velocities

Table 3 Specific wear rate and S/N ratio with control parameters for Taguchi L_{16} DOE

Sl. No.	B_4C wt.%	Load (N)	Velocity (m/s)	Sliding distance (m)	Specific wear rate (mm^3/Nm)	S/N ratio
1	2	10	1.0	500	0.0070	43.06
2	2	20	1.5	1000	0.0015	56.00
3	2	30	2.0	1500	0.0005	65.58
4	2	40	2.5	2000	0.0003	68.47
5	4	10	1.5	1500	0.0008	61.00
6	4	20	1.0	2000	0.0004	67.15
7	4	30	2.5	500	0.0004	66.31
8	4	40	2.0	1000	0.0010	59.42
9	6	10	2.0	2000	0.0034	49.23
10	6	20	2.5	1500	0.0040	47.80
11	6	30	1.0	1000	0.0015	56.11
12	6	40	1.5	500	0.0006	63.50
13	8	10	2.5	1000	0.0053	45.50
14	8	20	2.0	500	0.0020	53.71
15	8	30	1.5	2000	0.0004	66.08
16	8	40	1.0	1500	0.0002	71.06

Table 4 Response table: signal-to-noise ratios 'smaller is better'

Level	A	B	C	D
1	58.28	49.70	59.35	56.65
2	63.47	56.17	61.65	54.26
3	54.17	63.52	56.99	61.36
4	59.09	65.62	57.02	62.74
Delta	9.31	15.91	4.66	8.47
Rank	2	1	4	3

which helps in welding of debris material to the composites decreasing the wear rate [12].

3.2 *Effect of Sliding Distance*

With increase in the sliding distance (coded as D), the specific wear rate decreases. It decreases to a greater extent from 1000 to 1500 m due to the presence of the hard boron carbide reinforcement which acts as sharp asperities on the composite surface. At a lower sliding distance, the boron carbide particles protruded from the composite

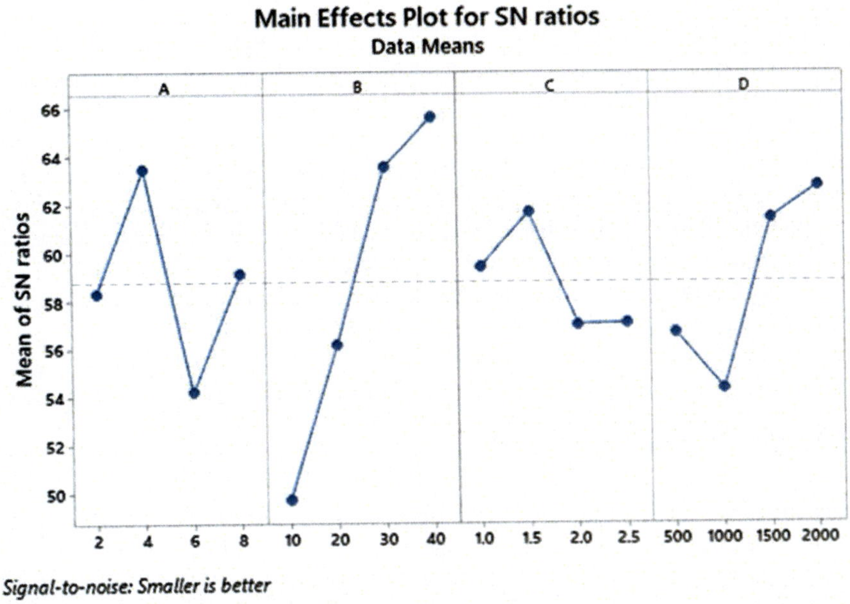

Fig. 2 Main effect plot for signal-to-noise ratio

surface. It decreases the interaction zone among the abrading wheel and the composites, resulting in increase of the wear rate. With an increment in the sliding distance, the sharp asperities become blunt due to the compaction between the abrading surfaces resulting in increasing the contact area. This leads to an improvement of the wear performance with the greater sliding distance.

3.3 *Effect of Sliding Velocity*

The specific wear rate was observed to be decreasing with the increase of sliding speed (coded as C). When sliding at higher velocity, the temperature increases over the contact surface, making the material to oxidize. This phenomenon leads to the transfer of materials, forming a mechanically mixed layer (MML), also called tribolayer. As the velocity increases, this tribolayer will act as a barrier or lubricant between the two surfaces decreasing the wear rate.

3.4 *B_4C and BN Reinforcements*

The enhancement of B_4C wt.% (coded as A) in the composite leads to decrease in the wear rate of the hybrid composites gradually. The enhancements in the strength and stiffness of the composites obtained were because of the grain refinement and the particle dispersion strengthening mechanism. Moreover, the B_4C and BN particles in the aluminum matrix are more resistant to wear due to the formation of an oxide layer between the sample and the rotating wheel. The lamellar crystalline structure of BN provides the lower shear stress between the material and the wheel, which is responsible for this increase in wear resistance. Moreover, due to the strong bond between the reinforced B_4C particles and the aluminum matrix, the deformation of the aluminum matrix at a higher load is not permissible. Also, the fractured oxide particles protruding from the composite surface limit the plastic flow by providing more wear resistance.

3.5 Microstructural Analysis

The microstructural analysis of the worn-out specimen was done by SEM and EDS. Figure 3a, b represents the worn-out surface of 2 wt.% and 8 wt.% of the boron carbide particles at a load of 10 N. It can be observed from the SEM images that the wear track has a layer of oxide particles. This oxide particle layer plays an important role in decreasing the wear rate of the hybrid composites by reducing the direct contact between the specimen and the abrading wheel [13]. No heavy flow of metal or delamination was observed on the worn-out surface of the hybrid composites [14]. As the load increases to 40 N, micro-cracks occur around the boron carbide particles due to high-stress concentration as shown in Fig. 3c, d. The oxidation layer and BN lubricant film reduce the wear rate with wear mechanisms such as abrasion and oxidation wear. The metal matrix hybrid composites show shallow grooves on the surface indicating it has better wear resistance. These findings suggest that between the two surfaces, a mechanically mixed layer (MML) is made on the worn surface, which works between the abrading wheel and the sample surface as a thin solid lubricant film. This MML reduces the wear rate by avoiding direct contact.

3.6 Confirmation Test

In this current study, the samples were subjected to three-body abrasion analysis. From the Taguchi results, A_2, B_4, C_2, D_4 parameters predicted the lowest specific wear rate of 0.00018 mm^3/Nm. The results obtained are presented in Table 5. To further confirm the selection of optimal parameters, experimental verification was done, and the errors were under 10% which confirms the accuracy of the models. However, by increasing the experimental runs, the error percent may be lowered further.

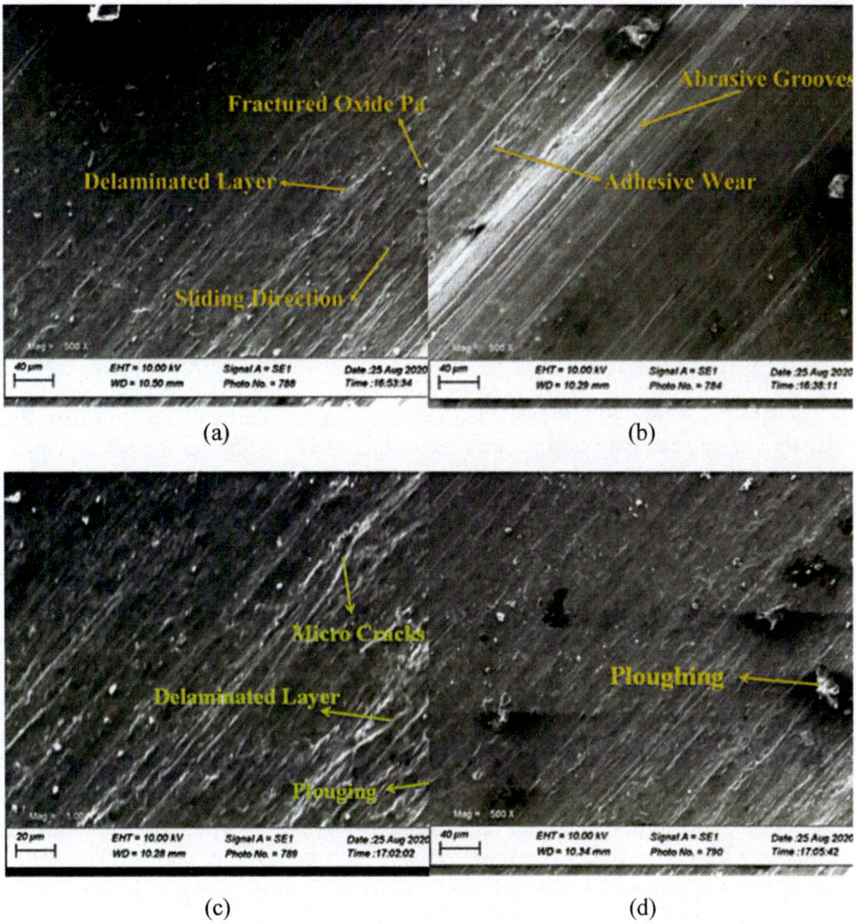

Fig. 3 SEM images of **a** 2 wt. % B_4C at 10 N load **b** 8 wt% B_4C at 10 N load **c** 2 wt.% B_4C at 40 N load **d** 8 wt.% B_4C at 40 N load

Table 5 Experimental and predicted response of the best sample

Optimal parameters						
Method used	A	B	C	D	Specific wear rate (mm^3/Nm)	Error (%)
Taguchi DOE	2	4	2	4	0.00018	–
Experimental	2	4	2	4	0.00017	5.5

4 Conclusion

In the present work, aluminum hybrid composites were manufactured using the mechanical alloying and powder metallurgy method with B_4C and BN as reinforcements. The following conclusions can be drawn from the test results.

- Among various process parameters used for the three-body abrasive wear test, the applied load was the utmost significant factor in reducing the specific wear rate.
- It was observed as the applied load was increased; there was a reduction in the specific wear rate. This is mainly due to uniformly distributed B_4C particles which enables the top surface of the composites to withstand the applied loads creating a solid tribo film.
- The specific wear rate decreased with increase in the sliding distance. This is due to the presence of hard boron carbide reinforcements on the surface which acts as sharp B_4C asperities reduces the contact surface increasing the wear rate. As the sliding distance increases, these get blunt due to compaction between the abrading surfaces, thus increasing the contact area. This leads to improvement of the wear behavior under the high sliding distance.
- As the temperature increases above the contact surface when sliding at a higher velocity, it creates a mechanically mixed layer (MML), or tribolayer, which serves as a barrier reducing the wear rate.
- With enhancements in the properties like strength and hardness, the specific wear rate of the hybrid composites decreases gradually with an increase in wt.% of B_4C particles reinforcement due to the grain refinement and uniform dispersion of particles.
- The optimal control parameters were found to be for the lowest specific wear rate was with 4 wt.% of B_4C, 40 N load, 1.5 ms^{-1} sliding velocity, and 2000 m sliding distance resulting in a specific wear rate of 0.00017 mm^3/Nm and also confirmed experimentally, with application area in nuclear industries.

References

1. Chen HS, Wang WX, Li YL, Zhou J, Nie HH, Wu QC (2016) The design, microstructure and mechanical properties of B_4C/6061Al neutron absorber composites fabricated by SPS. Mater Des 94:360–367
2. Zhang P, Li Y, Wang W, Gao Z, Wang B (2013) The design, fabrication and properties of B_4C/Al neutron absorbers. J Nucl Mater 437:350–358
3. Li YZ, Wang QZ, Wang WG, Xiao BL, Ma ZY (2015) Interfacial reaction mechanism between matrix and reinforcement in B_4C/6061Al composites, 154. https://doi.org/10.1016/j.matchemphys.2015.01.052
4. Kang PC, Cao ZW, Wu GH, Zhang JH, Wei DJ, Lin LT (2010) Phase identification of Al–B_4C ceramic composites synthesized by reaction hot-press sintering. Int J Refract Met Hard Mater 28:297–300

5. Chen H, Nie H, Zhou J, Wang W, Zhang P, Zhang Y, Wang J-F, Liu R (2019) Microstructure and mechanical properties of B_4C/6061Al neutron absorber composite tube fabricated by spark plasma sintering and hot spinning. J Nucl Mater 517:393–400
6. Chand S, Chandrasekhar P, Roy S, Singh S (2020) Influence of dispersoid content on compressibility, sinterability and mechanical behaviour of B_4C/BN reinforced Al6061 metal matrix hybrid composites fabricated via mechanical alloying. Met Mater Int
7. Chand S, Chandrasekhar P (2020) Influence of B_4C/BN on solid particle erosion of Al6061 metal matrix hybrid composites fabricated through powder metallurgy technique. Ceram Int https://doi.org/10.1016/j.ceramint.2020.04.064
8. Thuault A, Marinel S, Savary E, Heuguet R, Saunier S, Goeuriot D, Agrawal D (2013) Processing of reaction-bonded B_4C–SiC composites in a single-mode microwave cavity. Ceram Int 39:1215–1219
9. Domnich V, Reynaud S, Haber RA, Chhowalla M (2011) Boron carbide: structure, properties, and stability under stress. J Am Ceram Soc 94:3605–3628
10. Alihosseini H, Dehghani K, Kamali J (2017) Microstructure characterization, mechanical properties, compressibility and sintering behavior of Al-B_4C nanocomposite powders. Adv Powder Technol 28:2126–2134
11. Pani B, Chandrasekhar P, Singh S (2019) Application of box-behnken design and neural computation for tribo-mechanical performance analysis of iron-mud-filled glass-fiber/epoxy composite and parametric optimization using PSO. Polym Compos 40:1433–1449
12. Uthayakumar M, Aravindan S, Rajkumar K (2013) Wear performance of Al–SiC–B_4C hybrid composites under dry sliding conditions. Mater Des 47:456–464
13. Panwar RS, Pandey OP (2013) Analysis of wear track and debris of stir cast LM13/Zr composite at elevated temperatures. Mater Charact 75:200–213
14. Kumar PRS, Kumaran S, Rao TS, Natarajan S (2010) High temperature sliding wear behavior of press-extruded AA6061/fly ash composite. Mater Sci Eng A 527:1501–1509

Effect of Recast Layer Thickness of High-Carbon Alloy (EN-31) in Wire EDM Process by Varying Operating Parameters

P. C. Padhi and Bharat Chandra Routara

1 Introduction

Wire electrical discharge machining (WEDM) is the most popularly used unconventional material removal process in current manufacturing industries. It is an electro thermal machining technique in which the cutting takes place due to the occurrence of sequence of distinct sparks between the workpiece and wire electrode in presence of non-conductive dielectric fluid. The flushing is provided at the machining area continuously to wash out the eroded materials generated, also to avoid short circuit and improves the efficiency. The wire electrode movement is controlled by computer system to achieve the required shape and size of the workpiece. WEDM is generally used for cutting the profiles of intricate geometry and high hardness [1].

Selection of best combination of machining parameter to achieve optimal machining efficiency is difficult in wire-cut EDM due to the presence of huge number of control factors and stochastic process mechanism. The material removal rate, surface finish, and accuracy are the major responses to be achieved in a workpiece without having any surface defects like cracks due to the formation of recast layer. Several researches have already done experimentation on different materials in WEDM for optimization of the process.

The wire-cut surface is formed by a series of distinct sparks generated due to ionization between the wire electrode and the workpiece, which produces craters in the machined surface. WEDM is suitable for machining electrically conductive ferrous and non-ferrous materials regardless of its hardness and to achieve sharp corners.

P. C. Padhi
Central Institute of Petrochemicals Engineering and Technology (CIPET), Bhubaneswar, India

B. C. Routara (✉)
School of Mechanical Engineering, KIIT Deemed to Be University, Bhubaneswar, India
e-mail: bcroutray@gmail.com; broutrafme@kiit.ac.in

© The Author(s), under exclusive license to Springer Nature Singapore Pte Ltd. 2023

S. Revankar et al. (eds.), *Recent Advances in Thermofluids and Manufacturing Engineering*, Lecture Notes in Mechanical Engineering,
https://doi.org/10.1007/978-981-19-4388-1_43

The electrical energy develops spontaneous force by the dielectric fluid to erode the molten material and due to intense heating due to haphazard attack of the spark followed by cooling a multi-layered surface is generated on the workpiece. The three layers of WEDM-machined surface consist of recast layer on the upper most surface, heat-affected zone, and the transformation layer which changes the grain structure with respect to the base material [2]. The layer thickness depends on the input process variables and properties of the workpiece material and thermal conductivity [3]. During the cutting process, the material melted due to the spark is partially flushed away, and remaining material is solidified that forms the recast layer at the outer surface. It is the external part of the heat-affected region resulted from molten re-solidified work material having changed metallurgical structure due to exceedingly high thermal heating followed by dielectric quenching progression. Usually, the recast layer is a fine-grained hard and brittle structure form the martensite structure due to sudden cooling, and it is different from the base material. As the rate of cooling is faster from the top surface into the inside layers of the parent material, the martensites are directed perpendicular to the machined surface. The residual stresses developed due to sparking generate voids and micro-cracks which affect the functional behavior of the product. The pores and micro-cracks are generally limited to the recast layer. Walder and Richard [4] have studied the effects of ultrasonic vibrations of tool on machinability of EDM and recast layer. SEM was used to measure the machined surface and found that there were no remarkable changes between heat-affected zone and the base material structure. Shabgard et al. [5] has analyzed the effects of WEDM input variables; gap current and pulse time on RLT, heat-affected zone, and corrosion resistance of specimen. Liu et al. [6] have investigated the white layer and its crystallography, microstructure composition characteristics in WEDM machining of nitinol alloy by TEM analysis. Zhang et al. [7] have analyzed the WEDM machining input control factors on surface crack density and RLT relating to decrease the surface roughness of the workpiece. Klocke et al. [8] have studied the white layer composition and structure of cold worked tool steel in WEDM machining process. An improved quality of lean lamella was found in TEM characterizations and also studied the carbon and recast layers. Shunmugam [9] has studied the thickness of white layer on the WEDM machining of Ti6Al4V material, and it was established that structure of very fine grain lamellas is available. Zhang [10] has conducted the experimental investigation on the surface integrity of nanocomposite ceramics and studied the microstructure of WEDM-machined surface of the TiN/Si_3N_4 composite and TiN particles using TEM micrograph analysis. Huang et al. [11] have studied the recast layer of martensite-grade stainless steel machined in wire-cut EDM process. The TEM analysis was carried out and found the amorphous layer on few zones of cut surface with fine particles of crystalline structure randomly spread, spherical electrode deposits gas void in the workpiece. Punturat [12] studied the influence of control factors on the surface characteristics of workpiece in WEDM machining of silicon. Chaudhari [13] has studied the surface characteristics and RLT of wire EDM-machined nitinol workpiece. Gholipoor et al. [14] have investigated the effect of input variables on WEDM cut surface recast layer thickness and heat-affected zone of Inconel 617 material. The gap current and pulse on time are the key factors for development of RTL and HAZ. Kumar et al. [15] have analyzed the surface topography of Inconel 825 material in WEDM workpiece and found that the surface finish reduces with increase in pulse duration. The formation of surface crack, recast layer, and HAZ has relation with pulse duration and peak current.

To improve the end use application and to meet the functional requirements of the product, the recast layer needs to be removed as the part may be subjected tensile, shear, and fluctuating loads. The final workpiece surface should be defect free and corrosion resistant [16]. Therefore, the recast layer should be separate either by polishing or chemical etching to enhance the physical and mechanical properties. Generally, the polishing kit contains various grades of abrasive materials like emery paper, silicon carbide, aluminum oxide stones, and diamond paste are used to accomplish mirror polish on the surface texture of the WEDM-machined parts. Necessary care should be taken while removing the recast layer which sometimes causes damage to the part due to too much polishing may lead to excess material removal and loose the dimensional accuracy. However, these methods are additional and increases cost and time consuming. Therefore, it is essential to minimize the RLT appropriately by optimizing the process parameters during machining. It is obvious that achieving high cutting rate and good surface texture simultaneously is not possible in WEDM process. Hence, extensive research has been made by worldwide scientists to overcome this issue and still the study is in progress.

In this experimental investigation, an attempt has been made to visualize the effect of pulse on time, pulse off time, spark gap set voltage, and gap current on recast layer thickness of EN-31 high-carbon alloy steel in wire electrical discharge machining process using scanning electron microscopy (SEM). The machining has been performed as per Taguchi's experimental design concept of to study the effects of process variables on formation of recast layer thickness (RLT).

2 Experimental Methodology

The machining has been carried out in five axis CNC-WEDM, electronica make (Model: Ultracut-S2), and various settings of six controllable factors each at three levels were used as shown in Table 1. The experimental matrix adopted as per Taguchi's design (L_{27} orthogonal array) assuming the interaction effect of process variables is negligible.

The study was focused to analyze the effects of the process variables on recast layer thickness (RLT) of the WEDM-machined surface. EN-31 high-carbon alloy steel has been taken as workpiece material having size of 20 mm × 20 mm was cut. It achieves higher value of hardness after heat treatment with better tensile, compressive strength, and resistance to wear. Due to outstanding physical and mechanical properties, it has wide application in mold and die-making industries, manufacturing of jigs and fixtures, press tools, die-casting dies, punches, automobile and aerospace application. In the present experimental work, zinc-coated copper wire with a diameter of 0.25 mm

Table 1 Process variables and levels

Factors	Parameters	Levels		
		Level 1	Level 2	Level 3
1	Pulse on time (T_{on})	1 μs	1.1 μs	1.2 μs
2	Pulse off time (T_{off})	11.5 μs	14 μs	18 μs
3	Wire feed (WF)	8 m/min	10 m/min	12 m/min
4	Wire tension (WT)	1250 g	1450 g	1650 g
5	Spark gap set voltage (SV)	20 V	30 V	40 V
6	Servo feed (SF)	4 mm/min	6 mm/min	8 mm/min

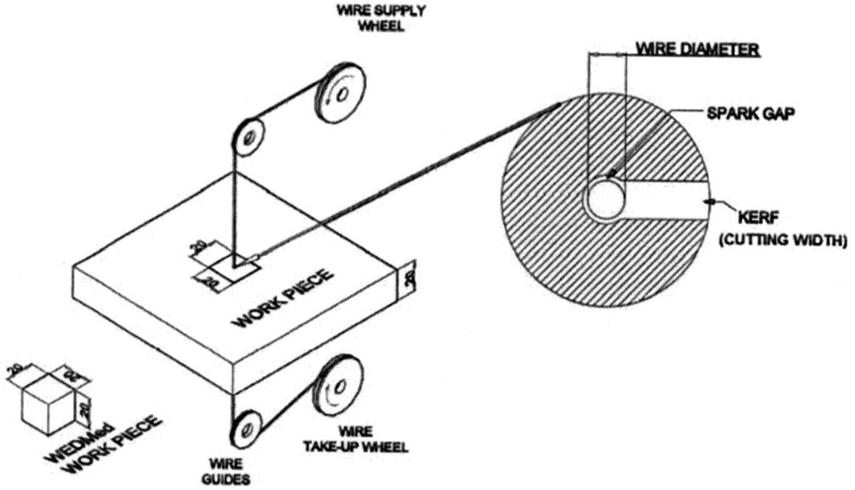

Fig. 1 Schematic diagram of WEDM cutting process

was used as electrode for cutting the workpieces. The schematic diagram cutting process is shown in Fig. 1.

3 Results and Discussion

Experiments were conducted in wire EDM machine as per Taguchi's design, and twenty-seven number of pieces were cut to study the influence of various process parameters like T_{on}, T_{off}, WF, WT, SV, and SF. Whereas, the gap current (I_g) is a measure of the power supplied to the discharge gap between the wire electrode, and workpiece is considered as an input variable. This is the actual value of gap current read on the WEDM machine ammeter during machining. The process parameter values along with the gap current input in the respective machining settings are shown in Table 2. From the experimental investigation, it was observed that, the input parameters, T_{on}, $T_{off,}$ SV, and I_g, have major effect on recast layer thickness (RLT) on WEDM-machined surface.

In order to measure the recast layer thickness of the cut workpieces, all the 27 numbers of pieces were transversely sectioned to a size of 20 mm × 20 mm × 5 mm thick. The transverse section area of the workpieces was polished successively with emery papers: silicon carbide and aluminum oxide of different grit sizes ranging from rough to finishing grade. Finally, the surface has been mirror polished using polishing machine with various grades of diamond paste. Later, these polished

Table 2 Observations of recast layer thickness (RLT)

Exp. No.	Process parameters						Gap current (I_g)	Recast layer thickness in micron
	T_{on}	T_{off}	WF	WT	SV	SF		
1	1	11.5	8	1250	20	4	3.5	6.895
2	1	11.5	8	1250	25	6	4.0	9.566
3	1	11.5	8	1250	30	8	4.3	11.60
4	1	14	10	1450	20	4	3.2	5.628
5	1	14	10	1450	25	6	3.3	6.248
6	1	14	10	1450	30	8	3.9	8.587
7	1	18	12	1650	30	4	2.3	1.615
8	1	18	12	1650	25	6	2.4	2.164
9	1	18	12	1650	20	8	3.0	3.5841
10	1.1	11.5	10	1650	20	6	5.4	28.960
11	1.1	11.5	10	1650	25	8	5.2	26.058
12	1.1	11.5	10	1650	30	4	4.5	12.538
13	1.1	14	12	1250	20	6	5.3	27.604
14	1.1	14	12	1250	25	8	4.8	16.706
15	1.1	14	12	1250	30	4	3.1	3.659
16	1.1	18	8	1450	20	6	4.5	12.677
17	1.1	18	8	1450	25	8	3.4	6.526
18	1.1	18	8	1450	30	4	2.8	2.772
19	1.2	11.5	12	1450	20	8	5.8	46.209
20	1.2	11.5	12	1450	25	4	5.2	26.725
21	1.2	11.5	12	1450	30	6	5.5	40.410
22	1.2	14	8	1650	20	8	5.6	42.539
23	1.2	14	8	1650	25	4	4.9	18.434
24	1.2	14	8	1650	30	6	5.0	20.081
25	1.2	18	10	1250	20	8	5.1	23.701
26	1.2	18	10	1250	25	4	4.2	11.008
27	1.2	18	10	1250	30	6	3.7	8.246

areas were etched with nital liquid containing 97% ethyl alcohol and 3% nitric acid for a duration of 20–25 s. After that, the metallographic sections of the surface were observed through a scanning electron microscope (Make: Carl Zeiss SMT, Germany). The micrographs of the etched workpieces were taken in scanning electron microscopy (SCM) machine for all the 27 numbers of samples. The depths of recast layer have been measured from the SCM micrographs, and the average measured values were taken for study. The magnitudes of recast layer thickness of WEDM-machined workpieces are shown in Table 2.

3.1 Imaging for Recast Layer Thickness (RLT)

The morphology of wire EDM surface generated because of high electrical discharge energy of spark with a temperature more than 10,000 °C at the spark gap between the wire electrode and workpiece removes the material in the form of hydro carbon by melting and vaporization. The part of the material is flushed away, and the remaining molten metal re-solidifies and quenched in presence of the dielectric fluid which are neither driven out nor eroded by pressure flushing of the dielectric. This re-solidified sticky layer is known as recast layer which contains numerous voids, pocks, and micro-cracks, and it depends on the process parameter combination of machining. Beneath the recast layer, other layers like HAZ and transformation zone are present, and the layer condition differs from material to material as reported by Kruth et al. [2]. The micrographs of WEDM-machined cross-sectional areas of EN-31 steel pieces were taken with high magnification scanning electron microscope to measure the recast layer thickness and surface morphology.

The SEM micrographs of the wire-cut EDMed surfaces for the machining parameter combinations (T_{on}: 1, T_{off}: 18, SV: 25), (T_{on}: 1, T_{off}: 14, SV: 20), (T_{on}: 1.1, T_{off}: 18, SV: 20) and (T_{on}: 1.2, T_{off}: 14, SV: 30) and (T_{on}: 1.2, T_{off}: 14, SV: 20), are shown in Figs. 2, 3, 4, 5, which reveal the increasing trend of recast layer thickness with gap current: $I_g = 2.4$ A, 3.2 A, 4.5 A, 5.0 A, and 5.6 A, respectively. The average recast layer thickness increases from 2.164 μm, 5.628 μm, 12.677 μm, 20.081 μm to 42.539 μm, respectively, due to increase in pulse on time, gap current, and reduced spark gap set voltage as shown in the micrographs. The increase in pulse duration and gap current causes rise in intensity of spark and produces more metal melting at the surface area. The dielectric fluid flushing pressure is insufficient to remove all the molten material produced by higher pulse current and during further quenching re-solidifies the sparked surface to form the added thickness of the recast layer.

3.2 Effect of Pulse on and Pulse off Time

As the pulse on time is directly proportional to the gap current intake which in turn generates higher thermal energy during spark, causes increase in pocks and cracks in

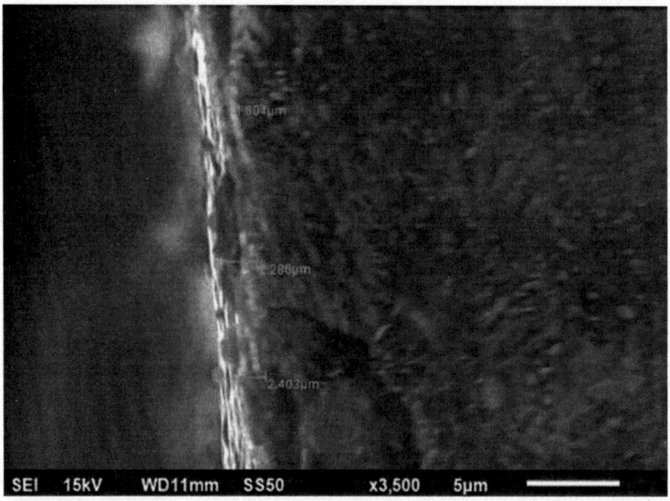

Fig. 2 SEM snap RLT at T_{on}-1, T_{off}-18, SV-25, I_g-2.4

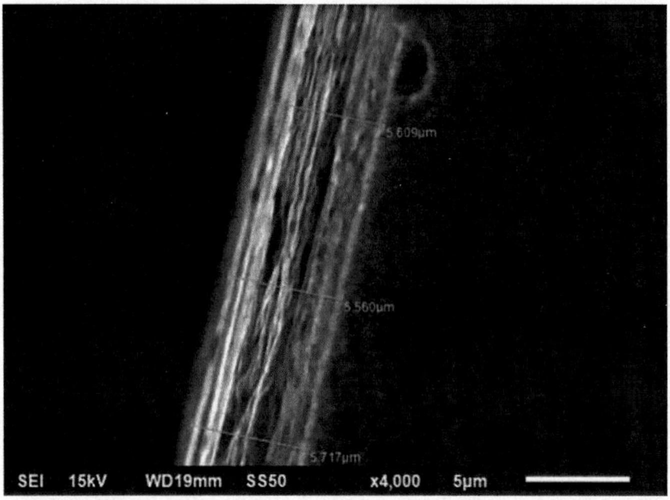

Fig. 3 SEM snap RLT at T_{on}-1, T_{off}-14, SV-20, I_g-3.2

the workpiece. The size of the crater increases with long-duration pulse energy and propagates the length of the surface cracks. In addition, the surface cracks penetrate into the recast layer, and its intensity depends on the pulse energy which is limited to the recast layer thickness only. The present experimental study is related to the remarks of Hascalyk and Caydas [17]. Thus, the research outcome suggests that, the surface cracking can be minimized by reducing the pulse on time and keeping the pulse off time at their highest level (Fig. 6).

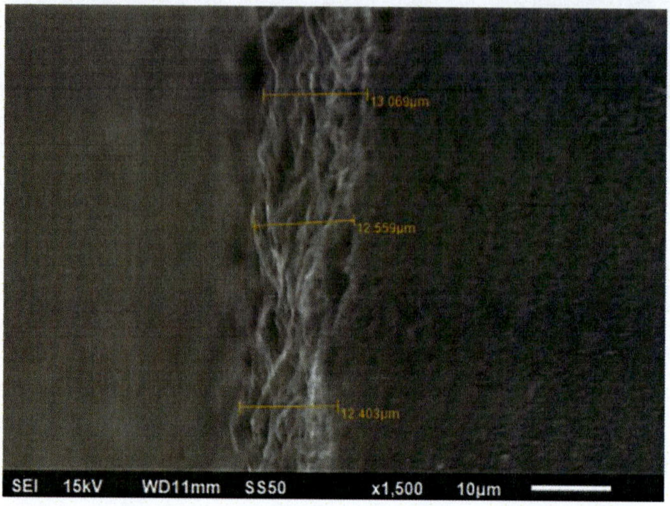

Fig. 4 SEM snap RLT at T_{on}-1.1, T_{off}-18, SV-20, I_g-4.5

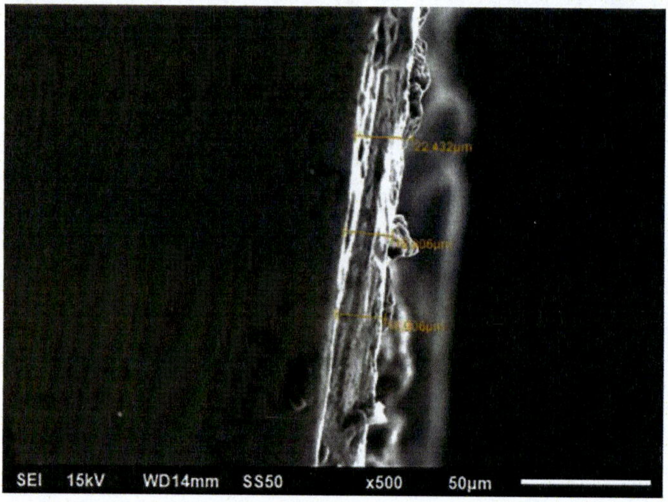

Fig. 5 SEM snap RLT at T_{on}-1.2, T_{off}-14, SV-30, I_g-5.0

Similarly, the SEM micrographs for levels at (T_{on}: 1, T_{off}: 14, SV: 25, I_g: 3.3), (T_{on}: 1.1, T_{off}: 14, SV: 25, I_g: 4.8), (T_{on}: 1.2, T_{off}: 14, SV: 25, I_g: 4.9) are shown in Figs. 7, 8, and 9, respectively. The average recast layer thickness is 6.248, 16.706, and 18.434 μm, where the increase of RLT can be seen from the SEM micrographs. The recast layer thickness increases with respect to the increase in pulse on time (T_{on}) and gap current. During the thermoelectrical machining process, the intensity of heat

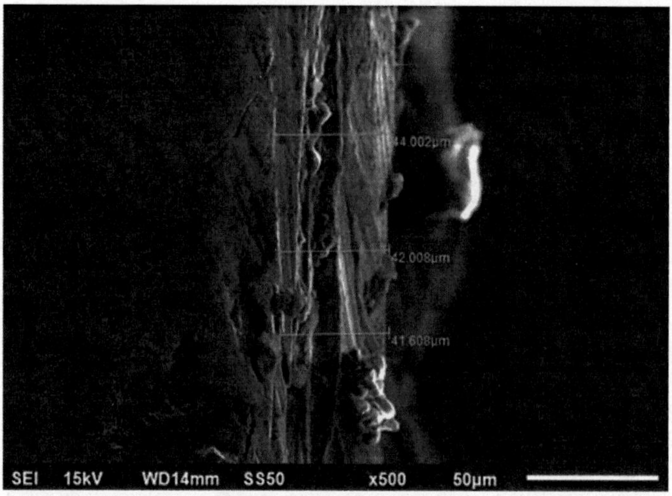

Fig. 6 SEM snap RLT at T_{on}-1.2, T_{off}-14, SV-20, I_g-5.6

increases with spark duration. The recast layer thickness increases with increase in T_{on} on as shown in the figures below.

It is evident from the SEM micrographs in Figs. 10, 11, and 12 that with increase in pulse off time (T_{off}) from 11.5 μs to 14 μs and 18 μs the corresponding recast layer thickness decreases from 11.60 μm to 8.587 μm and 3.584 μm, respectively. Recast layer thickness decreases with increase in T_{off} value. The pulse off time is the time lag between one spark to the next during which the flushing action takes place to remove

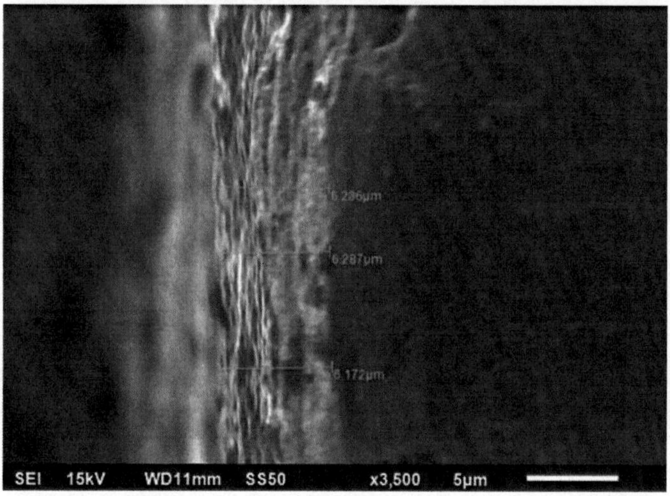

Fig. 7 SEM snap RLT at T_{on}-1, T_{off}-14, SV-25, I_g-3.3

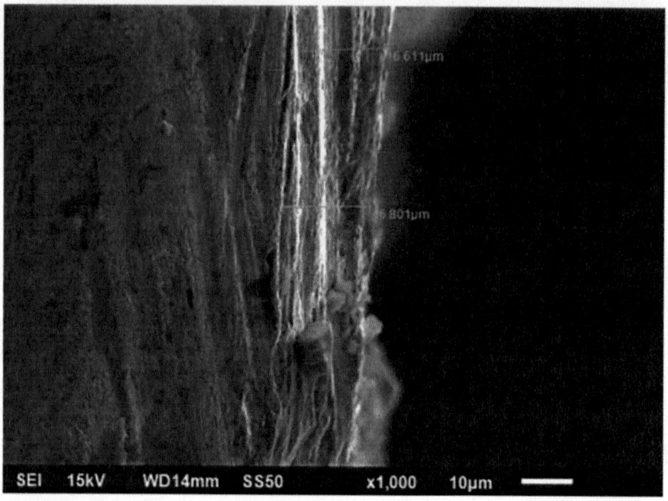

Fig. 8 SEM snap RLT at T_{on}-1.1, T_{off}-14, SV-25, I_g-4.8

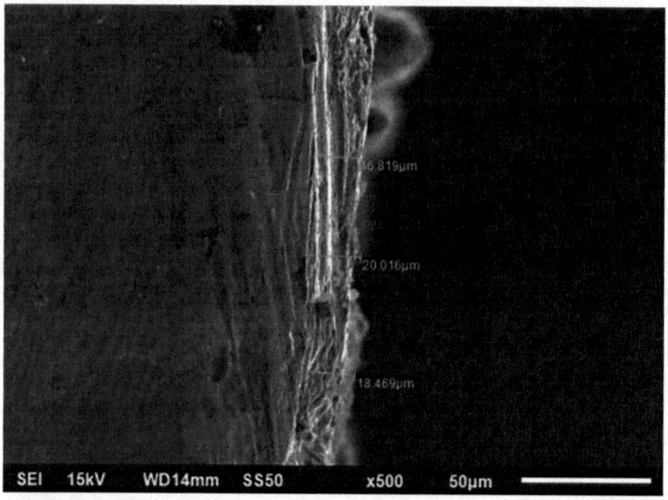

Fig. 9 SEM snap RLT at T_{on}-1.2, T_{off}-14, SV-25, I_g-4.9

the molten material in the form of hydro carbons from the workpiece. For smaller value of T_{off}, there is less time available to flush away the molten material, hence re-solidifies on the sparked surface due to quenching, resulting in higher thickness recast layer.

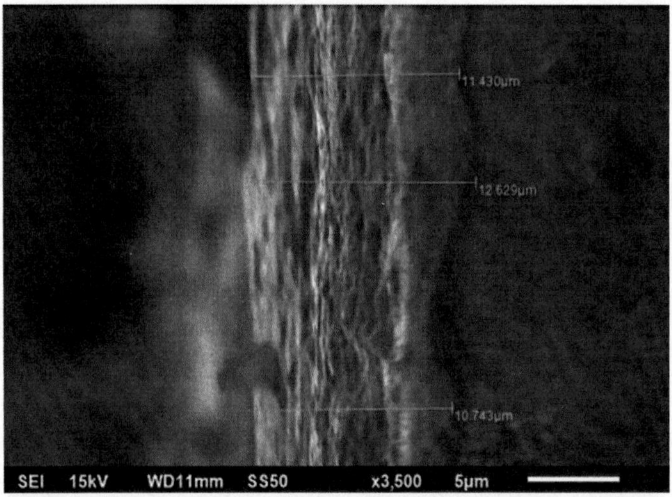

Fig. 10 SEM snap RLT at T_{on}-1, T_{off}-11.5, SV-30, I_{g}-4.3

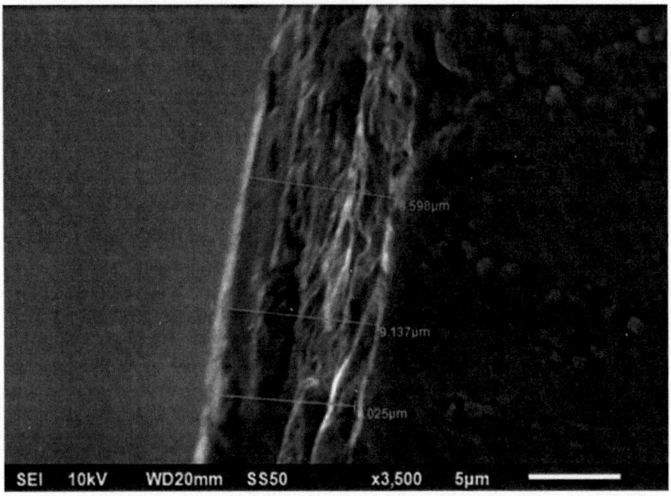

Fig. 11 SEM Snap RLT at T_{on}-1, T_{off}-14, SV-30, I_{g}-3.9

3.3 Recast Layer Thickness

The recast layer thickness of all 27 number of samples was measured in scanning electron microscopy (SEM), and the mean thickness of recast layer was obtained (observations are shown in Table 2 under column 'RLT').

It is evident that, the highest value of recast layer thickness of 46.209 μm is at process parameter level (T_{on}: 1.2, T_{off}: 11.5, SV: 20, I_{g}: 5.8) and lowest value of RLT

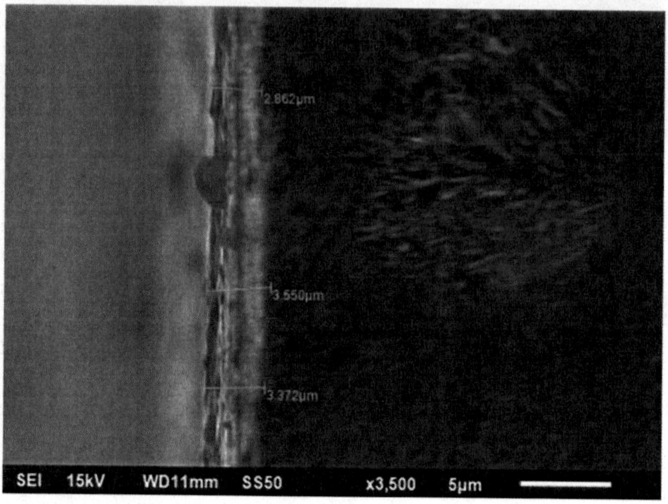

Fig. 12 SEM snap RLT at T_{on}-1, T_{off}-18, SV-30, I_g-3.0

1.615 µm at level (T_{on}: 1, T_{off}: 18, SV: 30, I_g: 5.8). The experimental investigation depicts, the recast layer thickness increases with increase in pulse on time, gap current, and at the lower value of pulse off time, spark gap set voltage. Similarly, the smallest size of recast layer thickness present at lowest value of pulse on time, gap current, and at highest value of pulse off time, spark gap set voltage.

4 Conclusions

The WEDM-machined workpiece sections were studied under scanning electron microscope to analyze the recast layer thickness. The influence of the most significant factors like pulse on time, pulse off time, spark gap set voltage, and gap current on recast layer thickness has been studied for EN-31 high-carbon alloy steel. It is observed that, the recast layer thickness increases with increase in pulse on time and gap current which is responsible for producing stronger thermal energy due to intense spark that increases the size of the craters during machining. The significance of the process parameters: wire feed, wire tension, and servo feed have negligible influence on formation of recast layer thickness. The present experimental study will help the tool room engineer to co-relate the input parameters with the output response (RLT) for optimization of WEDM process.

References

1. Ho KH (2004) state of the art in wire electrical discharge machining (WEDM). Int J Mach Tools Manuf 44:1247–1259
2. Kruth J, Stevens L, Froyen L, Lauwers B (1995) Study of the white layer of a surface machined by die-sinking electro-discharge machining. CIRP Ann Manuf Technol 44(1):169–172
3. Ramasawmy H, Blunt L, Rajurkar K (2005) Investigation of the relationship between the white layer thickness and 3d surface texture parameters in the die sinking EDM process. Precis Eng 29(4):479–490
4. Walder G, Richard J (2016) Removal of the heat affect zone created by EDM with pico-second LASER machining. Proc CIRP 42:475–480
5. Shabgard MR, Farzaneh S, Gholipoor A (2017) Investigation of the surface integrity characteristics in wire electrical discharge machining of Inconel 617. J Braz Soc Mech Sci Eng 39:857–864
6. Liu JF, Guo YB, Butler TM, Weaver ML (2016) Crystallography, compositions, and properties of white layer by wire electrical discharge machining of nitinol shape memory alloy. Mater Des 109:1–9
7. Zhang Z, Ming W, Huang H (2015) Optimization of process parameters on surface integrity in wire electrical discharge machining of tungsten tool YG15. Int J Adv Manuf Technol 81(5–8):1303–1317
8. Klocke F, Hensgen L, Klink A, Ehle L, Schwedt A (2016) Structure and composition of the white layer in the wire-EDM process. Proc CIRP 42:673–678
9. Jose JV, Shunmugam MS (2009) Investigation into white layer formed on wire electrical discharge machined Ti_6Al_4V surface. Int J Mach Mach Mater 6:234–249
10. Zhang C (2014) Effect of wire electrical discharge machining (WEDM) parameters on surface integrity of nanocomposite ceramics. Ceram Int 40:9657–9662
11. Huang CA, Hsu FY, Yao SJ (2004) Microstructure analysis of the martensitic stainless steel surface fine-cut by the wire electrode discharge machining (WEDM). Mater Sci Eng A 371:119–126
12. Punturat J, Tangwarodomnukun V, Dumkum C (2014) Surface characteristics and damage of monocrystalline silicon induced by wire-EDM. Appl Surf Sci 320:83–92
13. Chaudhari R, Vora JJ, Patel V, López de Lacalle LN, Parikh DM (2020) Surface analysis of wire-electrical-discharge-machining-processed shape-memory alloys. Materials 13:530
14. Gholipoor A, Shabgard M, Arghavanian R (2020) Experimental investigation of recast layer, heat affected zone and corrosion resistance in WEDM of Inconel 617. J Sci Ind Res 79:701–704
15. Kumar P, Guptab M, Kumar V (2021) Experimental investigation of surface crack density and recast layer thickness of WEDMed Inconel 825. J Comput Appl Res Mech Eng 11(1):205–216
16. Tai T, Lu S (2009) Improving the fatigue life of electro discharge machined SDK11 tool steel via the suppression of surface cracks. Int J Fatigue 31(3):433–438
17. Hascalyk A, Caydas U (2004) Experimental study of wire electrical discharge machining of AISI D5 tool Steel. J Mater Process Technol 148(3):362–367

Fabrication of Fly Ash-Based Refractory Brick Through Powder Metallurgy Technique

K. Balasubramanian, R. Vaira Vignesh, K. Periyaswamy, and M. Govindaraju

Abstract One of the major pollutants from thermal power plants is fly ash, which is a by-product of the combustion of pulverized coal. Fly ash consists of oxides of metals (primarily aluminum, silicon that constitute refractory compounds) heavy metals, sulfur, potassium, and sodium. It is generally dumped in the landfills, and hence, its leaching effect severely pollutes the environment. Hence, effective utilization of fly ash is the need of the hour. Also, the use of refractories is rapidly increasing in various applications because of their excellent properties and low-cost economic production. This work is on the development of technology to fabricate fly ash-based refractory bricks using 100% fly ash. The fabricated fly ash-based refractory bricks are characterized for their physical and mechanical properties. Also, the surface morphology, elemental composition, and phase composition are analyzed. The results demonstrate that the fabricated fly ash-based refractory bricks have desired physio-mechanical properties such as density of 1.7 cm^3, compression strength of 10 MPa, and less than 1% water absorption.

Keywords Fly ash · Sintering · Compaction · Ceramics · Refractory brick · Morphology

1 Introduction

Material is a substance or a mixture of a substance that makes up an object. The materials can be classified based on their physical or chemical or geological or biological property. The new materials can be instituted by synthesizing the raw materials by various processes and technology [1]. Also, the materials can be classified in a wide range according to their different use, property, and structure [2] as follows: metals, polymers, ceramics, and composites. Each classification of materials has its unique

K. Balasubramanian · R. Vaira Vignesh (✉) · K. Periyaswamy · M. Govindaraju
Department of Mechanical Engineering, Amrita School of Engineering, Amrita Vishwa Vidyapeetham, Coimbatore, India
e-mail: r_vairavignesh@cb.amrita.edu

© The Author(s), under exclusive license to Springer Nature Singapore Pte Ltd. 2023

S. Revankar et al. (eds.), *Recent Advances in Thermofluids and Manufacturing Engineering*, Lecture Notes in Mechanical Engineering,
https://doi.org/10.1007/978-981-19-4388-1_44

properties. In turn, metals are classified as ferrous and non-ferrous metals; polymers are classified as thermoplastic, thermosets, and elastomers; and composites are classified based on the base matrix as metal matrix composite, ceramic matrix composite, and polymer matrix composite. Ceramics are classified as glasses and crystalline ceramics.

The refractories are ceramic materials that exhibit high-melting points and corrosion resistance. Also, these refractories provide thermal insulation, good compression strength, wear resistance, good electrical resistance, and thermal shock resistance. The ceramic materials possess high-temperature stability between 800 and 2800 °C. The ceramic has low to medium thermal conductivity and high-electrical resistivity [3]. Examples of ceramic materials include SiO_2, Al_2O_3, mullite, ZrO_2, TiO_2, TiC, WC, SiC, etc. [4]. One of the cheap sources of SiO_2, Al_2O_3, and mullite is fly ash, which is produced as a by-product of the combustion of pulverized coal in thermal power plants. The annual fly ash generation of India stands at 112 million MT. Hence, the thermal power plants in India have been working on the development of effective waste management strategies for fly ash.

One of the convenient methodologies for the fabrication of ceramic components includes powder metallurgy. Powder metallurgy processing involves crushing, grinding, separating impurities, blending different powders, and compacting the homogeneous blend that is subsequently sintered to produce a finished product [3]. The powders are characterized by their size, flow, density, and compressibility. The powders are added with additives and mixed or blended to form a homogeneous mixture [5]. Then, the blended mixture is compressed by the hydraulic or mechanical press to compact. The compacted green pellet is then sintered in a furnace. The sintering process is a diffusion process that aids in the bonding of particles to form the final product. The sintering is performed in a controlled atmosphere to ensure good bond formation between the compacted powder particles. Typically, the sintering temperature range between 1000 and 1500 °C for ceramics (for metals it would be 0.7 Tm (melting temperature) [6].

An overview of the research works on the fabrication of fly ash-based ceramic refractory brick. Robert Kusiorowski [7] recycled magnesia–carbon refractories (MgO–C) from the waste steel plants to magnesia-zirconia refractories. The raw materials were mixed in the form of a fused magnesia-zirconia co-clinker to produce the component. The results indicated that MgO–C refractories could be recycled at the cost of quality. However, the addition of zirconia improved the cold crushing strength, reduced pores, and increased bulk density (2.8 g/cm^3). Xinyu Liu et al. analyzed the effect of rice husk powder binder on the properties of ZrO_2 hollow fiber refractories, as an alternative to Zr-sol binder [8]. The rice husk powder did not affect the density and thermal conductivity at 1300 °C. Also, the compressive strength (0.501 MPa) of the refractory increased about 285% compared to the Zr-sol binder. Shimizu et al. [9] produced a high-alumina refractory brick using the slurry and foaming method. The high-slurry foam was applied to the brick to increase the porosity and reduce the thermal conductivity. The fabricated material had a thermal conductivity of 0.12 W/mK (good thermal insulator) with a foam density of 0.1 g/cm^3. Gonzalez Otero et al. manufactured a refractory brick using fly ash and clay with the addition of a binder and foaming agent [10]. The specimen with 20 wt.% clay, 15 wt.% sodium silicate, and 5 ml of hydrogen peroxide had good mechanical strength of 6.03 MPa and low-thermal conductivity of 0.096 Wm^{-1} that indicated good insulating property.

Ravikishor et al. studied fly ash-rich bricks for high-temperature applications [11]. The fly ash was added with a small fraction of granite powder, alite cement, and black sand. The specimen with 25% of granite powder that was fired at 1000 °C showed better physical and mechanical properties (desirable density). Chakraborty et al. studied the effect of fly ash and quartz on earthen wear wall refractory [12]. The wall tile was fabricated with 70% sandstone and 30% fly ash or 30% quartz. The specimen with 30% of fly ash that was sintered at 1150 °C exhibited good thermomechanical properties. The mechanical strength increased from 8.0 to 18.6% when fly ash was used. Also, the increase in strength resulted in a gradual decrease in stresses because of less content of the sandstone. Tianyuan Wang et al. proposed a triple-layer microstructure for alumina fly ash-based ceramic where three unique layers of material were fabricated with 20, 30, and 40% fly ash [13]. The triple-layer microstructure of the ceramic was sintered at 1400 °C exhibited excellent physical properties such as the bulk density of 2.29 g/cm^3 and rupture modulus of 116.74 MPa. Also, the ceramic had zero water absorption and no pore formation. The fabricated triple microstructure alumina fly ash-based ceramic can be used as good waterproof material in a high temperature and corrosion environment. Ardalan Baradaran et al. experimented using refractory brick powder with portland cement and calcium aluminate cement [14]. The influence of firing temperature (100–800 °C) on the physio-mechanical properties of the specimen was evaluated. The specimen with calcium aluminate that was fired at 800 °C had improved residual strength. Also, the water absorption (5.17%) and porosity (11.93%) were low in the specimen. However, an increase in temperature decreased the elastic modulus of the specimen.

Jian Zhang et al. produced a glass-ceramic with 90% fly ash and 10% Na_2O [15]. The components were mixed and melted at a high temperature of about 1350 °C. Subsequently, the specimen was heated at 770 °C for 2 h to form glass-ceramic, which had good mechanical properties and chemical resistance. Ewais et al. used sludge from a water treatment plant and rice husk powder (RHP) to fabricate an insulating refractory brick [16]. For the fabrication, different batches of sludge with varying RHP were mixed and dried (batch 1–95% sludge and 5% RHP, batch 2–90% sludge and 10% RHP). The compacted specimens were fired at different temperatures ranging from 800–1200 °C, and physio-mechanical properties were evaluated. The results demonstrated that sludge with RHP could be successfully used to fabricate insulating bricks. Safeer Abbas et al. fabricated a sustainable clay brick utilizing the fly ash [17]. The fly ash was varied between 0 and 25% to study its influence on the mechanical and durability properties. After burning the bricks, the brick with 20% fly ash showed a compressive strength of about 11 MPa. The breaking load decreased with the increase of fly ash in the clay brick. The brick with 25% of fly ash satisfied the modulus of rupture requirement, i.e., >0.65 MPa. The literature survey shows that the prime requirements for fly ash-based refractory bricks are compressive strength, density, porosity, thermal conductivity, and microstructure.

Most of the research works was performed with a partial fraction of other materials in fly ash-based components and at a temperature ranging between 800 and 1350 °C. Also, most of the research work reported density ranging between 1 and 2.9 g/cm^3, compressive strength between 3 and 15 MPa, and less water absorption.

This research work attempted to fabricate the fly ash-based refractory (FAR) brick with 100% fly ash. The characterization tests were performed as per ASTM standards and Indian standards for determining the porosity, density, permanent linear change, and water absorption. Also, analytical instruments (SEM, EDS, XRD) were utilized to study the surface morphology, elemental composition, and phase composition of the FAR brick.

2 Materials and Method

The raw material fly ash was received from the coal-powered Mettur Thermal Power Plant, Tamil Nadu, India. The as-received fly ash was used in the fabrication of FAR brick. Distilled water was used for consolidating the fly ash powders in the compaction process. The wet blending process ensured a homogeneous mixing of fly ash particles without aggregation. The homogenous mixture was placed in a cylindrical die and was then compacted using a high-pressure hydraulic press. The compaction process increased the green strength of the specimen. The compacted specimens were positioned in the crucible and placed in the high-temperature furnace. The arrangement ensured the absence of thermal gradience in the course of the sintering process. The sintering was performed under normal atmospheric air and pressure. The crucible was heated in steps, as described in the sintering cycle (Fig. 1). The specimen was heated to 1400 °C at a constant heating rate of 10 °C/min for 16 h. Then, the specimen was furnace cooled.

The microstructure of the FAR brick was observed by using the high-resolution scanning electron microscope (SEM; Make: Zeiss Sigma). The surface morphology of the ceramic refractory brick was analyzed at an electron voltage of 5 and 10 kV at various magnification levels. An energy-dispersive x-ray spectroscope (EDS; Make: Oxford Instruments) was used to determine the elemental composition and elemental map of the specimen. The EDS analysis was performed at an electron acceleration voltage of 20 kV. The x-ray diffraction (XRD) analysis was used to determine the constitutional phase composition of the developed FAR brick. The XRD analysis was performed using Cu-Kα radiation at a scan rate of 0.5°/min between 10 and 90° in an x-ray diffractometer (Make: Rigaku).

2.1 Compression Test

The compression test was performed in a computerized universal testing machine to analyze the stress–strain characteristics of the FAR brick. Three specimens were

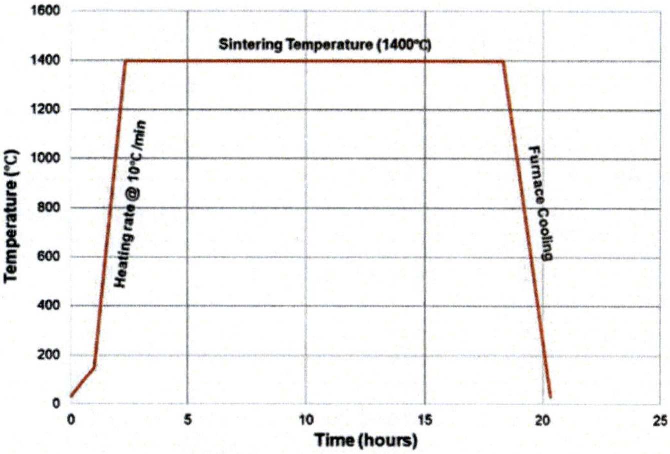

Fig. 1 Sintering cycle of the FAR brick

tested under normal room temperature. The stress–strain graph was obtained, and the maximum compression strength of the specimen was determined from the graph. The permanent linear change of the FAR brick (permanent change on dimension) after reheating was assessed as per the IS-1528 standard. Three specimens were heated to 1200 °C at the heating rate of 10 °C/min. A dwell time of 5 h was adopted at the peak temperature. Then, the specimens were furnace cooled. The dimensions of the specimens were measured using a digital vernier caliper with a least count of 0.01 mm (before and after the test). The change in dimensions of the specimens was reported.

2.2 *Water Absorption*

To analyze the water absorption, the specimens were suspended in the beaker such that the specimens did not touch the walls of the beaker. Distilled water was added to the beaker and ensured that the specimens were completely immersed in water. The setup was heated to 100 °C in a temperature-controlled oven. After 2 h, the heater was switched off. The specimen was left undisturbed for 12 h. Immediately upon removal from the setup, the saturated weight of the specimens was measured using a precision weighing balance with a readability of 0.001 g. Then, the specimens were dried in hot air over at 110 °C. Immediately, the dry weight of the specimens was determined. The water absorption of the specimens was calculated using Eq. (1).

$$\text{Water absorption (\%)} = \left[\frac{\text{Saturated Weight(g)} - \text{Dry Weight(g)}}{\text{Dry Weight (g)}}\right] \times 100 \quad (1)$$

2.3 Apparent Porosity

The apparent porosity was calculated by drying the specimen at 110 °C in the hot air oven. Then, the saturated weight was calculated by the boiling water method, which was performed for 12 h. Also, the volume of the specimen was calculated using Eq. (2), which in turn was used in Eq. (3) to determine the apparent porosity of FAR brick. The bulk density of the specimen was determined using the saturated weight and volume of the specimen, as given by Eq. (4).

$$\text{Volume}\left(\text{cm}^3\right) = \text{Saturated Weight (g)} - \text{Suspended Weight (g)} \tag{2}$$

$$\text{Apparent Porosity \%} = \left[\frac{\text{Saturated Weight (g)} - \text{Dry Weight (g)}}{\text{Volume}\left(\text{cm}^3\right)}\right] \times 100 \tag{3}$$

$$\text{Bulk Density}\left({}^{\text{g}}/_{\text{cm}^3}\right) = \frac{\text{Saturated Weight (g)}}{\text{Volume}\left(\text{cm}^3\right)} \tag{4}$$

The specific gravity of the specimen was determined using the dry weight and suspended weight of the specimen, as given by Eq. (5).

$$\text{Specific Gravity} = \frac{\text{Dry Weight (g)}}{\text{Dry Weight (g)} - \text{Suspended Weight (g)}} \tag{5}$$

3 Results and Discussion

3.1 Fabricated Specimen

The photographs of the typical compacted specimen and sintered specimens are shown in Figs. 2 and 3, respectively. The microstructure and physio-mechanical properties of the fabricated specimens are presented in the subsequent sections.

Fig. 2 Compacted specimen (FAR brick)

Fig. 3 Sintered specimen (FAR brick)

3.2 Microstructure

The microstructure of the FAR brick specimen was observed at various magnifications using the SEM. The FAR brick specimen exhibited the formation of pores, as shown in Fig. 4a. However, the pores were discontinuous. That is indicative of the absence of a networked pore structure. The magnified view of the pores is shown in Fig. 4b–d.

The phase diagram $Al_2O_3-SiO_2$ (alumina–silica) shows the low solubility for SiO_2 in the Al_2O_3 and vice versa. At the high temperature around 1700 °C, Al_2O_3 is transformed into a liquid state. However, the presence of SiO_2 reduces the liquidation temperature. In this study, the sintering temperature was 1400 °C. However, good bonding was observed between the fly ash particles. This could be attributed to the presence of other oxides that in turn lowered the liquidation temperature. The

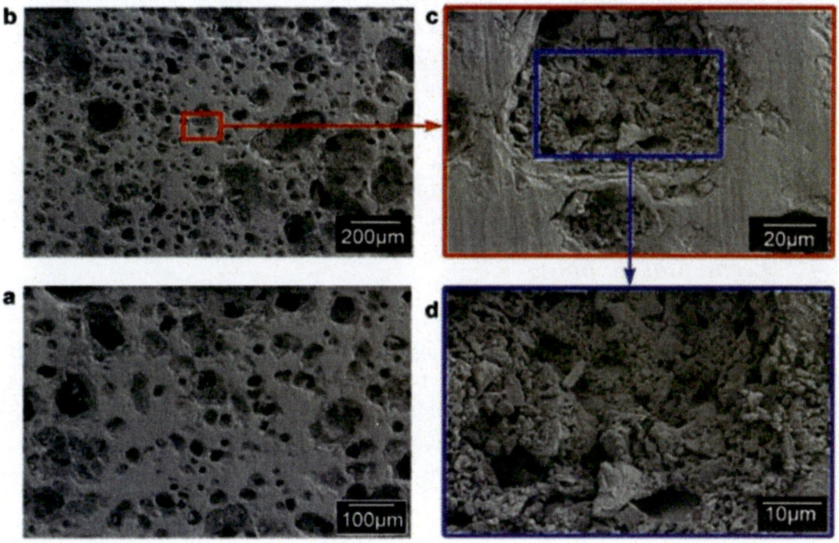

Fig. 4 SEM image of the specimen (FAR brick)

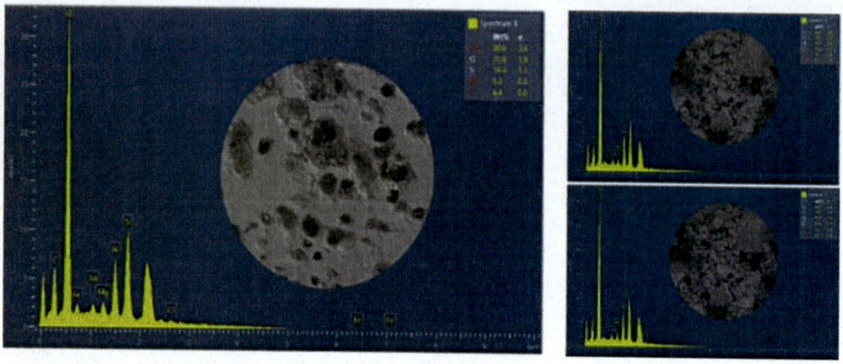

Fig. 5 Elemental composition of the specimen (FAR brick)

liquidation of oxides in the sintering temperature enabled good bonding between the fly ash particles.

3.3 *Elemental Composition Analysis*

The elemental composition of the FAR brick specimen was analyzed using EDS. The FAR brick specimen was sputtered with gold to avoid the charging effect. The EDS graph of the three regions is shown in Fig. 5.

The peak corresponding to gold could be neglected, as it was added deliberately through the sputtering process. Figure 5 shows the elements that were present in the ceramic refractory brick with their average composition. The O—oxygen concentration was about 24–45 wt.%, Si—silicon was about 12–28 wt.%, and Al—aluminum was about 5.3–13.7 wt.%. Also, the specimen had Mg—magnesium of 1–2 wt.%, and a small trace amount of Na—sodium of 0.5–1.7 wt.%.

3.4 *Elemental mapping*

The elemental mapping shown in Fig. 6 presents the visual of individual elements on the FAR brick. As observed in Fig. 6, Al, Si, and O are the major elemental concentration of FAR brick. The elemental map of aluminum (Al) and silicon (Si) overlaps with oxygen (O) and that indicates the formation of Al_2O_3 and SiO_2—mullite. The other element concentrations such as Mg, Na, Fe, and Cl were present in trace amounts.

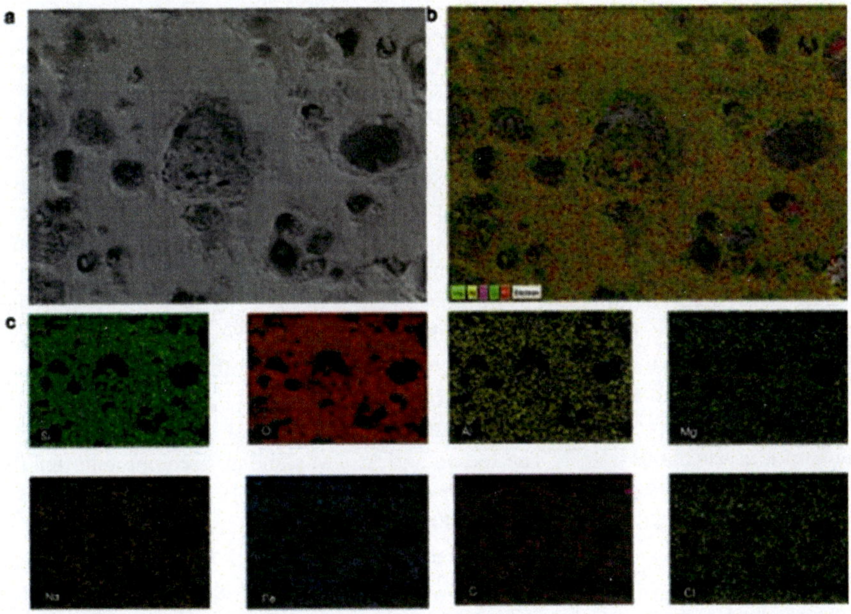

Fig. 6 Consolidated and individual elemental map of the specimen (FAR brick)

3.5 Phase Composition Analysis

The phase composition of the FAR brick specimen was analyzed using XRD. The XRD spectrum is shown in Fig. 7. The peak with maximum and second-maximum intensity corresponds to the mullite phase (* in Fig. 7) in the matrix. Besides, trace peaks corresponding to oxides of Ca, Na, Cr, Al, Si, K were found.

Fig. 7 Phase composition of the specimen (FAR brick)

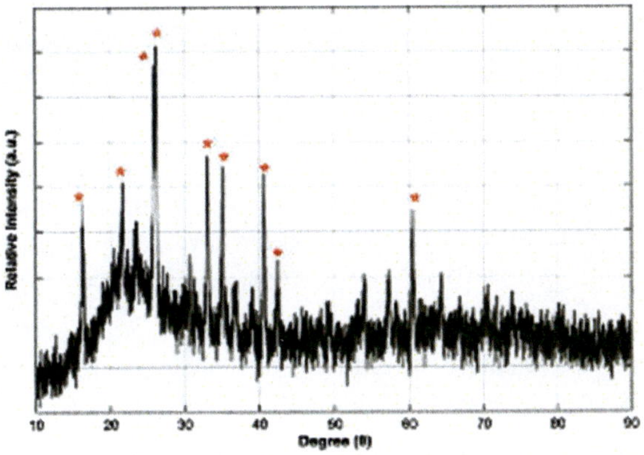

3.6 Physio-Mechanical characterization

The compression test was performed to determine the compressive strength of the FAR brick specimen. The deformation of the specimen against load was obtained. The average compressive strength of the specimen was 10 MPa. The compression strength was at par with the standard requirements [18]. The specimens experienced an average strain of 2.5% elongation during the compression test. The permanent linear change after reheating was determined by heating the FAR brick specimen at 1200 °C for 5 h. The dimensions of the refractory brick were measured before and after the heating cycle. The result indicated that there was no significant change in the dimensions of the specimens. Hence, it was concluded that the specimens had no linear change after reheating.

The water absorption test was performed on the FAR brick specimens. The water absorption was about 0.92%. The Indian and ASTM standards recommend that the water absorption should not exceed 15% of the dry weight [19]. The water absorbed was less than 1% in the fabricated FAR brick specimen. The absence of a continuous network of pores resulted in lesser water absorption in the FAR brick specimen. Also, the FAR brick specimen had a density of 1.7 g/cm^3. The bulk density was within the standard requirements of 1.7–2.3 g/cm^3. The pore fraction in the FAR brick specimen was calculated as 1.51%. The low porosity and absence of a continuous network of pores consequently improved the compressive strength of the specimen. The specific gravity of the FAR brick specimen was 1.71. The specific gravity of the specimen was lesser than the ASTM standard requirement. The results indicate that the developed FAR brick would be one of the potential refractory bricks.

4 Conclusion

A new and effective way to utilize fly ash was developed by fabricating a ceramic refractory brick. The refractory brick was fabricated by using 100% fly ash. The results demonstrated the following:

- The surface morphology confirmed the absence of networked pores.
- Discontinuous pore resulted in a density of 1.7 g/cm^3, reduced the water absorption (less than 1%), and enabled a good compressive strength of 10 MPa.
- The bulk density, specific gravity, water absorption, and porosity of the developed fly ash-based refractory brick were in line with the standard requirements of refractory brick.

The results demonstrated the development of promising technology for the fabrication of fly ash-based refractory brick.

References

1. Jagadeep R et al (2021) Mater Today Proc
2. Alex AJ et al (2020) Aust J Mech Eng 1
3. Govindaraju M et al (2020) Proc Inst Mech Eng Part C J Mech Eng Sci 234(12):2474
4. Guha Keshav M et al (2021) Mater Today: Proc 46:4493
5. Kannan KR et al (2019) Tribological performance of heavy-duty functionally gradient friction material (Cu–Sn–Fe–Cg–SiC–Al_2O_3) synthesized by PM route. In: Presented at International conference on materials, manufacturing and machining
6. Rajesh Kannan K et al (2020) Proc Inst Mech Eng Part J: J Eng Tribol 235(7):1463
7. Kusiorowski R (2020) Constr Build Mater 231
8. Liu X et al (2021) Ceram Int 47(6):8685
9. Shimizu T et al (2013) J Eur Ceram Soc 33(15–16):3429
10. Otero JG et al (2013) Br Ceram Trans 103(4):181
11. Kishor MSVR et al (2020) J Mater Eng Perform 29(10):6307
12. Chakraborty AK et al (2013) Adv Appl Ceram 106(4):196
13. Wang T et al (2020) Constr Build Mater 239
14. Baradaran-Nasiri A, Nematzadeh M (2017) Constr Build Mater 147:865
15. Zhang J et al (2007) J Hazard Mater 149(2):523
16. Ewais EMM et al (2017) Refract Ind Ceram 58(2):136
17. Abbas S et al (2017) J Build Eng 14:7
18. Standard test method for cold crushing strength (2015). https://doi.org/10.1520/c0133-97r15
19. Standard test methods for Refractory brick (2010). https://doi.org/10.1520/c0020-00r10

Turning Investigations of Al 7075 Alloy with ZrCN-Coated WC Inserts: Parametric Optimization and Cutting Temperature Prediction

Diptikanta Das, Santanu Kumar Dash, Roshan Kumar Rauniyar, Shah Md. Suaeb Ahemad, Ramanuj Kumar, and Chandrika Samal

1 Introduction

Distinctive amalgamation of properties, like high strength, reduced weight, high yield strength (≥500 MPa) [1], low density, and good thermal properties, along with excellent luster makes Al 7075 a versatile choice as a structural material in aviation, automobile, and marine industries [2]. Al 7075 is a strengthened wrought alloy with Zn, Cu, and Mg as major alloying elements, commonly referred as an aeronautical alloy. It is the strongest alloy among all Al alloys, and its strength is comparable to that of steel, but it weighs about 1/3rd of that of steel [3]. Machining of this alloy is very much essential for its structural application in various engineering sectors. Substitution of steel and cast iron with aluminum in automobile industries will lead to considerable reduction of vehicle weight and improvement of fuel economy [4]. Moreover, wide application of the Al alloys in aeronautical and automotive industries justifies their machining investigations [5–7]. But, the Al alloys provide machining difficulty due to the formation of long continuous chips, built-up edge (BUE), and built-up layer (BUL) on the tool [8, 9].

During high-speed milling of T6-conditioned Al 7075, Rao and Shin [10] reported generation of thin chips with large angle of shear, low thrust forces, and large chip flow angle. While investigating tool wear behavior during dry turning T6-treated UNS A97075 alloy with TiN-coated WC tools, Martín-Béjar et al. [11] reported increased BUL at higher feed and cutting depth levels. Okokpujie et al. [12] observed a reduced tool wear with increase of cutting speed while turning Al 1061 with high-speed steel (HSS) tool. While turning of Al 6061 alloy in dry and wet condition, Deepak and

D. Das (✉) · R. K. Rauniyar · S. Md. Suaeb Ahemad · R. Kumar
School of Mechanical Engineering, KIIT Deemed to Be University, Bhubaneswar 751024, India
e-mail: diptikantadas115@gmail.com

S. K. Dash · C. Samal
Department of Mechanical Engineering, GITA Autonomous College, Bhubaneswar 752054, India

© The Author(s), under exclusive license to Springer Nature Singapore Pte Ltd. 2023

S. Revankar et al. (eds.), *Recent Advances in Thermofluids and Manufacturing Engineering*, Lecture Notes in Mechanical Engineering,
https://doi.org/10.1007/978-981-19-4388-1_45

Rajendra [13] reported feed rate was more significant for the rate of material removal, taken after by cutting depth and cutting speed.

Cutting temperature is one of the significant machinability criteria, which directly affects life of the cutting tool, machined surface quality, and process economy [9, 14]. Tool-chip and tool-work interface frictions lead to high temperature generation in the machining zone, which needs to be optimized for different machining parameters. Higher heat generation during machining of Al alloys may lead to the risk of soft spot forming. Moreover, increased temperature at the tool-chip interface leads to higher tool wear, intensified residual stresses in the workpiece, and work-chip sticking [15]. While investigating the effect of chamfered cutting edges on heat generation during milling of Al-Li 2196 aerospace alloy, Denkena et al. [16] reported reduction of maximum workpiece temperature with the increase of feed per tooth. Palmai [17] reported an increase of cutting temperature on increasing the cutting speed up to a certain limit, but it reduced after attaining a maximum value for intermittent cutting. While turning Al 6063 alloy with WC inserts, Kannan et al. [14] observed minimum values of cutting temperature, cutting force, and surface roughness at 200 m/min of speed of cutting, 0.25 mm of cutting depth, and 0.05 mm rev^{-1} of feed rate. Patru et al. [18] compared the cutting temperature during milling of Al 6082, Al 5083, and Al 7075 at 2 mm of depth of cut, and observed maximum temperature (84.7 °C) for Al 6082, followed by that of Al 5083 (77.2 °C) and Al 7075 (75.6 °C). Vernaza-Pena et al. [19] used HgCdTe infrared detectors for the measurement of temperature distribution on the workpiece surface during orthogonal cutting of T6-Al 6061 alloy using D2 tool steel as cutting tool material through a modified Hopkinson bar technique at a speed of 15 m/s. It was reported that higher values of cutting depth increased the maximum cutting temperature. The temperature values were 138 °C, 207 °C, and 238 °C for cutting depths of 0.5 mm, 1.0 mm, and 1.5 mm, respectively. During turning of LM28 and LM13 alloys in dry condition, Dwivedi et al. [20] observed increased cutting temperature for LM28, which was due to its more silicon content, higher hardness, and lower value of thermal conductivity. Kishawy et al. [21] reported increased cutting temperature with increase of cutting speed during milling of A 356 alloy. Machining in dry environment causes a potential rise of temperature at the machining zone, leading to a substantial decrease in tool life and surface quality [22]. Tool wear, surface finish, and dimensional accuracy of the workpiece are directly influenced by the heat generation and its distribution in the cutting zone [23].

From open literature, it is evident that machining of Al alloys is essential for their potential application in automotive and aerospace industries. Though many literatures are available on study of machining characteristics of various Al alloys, further systematic investigation of cutting temperature during machining of the aerospace alloy Al 7075 is required to meet the challenges of tool wear, surface quality, and machining economy. Accordingly, this paper outlines an analysis on cutting temperature during turning Al 7075 alloy using ZrCN-coated WC tools in dry cutting condition, following L_{16} Taguchi orthogonal array. Then, the response was optimized for various machining parameters through Taguchi's signal-to-noise ratio approach, and then, regression model was developed for temperature of cutting tool.

2 Experimental Methodology

Al 7075 alloy heat treated to T6 condition was utilized as workpiece material for the turning trials. The alloy was procured from Bharat Aerospace Metals, Mumbai, and its chemical constituents by wt.% are Si: 0.1, Fe: 0.21, Cu: 1.5, Mn: 0.12, Mg: 2.0, Zn: 4.6, Ti: 0.04, Cr: 0.24, and the rest is Al. Density of the alloy was measured by conventional method of weight-to-volume ratio. A piece of material of previously measured weight was immersed in known volume of water in a measuring cylinder, and the difference in water level was noted. Hence, the density of material was obtained by dividing weight of alloy sample by volume of water displaced. The process was repeated for three times, and average density was calculated, which was found to be 2.81 gm/cm^3. Hardness of the alloy sample was measured by technomeasure Rockwell hardness tester, and it was 87.33 HRB. Ultimate strength (tensile), yield strength, and Young's modulus of the alloy were 530 MPa, 450 MPa, and 71 GPa, respectively. The turning workpiece was of dimension length, $l = 200$ mm and diameter, $D = 55$ mm. WIDIA makes TiN/TiCN/Al_2O_3/ZrCN-coated WC tools of ISO specification CNMG 120408 (80° rhombic) were utilized for assessment of their turning performance in dry environment. The cutting inserts were rigidly mounted in PCLNR 2525M12 right-hand tool holder during turning. HMT-NH 22 high-speed precision lathe was utilized to conduct the turning experiments. During machining, the workpiece was rigidly backed up by tailstock to reduce the vibration effects.

Cutting tool temperature (T) during turning was quantified by Fluke make Ti-32 infrared (IR) thermal imager, having temperature measurement range of −20 to 600 °C at accuracy ±2 °C and frequency 60 Hz. Each measurement was repeated for three times, and then, average of the measured values was utilized for data analysis. Figure 1a, b represents images of the workpiece for turning and the cutting tool mounted in the tool holder, whereas Fig. 2 portrays the experimental setups for cutting temperature measurement turning. Turning experiments were conducted with three machining parameters, associated with four levels to each parameter, as demonstrated in Table 1. Taguchi L_{16} standard orthogonal array (OA) was followed to design the experiments. Table 2 shows the design of experiments (DOE) as per Taguchi L_{16} OA.

3 Discussion of Experimental Results

3.1 *Influence of Cutting Parameters*

Results for cutting tool temperature (T) during machining T6-treated Al 7075 alloy with ZrCN-coated WC inserts at dry condition are presented in Table 3. T was minimum (43 °C) for Trial 1 and was maximum (115.93 °C) for Trial 13. A consistent rise of T was observed from Trial 1 to Trial 4, where spindle speed was constant (250 rpm), and both feed and cutting depth were increased. T increased by around 93% on increasing feed from 0.05 to 0.2 mm rev^{-1}, and cutting depth from 0.2 to

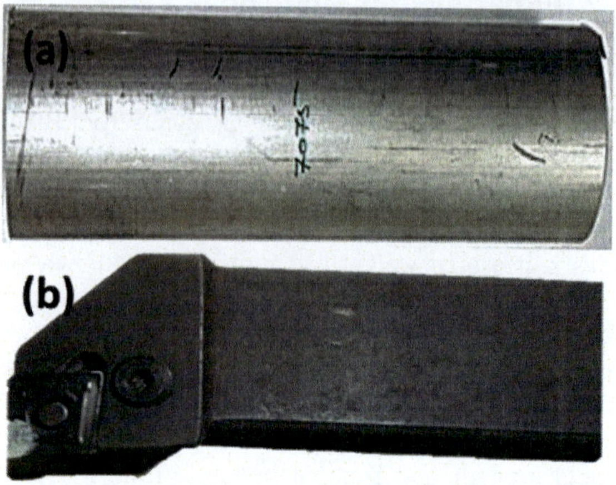

Fig. 1 Images of **a** Al 7075 workpiece for turning and **b** Cutting tool with tool holder

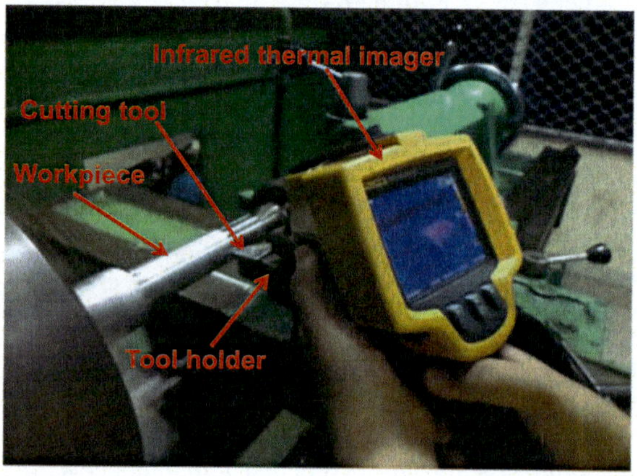

Fig. 2 Experimental setup for cutting temperature measurement during turning

Table 1 Machining parameter and levels

Parameters with unit	Notation	Levels of parameters			
		1	2	3	4
Spindle speed, rpm	*N*	250	715	1210	1575
Feed, mm/rev	*f*	0.05	0.1	0.16	0.2
Depth of cut, mm	*d*	0.2	0.3	0.4	0.5

0.5 mm. While T for Run 1 and Run 5 is compared, an increase of about 66% is observed for a consistent feed of 0.05 mm rev^{-1} and on increasing cutting depth from 0.2 to 0.3 mm, and spindle speed from 250 to 715 rpm. Response table for mean T (Table 4) reveals maximum delta value for d (Rank 1), taken after by N (Rank 2) and f (Rank 3). It implies that depth of cut (d) was the greatest influencing parameter for T, followed by speed (N) and feed (f). Main effects plot for mean T

Table 2 Design of experiments for turning trials

Trial No.	*N*	*f*	*d*
1	250	0.05	0.2
2	250	0.1	0.3
3	250	0.16	0.4
4	250	0.2	0.5
5	715	0.05	0.3
6	715	0.1	0.2
7	715	0.16	0.5
8	715	0.2	0.4
9	1210	0.05	0.4
10	1210	0.1	0.5
11	1210	0.16	0.2
12	1210	0.2	0.3
13	1575	0.05	0.5
14	1575	0.1	0.4
15	1575	0.16	0.3
16	1575	0.2	0.2

(Fig. 3) depicts an increase of T on enhancing either of the turning variables under consideration. Increase in spindle speed causes the workpiece to rotate at faster rpm, which in turn increases magnitude of tangential component of cutting force (P_z) on engagement of the tool with workpiece. It leads to heat generation at the tool-work interface zone, thus by rising the tool temperature. On increasing the feed, axial force component (P_x) and work-tool interface friction increases proportionately, leading to a more tool temperature. Similarly, on increasing the cutting depth, the cutting tool penetrates more into the workpiece in radial direction, causing more friction and increased magnitude of radial component of cutting force (P_y) that increases the tool temperature. Infrared thermal images of the cutting tools during dry machining the alloy are portrayed in Fig. 4a–d, for some experimental trials, as illustration.

3.2 *Optimization of Response*

Signal-to-noise ratio (SNR) is the power ratio between meaningful data and the background noise, having decibels (dB) as its unit. Better system performance is represented by the higher values of SNR, and its highest value constitutes optimal parametric combination for the desired response. In the present experiment, the cutting tool temperature (T) was optimized through the SNR concept of Taguchi method. The SNR values of T were evaluated considering "Lower the better" criterion through MINITAB 17 software, and these are included in Table 3. From the response

Table 3 Experimental results and signal-to-noise ratio for *T*

Trial No.	*N*	*f*	*d*	*T* (°C)	SNR of T (dB)
1	250	0.05	0.2	43.00	−32.6694
2	250	0.1	0.3	61.40	−35.7634
3	250	0.16	0.4	79.57	−38.015
4	250	0.2	0.5	83.10	−38.392
5	715	0.05	0.3	71.51	−37.0873
6	715	0.1	0.2	65.83	−36.3685
7	715	0.16	0.5	104.53	−40.3848
8	715	0.2	0.4	98.67	−39.8837
9	1210	0.05	0.4	96.97	−39.7327
10	1210	0.1	0.5	95.10	−39.5636
11	1210	0.16	0.2	65.63	−36.342
12	1210	0.2	0.3	94.00	−39.4626
13	1575	0.05	0.5	115.93	−41.2839
14	1575	0.1	0.4	101.70	−40.1464
15	1575	0.16	0.3	101.37	−40.1182
16	1575	0.2	0.2	80.63	−38.1299

Table 4 Response table for means of *T*

Level	*N*	*f*	*d*
1	66.7675	81.8525	63.7725
2	85.1350	81.0075	82.0700
3	87.9250	87.7750	94.2275
4	99.9075	89.1000	99.6650
Delta	33.1400	8.0925	35.8925
Rank	2	3	1

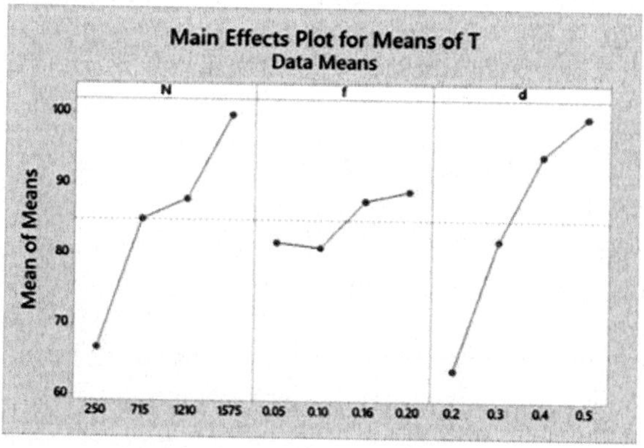

Fig. 3 Main effects plot for means of *T*

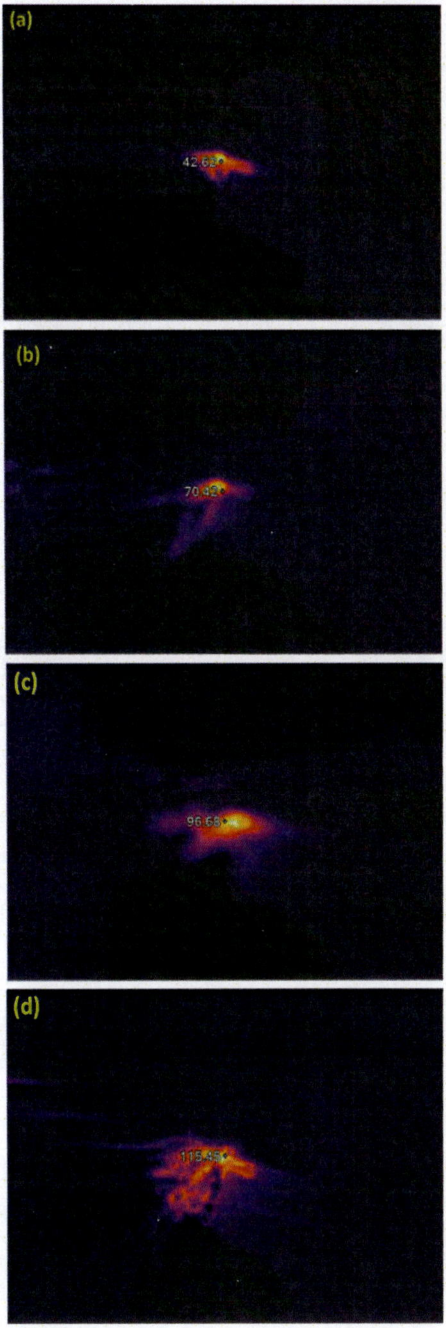

Fig. 4 Infrared thermal images of cutting tools during temperature measurement in **a** Trial 1; **b** Trial 5; **c** Trial 9; and **d** Trial 13

table (Table 5) and the main effects plot (Fig. 5), it is noticed that the SNR reduced consistently on increasing the levels of all the three parameters, and it was maximum for the 1st levels of speed, feed, and cutting depth, i.e., N_1-f_1-d_1. This combination can be regarded as the optimal parametric combination for T during dry turning of T6-conditioned Al 7075 alloy using ZrCN-coated WC inserts.

Significance of process parameters on the T was evaluated through analysis of variance at a confidence level of 95%, and the analyzed data is presented in Table 6. From the tabular data equal values (0.001) of probability of significance (P-value) were observed for spindle speed and depth of cut, which were less than 5%. It implies equal significance of both the parameters on T. However, contributions of these parameters on T were 39.55% and 53.33%, for spindle speed and cutting depth, respectively. Moreover, the statistical fisher consistency (F-value) for N and d are significantly more than the corresponding tabular value (4.76) [24] at confidence level of 95%. As the P-value for f is more than 5% and its statistical F-value is less as compared to its tabulated value, effect of feed was not significant on T.

Confirmation experiments were accomplished to predict and substantiate the enhancement of the performance characteristic utilizing the optimal combination of process parameters. The predicted SNR ($\hat{\gamma}$) was calculated using Eq. 1 for optimal parametric combinations.

Table 5 Response table for SNR T

Level	N	f	d
1	−36.2099	−37.6933	−35.8775
2	−38.4311	−37.9605	−38.1079
3	−38.7752	−38.7150	−39.4445
4	−39.9196	−38.9671	−39.9061
Delta	3.7097	1.2737	4.0286
Rank	2	3	1

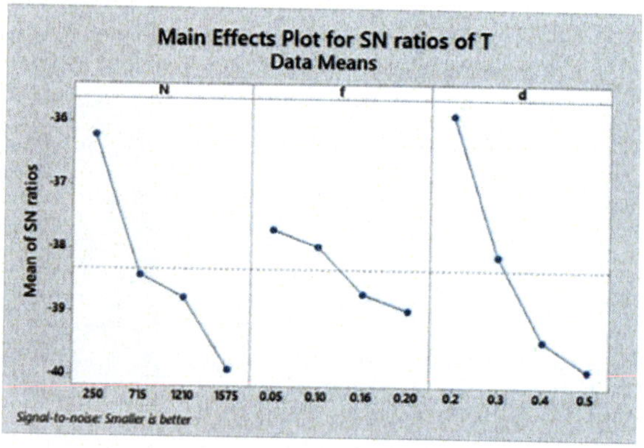

Fig. 5 Main effects plot for SNR T

Table 6 ANOVA for *T*

Source	DF	Adj SS	Adj MS	*F*-Value	*P*-Value	Contribution (%)
N	3	2252.86	750.95	22.07	0.001	39.55
f	3	201.36	67.12	1.97	0.220	3.54
d	3	3037.53	1012.51	29.76	0.001	53.33
Error	6	204.11	34.02			3.58
Total	15	5695.86				100

Table 7 Confirmation experimental results for *T*

Level	Initial process parameters	Optimal process parameters	
		Predicted	Experimental
	N_2-f_2-d_2	N_1-f_1-d_1	N_1-f_1-d_1
T (°C)	76.82		43
SNR for T (dB)	−37.7095	−33.1128	−32.6694

Improvement in SNR for $T = 5.0401$ (dB)

$$\hat{\gamma} = \gamma_m + \sum_{i=1}^{o} \left(\overline{\gamma}_i - \gamma_m\right) \tag{1}$$

where total mean SNR is γ_m, mean of SNR at optimal level is $\overline{\gamma}_i$, and number of design parameters affecting the performance characteristics is o. Confirmation experiment results for T using the optimal and initial process parameters are reported in Table 7. Good accordance existed between the predicted and optimal values. Improvement of SNR from the initial to the optimal parametric combination was 5.0401 dB.

3.3 Regression Model

Second-order regression model for T (Eq. 2) was obtained through response surface method (RSM) at 95% confidence level. The model presented was with high determination coefficient (R^2) near to 100%, and with rational concurrence with the adjusted R^2 value, which designates the righteousness of fit for the models and their high significance. Also, from normal probability (NP) plots of residuals (Fig. 6a), it is observed that the residuals fit fairly near to the normal probability line implicating that errors are normally dispersed, and the terms stated in the models are significant. Moreover, the residuals versus fits (RVF) plot (Fig. 6b) indicate a reasonable scatter of the residuals from the mean, mentioning the significance of the model.

$$\begin{aligned} T = & -7.4 + 0.0269 * N + 1 * f + 273 * d - 0.000007 * N^2 + 387 * f^2 \\ & - 321 * d^2 - 0.053 * N * f + 0.0462 * N * d + 133 * f * d \end{aligned}$$

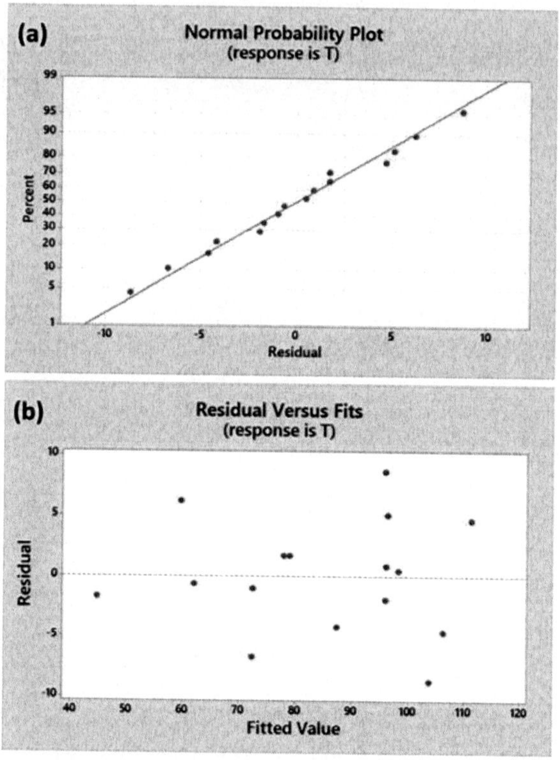

Fig. 6 **a** Normal probability plot of residuals; and **b** Residuals versus fits plot for T

$$R^2 = 94.11\%,\ R^2(\text{adj}) = 85.26\%,\ R^2(\text{pred}) = 43.12\% \quad (2)$$

In summary, during dry condition turning of T6-treated Al 7075 alloy with ZrCN-coated WC inserts, cutting tool temperature was minimum for the lowest levels of machining parameters under consideration. T was minimum (43 °C) for Trial 1 of the Taguchi L_{16} DOE, i.e., at the spindle speed 250 rpm, feed 0.05 mm rev^{-1}, and cutting depth 0.2 mm.

- T was optimized through SNR maximization method of Taguchi technique, which revealed N_1-f_1-d_1 was the optimal parametric combination, i.e., at 250 rpm of speed, 0.05 mm rev^{-1} of feed and 0.2 mm of cutting depth. Confirmation experiment results divulged a good concurrence between the predicted and experimental values, and the SNR for optimal parameters improved by 5.0401 dB than the initial parameters.
- Results of ANOVA at 95% confidence level divulged equal P-values for both speed and cutting depth, which were also less than 5%, mentioning both were equally significant for T. But, contribution of spindle speed on T was 39.55%, and that of depth of cut was 53.33%. However, effect of feed on T was not significant.

- Second-order regression model was generated for T using RSM. The model presented high determination coefficient (94.11%) with a reasonable agreement with its adjusted value. Residuals were close to normal probability line in the NP plots of residuals and were scattered reasonably from mean in the RVF plot, revealing good fitness, and high significance of the models.

4 Conclusions

The present work involves investigation of cutting temperature during turning Al 7075 alloy with ZrCN-coated WC tools in dry environment through L_{16} Taguchi orthogonal array, followed by optimization and regression analysis. However, the machining experiments have not been conducted in minimum quantity lubrication (MQL) or flood cooling environment, where the cutting temperature is expected to reduce compared to that in dry environment. So, a scope for this work is to investigate the cutting temperature during turning in MQL and flood cooling environment. Moreover, full factorial design may be adopted for experimentation to get more précised optimal results and regression model.

References

1. Davis JR (1993) Aluminum and aluminum alloys. ASM International, pp 351–416.https://doi.org/10.1361/autb2001p351
2. Imran M, Khan AA (2019) Characterization of Al-7075 metal matrix composites: a review. J Market Res 8(3):3347–3356
3. Web link 1. https://www.aluminiumleader.com/application/transport/. Accessed on 16 Oct 2021
4. Miller WS, Zhuang L, Bottema J, Wittebrood AJ, De Smet P, Haszler A, Vieregge A (2000) Recent development in aluminium alloys for the automotive industry. Mater Sci Eng 280:37–49
5. Santos MC, Machado AR, Barrozo MA (2018) Temperature in machining of aluminum alloys, temperature sensing. Intech Open, pp 71–89. https://doi.org/10.5772/intechopen.75943
6. Hovsepian PE, Luo Q, Robinson G, Pittman M, Howarth M, Doerwald D, Tietema R, Sim WM, Deeming A, Zeus T (2006) TiAlN/VN superlattice structured PVD coatings: a new alternative in machining of aluminium alloys for aerospace and automotive components. Surf Coat Technol 201:265–272
7. Demir H, Gündüz S (2009) The effects of aging on machinability of 6061 aluminium alloy. J Mater Des 30:1480–1483
8. Gangopadhyay S, Acharya R, Chattopadhyay AK, Sargade VG (2010) Effect of cutting speed and surface chemistry of cutting tools on the formation of BUL or BUE and surface quality of the generated surface in dry turning of AA6005 aluminium alloy. Mach Sci Technol 14(2):208–223
9. Soren TR, Kumar R, Panigrahi I, Sahoo AK, Panda A, Das RK (2019) Machinability behavior of aluminium alloys: a brief study. Mater Today: Proc 18:5069–5075
10. Rao B, Shin YC (2001) Analysis on high-speed face-milling of 7075–T6 aluminum using carbide and diamond cutters. Int J Mach Tools Manuf 41(12):1763–1781
11. Martín-Béjar S, Trujillo FJ, Sevilla L, Marcos M (2017) Indirect adhesion wear parametric analysis in the dry turning of UNS A97075 alloys. Proc Manuf 13:418–425

12. Okokpujie IP, Ohunakin OS, Bolu CA, Okokpujie KO (2018) Experimental data-set for prediction of tool wear during turning of Al-1061 alloy by high speed steel cutting tools. Data Brief 18:1196–1203
13. Deepak D, Rajendra B (2015) Studies on material removal rate of Al 6061 while turning with coolant and without coolant using Taguchi method. Int J Res Eng Technol 4(09):75–78
14. Kannan A, Esakkiraja K, Nataraj M (2013) Modeling and analysis for cutting temperature in turning of aluminium 6063 using response surface methodology. IOSR J Mech Civil Eng 9(4):59–64
15. Chandrasekhar P, Chand S, Sarangi RK, Kar SP, Swain A (2021) Thermal analysis of Al 7075-T651 during high-speed machining. In: Proceedings of international conference on thermofluids. Springer, Singapore, pp 643–653. https://doi.org/10.1007/978-981-15-7831-1_60
16. Denkena B, Brüning J, Niederwestberg D, Grabowski R (2016) Influence of machining parameters on heat generation during milling of aluminum alloys. Proc CIRP 46:39–42
17. Palmai Z (1987) Cutting temperature in intermittent cutting. Periodica Polytechnica Mech Eng 31(1):61–78
18. Patru EN, Craciunoiu N, Panduru D, Bica M (2018) Study on cutting temperature and surface roughness during the milling process of aluminium alloys. Earth Environ Sci 172(1):0120181–0120188
19. Vernaza-Pena KM, Mason JJ, Li M (2002) Experimental study of the temperature field generated during orthogonal machining of an aluminum alloy. Exp Mech 42(2):221–229
20. Dwivedi DK, Sharma A, Rajan TV (2008) Machining of LM13 and LM28 Cast a aluminium alloys: part I. J Mater Process Technol 196:197–204
21. Kishawy HA, Dumitrescu M, Ng EG, Elbestawi MA (2005) Effect of coolant strategy on tool performance, chip morphology and surface quality during high-speed machining of A356 aluminum alloy. Int J Mach Tools Manuf 45:219–227
22. Sharma SK, Panda A, Kumar R, Sahoo AK, Routara BC (2021) Embedded heat pipe-assisted cooling in machining process: a comprehensive review. In: Proceedings of international conference on thermofluids. Springer, Singapore, pp 665–676
23. Kumar R, Panda A, Sahoo AK, Singhal D (2021) Analysis of heat transfer coefficient in turning process. In: Proceedings of international conference on thermofluids. Springer, Singapore, pp 655–663. https://doi.org/10.1007/978-981-15-7831-1_61
24. Kothari CR (2012) Research methodology. Second revised edition: Reprint, New Age International Publishers

Influence of Roughness on Micropolar Lubricated Finite Porous Journal Bearing

Ujjal Baidya, Sanjoy Das, and Santanu Das

Notations

c	Bearing clearance in radial direction
$l, \bar{l}$	Formal film thickness, $\bar{l} = {}^{l}/_{c}$
l_s	Fluid film thickness in stochastic condition
L_m	Dimensionless micro-additives characteristics length, $L_m = c/\Lambda$
$\delta, \overline{\delta}$	Hydrodynamic fluid film thickness, $\overline{\delta} = \delta/c$
Ω_j	Porous matrix permeability, $j = x, y, z$
Π_i	Non-dimensional porous matrix permeability, $\Pi_i = \Omega_i/\Omega_y$ where $i = \theta, \overline{z}$
$\overline{\sigma}_k$	Non-dimensional mean, standard deviation and skewness roughness pattern, where $k = M, SD, SK$
α	Bearing slip coefficient
β	Feeding parameter of porous bearing
ω, γ	Micropolar fluid viscosity parameter
μ	Viscosity parameter of Newtonian fluid
ν	Journal angular velocity
Λ	Characteristics length of the micro-additives
ϕ_0	Attitude angle in steady state condition

U. Baidya (✉)
Department of Mechanical Engineering, Medinipur Sadar Government Polytechnic College, Khasjungle, Medinipur, India
e-mail: baidya.ujjal@gmail.com

Kalyani Governtment Engineering College, Kalyani 741235, India

S. Das
Department of Engineering and Technological Studies, Kalyani University, Kalyani, India

S. Das
Department of Mechanical Engineering, Kalyani Government Engineering College, Kalyani, India

© The Author(s), under exclusive license to Springer Nature Singapore Pte Ltd. 2023

S. Revankar et al. (eds.), *Recent Advances in Thermofluids and Manufacturing Engineering*, Lecture Notes in Mechanical Engineering,
https://doi.org/10.1007/978-981-19-4388-1_46

1 Introduction

Surface roughness plays an important role in any moving parts of different kinds of plants and machines to influence friction characteristics as well as related loss of energy. The effect of friction has been reported to have resulted in about 22.5% loss of power of an engine [1], and approximately 1% loss of GDP of a developed country [2].

Effect of 2D Sine curve type of roughness was first tried [3] to explore on lubrication. Later on, a model considering stochastic nature of isotropic roughness was introduced [4]. A novel consideration was also made [5] to explore its effect on partially lubricated surface of hydrodynamic type sleeve journal bearing.

In a journal bearing, when the bush is having porosity, it serves as to supply fluid when necessary. This fluid supply interacts with hydrodynamic fluid film generating occurrence of slip [6–8]. Stochastic roughness model [4] was also utilized [9, 10] to model the nature of porous hydrodynamic journal bearing. Another work was done [11] on isotropic random nature of roughness present in hydrodynamic porous journal bearing with different slip conditions. Up to this, research works were confined mainly on the lubricants following Newton's law.

Moving parts operating at a high velocity involve large friction, subsequent heat generation and wearing. Particles coming out of wearing get distributed within lubricating fluid; temperature rise lowers fluid viscosity and hence, the lubricating effect. These phenomena are not possible to be explained by classical theories [12]. In a recent development, long chain carbon polymer has been put inside the prime lubricating fluid and is used effectively in high speed mechanisms. Polymers added are subject to linear as well as angular (spin) motion while in operation within moving bodies. This type of embedded lubricants is called micropolar lubricant or fluid. Eringen [13, 14] presented the complete theory of these suspended lubricant particles. A number of investigations was performed on micropolar lubrication [15, 16], and micropolar lubrication in bearing was reported to have superior performance over Newtonian counterpart. Change in viscous nature of fluid having some additives in it while moving around rough bearing surface resulted in [17] improvement in load bearing capability as well as lowering of friction parameter. Another recent work [18] noted that steady state bearing pressure was reduced by increasing values of roughness characteristics.

Till today, these authors could not find out any work done on application of statistics to explore effect of roughness and slip effect occurring in finite porous hydrodynamic journal bearing with the use of micropolar lubricating fluid. Therefore, this objective of this work is drawn up so as to explore applicability of stochastic nature of surface roughness while carrying out analysis on finite porous hydrodynamic journal bearing at steady state situation using micropolar lubricating condition using some statistics.

2 Analytical Method

Bearing system considered in this work is depicted in Fig. 1. Lubricant used in porous hydrodynamic journal bearing is assumed to be iso-viscous, fluid flow is taken laminar and mean size of additives is taken bigger than average size of uniformly distributed pores. Inertia of lubricating fluid in porous bush is not considered. Darcy's as well as Continuity equations are combined to make governing Eq. (1):

$$\Omega_x \frac{\partial^2 \dot{p}}{\partial x^2} + \Omega_y \frac{\partial^2 \dot{p}}{\partial y^2} + \Omega_z \frac{\partial^2 \dot{p}}{\partial z^2} = 0 \quad (1)$$

where Ω_i and $\dot{p}$ are the permeability and pressure in porous matrix, $i = x, y$ and z directions, respectively.

Following the basic assumptions related to micropolar lubrication, Reynolds equation [19] for finite sleeve journal bearing within bearing clearance area can be expressed as given in Eq. (2).

$$\frac{\partial}{\partial x}\left[\frac{\delta^3}{\mu}\pi(\delta, \Lambda, N)\frac{\partial p}{\partial x}\right] + \frac{\partial}{\partial z}\left[\frac{\delta^3}{\mu}\pi(\delta, \Lambda, N)\frac{\partial p}{\partial z}\right] = 6\frac{\partial}{\partial x}(u\delta) + 12\frac{\partial \delta}{\partial t} \quad (2)$$

where

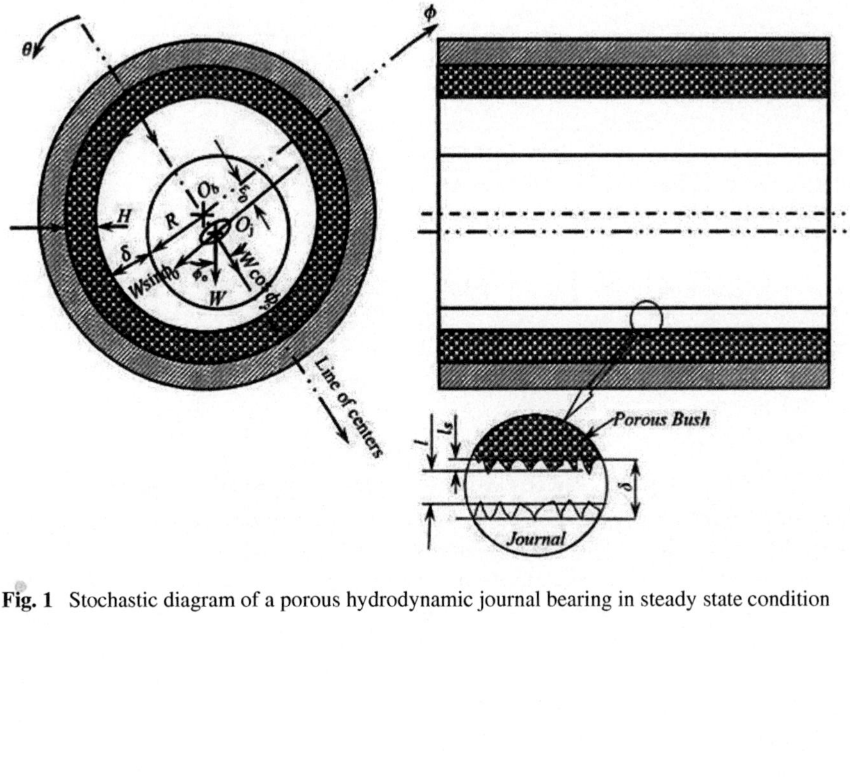

Fig. 1 Stochastic diagram of a porous hydrodynamic journal bearing in steady state condition

$$\pi(\delta, \Lambda, N) = \left\{1 + 12\frac{\Lambda^2}{\delta^2} - \frac{6N\Lambda}{\delta}\coth\left(\frac{N\delta}{2\Lambda}\right)\right\}$$

in which

$$N = \left(\frac{\omega}{2\mu + \omega}\right)^{1/2} \quad \text{and} \quad \Lambda = \left(\frac{\gamma}{4\mu}\right)^{1/2}.$$

Linear Newtonian viscosity of the substrate lubricating fluid is denoted as μ. ω and γ are external microrotational viscosity characteristics of micropolar fluid. N is coupling number, an indexed dimensionless value that combines linear and angular momentum relationships considering spinning of suspended particles. Λ is length characteristic related to average molecular size of suspended micro particles and δ the gap within the journal and porous bush. Had linear and angular momentum been separated, ω and N may approach 0 value making Eq. (2) becoming classical Newton's equation.

Making assumptions of porous surface and micropolar lubrication, Beavers-Joseph criterion of velocity slip within the interface of lubrication film boundary and porous bush are used to modify Reynold's equation [15] as in Eq. (3).

$$\begin{aligned}&\frac{\partial}{\partial x}\left[\frac{\delta^3}{\mu}\psi(\delta, S_i, \Lambda, N, \alpha)\frac{\partial p}{\partial x}\right] + \frac{\partial}{\partial z}\left[\frac{\delta^3}{\mu}\psi(\delta, S_i, \Lambda, N, \alpha)\frac{\partial p}{\partial z}\right] \\ &= 12\mu\left.\frac{\partial \delta}{\partial t}\right|_r + 6\mu\frac{\partial}{\partial x}(u\delta\tau_{0x}) - 12\mu\frac{\partial\varphi}{\partial t}\frac{\partial\delta}{\partial x} + \left.\frac{12\Omega_y}{\mu}\frac{\partial \dot{p}}{\partial y}\right|_{y=0} \end{aligned} \tag{3}$$

where

$$\begin{aligned}\psi(\delta, S_i, \Lambda, N, \alpha) = &\left[\delta^3 6\Lambda\delta N\left\{\delta\coth\frac{N\delta}{2\Lambda} - \frac{2\Lambda}{N}\right\} + 3\delta S_i\delta(1 - N^2) + 2\alpha^2 S_i\right. \\ &\left.\cdot\left\{1 - \frac{(1 - N^2)}{\delta S_i} - \frac{2N\Lambda(1 - N^2)}{\delta^2 S_i}\tanh\left(\frac{N\delta}{2\Lambda}\right)\right\}\right]\end{aligned}$$

$$\tau_{0x} = \left[\left\{1 - \frac{(1 - N^2)}{\delta S_i} - \frac{2N\Lambda(1 - N^2)}{\delta^2 S_i}\tanh\left(\frac{N\delta}{2\Lambda}\right)\right\}\right]$$

and

Randomized roughness can be classified as isotropic, longitudinal and transverse. Resistance to flow of fluid under hydrodynamic system within bearing housing has been affected the least with longitudinal roughness. Transverse roughness largely resists fluid flow as well as induces pressure during bearing operation. Isotropic roughness having considered enough influence is incorporated in this work along with stochastic approach of surface roughness of the bearing. The journal is considered to rotate at an angular velocity ω. Hydrodynamic film thickness, δ along with randomized roughness is considered to remodel the relations as below:

$$\delta = l(\theta) + l_s(\sigma_M, \sigma_{SD}, \sigma_{SK}) \tag{4}$$

where $l(\theta) = 1 + \varepsilon_0 \cos\theta$.

When, $l(\theta)$ and l_s are mean and stochastic film thickness of hydrodynamic fluid film while ε_0 is eccentricity of journal bearing.

Stochastic film thickness has probability density function distributing over a range $[-M_s \le l_s \le +M_s]$ when M_s is the largest variance from mean thickness of fluid film. Randomized variation of stochastic film thickness is represented as Mean (σ_M), Standard Deviation (σ_{SD}) and Skewness (σ_{SK}) that are given by:

$$\left.\begin{array}{rl} \text{Mean,} & \sigma_M = \mathcal{E}(l_s) \\ \text{Standard Deviation,} & \sigma_{SD} = \sqrt{\mathcal{E}(l_s - \sigma_M)^2} \\ \text{Skewness,} & \sigma_{SK} = \mathcal{E}(l_s - \sigma_M)^3 \end{array}\right\} \tag{5}$$

where $\mathcal{E}()$ is the expectancy operator given as Eq. (6).

$$\mathcal{E}[F(x)] = \int_{-\infty}^{+\infty} F(x) l_s dl_s = \int_{-M_s}^{+M_s} F(x) l_s dl_s \tag{6}$$

Equations (1) and (3) are multiplied by $F(l_s)$ and they are integrated over variable l_s within the range of $-M_s$ to $+M_s$ to be modified as Eqs. (7) and (8).

Within porous bush,

$$\Omega_x \frac{\partial^2 \mathcal{E}(\dot{p})}{\partial x^2} + \Omega_y \frac{\partial^2 \mathcal{E}(\dot{p})}{\partial y^2} + \Omega_z \frac{\partial^2 \mathcal{E}(\dot{p})}{\partial z^2} = 0 \tag{7}$$

Within bearing clearance,

$$\begin{aligned} &\frac{\partial}{\partial x}\left[\frac{\delta^3}{\mu}\xi(\delta, S_x, \Lambda, N, \alpha)\frac{\partial \mathcal{E}(p)}{\partial x}\right] + \frac{\partial}{\partial z}\left[\frac{\delta^3}{\mu}\xi(\delta, S_z, \Lambda, N, \alpha)\frac{\partial \mathcal{E}(p)}{\partial z}\right] \\ &+ 6\mu.u\frac{\partial}{\partial x}(\eta_{0x}) - 12\mu\frac{\partial \varphi}{\partial t}\frac{\partial \mathcal{E}(\delta)}{\partial x} + \frac{12\Omega_y}{\mu}\frac{\partial \mathcal{E}(\dot{p})}{\partial y}\bigg|_{y=0} \end{aligned} \tag{8}$$

where

$$\xi(\delta, S_x, \Lambda, N, \alpha) = \mathcal{E}[\psi(\delta, S_i, \Lambda, N, \alpha)] = \int_{-M_s}^{+M_s} \psi(\delta, S_i, \Lambda, N, \alpha) F(l_s) dl_s$$

$$= \int_{-M_s}^{+M_s} \left[\delta^3 - 6\Lambda\delta N\left\{\delta \coth\left(\frac{N\delta}{2\Lambda}\right) - \frac{2\Lambda}{N}\right\} + \frac{3\delta}{S_i}\left\{\delta(1 - N^2) + \frac{2\alpha^2}{S_i}\right\}.\right.$$

$$\left\{1-\frac{(1-N^2)}{\delta S_i}-\frac{2N\Lambda(1-N^2)}{\delta^2 S_i}\tanh\left(\frac{N\delta}{2\Lambda}\right)\right\}\Bigg]F(l_s)dl_s$$

and,

$$\eta_{0x}=\mathcal{E}(\delta\tau_{0x})=\int_{-M_s}^{+M_s}\left[\delta\left\{1-\frac{(1-N^2)}{\delta S_i}-\frac{2N\Lambda(1-N^2)}{\delta^2 S_i}\tanh\left(\frac{N\delta}{2\Lambda}\right)\right\}\right]F(l_s)dl_s$$

in which, $i = x, z$.

Equations (7) and (8) are made in dimensionless form from Laplace equation to apply to porous bush and within clearance zone with following substitutions:

$x = R\theta, \overline{z} = \frac{2z}{L}, \overline{l} = \frac{\delta}{c}, L_m = \frac{c}{\Lambda}, \overline{y} = \frac{y}{H}, \overline{p} = \frac{pc^2}{\mu\chi R^2}, \overline{\dot{p}} = \frac{pc^2}{\mu\chi R^2}.$

$\Pi_\theta = \frac{\Omega_x}{\Omega_y}, \Pi_{\overline{z}} = \frac{\Omega_z}{\Omega_y}\ \Gamma = \chi t, \overline{\sigma}_M = \frac{\sigma_M}{c}, \overline{\sigma}_{SD} = \frac{\sigma_{SD}}{c^2}, \overline{\sigma}_{SK} = \frac{\sigma_{SK}}{c^3}.$

$\overline{S}_i = \alpha\gamma_i = S_i c = \frac{\alpha}{\sqrt{\Omega_i}}$, where, $i = x, z$ and $\beta = \frac{12\Omega_y R^2}{c^3 H}$.

Accordingly, steady state dimensionless form of Eq. (7) becomes:

$$\Pi_\theta\frac{\partial^2\overline{\dot{p}}}{\partial\theta^2}+\left(\frac{R}{H}\right)^2\frac{\partial^2\overline{\dot{p}}}{\partial\overline{y}^2}+\Pi_{\overline{z}}\left(\frac{D}{L}\right)^2\frac{\partial^2\overline{\dot{p}}}{\partial\overline{z}^2}=0 \tag{9}$$

Similarly, dimensionless form of Eq. (8) for a finite width of journal bearing becomes:

$$\begin{aligned}&\frac{\partial}{\partial\theta}\left[\frac{l^3}{\mu}\overline{\xi}(\overline{l},\overline{S}_\theta,L_m,N,\alpha,\overline{\sigma}_M,\overline{\sigma}_{SD},\overline{\sigma}_{SK})\frac{\partial\overline{p}}{\partial\theta}\right]\\&+\frac{\partial}{\partial\overline{z}}\left[\frac{l^3}{\mu}\overline{\xi}(\overline{l},\overline{S}_z,L_m,N,\alpha,\overline{\sigma}_M,\overline{\sigma}_{SD},\overline{\sigma}_{SK})\frac{\partial\overline{p}}{\partial\overline{z}}\right]\\&=\beta\left.\frac{\partial\overline{\dot{p}}}{\partial\overline{y}}\right|_{y=0}+6U_s\frac{\partial\overline{\eta}_{0\theta}}{\partial\theta}\end{aligned} \tag{10}$$

where

$$\overline{\xi}(\overline{l},\overline{S}_i,L_m,N,\alpha)=\overline{\xi}_1-\overline{\xi}_2+\overline{\xi}_3$$

here,

$$\overline{\xi}_1=\left(\overline{l}^3+\overline{\sigma}_M^3+3\overline{l}^2\sigma_M+3\overline{l}\overline{\sigma}_M^2+3\overline{l}\overline{\sigma}_{SD}^2+3\overline{\sigma}_M\overline{\sigma}_{SD}^2+\overline{\sigma}_{SK}\right)$$

$$\overline{\xi}_2=\left[6NL_m\left(\overline{l}^2+\overline{\sigma}_M^2+\overline{\sigma}_{SD}^2+2\overline{l}\overline{\sigma}_M\right)\coth\left(\frac{NL_m(\overline{l}+\overline{\sigma}_M)}{2}\right)-\frac{12(\overline{l}+\overline{\sigma}_M)}{L_m^2}\right]$$

$$\bar{\xi}_3 = \frac{3}{\bar{S}_i}\left[\left\{(\bar{l}+\bar{\sigma}_M)(1-N^2)+\frac{2\alpha^2}{\bar{S}_i}\right\}\left\{(\bar{l}+\bar{\sigma}_M) - \frac{(1-N^2)}{\bar{S}_i} - \frac{2N(1-N^2)(\bar{l}-\bar{\sigma}_M)}{L_m\bar{l}^2\bar{S}_i}\tanh\left(\frac{NL_m(\bar{l}+\bar{\sigma}_M)}{2}\right)\right\}\right]$$

$$\overline{\eta}_{0\theta} = \left\{(\overline{l}+\overline{\sigma}_M) - \frac{(1-N^2)}{\overline{S}_\theta} - \frac{2N(1-N^2)(\overline{l}-\overline{\sigma}_M)}{L_m\overline{l}^2\overline{S}_\theta}\tanh\left(\frac{NL_m(\overline{l}+\overline{\sigma}_M)}{2}\right)\right\}$$

$\mathcal{E}(\overline{\delta}) = (\overline{l}+\overline{\sigma}_M)$, $\mathcal{E}(\overline{p}) = \overline{p}$ and $\mathcal{E}(\overline{\dot{p}}) = \overline{\dot{p}}$ here, $i = \theta, z$.

2.1 Boundary Conditions

A. Boundary conditions corresponding to porous bush portion are:

Porous Bush Region Boundary Conditions	Bearing Positions
$\overline{\dot{p}}(\theta, \bar{y}, \pm 1) = 0$	(for open ends)
$\frac{\partial \overline{\dot{p}}(\theta, -1, \bar{z})}{\partial \bar{y}} = 0$	(for the porous matrix)
$\frac{\partial \overline{\dot{p}}(\theta, \bar{y}, 0)}{\partial \bar{z}} = 0$	(At mid - span of the bearing)

(11)

B. Boundary conditions related to bearing clearance portion are:

Bearing Clearance Region Boundary Conditions	Bearing Positions
$\overline{p}(\theta, \pm 1) = 0$	(for open ends)
$\frac{\partial \overline{p}(\theta, 0)}{\partial \bar{z}} = 0$	(for the mid-span of the bearing)
$\frac{\partial \overline{p}(\theta_2, \overline{z})}{\partial \theta} = 0$	(Reynolds boundary condition)
$\overline{p} = 0$ and $\overline{p}(\theta, \overline{z}) = 0$, for, $\theta_2 \le \theta \le \theta_1 + 2\pi$	(Cavitation condition)

(12)

θ_1 and θ_2 are angular positions at which formation of hydrodynamic fluid film occurs and cavitation begins.

C. The boundary condition within the interface between porous bush and lubricating fluid is:

$$\overline{\dot{p}}(\theta, 0, \overline{z}) = \overline{p}(\theta, \overline{z}) \tag{13}$$

3 Numerical Solution

Employing the boundary conditions stated in Eqs. (11–13), finite difference analysis using over relaxation is performed to solve Eqs. (9 and 10). Convergence criterion is given by Eq. (14):

$$\frac{\left|\left(\sum\left(\overline{p}_{i,j,k}\right)^{k+1} - \sum\left(\overline{p}_{i,j,k}\right)^{k}\right)\right|}{\left|\sum\left(\overline{p}_{i,j,k}\right)^{k+1}\right|} \leq 0.001 \tag{14}$$

Steady state properties (load bearing capacity and friction) are analyzed in the following sections.

3.1 *Load Bearing Capacity*

Non-dimensional load bearing capacity can be radial ($\overline{W}_r$) as well as tangential ($\overline{W}_t$). These are computed as:

$$\overline{W}_r = -2\int_0^1\int_{\theta_1}^{\theta_2} \overline{p}\cos\theta \mathrm{d}\theta \mathrm{d}\overline{z} \quad \text{and} \quad \overline{W}_t = 2\int_0^1\int_{\theta_1}^{\theta_2} \overline{p}\sin\theta \mathrm{d}\theta \mathrm{d}\overline{z} \tag{15}$$

Hence, dimensionless total load can be found out using Eq. (16).

$$\overline{W} = \sqrt{\overline{W}_r^2 + \overline{W}_t^2} \tag{16}$$

3.2 *Friction Characteristic*

Non-dimensional friction force can be found out using Eq. (17).

$$\overline{f}_s = \left[\int_0^1\int_0^{\theta} A\mathrm{d}\theta \mathrm{d}\overline{z} + \int_0^1\int_0^{\theta} A\left(\frac{\overline{\delta}_{\mathrm{cav}}}{\overline{l}}\right)\mathrm{d}\theta \mathrm{d}\overline{z}\right] \tag{17}$$

where $\overline{\delta}_{\mathrm{cav}}$ is film thickness of cavitated lubricating fluid,

$$\overline{\delta}_{\mathrm{cav}} = \delta_{\mathrm{cav}}/c$$

$$A = \frac{1}{2}\Bigg[\left(\bar{l} + \bar{\sigma}_M\right)\left(1 - N^2\right)\frac{\partial \bar{p}}{\partial \theta} - \bar{S}_\theta\left(1 - \bar{\zeta}_{0\theta}\right) - \left\{\left(\bar{l} + \bar{\sigma}_M\right)\left(1 - N^2\right) + \frac{2\alpha^2}{\bar{S}_\theta}\right\}\left(1 - \bar{\zeta}_{0\theta}\right)\frac{\partial \bar{p}}{\partial \theta}\Bigg]$$

and

$$\overline{\zeta}_{0\theta} = \left[1 - \frac{\left(1 - N^2\right)\left(\bar{l} - \overline{\sigma}_M\right)}{\overline{S}_\theta \bar{l}^2} - \frac{2N\left(1 - N^2\right)\left(\bar{l}^2 - \overline{\sigma}_M^2\right)}{L_m \overline{S}_\theta \bar{l}^2}\tanh\left(\frac{N L_m\left(\bar{l} + \overline{\sigma}_M\right)}{2}\right)\right]$$

The friction parameter is found by:

$$F\left(\frac{R}{c}\right) = \frac{\overline{f}_s}{\overline{W}} \tag{18}$$

4 Results and Discussion

The stochastic schematic diagram of finite porous journal bearing is shown in Fig. 1. Since, the steady state characteristics like $\overline{W}$ and $F(R/c)$ of journal are depended on steady state film pressure $\overline{p}$. The steady state pressures are estimated by combining stochastic surface roughness properties, i.e., $\overline{\sigma}_M, \overline{\sigma}_{SD}$ and $\overline{\sigma}_{SK}$ of journal and porous bush with bearing variable properties like β & α, also constant properties such as $R/c = 500.0$, $L/D = 1.0$, $\varepsilon_0 = 0.5$ and $\Pi_\theta = \Pi_{\bar{z}} = 1.0$. It is assumed that the bearing is operated with micropolar lubricant. The micropolar lubricant characteristics parameter like $L_m = 20.0$ and $N^2 = 0.5$ are considered as constant in this estimation.

4.1 Combined Influence of $\overline{\sigma}_M$, $\overline{\sigma}_{SD}$ and $\overline{\sigma}_{SK}$ with β on $\overline{W}$

Effect of surface roughness, i.e., $\overline{\sigma}_M, \overline{\sigma}_{SD}$ and $\overline{\sigma}_{SK}$ on $\overline{W}$ are shown in Figs. 2, 3 and 4, for different values of β. It can be scrutinized from the figures that an increase of any types of roughness, i.e., $\overline{\sigma}_M, \overline{\sigma}_{SD}$ and $\overline{\sigma}_{SK}$ decreases the magnitude of $\overline{W}$. In Fig. 2 shows that the mean roughness, i.e., $\overline{\sigma}_M$ are varying from negative quantity to positive one. $\overline{\sigma}_M$ value negative signify high asperity and narrow clearance gap among the surface asperities. This geometrical configuration is generated high flow resistance, results high pressure produced in the bearing house during in hydrodynamic lubrication. So, the above physical phenomenon is responsible to increase value $\overline{W}$ at negative values $\overline{\sigma}_M$. An increases of $\overline{\sigma}_M$ revel lower down the asperity size, thus reduce the flow resistance and pressure generation, consequently the $\overline{W}$. Similar types of natures are observed with an enhancement of different values of $\overline{\sigma}_{SD}$ and $\overline{\sigma}_{SK}$ in the Figs. 3 and 4.

It is found from the Figs. 2, 3 and 4 that the micropolar lubricated bearing is generated higher values $\overline{W}$ compared to Newtonian lubricant. Micropolar lubricant is categorized with micro-element parameters viz. N and L_m. Parameter N is a function of two quantities, i.e., ω and μ. ω is signified as microrotation viscosity

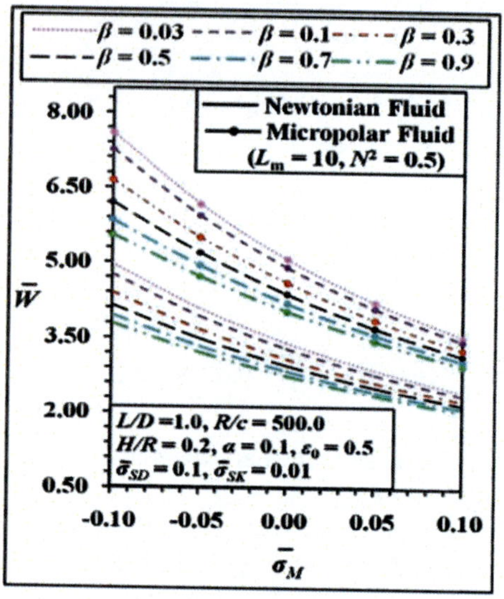

Fig. 2 Variation of $\overline{W}$ with respect to $\overline{\sigma}_M$ at different β

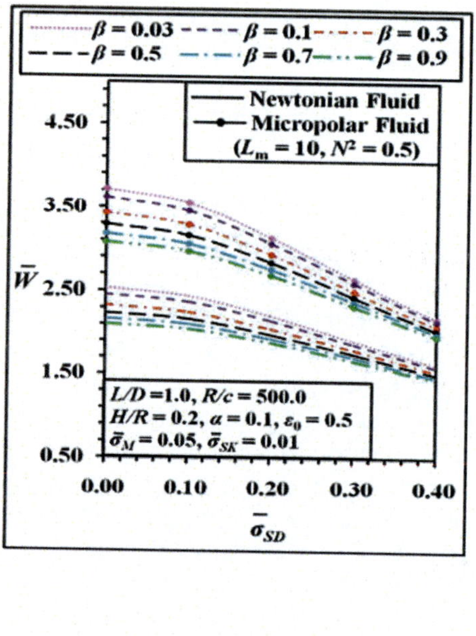

Fig. 3 Variation of $\overline{W}$ with respect to $\overline{\sigma}_{SD}$ at different β

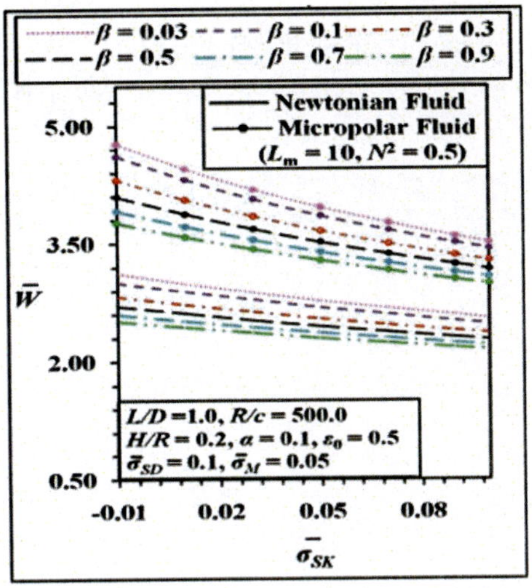

Fig. 4 Variation of $\overline{W}$ with respect to $\overline{\sigma}_{SK}$ at different β

of additives, which arise from microrotational motion of the local lubricant structure. The micropolar lubricant variable N adjoins the linear and angular momentum equations and becomes vanish as $\omega = 0.0$, finally the lubricant behaves as Newtonian lubricant. Other micropolar lubricant variable L_m identify with the ratio of c to Λ. Parameter c narrates as clearance gap between journal and bearing and Λ defines the average molecular size of the micro-additives. The L_m value tends to become zero as c value between journal bearing is small and $\Lambda \to \infty$, means the micro-additives enhances the effective viscosity with a factor of $\{(\mu + \omega\frac{1}{2})/\mu\}$. The parameter L_m tends to infinity revels Λ value become zero, so the micropolarity of lubricant is lost and all curves mapping into Newtonian one.

It is further concluded from the Figs. 2, 3 and 4 that $\overline{W}$ values are dipped with an increase values of β [20]. The higher value of β signify, higher rate of lubricant trap into the porous bush, results low pressure is generated in the bearing house, simultaneously the $\overline{W}$ decrease.

4.2 *Combined Influence of $\overline{\sigma}_M$, $\overline{\sigma}_{SD}$ and $\overline{\sigma}_{SK}$ with α on $\overline{W}$*

Variation of $\overline{W}$ of the bearing with different kind surface roughness, i.e., $\overline{\sigma}_M$, $\overline{\sigma}_{SD}$ and $\overline{\sigma}_{SK}$ for different values of α is presented in the Figs. 5, 6 and 7. The $\overline{W}$ is found to decrease with increase in the values of all category of surface roughness, i.e., $\overline{\sigma}_M$, $\overline{\sigma}_{SD}$ and $\overline{\sigma}_{SK}$. The physics of reduction of $\overline{W}$ already explain in the above section. It is also noticed from the Figs. 5, 6 and 7 that the magnitude of $\overline{W}$ raises with an increase the values of α [20]. Bearing porous bush permeability factor is a linear function of α. The α value of the bearing increases as permeability of porous bush increases, results reduces the availability of lubricant in the bearing. This phenomenon is reduce increment rate of $\overline{W}$ at higher of α. It is further depicted from the Figs. 5, 6 and 7 that $\overline{W}$ of micropolar lubricated bearing is higher as compared to Newtonian one.

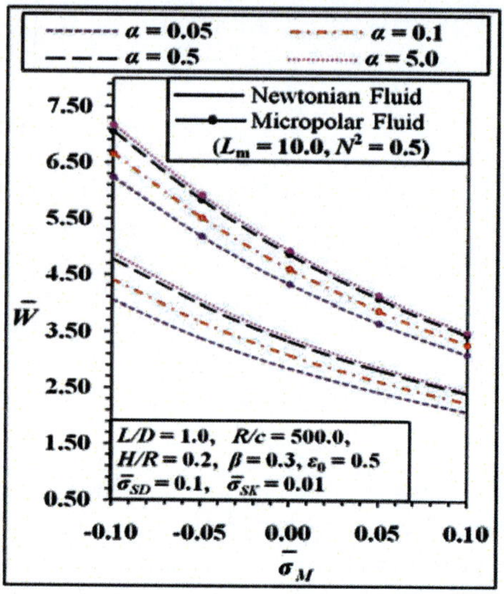

Fig. 5 Variation of $\overline{W}$ with respect to $\overline{\sigma}_M$ at different α

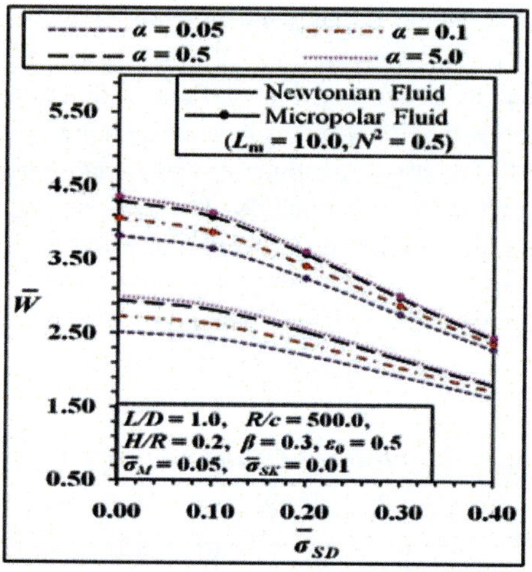

Fig. 6 Variation of $\overline{W}$ with respect to $\overline{\sigma}_{SD}$ at different α

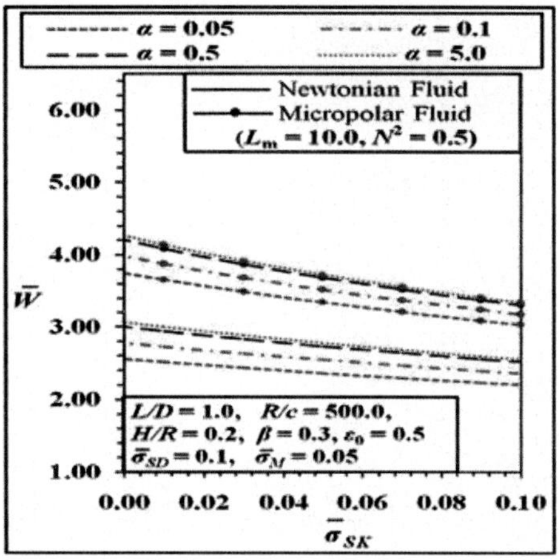

Fig. 7 Variation of $\overline{W}$ with respect to $\overline{\sigma}_{SK}$ at different α

4.3 Combined Influence of $\overline{\sigma}_M$, $\overline{\sigma}_{SD}$ and $\overline{\sigma}_{SK}$ with β on $F(R/c)$

The $F(R/c)$ of porous bearing are varying with different magnitudes of stochastic roughness pattern, i.e., $\overline{\sigma}_M$, $\overline{\sigma}_{SD}$ and $\overline{\sigma}_{SK}$ at variance of β, are exhibited in Figs. 8, 9 and 10. It can be summarized from the Figs. 8, 9 and 10 that $F(R/c)$ of bearing enhance with raising values of any types of stochastic surface roughness. $F(R/c)$ of bearing is linearly vary with $\overline{f}_s$ and inversely proportional with $\overline{W}$. As $\overline{\sigma}_M$ approaches to zero or negative quantities, thus improve flow resistance in the bearing house, consequently increase $\overline{W}$. This phenomenon reduces $F(R/c)$ at lower values of $\overline{\sigma}_M$ and opposite trend is also observed an increment quantities mean roughness. Again, $F(R/c)$ of porous bearing enhances with increasing quantities of $\overline{\sigma}_{SD}$ and $\overline{\sigma}_{SK}$. The physical explanation is already discussed in the above.

It is also seen from Figs. 8, 9 and 10 that $F(R/c)$ of the bearing raises with increment value of β. Increment quantities of β denote, higher amount of lubricant leaving from bearing clearance zone to porous bush. This incident is reduced the availability of the lubricant in clearance zone, results $\overline{W}$ magnitude dips. Finally, the $F(R/c)$ quantities are increasing.

It is further noticed from the Figs. 8, 9 and 10 that $F(R/c)$ of micropolar lubricant lower as compared to Newtonian lubricant. The micropolar lubricant is generated higher values of $\overline{W}$, simultaneously reduce the $F(R/c)$ quantities.

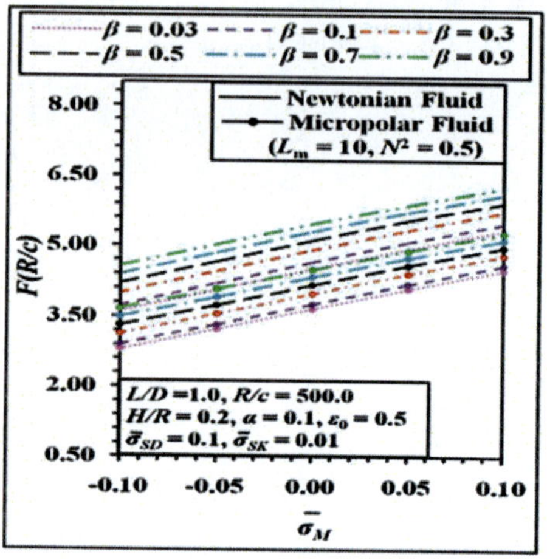

Fig. 8 Variation of $F(R/c)$ with respect to $\overline{\sigma}_M$ at different β

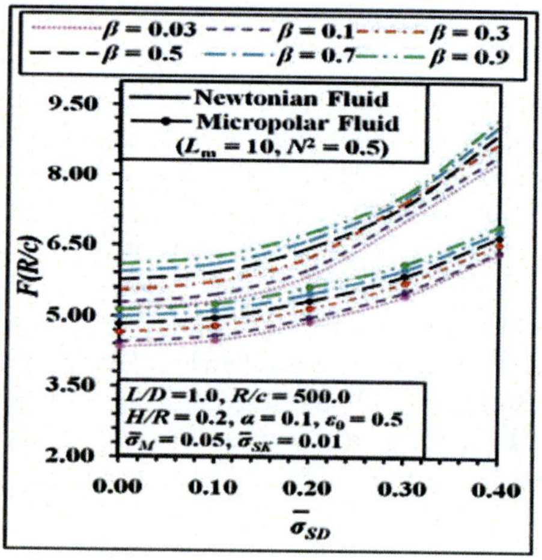

Fig. 9 Variation of $F(R/c)$ with respect to $\overline{\sigma}_{SD}$ at different β

4.4 Combined Influence of $\overline{\sigma}_M$, $\overline{\sigma}_{SD}$ and $\overline{\sigma}_{SK}$ with α on $F(R/c)$

Figures 11, 12 and 13 show variations $F(R/c)$ of porous bearing with respect to different types of stochastic surface roughness, i.e., $\overline{\sigma}_M, \overline{\sigma}_{SD}$ and $\overline{\sigma}_{SK}$ for parametric variation of α. It can be discerned from the Figs. 11, 12 and 13 that $F(R/c)$ quantities of bearing enhances with an increase values of $\overline{\sigma}_M, \overline{\sigma}_{SD}$ and $\overline{\sigma}_{SK}$. The $F(R/c)$ values is ratio two quantities, i.e., $\overline{f}_s$ and $\overline{W}$. At lower values of $\overline{\sigma}_M$, $\overline{\sigma}_{SD}$ and $\overline{\sigma}_{SK}$ revels higher values of $\overline{W}$ generation, results lower down the $F(R/c)$ of the bearing. Higher values of surface roughness reduce $\overline{W}$ quantities, hence raise the values of $F(R/c)$.

It can be derived from the Figs. 11, 12 and 13 that $F(R/c)$ values of porous bearing reduce with enhancement values of α. The increment values of α, is increased the

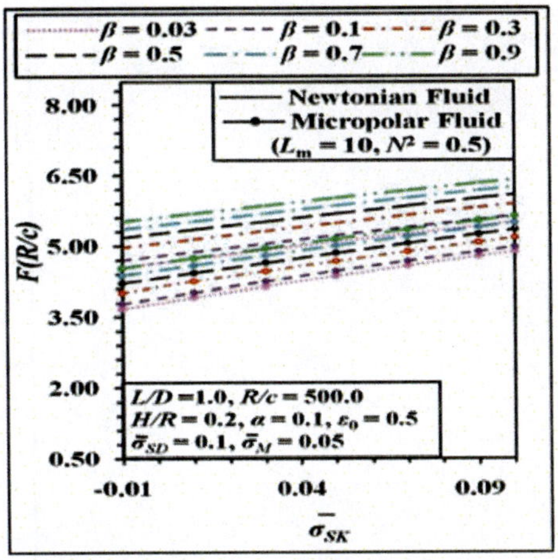

Fig. 10 Variation of $F(R/c)$ with respect to $\overline{\sigma}_{SK}$ at different β

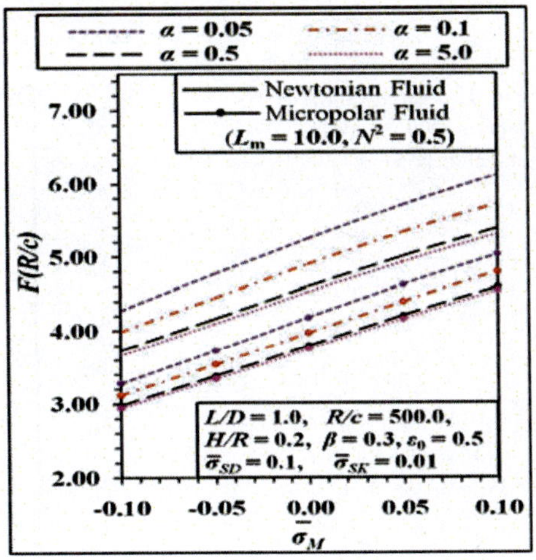

Fig. 11 Variation of $F(R/c)$ with respect to $\overline{\sigma}_M$ at different α

quantities of $\overline{W}$, thus reduce the aforesaid parameter. Similar types of nature of micropolar fluid are seen as previously mentioned in the Sect. 4.3.

5 Conclusions

From this research analogy, following statements can be outlined-

1. Increasing values of stochastic surface roughness, i.e., $\overline{\sigma}_M$, $\overline{\sigma}_{SD}$ and $\overline{\sigma}_{SK}$ reduce the quantities of $\overline{W}$. Lower values of any types of roughness leads to higher value of bearing steady state characteristics, i.e., $\overline{W}$.

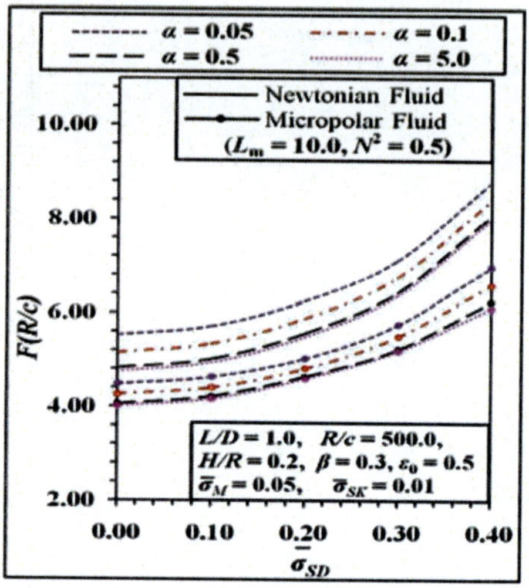

Fig. 12 Variation of $F(R/c)$ with respect to $\overline{\sigma}_{SD}$ at different α

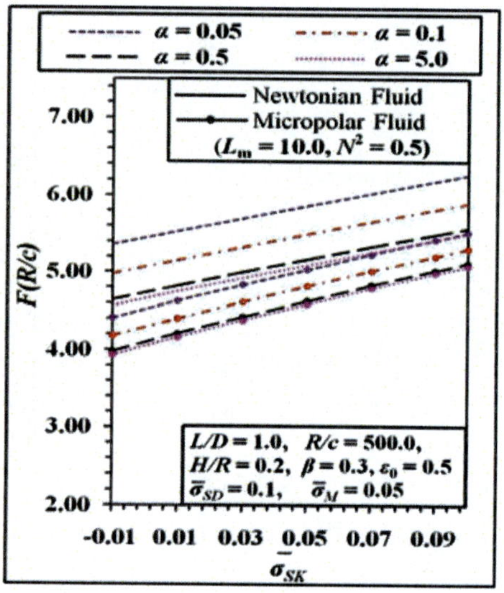

Fig. 13 Variation of $F(R/c)$ with respect to $\overline{\sigma}_{SK}$ at different α

2. Porous bearing parameter like β, reduce the values of $\overline{W}$ of porous bearing with an increment values aforesaid parameter.
3. The steady state nature, i.e., $\overline{W}$ of bearing raises with an increasing values of α.
4. Micropolar fluid load carrying ability is higher as compared to Newtonian fluid.
5. $F(R/c)$ values of porous bearing uplifts with an increment quantities of $\overline{\sigma}_M$, $\overline{\sigma}_{SD}$ and $\overline{\sigma}_{SK}$ surface roughness patterns.
6. The steady state property like $F(R/c)$ of porous bearing hikes with the raising quantities of β.
7. All increment values α of porous bearing, reduces the $F(R/c)$ parameter.

8. All $F(R/c)$ value of micropolar lubricant is lower as compared to Newtonian lubricant.

References

1. Allmaier H, Priestner C, Six C, Priebsch HH, Forstne C, Novotny-Farkas F (2011) Predicting friction reliably and accurately in journal bearings; a systematic validation of simulation results with experimental measurements. Tribol Int 44:1151–1160
2. Jost PH, Schofield J (1981) Energy saving through tribology: a techno economic study. Proc Ins Mech Eng 195:151–173
3. Burton RA (1963) Effects of two dimensional, sinusoidal roughness on the load support characteristics of a lubricant film. Trans ASME 85(2):258–262
4. Christensen H (1969) Stochastic model for hydrodynamic lubrication of rough surfaces. Proc Ins Mech Eng 184(55):1013–1025
5. Patir N, Cheng HS (1978) An average flow model for determining effects of three-dimensional roughness on partial hydrodynamic lubrication. Trans ASME J Lubr Technol 100(1):12–17
6. Ishizawa S, Hori E (1966) The flow of a viscous fluid through a porous wall into a narrow gap (a consideration of the slip of fluid on a porous wall surface). Bull JSME 9(36):719–730
7. Beavers GS, Joseph DD (1967) Boundary conditions at a naturally permeable wall. J Fluid Mech 30(1):197–207
8. Beavers GS, Sparrow EM, Magnuson RA (1970) Experiments on coupled parallel flows in a channel and a bounding porous medium. J Basic Eng Trans ASME 92(4):843–848
9. Prakash J, Gururajan K (1999) Effect of velocity slip in an infinitely long rough porous journal bearings. Tribol Trans 42(3):661–667
10. Gururajan K, Prakash J (2003) Effect of velocity slip in a narrow rough porous journal bearing. Proc Ins Mech Eng Pt-J 217(1):59–70
11. Kalavathi GK, Dinesh PA, Gururajan K, Gurubasavaraj G (2014) Effect of surface roughness in narrow porous journal bearing with a heterogeneous slip/No slip surface. Int J Sci Innov Math Res 2(12):944–959
12. Balaram M (1975) Micropolar squeeze films. J Lub Tech ASME 97(4):635–637
13. Eringen AC (1964) Simple microfluids. Int J Eng Sci 2(2):205–217
14. Eringen AC (1966) Theory of micropolar fluids. J Math Mech 16(1):1–18
15. Baidya U, Das S, Das S (2019) Porous hydrodynamic bearing analysis using micropolar fluid in steady state condition. Tribol Online 14(2):24–31
16. Bhattacharjee B, Chakraborti P, Choudhuri K (2019) Evaluation of the performance characteristics of double-layered porous micropolar fluid lubricated journal bearing. Tribo Int 138:415–423
17. Siddangouda A, Biradar TV, Naduvinami NB (2013) Combined effect of surface roughness and viscosity variation due to additives on long journal bearings. Tribol-Mater Surfaces Interfaces 7(1):21–35
18. Kumar G, Jha KS, Baidya U, Das S, Das S (2016) Effect of surface roughness on the pressure generation in a finite rough hydrodynamic journal bearing under micropolar lubrication in steady-state. Reason Tech J 15:39–53
19. Prakash J, Sinha P (1975) Lubrication theory for micropolar fluids and its application to a journal bearing. Int J Eng Sci 13(3):217–232
20. Haque R, Guha SK (2005) On the steady state performance of isotropically rough Porou hydrodynamic journal bearings finite width with slip flow effect. Proc Inst Mech Eng Pt-J 219:1249–1267

Parametric Optimization and Prediction of Material Removal Rate During Turning Al 7075 Alloy with ZrCN-Coated WC Inserts

Ramanuj Kumar, Rakesh Kumar Dikshit, Diptikanta Das, Shah Md. Suaeb Ahemad, Achinta Sarkar, and Chandrika Samal

1 Introduction

Superlative coalescence of properties, like excellent luster, age hardening behavior, high strength, high yield strength, low density, and good thermal properties, renders the Zn–Mg–Cu-alloyed Al (Al 7075) as a versatile choice for structural applications in aviation, automobile, and marine industries [1–4]. Machining investigations of this alloy are indispensable for its structural application in the abovementioned industries focusing the productivity and economy aspects. Reduction of vehicle weight and improvement of fuel economy can be substantially attended through substitution of steel and cast iron with aluminum in automobile industries [5]. Wide application of the Al alloys for intricate-shaped products in aeronautical and automotive industries also justifies their machining investigations [6–8]. But, formation of long continuous chips, built up edge (BUE), and built up layer (BUL) on the tool provides difficulty in machining of the Al alloys [9, 10].

Increased productivity is one of the major objectives of every manufacturing industry, and material removal rate (*M*) is the paradigmatic of productivity during machining process [11–14]. Rate of material removal must be improvised without compromising with product quality while machining. Das et al. [15], while turning Al 6063 alloy with uncoated and multilayer-coated WC tools reported that there was around 8.75 times more volume of chip removal and around 70% reduction of machining cost for coated tools as compared to the uncoated tools. Alagarsamy et al. [12] reported speed was the most influencing parameter for material removal, and the optimal parametric setting for the material removal rate was 1500 m/min speed,

R. Kumar · D. Das (✉) · S. Md. Suaeb Ahemad · A. Sarkar
School of Mechanical Engineering, KIIT Deemed to Be University, Bhubaneswar 751024, India
e-mail: diptikantadas115@gmail.com

R. K. Dikshit · C. Samal
Department of Mechanical Engineering, GITA Autonomous College, Bhubaneswar 752054, India

© The Author(s), under exclusive license to Springer Nature Singapore Pte Ltd. 2023

S. Revankar et al. (eds.), *Recent Advances in Thermofluids and Manufacturing Engineering*, Lecture Notes in Mechanical Engineering,
https://doi.org/10.1007/978-981-19-4388-1_47

0.20 mm/rev feed, and of 0.8 mm cutting depth for turning Al 7075 with coated carbide tools. While turning of Al 6061 alloy in dry and wet condition, Deepak and Rajendra [13] reported feed rate was more significant for the rate of material removal, taken after by cutting depth and cutting speed. Das et al. [16], while turning Al 7075 alloy with uncoated WC inserts, reported increment of rate of material removal with the increase of speed, feed, or depth of cut.

Though many literatures are available in open literature on study of machining characteristics of various Al alloys, further systematic investigation of material removal rate (M) during machining of the aerospace alloy Al 7075 is required for enhancement of productivity during machining. Accordingly, this paper outlines an analysis of M during turning Al 7075 alloy using ZrCN-coated WC tools in dry cutting condition, following L_{16} Taguchi orthogonal array. Then, the response was optimized for various machining parameters through Taguchi's signal-to-noise ratio approach, and then, regression model was developed for the rate of material removal during turning the material.

2 Experimental Methodology

Thermally, T6-treated Zn–Mg–Cu-alloyed Al (Al 7075) as procured from Bharat Aerospace Metals, Mumbai, was deployed for the turning trials. The workpiece material contained 4.6 wt.% of Zn, 2 wt.% of Mg, and 1.5 wt.% of Cu as major alloying elements. Conventional weight-to-volume ratio method was adopted for density measurement, which was 2.81 gm/cm^3. Hardness, ultimate tensile strength, yield strength, and Young's modulus of the alloy sample were 87.33 HRB, 530 MPa, 450 MPa, and 71 GPa, respectively. Dimension of the turning workpiece sample was 55 mm diameter × 200 mm length. WIDIA makes TiN/TiCN/Al_2O_3/ZrCN-coated WC tools of ISO specification CNMG 120,408 (80^0 rhombic) were utilized for assessment of their turning performance in dry environment. The cutting inserts were rigidly mounted in PCLNR 2525M12 right-hand tool holder during turning. HMT-NH 22 high-speed precision lathe was utilized to conduct the turning experiments. During machining, the workpiece was rigidly backed up by tailstock to reduce the vibration effects. Figure 1 presents the experimental setup for measurement of material removal rate during turning. Turning experiments were conducted with three machining parameters, associated with four levels to each parameter, as demonstrated in Table 1. Taguchi L_{16} standard orthogonal array (OA) was followed to design the experiments. Table 2 shows the design of experiments (DOEs) as per Taguchi L_{16} OA. Assuming D_1 and D_2 as the un-machined and machined workpiece diameters (mm), f as feed (mm/rev) and N as spindle speed (rpm), material removal rate (M) (mm^3/min) during turning was quantified using Eq. (1). Each measurement was repeated for three times, and then, average of the measured values was utilized for data analysis.

Fig. 1 Experimental setup during turning of T6-Al 7075 alloy

Table 1 Machining parameter and levels

Parameters with unit	Notation	Levels of parameters			
		1	2	3	4
Spindle speed, rpm	*N*	250	715	1210	1575
Feed, mm/rev	*f*	0.05	0.1	0.16	0.2
Depth of cut, mm	*d*	0.2	0.3	0.4	0.5

$$M = \frac{\pi}{4} * (D_1^2 - D_2^2) * f * N \tag{1}$$

3 Results and Discussion

3.1 Influence of Cutting Parameters

Results for material removal rate (M) during machining T6-treated Al 7075 alloy with ZrCN-coated WC inserts at dry condition are presented in Table 3. M was minimum (397.65 mm^3/min) for Trial 1 and was maximum (9746.28 mm^3/min) for Trial 12. A consistent rise of M was observed from Trial 1 to Trial 4, where spindle speed was constant (250 rpm), and both feed and cutting depth were increased. M increased by around 860% on increasing feed from 0.05 to 0.2 mm rev^{-1} and cutting depth from 0.2 to 0.5 mm. While M for Run 1 and Run 5 is compared, an increase of about 304% is observed for a consistent feed of 0.05 mm rev^{-1}, and on increasing cutting depth from 0.2 to 0.3 mm and spindle speed from 250 to 715 rpm. Response table for mean M (Table 4) reveals maximum delta value for N (Rank 1), taken after by f (Rank 2) and d (Rank 3). It implies that spindle speed (N) was the greatest influencing

Table 2 Design of experiments for turning trials

Trial No.	*N*	*f*	*d*
1	250	0.05	0.2
2	250	0.1	0.3
3	250	0.16	0.4
4	250	0.2	0.5
5	715	0.05	0.3
6	715	0.1	0.2
7	715	0.16	0.5
8	715	0.2	0.4
9	1210	0.05	0.4
10	1210	0.1	0.5
11	1210	0.16	0.2
12	1210	0.2	0.3
13	1575	0.05	0.5
14	1575	0.1	0.4
15	1575	0.16	0.3
16	1575	0.2	0.2

parameter for *M*, followed by feed (*f*) and cutting depth (*d*). Main effects plot for mean *M* (Fig. 2) depicts an increase of *M* on enhancing either of the turning variables under consideration. Increase in spindle speed causes the workpiece to rotate at faster rpm, leading to more material removal at the shortest possible time. On increasing the feed, linear traverse of the cutting tool past the workpiece increases, causing removal of more volume of material from the workpiece in less time, and therefore, *M* increases. Similarly, on increasing the cutting depth, the cutting tool penetrates more into the workpiece in radial direction, causing more volume of material removal at an increased magnitude of radial component of cutting force, which increases the rate of material removal.

3.2 *Optimization of Response*

Signal-to-noise ratio (SNR) is the power ratio between meaningful data and the background noise, having decibels (dB) as its unit. Better system performance is represented by the higher values of SNR, and its highest value constitutes optimal parametric combination for the desired response. In the present experiment, the material removal rate (*M*) was optimized through the SNR concept of Taguchi method. The SNR values of *M* were evaluated considering "Larger the better" criterion through MINITAB 17 software, and these are included in Table 3. From the response table (Table 5) and the main effects plot (Fig. 3) for SNR of *M*, it is noticed that the SNR

Table 3 Experimental results and signal-to-noise ratio for *M*

Trial No.	*N*	*f*	*d*	*M* (mm^3/min)	SNR for *M* (dB)
1	250	0.05	0.2	397.65	51.99
2	250	0.1	0.3	1181.17	61.4462
3	250	0.16	0.4	2484.66	67.9053
4	250	0.2	0.5	3811.6	71.6221
5	715	0.05	0.3	1608.23	64.127
6	715	0.1	0.2	2121.85	66.5343
7	715	0.16	0.5	8361.62	78.4458
8	715	0.2	0.4	8199.92	78.2762
9	1210	0.05	0.4	3408.39	70.651
10	1210	0.1	0.5	8349.94	78.4337
11	1210	0.16	0.2	5258.83	74.4178
12	1210	0.2	0.3	9746.28	79.7768
13	1575	0.05	0.5	5187.01	74.2983
14	1575	0.1	0.4	8121.12	78.1923
15	1575	0.16	0.3	9579.12	79.6265
16	1575	0.2	0.2	7883.66	77.9346

Table 4 Response table for means of *M*

Level	*N*	*f*	*d*
1	1969	2650	3915
2	5073	4944	5529
3	6691	6421	5554
4	7693	7410	6428
Delta	5724	4760	2512
Rank	1	2	3

increased consistently on increasing the levels of all the three parameters, and it was maximum for the 4th levels of speed, feed, and cutting depth, i.e., N_4-f_4-d_4. This combination can be regarded as the optimal parametric combination for *M* during dry turning of T6-conditioned Al 7075 alloy using ZrCN-coated WC inserts.

Significance of process parameters on *M* was evaluated through analysis of variance (ANOVA) at a confidence level of 95%, and the analyzed data are presented in Table 6. The tabular data reveal the probability of significance (*P*-value) was less than 5% for *N* and *f*. It was minimum (0.015) for *N* with the highest percentage of contribution (47.52%), which signifies its highest significance for *M*. Next to *N*, *f* was significant parameter for *M* with 32.48% of contribution. Moreover, the statistical fisher consistency (*F*-value) for *N* and *f* is significantly more than the corresponding tabular value (4.76) [17] at confidence level of 95%. Because of the *P*-value of *d* was

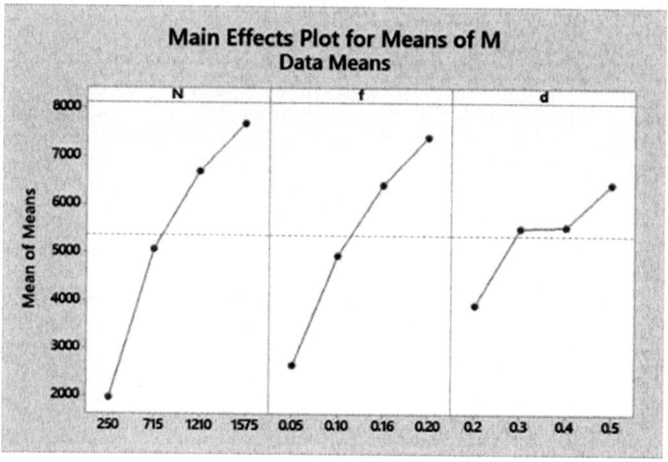

Fig. 2 Main effects plot for means of *M*

Table 5 Response table for SNR *M* (Larger is better)

Level	*N*	*f*	*d*
1	63.24	65.27	67.72
2	71.85	71.15	71.24
3	75.82	75.1	73.76
4	77.51	76.9	75.7
Delta	14.27	11.64	7.98
Rank	1	2	3

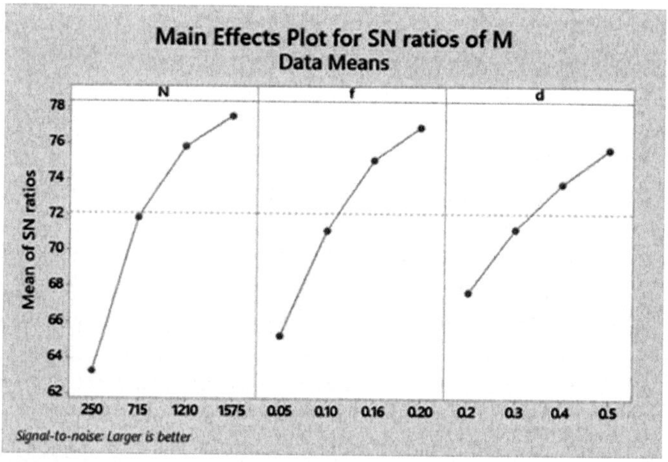

Fig. 3 Main effects plot for SNR *M*

Table 6 ANOVA for *M*

Source	DF	Adj SS	Adj MS	*F*	*P*	Contribution (%)
N	3	75,182,464	25,060,821	8.14	0.015	47.52
f	3	51,382,427	17,127,476	5.56	0.036	32.48
d	3	13,168,363	4,389,454	1.43	0.325	8.32
Error	6	18,477,204	3,079,534			11.68
Total	15	158,210,458				100

Table 7 Confirmation experimental results for *M*

Level	Initial process parameters	Optimal process parameters	
		Predicted	Experimental
	N_2-f_2-d_2	N_4-f_4-d_4	N_4-f_4-d_4
M (mm^3/min)	3176.03		19,560.73
SNR for *M* (dB)	70.0377	85.9056	85.8274

Improvement in SNR ratio for $M = 15.7897$ (dB)

more than 5% and its statistical *F*-value is less as compared to its tabulated value, it was not significant for *M*.

Confirmation experiments were accomplished to predict and substantiate the enhancement of the performance characteristic utilizing the optimal combination of process parameters. The predicted SNR ($\hat{\gamma}$) was calculated using Eq. (2) for optimal parametric combinations.

$$\hat{\gamma} = \gamma_m + \sum_{i=1}^{o} \left(\overline{\gamma}_i - \gamma_m\right) \tag{2}$$

where total mean SNR is γ_m, mean of SNR at optimal level is $\overline{\gamma}_i$, and number of design parameters affecting the performance characteristics is o. Confirmation experiment results for *M* using the optimal and initial process parameters are reported in Table 7. Good accordance existed between the predicted and optimal values. Improvement of SNR from the initial to the optimal parametric combination was 15.7897 dB.

3.3 Regression Model

Second-order regression model for *M* (Eq. 3) was obtained through response surface method (RSM) at 95% confidence level. The model presented was with high determination coefficient (R^2) near to 100%, and with rational concurrence with the adjusted R^2 value, which designates the righteousness of fit for the models and their high significance. Also, from normal probability (NP) plots of residuals (Fig. 4a), it is observed that the residuals fit fairly near to the normal probability line implicating that errors are normally dispersed, and the terms stated in the models are significant. Moreover, the residuals versus fits (RVF) plot (Fig. 4b) indicate a reasonable scatter of the residuals from the mean, mentioning the significance of the model developed.

$$M = 550 - 2.625 * N - 16236 * f - 1155 * d - 0.002076 * N^2 - 90122 * f^2 - 18762 * d^2 + 45.25 * N * f + 17.416 * N * d + 117767 * f * d$$
$$R^2 = 99.97\%, R^2(\text{adj}) = 99.94\%, R^2(\text{pred}) = 99.77\% \quad (3)$$

In summary, during dry condition turning of T6 thermally treated Al 7075 alloy with ZrCN-coated WC inserts, material removal rate was maximum (9746.28

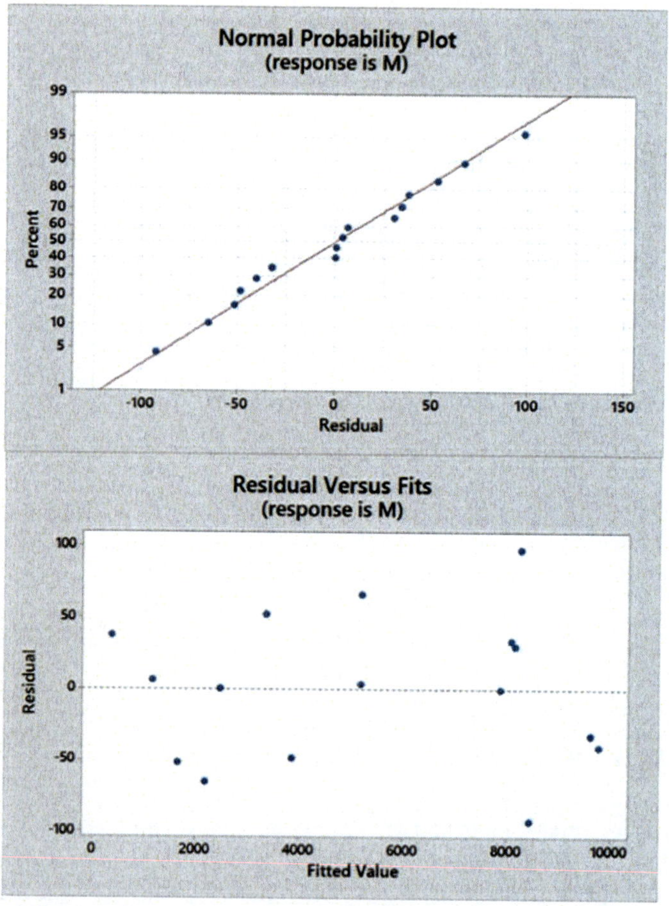

Fig. 4 **a** Normal probability plot of residuals; and **b** Residuals versus fits plot for *M*

mm^3/min) for the Trial 12 of the Taguchi L_{16} DOE, i.e., at the spindle speed 1210 rpm, feed 0.2 mm rev^{-1}, and cutting depth 0.3 mm.

- *M* was optimized through SNR maximization method of Taguchi technique, which revealed N_4-f_4-d_4 was the optimal parametric combination, i.e., at 1575 rpm of speed, 0.2 mm rev^{-1} of feed, and 0.5 mm of cutting depth. Confirmation experiment results divulged a good concurrence between the predicted and experimental values, and the SNR for optimal parameters improved by 15.7897 dB than the initial parameters.
- Results of ANOVA at 95% confidence level divulged minimum *P*-value for speed, taken after by feed, which were also less than 5%, mentioning the highest significance of *N*, followed by *f* on *M*. Contribution of spindle speed on *M* was 47.52%, and that of feed was 32.48%. However, effect of cutting depth on *M* was not significant.
- Second-order regression model was generated for *M* using RSM. The model presented high determination coefficient (99.97%) with a reasonable agreement with its adjusted value (99.94%). Residuals were close to normal probability line in the NP plots of residuals and were scattered reasonably from mean in the RVF plot, revealing good fitness and high significance of the models.

4 Conclusions

The present work involves investigation of material removal rate during turning Al 7075 alloy with ZrCN-coated WC tools in dry environment through L_{16} Taguchi orthogonal array, followed by optimization and regression analysis. However, the machining experiments have not been conducted in minimum quantity lubrication (MQL) or flood cooling environment, where the machining performance of the alloy is expected to be better compared to dry environment. So, a scope for this work is to investigate the machining performance of the alloy during turning in MQL and flood cooling environment. Moreover, full factorial design may be adopted for experimentation to get more précised optimal results and regression model.

References

1. Imran M, Khan AA (2019) Characterization of Al-7075 metal matrix composites: a review. J Market Res 8(3):3347–3356
2. Lei Z, Bi J, Chen Y, Chen X, Tian Z, Qin X (2020) Effect of TiB_2 content on microstructural features and hardness of TiB_2/AA7075 composites manufactured by LMD. J Manuf Process 53:283–292
3. Sahoo BP, Das D (2021) Investigation on reinforcement incorporation factor and microstructure of Al 7075/submicron-TiB_2 metal matrix composites processed through a modified liquid metallurgy technique. Exp Tech 45(2):179–193

4. Sahoo BP, Das D, Chaubey AK (2021) Strengthening mechanisms and modelling of mechanical properties of submicron-TiB_2 particulate reinforced Al 7075 metal matrix composites. Mater Sci Eng A 141873
5. Miller WS, Zhuang L, Bottema J, Wittebrood AJ, De Smet P, Haszler A, Vieregge A (2000) Recent development in aluminium alloys for the automotive industry. Mater Sci Eng 280:37–49
6. Santos MC, Machado AR, Barrozo MA (2018) Temperature in machining of aluminum alloys, temperature sensing. Intech Open, pp 71–89. https://doi.org/10.5772/intechopen.75943
7. Hovsepian PE, Luo Q, Robinson G, Pittman M, Howarth M, Doerwald D, Tietema R, Sim WM, Deeming A, Zeus T (2006) TiAlN/VN superlattice structured PVD coatings: a new alternative in machining of aluminium alloys for aerospace and automotive components. Surf Coat Technol 201:265–272
8. Demir H, Gündüz S (2009) The effects of aging on machinability of 6061 aluminium alloy. J Mater Des 30:1480–1483
9. Gangopadhyay S, Acharya R, Chattopadhyay AK, Sargade VG (2010) Effect of cutting speed and surface chemistry of cutting tools on the formation of BUL or BUE and surface quality of the generated surface in dry turning of AA6005 aluminium alloy. Mach Sci Technol 14(2):208–223
10. Soren TR, Kumar R, Panigrahi I, Sahoo AK, Panda A, Das RK (2019) Machinability behavior of aluminium alloys: a brief study. Mater Today: Proc 18:5069–5075
11. Mishra P, Das D, Ukamanal M, Routara B, Sahoo A (2015) Multi-response optimization of process parameters using Taguchi method and grey relational analysis during turning AA 7075/SiC composite in dry and spray cooling environments. Int J Ind Eng Comput 6(4):445–456
12. Alagarsamy SV, Raveendran P, Arockia VS, Tamil V (2016) Optimization of machining parameters for turning of aluminium alloy 5 using Taguchi method. Int Res J Eng Technol 3(1):316–322
13. Deepak D, Rajendra B (2015) Studies on material removal rate of Al 6061 while turning with coolant and without coolant using Taguchi method. Int J Res Eng Technol 4(09):75–78
14. Pant G, Kaushik S, Rao DK, Negi K, Pal A, Pandey DC (2017) Study and analysis of material removal rate on lathe operation with varing parameters from CNC Lathe machine. Int J Emerg Technol 8(1):683–689
15. Das D, Mishra PC, Sahoo AK, Ghosh D (2015) Experimental investigation on cutting tool performance during turning AA 6063 using uncoated and multilayer coated carbide inserts. Int J Mach Mach Mater 17(3–4):277–294
16. Das D, Sahoo BP, Bansal S, Mishra P (2018) Experimental investigation on material removal rate and chip forms during turning T6 tempered Al 7075 alloy. Mater Today: Proc 5(2):3250–3256
17. Kothari CR (2012) Research methodology. Second revised edition: Reprint, New Age International Publishers

Supply Chain Transformation Through Digital Servitization in Manufacturing Sector

Vivek Mishra, Bhubaneswari Bisoyi, Biswajit Das, Ipseeta Satpathy, and Abhishek Gupta

1 Introduction

The industrial activity has been largely changed by the proliferation of emerging technology and increased levels of digitization. Propelled by Big Data, advanced analytics and Internet of things, industry 4.0 has transformed manufacturing into a service oriented mechanism [1]. Operating both at the front and back end, digitization has completely altered the entire value chain by higher levels of connectivity across products and firms [2]. Such disruptions are not limited to the scope of manufacturing, they are also influencing other economies such as maritime and publishing. The emergence of technology as a facilitator of servitization has caught the fancy of researchers and managers alike and they have exploring the same since the last decade. Considering the example of KONE, one of the biggest manufacturers of elevators globally, by partnering with IBM, they developed advanced condition monitoring and predictive maintenance services. Based on IoT technology these services have led to higher efficiency by reduced downtime and faster restoration of equipment [3]. The perfect marriage of digital manufacturing technologies (DMT) offers scalability in delivery tailored solutions, whereas new ways for increased connectivity with customers is facilitated through digital information technologies (DIT) [4]. Aimed at implementing smart autonomous ecosystems, digital servitization also requires the blurring of boundaries across firms as smart solutions merge with product-service-software systems of other companies [5]. Undeniably, in the process of creating

V. Mishra
BIITM, BPUT, Rourkela , Odisha, India

B. Bisoyi (✉)
Faculty of Management Studies, Sri Sri University, Godisahi, Odisha, India
e-mail: bhubaneswari.b@srisriuniversity.edu.in

B. Das · I. Satpathy · A. Gupta
KIIT School of Management, KIIT Deemed to Be University, Bhubaneswar, Odisha, India

© The Author(s), under exclusive license to Springer Nature Singapore Pte Ltd. 2023
S. Revankar et al. (eds.), *Recent Advances in Thermofluids and Manufacturing Engineering*, Lecture Notes in Mechanical Engineering,
https://doi.org/10.1007/978-981-19-4388-1_48

autonomous products/services companies such as Tesla, Rolls-Royce, Caterpillar, and many others, have avoided operating separately from customers and have moved beyond the boundaries of the firm [6]. The core philosophy behind the designing of smart solutions include; interaction with other manufacturers, increased customer access, collaboration between different service partners, and third part [7]. However, such rapid transformation through integration of smart solutions requires substantial changes in the business model followed by collaborative innovations, an area where majority of manufacturers struggle.

The aim of this study is to contribute to theory and practice by identifying the strategic antecedents essential for digital servitization transformation. By adopting a macro-level approach based on the competencies framework we determine the key digital competencies essential for digital servitization transformation, in order to enhance firm performance. This research paper also explores the inter-relationships between these competencies that can be pivotal in improving a firm's ability to upgrade or re-define existing competences while simultaneously experimenting with new ideas, to find out which core-areas are preconditions for firms pursuing digital servitization. The structure of our research article is as follows, at the outset, it has concisely reviewed the available literature on digital servitization and digital competencies to excavate prior research, uncover current knowledge gaps and develop hypotheses. In subsequent section, the research methodology adopted in the study is discussed, comprising scale development, data collection and analysis. Next, based on the output obtained in our structural model, it has also explored the relationship between constructs and relate the evidence to our hypotheses. Lastly, the major findings and their managerial implications are also highlighted.

2 Defining the Competencies

2.1 Data and Analysis Management

This is the first category and it consists of (1) Research, position, process and analysis of data, which comprises the ability of (a) finding out needs, filtering the needs, elaborating and organizing everything (with the accurate tools and proper format) a huge set of both unstructured and structured data in order to finalize purple specific analysis (e.g., using pivot tables and SQL script to segment data and calculate the total cost) [8]. (b) Analysis, comparison and critical evaluation of the credibility and reliability of data sources, digital content and received information [9]. (c) Organizing the present data, storing it and then retrieving it, content and various information in digital environment, and (d) organizing received information and then processing them in the environment which is more structured. (2) Data security management and protection of digital assets that includes the use of right technologies to treat assets in a secure way [10] digitally, protection from leakages, hosts and viruses, online fraud, hacking, and other cyber-attacks by cyber criminals.

H5: Data Analysis and management has a positive impact on a firms digital servitization transformation

H6: Data Analysis and management has a significant impact on digital content creation.

2.2 Digital Content Creation

In the next category, the main areas are: (1) creating and editing the contents digitally in technical form by which one can express oneself easily on the digital platform [11] (2) finding solutions to problems, creativity, finding solutions, logical way of thinking, and developing of quantitative and qualitative models (statistical, logical, analytical, heuristic models), developing applications to study a specific business or technical problem, comparing in finding the best solution, having proper insights to take decisions, developing tools for proper simulation, using various techniques for analysis etc. [12] (3) designing and development of software and hardware that includes (a) Coding, (b) development of applications digitally, interfaces, various scripts, different algorithms and prototypes. (c) finding solutions to specific business and technical problems (developing applications for demand forecasting), simulation techniques and modeling [13].

H7: Digital content creation has a positive impact on a firm's digital servitization transformation.

2.3 Soft Skills

This is the third category and it contains (1) communication skills: it includes choosing the right platform/channel for communication (email, text messaging, social media messaging applications), creation of proper communication content, usage of language properly (business, analytical, spoken and written, technical languages), getting the right replies from the people receiving the messages (behaviors, actions, and solicitations) [14, 15] (2) attitude for teamwork: it means the ability of involving with the colleagues, achieving results by working as a team, liking social relationships established in the work, knowledge sharing, and creation of collective intelligence [16]. (3) leadership skill: The skill of leading people toward goal achievement, having enough emotional intelligence, getting in touch with the requirements of the colleagues and investors, fulfilling expectations, creating new social relationships and maintaining them properly in the work environment, creation of a super strong network, and being a leader and entrepreneur of the company [17]. (4) Time, stress management, and results: the ability to get the work done in various innovative ways, being reliable and responsible, having a control on time and being able to work properly under pressure [18].

H1: Soft skills have a positive impact on a firm's digital servitization transformation
H2: Soft skills have a significant impact on a firm's innovation appetite.

2.4 Innovation Appetite

This is the last area and its components include: (a) thorough knowledge of the present technologies, assessment of impact, benefits, risks and costs which also includes the ability of knowing about the potential impact of technology on various business models, operations, channels and acting as a medium to create common consensus on the present opportunities and introducing new technologies [19] (b) project management for interruption of new technologies, services and products, different processes that involve the part dissipation by the management for innovation (creating the technology, definition requirements, planning, organizing, developing and executing, testing reviewing and controlling), innovating new technologies and analysis of processes [20].

H3: Innovation Appetite has a positive impact on a firm's digital servitization transformation
H4: Innovation Appetite has a significant impact on the data analysis and management abilities of a firm.

3 Problem Formulation

3.1 Servitization and Firm's Performance

According to different authors, digital servitization covers different models which are technology oriented and platform-based. These business models facilitate exchange between the service provider and the service takers (customers) by helping the firms and reaching a stage of competitive advantage while delivering the prime consumers with knowledge-based services during the entire life-cycle of the product [21] and by leveraging complexity of network. As seen and indicated by various researchers, digital servitization creates new business opportunities, either for industries to set up service-based business models to increase their competitive edge [22] or for the new competitors that enter the market. The study conducted by companies create competitive advantage with the combination of digital systems and PSSs and also harvest value this way. However, large-scale improvement in the effectiveness and efficiency of maintenance have been observed by many researchers (Fig. 1).

H8: Digital servitization transformation has a positive impact on a firm's performance.

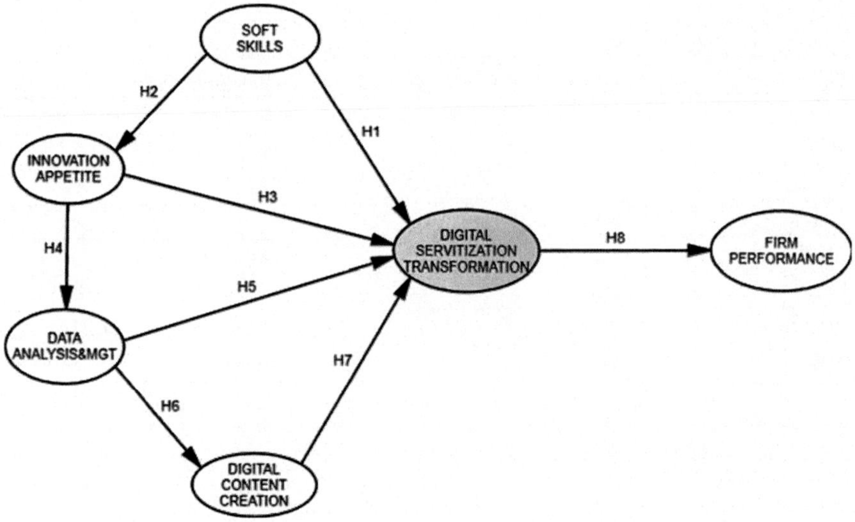

Fig. 1 The proposed research model. *Source* Authors own depiction

3.2 Evaluation of the Structural Model

The SEM output obtained from the analysis done by using Amos 23.0 was evaluated based on the recommended indices for model fit. To improve the overall model fit, it utilized the function of modification indices. Based on the suggestions obtained in the output, direct covariates paths were added between some error terms, which improved the overall model fit to an agreeable extent. The results illustrated in Table 1 given below suggest that the proposed model was an excellent fit, the Relative χ^2 (CMIN/d*f*) statistic was 1.416, which was well below the recommended value of <0.03 [23]. The goodness of fit index (GFI) and comparative fit index (CFI) indices obtained had a value of 0.905 and 0.979, respectively, which surpassed the cut-off value of >0.90 [24]. The RMSEA index was 0.039 and was in the excellent range [25] and the indices obtained for NFI was 0.934, which was above the recommended value of >0.90.

4 Results of Hypotheses

After the evaluation of model fit based on the recommended indices, the hypothesized relationships has been examined. The path coefficients obtained through SEM are illustrated in the table below (Table 2), as seen in the table all paths display a significant relationship between the constructs. As evident in the table, the results of the path coefficient indicated that all hypotheses including H1, H2, H3, H4, H5, H6, H7, and H8 were supported. Specifically, all the four antecedents had a significant

Table 1 Model fit measures

Fit index	Recommended value	Structural model
χ^2	NS at $p < 0.05$	366.770
df	n/a	259
χ^2/df	<3	1.416
GFI	>0.90	0.905
AGFI	>0.80	0.881
CFI	>0.90	0.979
RMSEA	<0.08	0.039
NFI	>0.90	0.934

GFI goodness of fit index, *AGFI* adjusted goodness of fit index, *CFI* comparative fit index; df degrees of freedom; *NFI* normed fit index; *RMSEA* root mean square error of approximation
Sources Hu and Bentler [26], and Hair et al. [27]

Table 2 Results of hypothesis

Hypothesis	Path			Estimate	S.E	C.R	*P*	Results
H2	IA	←	SO	0.600	0.078	7.698	***	Supported
H4	DA	←	IA	0.250	0.067	3.761	***	Supported
H6	DC	←	DA	0.143	0.044	3.252	0.001**	Supported
H7	DS	←	DC	0.389	0.084	4.656	***	Supported
H1	DS	←	SO	0.230	0.071	3.234	0.001**	Supported
H3	DS	←	IA	0.215	0.063	3.437	***	Supported
H5	DS	←	DA	0.152	0.053	2.869	0.004**	Supported
H8	FP	←	DS	0.644	0.077	8.378	***	Supported

*** $p = < 0.001$,** $p = < 0.005$
SO soft skills, *DC* digital content creation, *FP* firm performance, *DA* data analysis and management, *IA* innovation appetite, *DS* digital servitization transformation

impact on DS; SO ($\beta = 0.600, p = < 0.001$), IA ($\beta = 0.215, p = < 0.001$), DA ($\beta = 0.152, p = < 0.004$), and DC ($\beta = 0.389, p = < 0.001$). A significant observation was that DS was positively related to FP ($\beta = 0.644, p = < 0.001$).

4.1 Managerial Implications and Recommendations

A large number of manufacturing concerns are attempting to keep pace with the digital revolution and technological advancement, yet the contribution of academic research toward digital competencies is at a nascent stage. Additionally, although research is focused on the organizational transformation for increased level of digitization, a concrete approach toward the linkage between competencies and firm capabilities was lacking. Directionally, our study provides a macro-view of the antecedents essential for digital servitization transformation, which ultimately has a bearing on firm performance [28, 29]. Based on the four antecedents/pre-requisites of digital servitization transformation, our study recommends the following:

(i) The proposed framework can be effectively utilized by managers, as it defines the requisite skills to support the complex process of digital servitization transformation.
(ii) The results obtained can be used for training need identification or re-skilling of existing employees as well as the core competencies to be looked for while recruiting new ones.
(iii) The strategic decisions taken at the top level can be aligned with these competencies, to ensure optimal utilization of available human resources.
(iv) Based on their leadership capabilities managers can increasingly focus on data-driven decisions based on proper data management and analysis for effective forecasts.
(v) Marketers have an important role to play, when choosing the appropriate communication channel and a bottom-up approach is suggested in designing the communication mix.
(vi) The innovation appetite can be enhanced by proper collaboration between other firms and academia, as it would provide a proper knowledge sharing platform.
(vii) Another important aspect is the creation and curation of digital, a proper balance is essential between by considering the pros and cons of both.
(viii) While focusing on the alignment of the strategy with competencies, firms need to adopt a contingency approach in order to deal with the turbulence often caused by rapid technology change and intensifying competition.
(ix) Alterations in the business model are suggested, however they are to be based on the gradual and slow incorporation of digitization and a continuous monitoring tracking its effectiveness.
(x) While entering foreign markets more emphasis should laid on the digital content, which can potentially attract and retain the interest of clients and a value co-creation approach is advised based on increased interaction with the target segment.

5 Conclusions

The study was conducted with the objective of identifying and validating the antecedents in the form of key competencies essential for digital servitization transformation, the study is significant from a manufacturing firm's standpoint. The major findings of our study include the following (i) soft skills, innovation appetite, data analysis and management, and digital content creation are the macro-level antecedents of digital servitization transformation (ii) there is a significant inter-correlation between; soft skills-innovation appetite, innovation appetite-data analysis and management, and data analysis and management-digital content creation (iii) digital servitization transformation has a significant and positive impact on firm performance.

References

1. Adolph KE, Tamis-Lemonda CS (2014) The costs and benefits of development: the transition from crawling to walking. Child Dev Perspect 8(4):187–192. https://doi.org/10.1111/cdep.12085
2. Adrodegari F, Saccani N, Kowalkowski C, Vilo J (2017) PSS business model conceptualization and application. Prod Plan Control 28(15):1251–1263. https://doi.org/10.1080/09537287.2017.1363924
3. Ardolino M, Rapaccini M, Saccani N, Gaiardelli P, Crespi G, Ruggeri C (2018) The role of digital technologies for the service transformation of industrial companies. Int J Prod Res 56(6):2116–2132. https://doi.org/10.1080/00207543.2017.1324224
4. Arntz M, Gregory T, Zierahn U (2016) The risk of automation for jobs in OECD countries: a comparative analysis. OECD social, employment and migration working paper. https://doi.org/10.1787/5jlz9h56dvq7-en
5. Bagozzi RP, Yi Y (2012) Specification, evaluation, and interpretation of structural equation models. J Acad Mark Sci 40(1):8–34. https://doi.org/10.1007/s11747-011-0278-x
6. Baines T, Lightfoot H, Smart P, Fletcher S (2013) Servitization of manufacture: exploring the deployment and skills of people critical to the delivery of advanced services. J Manuf Technol Manag 24(4):637–646. https://doi.org/10.1108/17410381311327431
7. Benedettini O, Swink M, Neely A (2017) Examining the influence of service additions on manufacturing firms' bankruptcy likelihood. Ind Mark Manage 60:112–125. https://doi.org/10.1016/j.indmarman.2016.04.011
8. Bentler PM (1990) Comparative fit indexes in structural models. Psychol Bull 107(2):238–246. https://doi.org/10.1037/0033-2909.107.2.238
9. Bentler PM, Bonett DG (1980) Significance tests and goodness of fit in the analysis of covariance structures. Psychol Bull 88(3):588–606. https://doi.org/10.1037/0033-2909.88.3.588
10. Bilgeri D, Gebauer H, Fleisch E, Wortmann F (2019) Driving process innovation with IoT field data. MIS Q Executive 18(3):191–207. https://doi.org/10.17705/2msqe.00016
11. Böhm E, Eggert A, Thiesbrummel C (2017) Service transition: a viable option for manufacturing companies with deteriorating financial performance? Ind Mark Manage 60:101–111. https://doi.org/10.1016/j.indmarman.2016.04.007
12. Boldosova V (2020) Telling stories that sell: the role of storytelling and big data analytics in smart service sales. Ind Mark Manage 86:122–134. https://doi.org/10.1016/j.indmarman.2019.12.004
13. Bowen DE, Siehl C, Schneider B (1989) A framework for analyzing customer service orientations in manufacturing. Acad Manag Rev 14(1):75–95. https://doi.org/10.5465/amr.1989.4279005
14. Bustinza OF, Gomes E, Vendrell-Herrero F, Tarba SY (2018) An organizational change framework for digital servitization: evidence from the Veneto Region. Strateg Chang 27(2):111–119. https://doi.org/10.1002/jsc.2186
15. Carretero S, Vuorikari R, Punie Y (2017) The digital competence framework for citizens with eight. Publications Office of the European Union.https://doi.org/10.2760/38842

16. Cenamor J, Rönnberg Sjödin D, Parida V (2017) Adopting a platform approach in servitization: leveraging the value of digitalization. Int J Prod Econ 192:54–65. https://doi.org/10.1016/j.ijpe.2016.12.033
17. Coreynen W, Matthyssens P, Van Bockhaven W (2017) Boosting servitization through digitization: pathways and dynamic resource configurations for manufacturers. Ind Mark Manage 60:42–53. https://doi.org/10.1016/j.indmarman.2016.04.012
18. Cotet GB, Balgiu BA, Negrea VCZ (2017) Assessment procedure for the soft skills requested by industry 4.0. In: MATEC web of conferences, vol 121. https://doi.org/10.1051/matecconf/201712107005
19. Demirkan H, Bess C, Spohrer J, Rayes A, Allen D, Moghaddam Y (2015) Innovations with smart service systems: analytics, big data, cognitive assistance, and the internet of everything. Commun Assoc Inform Syst 37(1):733–52. https://doi.org/10.17705/1cais.03735
20. Dworschak B, Zaiser H (2014) Competences for cyber-physical systems in manufacturing-first findings and scenarios. Proc CIRP 25:345–350. https://doi.org/10.1016/j.procir.2014.10.048
21. Eloranta V, Turunen T (2015) Seeking competitive advantage with service infusion: a systematic literature review. J Serv Manag 26(3):394–425. https://doi.org/10.1108/JOSM-12-2013-0359
22. Fischer T, Gebauer H, Gregory M, Ren G, Fleisch E (2010) Exploitation or exploration in service business development?: Insights from a dynamic capabilities perspective. J Serv Manag 21(5):591–624. https://doi.org/10.1108/09564231011079066
23. Fliess S, Lexutt E (2019) How to be successful with servitization—guidelines for research and management. Ind Mark Manage 78:58–75. https://doi.org/10.1016/j.indmarman.2017.11.012
24. Fornell C, Larcker DF (1981) Evaluating structural equation models with unobservable variables and measurement error. J Mark Res 18(1):39. https://doi.org/10.2307/3151312
25. Gebauer H, Fleisch E, Friedli T (2005) Overcoming the service paradox in manufacturing companies. Eur Manag J 23(1):14–26. https://doi.org/10.1016/j.emj.2004.12.006
26. Hu L, Bentler PM (1999) Cutoff criteria for fit indexes in covariance structure analysis: conventional criteria versus new alternatives. Struct Eqn Model: Multi J 6(1):1–55
27. Hair J, Black W, Babin BYA, Anderson R, Tatham R (2010) Multivariate data analysis, 7th edn. Pearson Prentice Hall, New Jersey
28. Ferreira FNH, Proença JF, Spencer R, Cova B (2013) The transition from products to solutions: external business model fit and dynamics. Ind Mark Manage 42(7):1093–1101
29. Gebauer H (2008) Identifying service strategies in product manufacturing companies by exploring environment-strategy configurations. Ind Mark Manage 37(3):278–291